Concepts of Ecology

Fourth Edition

Edward J. Kormondy

PRENTICE HALL, Upper Saddle River, New Jersey 07458

Library of Congress Cataloging-in-Publication Data

Kormondy, Edward John
Concepts of ecology / Edward J. Kormondy. — 4th ed.
p. cm.
Includes bibliographical references and index.
ISBN 0-13-478116-3 (pbk.)
1. Ecology. I. Title.
QH541.K59 1996
574.5—dc20 95-37090
CIP

Senior editor: Sheri L. Snavely
Editorial/production supervision and interior design: ETP/Harrison
Copy editor: Jane Loftus
Cover director: Jayne Conte
Cover photographer: Heather Scott
Manufacturing manager: Trudy Pisciotti

Printed in the United States of America

ISBN 0-13-478116-3

Prentice-Hall International (UK) Limited,London
Prentice-Hall of Australia Pty. Limited, Sydney
Prentice-Hall Canada Inc., Toronto
Prentice-Hall Hispanoamericana, S.A., Mexico
Prentice-Hall of India Private Limited, New Delhi
Prentice-Hall of Japan, Inc., Tokyo
Pearson Education Asia Pte. Ltd., Singapore
Editora Prentice-Hall do Brasil, Ltda., Rio de Janeiro

TO ALL ECOLOGISTS—IN SPIRIT AND IN FACT

. . . concerning the estate of the sons of men,
that God might manifest them,
and that they may see themselves are beasts.
For that which befalleth the sons of men befalleth beasts;
even one thing befalleth them: as the one dieth so dieth the other;
yea, they have all one breath;
so that a man hath no preeminence above a beast;
for all *is* vanity.

Ecclesiastes 3:18–19

Nearer the coasts are the giant forests of the redwoods extending from near the Oregon line to Santa Cruz. . . . these withdrawn gardens of the woods—long vistas opening to the sea—sunshine sifting and pouring upon the flowery ground in a tremulous, shifting mosaic, as the light-ways in the leafy wall open and close with the swaying breeze . . .

—John Muir

Lying out at night under those giant sequoias was lying in a temple built by no hand of man, a temple grander than any human architect could by any possibility build, and I hope for the preservation of the groves of giant trees simply because it would be a shame to our civilization to let them disappear. They are monuments in themselves.

—Theodore Roosevelt

I went to the woods because I wished to live deliberately, to front only the essential facts of life, and see if I could not learn what it had to teach, and not, when I came to die, discover that I had not lived.

—Henry David Thoreau

Contents

Preface

I have tried to present the significant concepts of modern ecology in a readable and intelligible way, and to develop these concepts in a manner that reflects the way in which they have in fact developed—not as truth revealed but as a searching into the nature of things, asking questions, analyzing data, generalizing, and predicting. I have not been completely consistent in this approach, any more than any scientist is in his day-to-day research. A scientist builds on the store of previous work and hence must master established facts and formulas. It would be imprudent and ludicrous to search out everything from scratch; little, if any, progress would ever be made. Thus, in treating some ideas I have been more descriptive than developmental, more assertive than investigatory. At the least, however, I hope I have portrayed ecology as both very old and very new, very secure on some concepts and very open on others, and very much of an exciting cross- or inter-disciplinary field, as is much of contemporary biology.

I have aimed for an audience of post-general biology students but there is little reason for students to avoid the book just because they have not been exposed to Mendelism, Darwinism, fetal pigs, or glycolysis. I have consciously minimized jargon, hopefully without sacrificing accuracy, and I have tried to write about ideas rather than things. Nonetheless, I have assumed some scientific sophistication on the part of the reader, and this may necessitate correlative reading in biology and other sciences.

At the suggestion of users of the earlier editions, I have identified the investigators and sources of most of the studies that form the basis of this book, doing so in the style common to professional journals. However, in recognition of the wide disparity among libraries, I have, for the most part, limited those sources to journals that are most likely to be found in most college or university libraries—

American Scientist, Ecology, Ecological Monographs, Science, and *Scientific American.*

ACKNOWLEDGMENTS

I would suppose that investigators would pass beyond the bounds of ecology when they were no longer directly concerned with interrelationships of the physical and the biological but with any particular process in its own right. It has been my intent here to delve into these processes but always in the context of the whole picture of the ecological system, that is a holistic point of view. If I have not succeeded, it is no reflection on the many upon whom I've depended for the development of my ideas and the execution of this book. Like any venture, this book is of the "OTSOG" sort because I have stood "*o*n *t*he *s*houlders *o*f *g*iants"—the many investigators who have given the substance from which I've drawn heavily, the few really creative scientists who have given direction to the discipline, my former students at Oberlin College and The Evergreen State College who provided a critical audience against which my thoughts and presentations have been tested. Ecologist colleagues Francis Evans, Gene Likens, Robert Whittaker, and George Woodwell reviewed the manuscript of the first edition; Francis Evans, Frank Golley, Monte Lloyd, Frank McCormick, David Milne, Oscar Soule, Robert Whittaker, and Ronald Wolff reviewed the draft of the second edition; Robert Burgess, Clark Grier, Robert McIntosh, and David Milne reviewed the draft of the third edition. John Beaver, The University of Akron; Lloyd C. Fitzpatrick, University of North Texas; Kipp Kruse, Eastern Illinois University; Don Moll, Southwest Missouri State University; Shirley Porteous, Fresno City College; Brian Reeder, Morehead State University; Lee Rockett, Bowling Green State University; Daniel Simberloff, Florida State University; and Donald R. Young, Virginia Commonwealth University; reviewed the draft of this fourth edition. Their critiques were most helpful to me, and thereby, I trust, to the reader. In no way, however, are they culpable for the final product—that is my burden. I am especially grateful to my nonecologist colleague, Thomas F. Sherman of Oberlin College, who read the entire manuscript of the first edition, and to the many users who were kind enough to send me both their compliments and condemnations on the earlier editions and the draft of this one.

Finally I am deeply indebted to my colleagues at the University of Hawaii-Hilo—Ronald Amundson, Dan Brown, Don Hemmes, and Ronald Terry—who enlarged my perspectives on aspects of ecology, and especially to Librarian Helen Rogers, whose invaluable reference assistance enabled me to complete this project in a timely fashion. I also wish to acknowledge the technical assistance of Eric Flower, Librarian, at the University of Hawaii-West Oahu.

In spite of all this indispensable assistance, I alone am responsible for errors of fact or interpretation and will appreciate comments to that effect in order to improve the quality of this publication in the future.

Edward J. Kormondy
Los Angeles, California

PART I

Ecology and Ecosystems

Ecology as a Science

Tracing the historical development of any science is beset with many difficulties, not the least of which is the precise definition of the science in question and of "science" itself. If one accepts the concept of science as the model of classical physics, ecology doesn't fit. Insofar as ecology describes, classifies, hypothesizes, and tests hypotheses, it fits the classical mold of science. However, because ecology extends to human interactions, including aesthetics, ethics, politics, and economics, it doesn't fit that mold. Further, because ecology is a multidisciplinary enterprise, it doesn't fit precisely into one channel of scientific inquiry; instead it ranges from reductionism in the study of individual species populations, through less reductionist approaches in the study of communities, to the holistic in studies of the totality of communities on Earth. The result is sometimes strong disagreement on the methods and viewpoints used by different ecologists investigating different phenomena at different levels of complexity and organization (Pomeroy et al. 1988). Further, there are some who contend that much of ecology cannot be regarded as science because of the paucity of information about nature it provides and that what information it does provide is of such poor quality that ecology can at best be regarded as a "soft" science (Peters 1991).

THE ANTECEDENTS OF ECOLOGY

In one of the few excellent historical reviews of the development of ecology, ecologist Robert McIntosh (1985) noted that the polymorphic nature of ecology as a science, the widespread distortion of its content and competence during the environmental crisis of the 1960s and 1970s, coupled with the lack of historical studies, contribute to diverse and even contradictory opinions about the origins of ecology. He notes further that ecology is "more a bush with multiple stems and a diffuse rootstock than a tree with a single, well-defined trunk and roots." This botanical metaphor fits the model suggested by the seminal philosopher of science, Thomas Kuhn (1970), who proposed that a developing scientific discipline may represent a fusion of several separate trunks lacking a common initial rootstock.

These caveats aside, the antecedents of modern ecology and of ecology as a science certainly extend to the origins of humanity itself. Our human forebears must have learned about their environment, else *Homo sapiens* would not have survived. Humans' conscious observations of their natural surroundings can be traced to ancient civilizations, especially in matters of agriculture and aquaculture. As with the beginnings of what became the different sciences, more formal and systematic study of the environment began in Greece in the third and fourth century B.C. Among the early natural historians were Aristotle and, notably, his student and successor as head of the Lyceum, Theophrastus.

The tradition of scientific observations of nature, which began with Theophrastus, reached its zenith in the eighteenth and nineteenth centuries in the works of such great natural historians as Buffon, Linnaeus, Reaumur, Darwin, and von Humboldt, among others, and extends into the present day. The best of these traditions is characterized by careful attention to detail, precise measurement and recording of information, recognition and interpretation of variables, awareness and questioning of previous contributions to theory, and development of new tools of analysis. By the same token, many natural history studies lacked, and still lack, the unifying focus that is critical to the development of concepts and theories.

DEFINITIONS OF ECOLOGY

While definitions of a field of inquiry do not necessarily eliminate all ambiguities, they do set boundaries, sharpen perspective, and give direction. The introduction of the term *ecology* and the evolution of its meaning have done just that.

Although the German biologist Ernst Haeckel is generally credited with having done so, Hanns Reiter (1885) appears to have been the first to combine the Greek words *oikos* (house) and *logos* (study of) to form the term ecology (Egerton 1977). There is consensus, however, that Haeckel gave definition and substance to the term, which he first used in 1866, in the following statement written in 1870:

> By ecology we mean the body of knowledge concerning the economy of nature—the investigation of the total relations of the animal both to its inorganic and to its organic environment; including above all, its friendly and inimical relation with those animals and plants with which it comes directly or indirectly into contact—in a word, ecology is the study of all the complex interrelations referred to by Darwin as the conditions of

> the struggle for existence. The science of ecology, often inaccurately referred to as "biology" in a narrow sense, has thus far formed the principal component of what is commonly referred to as "Natural History." As is well shown by the numerous popular natural histories of both early and modern times, this subject has developed in the most close relations with systematic zoology. The ecology of animals has been dealt with quite uncritically in natural history; but natural history has in any case had the merit of keeping alive a widespread interest in zoology.

Some seven years previously, the French zoologist Isodore Geoffroy St. Hilaire had proposed the term *ethology* for "the study of the relations of the organisms within the family and society in the aggregate and in the community." At about the same time the English naturalist St. George Jackson Mivart coined the term *hexicology*, which he defined in 1894 as "devoted to the study of the relations which exist between the organisms and their environment as regards the nature of the locality they frequent, the temperatures and the amounts of light which suit them, and their relations to other organisms as enemies, rivals, or accidental and involuntary benefactors."

It is curious to note that, although coined in 1866, the term ecology was not widely used nor recognized at the end of the nineteenth century. It first began to catch the fancy of botanists at the time that some zoologists, including American entomologist William Morton Wheeler, preferred St. Hilaire's *ethology* (McIntosh 1985). By 1913, the term became institutionalized with the formation of the British Ecological Society and shortly thereafter, in 1915, with the formation of the Ecological Society of America.

Institutionalization of the term ecology, however, did not necessarily mean there was consensus on what ecology's purview was. Haeckel's definition, involving the concept of interrelationships of organisms and environment and having a strong physiological orientation, sometimes has had quite different interpretations placed upon it. The British ecologist, Charles Elton (1927), for example, defined ecology as "scientific natural history" concerned with the "sociology and economics of animals." American plant ecologist Frederick Clements (1905) considered ecology to be "the science of the community," while American animal ecologist Victor Shelford (1937) regarded it as "that branch of general physiology which deals with the organism as a whole, with its general life processes as distinguished from the more special physiology of organs." By contrast, the German ecologist Karl Friederichs (1958) regarded ecology as "the science of the environment" (*Umweltlehre*). Eugene P. Odum (1959), the contemporary American ecologist who has had the most influence in defining the parameters of the discipline, has defined it as "the study of the structure and function of nature" and later (1962) as "the study of the structure and function of ecosystems."

Regardless of precise definition, the substance of ecology is found in the multitude of nonliving and living structures, processes, and interrelations involved in moving energy and nutrients and in regulating population and community structure and dynamics. Like many fields of contemporary biology, ecology is multidisciplinary and almost boundless in its concern. This point has been well stated by the British ecologist A. Macfadyen (1957):

> Ecology concerns itself with the interrelationships of living organisms, plant or animal, and their environments; these are studied with a view to discovering the principles which govern the relationships. That such principles exist is a basic assumption—and an act of faith—of the ecologist. His field of inquiry is no less wide than the totality of the living conditions of the plants and animals under observation, their systematic position, their reactions to the environment and to each other, and the physical and chemical nature of their inanimate surroundings.... It must be admitted that the ecologist is something of a chartered libertine. He roams at will over the legitimate preserves of the plant and animal biologist, the taxonomist, the physiologist, the behaviourist, the meteorologist, the geologist, the physicist, the chemist and even the sociologist; he poaches from all these and from other established and respected disciplines. It is indeed a major problem for the ecologist, in his own interest, to set bounds to his divagations.

This broad definition contributed in part to compounding the precise determination of the field of ecology during the 1960s when popular usage of the term became commonplace. In this context, not only did it seem that everyone had a particular, if not peculiar, use of the term, but its usage was extended politically to encompass a philosophy that broadly incorporated a variety of environmental concerns in a yet more ambiguous *environmental science.*

ECOSYSTEM ECOLOGY

As we embark here on describing the central concepts of the seemingly elusive term, ecology, it is necessary to put some constraints on our discussion. Thus, for the purposes of our discourse, we define ecology in Odum's (1962) somewhat tautological terms as the scientific study of the structure and function of ecological systems, or *ecosystems.* The term ecosystem was coined and defined by British ecologist Arthur Tansley (1935) as follows:

> ...the whole system (in the sense of physics) including not only the organism-complex, but also the whole complex of physical factors forming what we call the environment of the biome—the habitat factors in the widest sense. Though the organisms may claim our primary interest, when we are trying to think fundamentally we cannot separate them from their special environment, with which they form one physical system. It is the systems so formed which, from the point of view of the ecologist, are the basic units of nature on the face of the Earth.

A system consists of two or more components that interact and it is surrounded by an environment with which it may or may not interact (O'Neill et al. 1986). A system is thus an arbitrary unit of the universe selected for study because it is a construct of the human mind. However, some logic is applied in deciding what to include and what to exclude in the system. In the case of ecological systems, the two major components are nonliving, or *abiotic*, and living, or *biotic*. The abiotic component is a setting of physicochemical substances and gradients including basic inorganic elements and compounds such as calcium and oxygen, water and carbon dioxide, carbonates and phosphates, and an array of organic compounds, the byproducts of organism activity. The abiotic component also includes such physical factors and gradients as moisture, winds, currents, tides, and solar radiation with its concomitant light and heat. The biotic component is the particular assemblage of plants, animals, and microbes existing in the abiotic setting.

Ecosystems are real—like a pond, a field, a forest, an ocean, or even an aquarium. They are also abstract in the sense of being conceptual schemes developed from a knowledge of real systems. There is great diversity in the types of actual ecosystems—from small to large, terrestrial to freshwater to marine, field to laboratory. There are also unique combinations of particular abiotic and biotic components in any particular ecosystem. Nonetheless, they have in common certain general structural and functional attributes that are recognizable, analyzable, and predictable.

Most historians of science mark the rise of modern ecology with the introduction and elaboration of the ecosystem concept (Kormondy and McCormick 1981). This "new" ecology has reached highly sophisticated levels of abstraction and developed increasingly powerful theoretical constructs and methodologies. The worldwide emphasis on ecosystem ecology derives in some large measure from the source of educational preparation and advanced training of ecologists. Until quite recently, most ecologists received their training in English in American universities in which ecosystem ecology has predominated. No less important has been the omnipresence of Anglo-American textbooks and major international initiatives in research (Kormondy and McCormick 1981).

In the following chapter, we will begin to explore in greater depth the basic nature of ecosystems.

SUMMARY

VOCABULARY

abiotic
biotic
ecology
ecosystem
environmental science
ethology
hexicology

KEY POINTS

- Ecology is a multidisciplinary field extending across the physical, biological, and social sciences.
- Environmental observations made by agriculture-based societies were systematized first by early Greek naturalists, then by western natural historians, and currently by ecologists.
- Various definitions of ecology have been advanced; the contemporary definition focuses on ecosystems as the major unit of study.

The Nature of Ecosystems

One characteristic of any system is organization, that is, a unified group of components forming a systemized whole. O'Neill et al. (1986) note other properties of a biological organization, including ecosystems, as follows.

- Ecosystems exist independently of specific components (e.g., an individual tree may die, but the forest's organization remains)
- Its components are interdependent (e.g., when removed from its colony, a social insect does not often survive)
- An ecosystem has a function (e.g., the component parts each have functions that, together, produce a function of the whole)
- It is active; something dynamic, past or present, is implied (e.g., change occurs or has occurred)
- A sliding scale of organization exists (e.g., two populations may independently coexist in an area or they may be intertwined in a complex relationship).

These attributes of organization apply fully to ecosystems. This will become evident as we explore in increasing detail the various structural and functional features of ecosystems in this and later chapters.

PRODUCERS, CONSUMERS, AND DECOMPOSERS

Producers

Because energy relationships underlie ecological kinships in so many ways, the nature of the ecosystem is best studied by beginning with the influx of energy. Radiant energy, in the form of sunlight, is the ultimate and only significant source of energy for any ecosystem. It is used in the photosynthetic process whereby carbon dioxide is assimilated into energy-rich carbon compounds. The organisms that perform this vital function are the *producers*. Typically they are the chlorophyll-bearing plants, the algae of a pond, the grass of a field, the trees of a forest. Of considerably less significance in most ecosystems are the purple bacteria that also assimilate carbon dioxide from inorganic compounds with the energy of sunlight. In certain stagnant lakes rich in hydrogen sulfide these photosynthetic bacteria may account for 25 percent of the total photosynthesis. Finally, producers also include chemosynthetic bacteria, all of which obtain energy by oxidizing simple inorganic compounds. Chemosynthetic producers are relatively insignificant in the energy relations of most ecosystems, except as the foundation for energy flow in such environments as deep-sea geothermal vents in the Pacific Ocean. They also play a substantial role in the movement of mineral nutrients in ecosystems. Under anaerobic conditions these chemosynthesizers "rescue" energy in the form of organic compounds, energy that would otherwise be "lost" through storage in sediments.

The term producer, in an energy context, is somewhat misleading and misrepresentative. Producers create carbohydrates, not energy. Because they convert or transduce radiant energy into a chemical form, it might be better to refer to them as converters, transformers, or transducers. Yet "producer" is so widely used and so firmly entrenched in the ecological literature that it would be futile to try to dislodge it.

Consumers

Because the energy incorporated in the producer by photosynthesis is subsequently synthesized into molecules that serve the nutritional requirements of the producer's own growth and metabolism, we can speak of the producer as being *autotrophic* (self-feeding). In the same way, organisms whose nutritional needs are met by feeding on other organisms are referred to as *heterotrophic* (other feeding). A *primary consumer*, or, more commonly, a *herbivore*, is a heterotroph that derives its nutrition directly from plants. A *carnivore*, or *secondary consumer*, is a heterotroph deriving its energy indirectly from the producer by way of the herbivore. Some ecosystems contain tertiary consumers—carnivores that feed on other carnivores. Omnivores are consumers that derive their energy from both producers and herbivores.

A simple but nonetheless actual example of this autotroph-heterotroph relationship is the reindeer moss → reindeer → human *food chain* in Lapland (Figure 2-1). Energy captured by the photosynthesizing reindeer moss (which is actually a lichen,

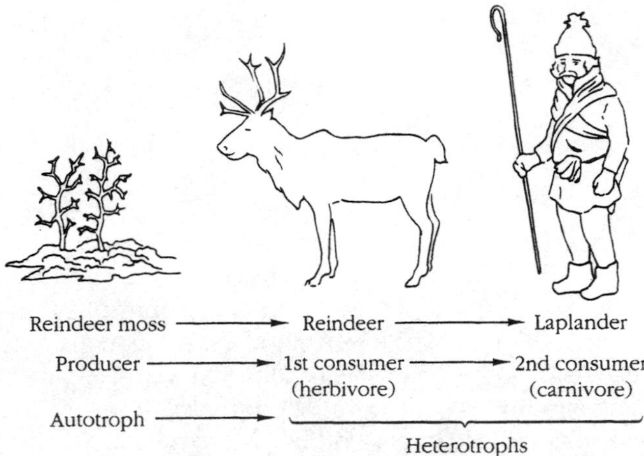

Figure 2-1 Energy movement in a human food chain.

not a moss, a lichen being a composite organism consisting of a fungus and an alga) serves as the food base for reindeer and the latter, in turn, for the Laplander. In this chain the reindeer moss is the producer and is an autotroph; the reindeer and the Laplander are heterotrophs, the former a primary consumer or herbivore, the latter a secondary consumer or carnivore.

Implicit in this autotroph → heterotroph, producer → consumer or producer → herbivore → carnivore relationship is the direction of energy movement through the ecosystem. It is unidirectional and noncyclic. In part, we could reason this a priori, for plants do not consume animals except for the unusual carnivorous plants, such as sundew, pitcher plant, Venus flytrap, and a number of bromeliads. The explanation for the noncyclic, unidirectional flow of energy, however, is to be found in the energy losses that occur at each transfer along the chain and in the efficiency of energy utilization that occurs within each link of the chain. Thus, the amount of energy at each successive step is less and less, imposing an ultimate limit on the number of transfers, hence, levels. These matters will be considered at length in Part III. For the moment, however, it is enough to say that the one-way flow of energy constitutes a cardinal principle of ecosystems.

Decomposers

In addition to herbivores and carnivores, there is another major group of heterotrophs in ecosystems. This group of organisms is collectively referred to as *decomposers*, or *reducers*, and consists chiefly of bacteria and fungi. For the most part, decomposers do not consume food in the ingestive manner of a herbivore or carnivore; they do so by absorption. Enzymes produced within their bodies are released into dead plant and animal material, and some of the degraded and digested products are then absorbed. Generally in terrestrial ecosystems bacteria act on animal tissues, and fungi act on plant tissues. In some aquatic ecosystems consumers play a more significant role than either bacteria or fungi in degrading and digesting organic products. As a byproduct of supplying their own growth and metabolism requirements, however, decomposers perform an invaluable service to the ecosystem—the mineralization of organic matter. Through their exoenzyme digestive activity basic elements bound in protoplasm are released to the environment and are thereby made available for reuse, primarily by producers, but also by other organisms.

ENERGY FLOW AND MINERAL CYCLING

It should now be evident that two processes proceed concurrently in ecosystems, the unidirectional flow of energy and the cycling of nutrient elements. The former activity has been described as unidirectional and noncyclic, whereas the implication of decomposer mineralization activity is that the movement of nutrients is cyclic. This point needs further clarification.

In the process of converting radiant energy into chemical energy by photosynthesis, the green plant incorporates into its protoplasm various inorganic elements and compounds. Among the important ones are the direct components of the photosynthetic reaction, carbon dioxide and water, and those that are critical to protoplasmic synthesis—notably nitrogen, phosphorous, sulfur, and magnesium, as well as 15 or more other essential nutrients. As the green plant is grazed on, not only is chemical energy in the form of carbohydrates, fats, and proteins transferred to the herbivores but so are a host of nutrients as well. Similarly, there is a transfer of both energy and nutrients from herbivore to carnivore and from all the preceding levels to the decomposers. Thus energy flow must accompany nutrient cycling, and nutrient cycling must accompany energy flow.

Although, as noted, there is a progressive diminution of energy in this trophic or feeding chain, the nutrient component is not diminished. In fact, as we will see in Chapter 8, some may even become concentrated in certain steps of the chain. In any event, nutrients are not lost in the manner of energy. When nutrient-containing protoplasm is eventually subjected to decomposer activity, the nutrients are released to the environment where they are available for reuse, or recycling. The nature of the nutrient cycling process and the interaction of physical, chemical, and biotic factors in it are discussed in detail in Part IV.

These two ecological processes of energy flow and mineral cycling, involving interaction between the physicochemical environment and the biotic assemblage, lie at the heart of ecosystem dynamics. They are fundamental to the structure of ecosystems and to the myriad complex processes that take place within ecosystems. For the moment, however, we can represent these two processes by an oversimplified graphic model (Figure 2-2) in order to codify the discussion up to this point. While this model does indicate the movement of energy and matter within the biotic component and between it and the abiotic nutrient pool, it is quite inadequate in

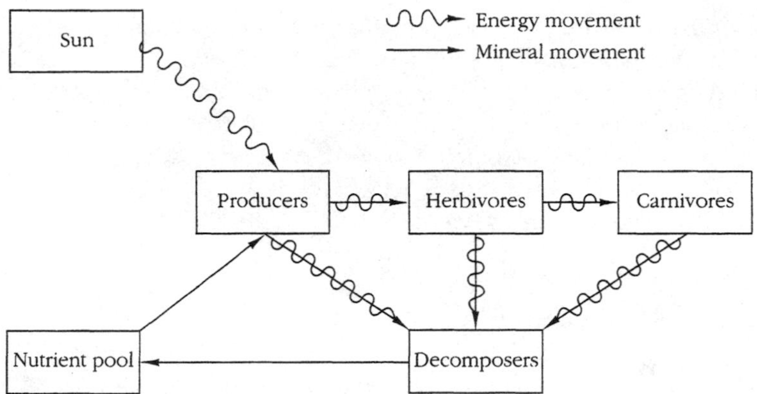

Figure 2-2 An overly simplified model of energy and mineral movement in ecosystems. Note that energy flow is unidirectional and noncyclic, whereas nutrient movement is cyclic.

describing an ecosystem, either structurally or functionally. That description and the building of a model that both organizes knowledge about ecosystems and permits prediction are prime objectives of this book. We will be accruing the kinds of information needed to conceptualize an ecosystem more accurately. At the start, however, several additional qualities of ecosystems and several critical characteristics of energy flow and mineral cycling need to be identified.

First, energy not only flows unidirectionally but is also lost irretrievably from the system in several ways. This loss occurs largely in the form of heat as a byproduct of metabolic reactions and also by export to other ecosystems through such agencies as wind, floods, and animal movements. Secondly, although minerals circulate within an ecosystem, there is a net loss in several ways: by being bound up in the sediments in the system or by export through such actions as wind, floods, and animal activities. But these serious and acknowledged deficiencies of the present model do not obviate its immediate usefulness in providing a general overview.

These two major processes—energy flow and nutrient cycling—have considerable significance in today's society. Meeting the food requirements of an expanding human population, for example, demands a more intimate knowledge of the efficiency and output of producers. Of no less importance is the efficiency of energy flow through human ecosystems and the dependence of those systems on the availability and abundance of essential nutrients. Moreover, the introduction into the environment of toxic substances, such as certain wastes and pesticides, and of radioactive forms of nutrients, typically results in their incorporation into human food chains, thereby constituting one of the major problems facing humanity today. Such considerations, as we shall see, involve a thorough understanding of these two major ecological processes.

INDIVIDUALS AND SPECIES

Although the processes of energy flow and mineral cycling are indeed fundamental, they are not the whole of ecology. The manifestation of these processes, as we have already seen, is through the vehicle of living organisms—plants, animals, and microbes. These process terms, like the term ecosystem, have objectivity in actual populations of specific organisms, each of which plays a role that is characteristic of the species and unique in any given ecosystem. Each species not only has unique morphological, physiological, and behavioral attributes but, because of them, unique ecological roles in the processes in an ecosystem as well. Some species are an energy source only to certain other species whereas others serve a wider range of consumers. Some play a crucial role in one specific phase of a nutrient cycle while others may play no evident role whatsoever. Given plant species, as we shall see, are considerably more efficient in energy capture at particular altitudes or at particular depths in aquatic systems than are other species. Each species is more tolerant of excesses and minima of particular nutrients and less tolerant of extremes in others, and each is most successful within a range of particular gradients. Consequently, we expect a priori that no two ecosystems would have precisely the same biological composition but rather would have ecological equivalents performing comparable functions.

POPULATIONS AND ECOSYSTEMS

In addition, it should follow that as physical and chemical factors show gradation, species *populations* (that is, a group of organisms, all of the same species, that occupies a particular area) sensitive to these gradations should correspondingly increase or decrease in abundance, as well as importance. And it should also follow that these shifts in any given population would affect the relationships that become established between populations of different species. The consequences of varying specificities of organisms and environment ramify quickly and affect the growth, size, and distribution of populations. In Part V we shall consider in some detail both general patterns of growth and regulation of populations and some of the unique aspects of particular species populations.

COMMUNITIES AND ECOSYSTEMS

These comments on environmental tolerances of species populations, when coupled with the acknowledged differences in physical and chemical parameters of the environment, raise several significant implications. In the first instance, although no two ecosystems would be expected to be exactly alike, some, like members of the same species population, would be expected to be more alike than others. Thus we group together fields or ponds or forests; but, in turn, evergreen forests are recognized as distinct from deciduous forests and, within the latter, oak-hickory forests as different from beech-maple forests. The ecologist can have as an object of study a general kind of ecosystem (e.g., forest) or any of its subdivisions (beech-maple or oak-hickory forests) or any of its particular manifestations (the beech-maple woods north of town). These assemblages of populations of different organisms in a particular environment constitute an *ecological community.* Large or small, geographically limited or expansive, these communities are the systems in which energy flows, minerals cycle, and populations grow and interact. These interactive processes of organism and environment are reflected not only in the structure and function of ecological communities but also in their dynamics, in growth and development, homeostasis and adaptation. These attributes are an undercurrent throughout this book but will receive special focus in Part VI.

HIERARCHY OF ECOSYSTEM ORGANIZATION

Organizationally, ecosystems are constituted hierarchically. Individuals, which are at the base of the hierarchy, in turn compose a species, and populations are the next order of complexity in the hierarchy. Communities constitute the next order, and those communities existing in particular physicochemical environments constitute an ecosystem. The aggregation of all ecosystems on Earth is sometimes referred to as the *ecosphere*, the ecosystem of the whole planet. In this pyramidal hierarchy, there are many individuals at the base, a smaller number of species than individuals at the next level, fewer populations than species at the next, and fewer communities than populations of the following level, and so on up to the top of the pyramid, the ecosphere.

SPACE, TIME, AND ECOSYSTEMS

Yet two additional attributes of ecosystems are of fundamental significance to their structure and function: the dimensions of space and time and the interrelatedness with other ecosystems. Ecosystems occur in space and exist in time—they have width, depth, and height, plus a past as well as a present and a future. These dimensional attributes pose continuing problems in studying ecosystems—in sampling their nooks and crannies and in determining their status yesterday and tomorrow. If we sample a pond only in the very deepest part, for instance, a substantial component of shallow-water characteristics would be missed. Similarly, a summer sample is usually very different from a winter sample, especially in areas subject to freezing conditions, and samples taken 10 to 20 years apart could be very different, especially in young ecosystems.

The spatial aspect of ecosystems is real, but precise delimitation is arbitrary, for one ecosystem is interrelated with other ecosystems. As no organism is sufficient unto itself, neither is an ecosystem. Ecosystems are not discrete entities delimited sharply from other ecosystems. No pond exists that is not surrounded by another ecosystem, perhaps a field, from which organic matter may be added, or that is not connected by a stream to another pond to which it contributes organic or nutrient

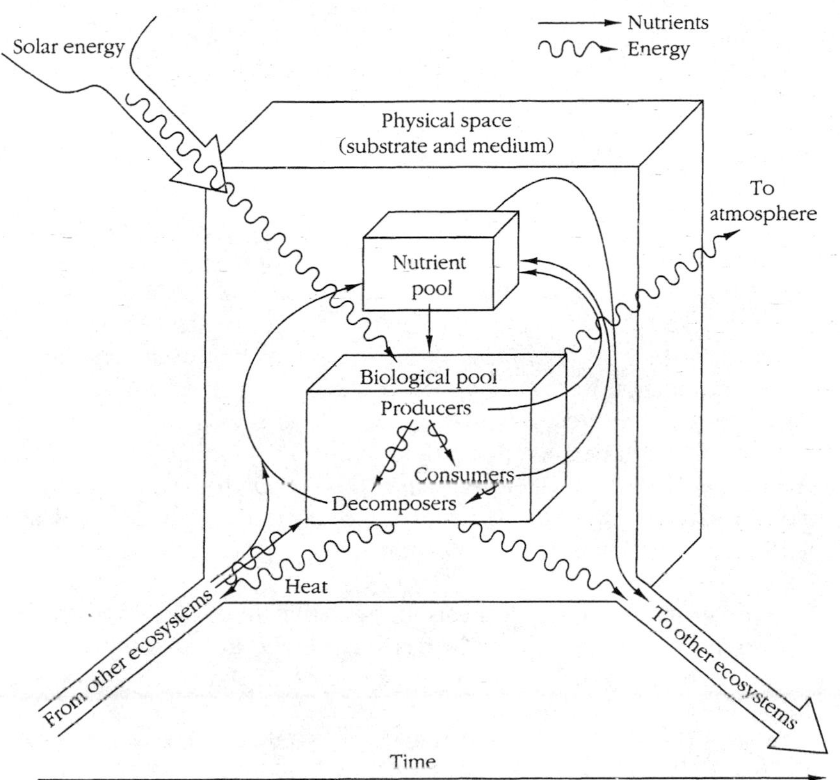

Figure 2-3 Model of an ecosystem reflecting its spatial and temporal dimensions, its external open but internal cybernetic system, its relationship to other ecosystems, the unidirectional, noncyclic flow of energy, and the cyclic flow of nutrients. (Modified by permission from E. J. Kormondy. 1974. In F. Sargent II (ed.). *Human Ecology.* Amsterdam: North Holland Publishing Co.)

matter. Complex feeding relationships often result in biological transfer of energy and nutrients from one system to another. The existence of contiguity and/or continuousness complicates the study of ecosystems, requiring the investigator to take into account the influences of surrounding and connecting systems. Although we shall explore this interrelatedness in more detail subsequently, the basic point should now be clear enough to warrant a substantial redrafting of our model (Figure 2-3).

HE INTERCONNECTEDNESS OF ECOSYSTEMS

Yet another quality to be noted from the model (Figure 2-3) is that each ecosystem is an open one. It depends on factors outside it for its sustenance, chiefly in the form of radiant energy from the sun. But as we will see, for some ecosystems the input is entirely organic (e.g., a cave ecosystem). It has an output, and that very output necessitates continual input. But although externally it is an open system receiving and delivering, internally it is largely a self-regulating one, as we will see in subsequent chapters.

The Gaia Hypothesis

From the larger perspective of the biosphere there is considerable interest in and controversy about the *Gaia hypothesis* (for Gaia, the goddess Earth in ancient Greece), which states that the Earth's atmosphere would not support life if it were not regulated by the totality of life in the biosphere (Lovelock and Margulis 1974; Lovelock 1979, 1988; Mann 1991). Originally Lovelock and Margulis seemed inclined to accept the notion that some purposeful design and foresight organized living things to stabilize their atmosphere and climate, but later came to believe this is achieved through negative feedback. For example, they suggest that a planet covered by light- and dark-colored daisies could control solar heat. The dark daisies would absorb sunlight and warm the planet until it became too warm for the dark daisies, thereby favoring the expansion of the light daisies, which would reflect light, leading to a cooling effect.

At heart, the Gaia hypothesis suggests that the evolution of organisms and their environment are tightly coupled as a single process through self-regulation. Lovelock and Margulis maintain, among other aspects of the hypothesis, that the decrease in carbon dioxide over the course of history has been a direct consequence of biological intervention and that without living organisms the Earth's climate could well have gone the way of icy Mars (–60°C) or torrid Venus (460°C).

While acknowledging that the biota are important in contributing to the regulation of the Earth's atmosphere, and of carbon dioxide in particular, Kasting et al (1988) argue that the fundamental controls on atmospheric carbon dioxide levels are physical rather than biological. They contend that even without organisms the atmospheric conditions of the Earth, Mars, and Venus are due to each planet's differing ability to cycle carbon between its crust and its atmosphere. Their argument will best be understood after the discussion on the carbon cycle (see Chapter 8).

The Gaia hypothesis poses a radical departure from the central concept of modern evolutionary theory, namely, that life on Earth is surrounded by and adapts to an environment in which conditions change by chance. Further, Gaia seems to require that some organisms restrain their reproduction in order to benefit the larger community, in spite of the generally accepted evolutionary concept that natural selection favors genes that increase their frequency. In this view, there is no mechanism for the evolution of organisms that will altruistically sacrifice immediate advantage for some future benefit (Beardsley 1989).

Although a number of ecologists, Eugene Odum (1989) among them, are inclined to accept the Gaia hypothesis, most do not cater to its notion of a superorganism, that of a giant system controlling temperature and chemical composition such that they are conducive to life on Earth.

While the Gaia hypothesis is certainly challenging, and, from a cybernectic viewpoint, intriguing, it remains to be seen if, as Beardsley (1989) noted, "the goddess of their temple is nowhere to be found."

CYBERNETICS AND ECOSYSTEMS

The implication of the Gaia hypothesis is that individually as well as collectively ecosystems are self-regulating entities. To maintain the homeostatic state, the feedback must be negative. Negative feedback is feedback that regulates a process or set of events by turning it off or slowing it down. Such negative feedback is the basic principle of the home thermostat that turns off the heating unit when the temperature exceeds a preset level, but allows it to operate when the temperature is below the set level. Similar feedback occurs at sublevels in ecosystems: when nutrient release exceeds a certain level, feedback, largely through chemical equilibria, inhibits further release; when a given population exceeds a certain size, various events are triggered that curtail further reproduction. We shall be exploring some of these feedback mechanisms and their role as we gain some understanding of the *cybernetics*, or science of control, of ecosystems.

Note that in a cybernetic system the emphasis is on stability and negative feedback, as described briefly in the previous paragraph. As O'Neill et al. (1986) noted, however, not everything about ecosystem behavior is well regulated, and many phenomena are more easily conceptualized as unstable, positive-feedback systems. For example, populations tend to grow exponentially until they meet some constraint imposed on them from outside. DeAngelis et al. (1986) discuss the widespread presence of positive-feedback processes in ecosystems.

STUDYING ECOSYSTEMS

Whereas the biologist likes to have as an object of study something discrete, like a cell or cell organelle that can be isolated into a test tube, the ecologist, in studying natural systems, is confronted by the complexities of almost unlimited variables. Ecologists seek, then, systems that are either more delimited or in which interrelationships can be more sharply defined—or they move into the laboratory or to the computer. A good deal of very sound ecology has, in fact, taken place in the laboratory as we will see in subsequent chapters. Here ideas and concepts can be tested with a limitation of variables and variability that nature seldom provides. The ultimate test of laboratory pronouncement, however, is to be found in the reality of nature.

The dynamics of these open, cybernetic systems of complex interaction are appropriate to study using computers. Simulation models allow the programming of dynamic interaction and, with increasing refinement of the input data, ecologists are able to manipulate variables and subcomponents almost limitlessly. A systems model organizes the knowledge of an ecosystem and describes its dynamics. The more accurately it mimics the real system, the more it serves as a tool of prediction and, in turn, as a means of optimization—that is, how best to manipulate the system to achieve a stated goal. Moreover, with large-scale systems, such as Lake Michigan or even the biosphere, mathematical modeling and computer simulation may constitute the only feasible approach to prediction. Competent treatment of these approaches is beyond the scope of this book, although we shall be considering their general results.

SUMMARY

VOCABULARY

autotrophic
carnivore
consumer
cybernetics
decomposers
ecological community
ecosphere
food chain
Gaia hypothesis
herbivore
heterotrophic
populations
primary consumer
producers
reducers
secondary consumer

KEY POINTS

- Ecosystems are organizations consisting of a unified group of components forming a systematized whole.
- Energy moves unidirectionally through ecosystems from producers to consumers and from both of these groups to decomposers along a food chain and ultimately is lost irretrievably.
- Minerals cycle, and recycle, through ecosystems.
- Energy flows and minerals cycle through individual organisms that are members of species whose populations are assembled into ecological communities; these form the biological component of ecosystems.
- The hierarchy of ecosystems (individuals, species, populations, communities) culminates in the aggregation of the world's ecosystems as the ecosphere.
- Ecosystems exist in space as well as time, and all are ultimately interconnected.
- The Gaia hypothesis proposes that the Earth's biota regulates its atmospheric environment; contrariwise, physicochemical processes independent of biota may be responsible.
- Both positive and negative (cybernetic) feedback operate in ecosystems.
- Because of their complexity, ecosystems study is beset with difficulty; computer simulations provide another route for their study.

PART II

The Abiotic Environment

Minimums, Tolerances, and the Medium

In Chapter 2 it was noted that ecosystems exist in time, an attribute that results in their being subject to change in both their biotic and abiotic components. These changes, as we shall see, come about because of the activities of organisms, interactions of organisms and their abiotic environment, and interactions of abiotic factors that, in turn, affect organisms. For example, the freshwater alga *Cladophora*, excretes calcium carbonate as a byproduct of its general metabolism. Over time, the calcium carbonate modifies the chemistry of the water as well as the substrate to the point that the alga can no longer survive, but other species not initially present can be established. The end result is not only an abiotic chemical change but a change in the biotic assemblage as well.

In this simple example, the change in water chemistry became a *limiting factor* for the alga. American ecologist George Clarke (1954) defined a limiting factor as that which "first stops the growth or spread of an organism." The origin of the concept, however, is generally attributed to the German chemist Justus von Liebig in his 1840 treatise, *Organic Chemistry and Its Application to Agriculture and Physiology*, wherein he noted, "The crops of a field diminish or increase in exact proportion to the diminution or increase of the mineral substances conveyed to it in nature."

LIEBIG'S LAW OF THE MINIMUM

Liebig had determined that phosphorus was not only an important element to agriculture, but that if it were present in small amounts plant growth was poor and that in its absence there was no plant growth at all. Thus, growth was limited by the amount of phosphorus present. In formulating his *law of the minimum*, Liebig indicated that each plant requires certain kinds and quantities of nutrients. If the quantity is minimal, plant growth will be minimal. In his own words,

> the perfect development of a plant, according to this view, is dependent on the presence of alkalies or alkaline earths; for when these substances are totally wanting its growth will be arrested, and when they are only deficient it must be impeded.

As originally conceptualized, Liebig's law of the minimum applied to inorganic nutrients. With the passage of time, the concept expanded to include a wide spectrum of physical factors as well, temperature and rainfall (e.g., a minimum of either or both) being among the most obvious.

As we proceed through the text, numerous examples of the law of the minimum will be evident in energy flow and mineral cycling, as well as in population growth and community structure and function.

Distinguishing a limiting factor is not always easy. As in the instance of Liebig's study of phosphorus, it appears there is a direct cause and effect, low phosphorus level resulting in low plant growth. However, further analysis might show that the relationship is indirect and that both factors are controlled by or through a third factor. Suppose that nitrogen availability affected the water needs of the plant and also contributed to making the phosphorus available in a form in which it could be assimilated. Then nitrogen would be a third factor in the cause-effect sequence. These are instances of *factor interaction*, which is discussed below.

LAW OF LIMITING FACTORS

F. F. Blackman (1905) extended Liebig's law of the minimum to encompass the limiting effects of the maximum as well. In what has become known as the *law of limiting factors*, Blackman noted

> in treating physiological phenomena, assimilation, respiration, growth, and the like, which have a varying magnitude under varying external conditions of temperature, light, supply of materials, etc., it is customary to speak of three cardinal points, the *minimal* condition below which the phenomenon ceases altogether, the *optimal* condition at which it is exhibited to its highest observed degree, and the *maximal* condition above which it ceases again.

It is relatively easy to see the applicability of *minimal, optimal,* and *maximal conditions* in organism growth. For example, if the temperature or availability of water is below the minimal or above the maximal, growth ceases, and the organism will likely expire. Also, the optimum for a given factor might not be achievable without at least a suboptimal amount of another factor, for example light and heat for plants. Again, as we proceed, numerous examples of the law of limiting factors will be evident.

In this same paper, Blackman considered the case of a given chloroplast engaged in photosynthesis, noting that there are five controlling factors: the amounts of CO_2, H_2O, and chlorophyll present; intensity of available radiant energy; and temperature in the chloroplast. This led him to conclude that, "When a process is conditioned as to its rapidity by a number of separate factors, the rate of the process is limited by the pace of the 'slowest' factor." This could be interpreted as an extension of the law of the minimum, but in a larger context it laid the foundation for what ecologists subsequently developed as the concept of *factor interaction* (see below).

LAW OF TOLERATION

Blackman's concept of limiting factors was developed from the perspective of control exerted by the environment. It is the organism, however, which responds to minimum and maximum factors, and thus the ecological and physiological attributes of organisms assume significance in explaining their geography. Building on the concept of the law of the minimum and the law of limiting factors, one of American ecology's outstanding contributors, Victor E. Shelford, incorporated both the geographic environment and the ecological physiology of organisms in his *law of toleration.* Shelford (1911) noted that success of a particular physiological activity (e.g., egg-laying) depends "upon the qualitative and quantitative *completeness* of the complex of conditions," and stated that

> ...the *law of the minimum* is but a special case of the *law of toleration.* Combinations of the factors which fall under the law of the minimum may be made, which make the law of toleration apply quite generally.

In applying the law of toleration to geographic distribution, Shelford pointed out that "centers of distribution are often only areas in which conditions are optimum for a considerable number of species." He depicted the relation of the law of toleration to centers of distribution as shown in Figure 3-1.

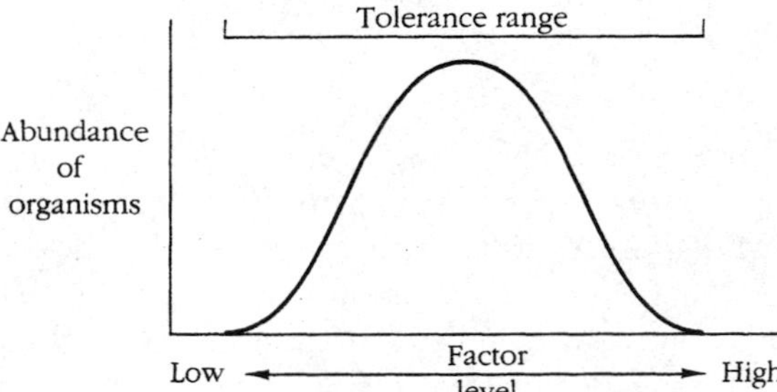

Figure 3-1 The law of toleration.

FACTOR INTERACTION

As might be expected, a given species may have a wide tolerance for one factor and a narrow tolerance for another. To describe a wide tolerance to a particular factor, the prefix *eury* is applied; *steno* is the prefix applied to indicate a narrow or limited tolerance (Ruttner 1963). In the instance of temperature, a species with a wide tolerance would be described as eurythermal and one with narrow tolerance as stenothermal (Figure 3-2). Euryhaline would describe a species with wide tolerance to salinity and stenohaline a species with narrow or limited salinity tolerance. A given species might well be eurythermal but stenohaline; similarly, a given species has different tolerances to other environmental factors. For various physiological processes to be optimally functional, at least suboptimal amounts of all pertinent factors must be present.

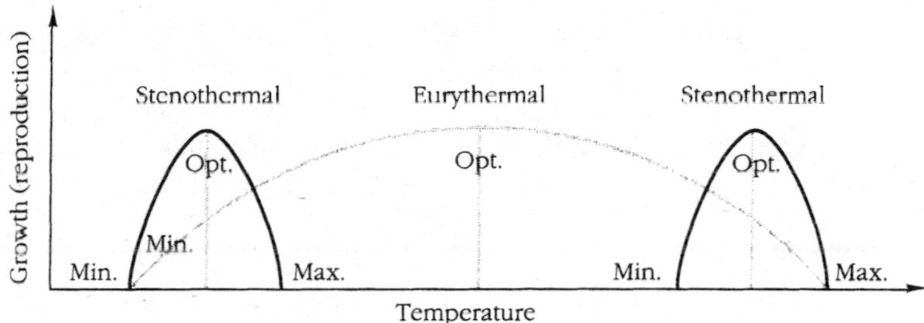

Figure 3-2 The ranges of temperature tolerance in stenothermal and eurythermal animals. (Redrawn with permission from Ruttner 1963).

The conclusion that organisms with a wide range of tolerance to many factors are likely to be very widely distributed should not be surprising. What may not be immediately self-evident is that when one factor becomes less than optimum the limits of tolerance for other factors may become limiting. For example, more water is required for healthy plant growth at low levels of nitrogen than at high nitrogen levels.

It is also the case that age, season, and physiological or behavioral state affect tolerance. The stage of life that is usually least tolerant to all factors is the egg or very young. For example, salmon eggs hatch and the young begin development in freshwater; they can be described as being stenohaline. Adult salmon, by contrast, are euryhaline, spending the majority of the adult period in salt water. Also, in 83.5 percent of 119 studies, increased temperature during rearing of a wide variety of animals, plants, protists, and a bacterium led to reduced size at a given stage of development (Atkinson 1994).

ADAPTATIONS TO LIMITING FACTORS

Another characteristic of environmental factors is that in one environment a given factor may be limiting, while not being so in another. For example, oxygen is often limiting in the deeper recesses of fresh- and saltwater environments, whereas it is rarely limiting on land, except at high altitudes. Adaptations to low oxygen are many, including chemical modifications of hemoglobin, structural modifications of lungs and gills, physiological changes in respiration rates, and behavioral activities. Tubificid worms, often called blood worms because of their red color (the result of numerous blood vessels just below the cuticle), are commonly found in regions of ponds and lakes that have low oxygen levels. With their heads buried in the sediment, they vigorously wave the distal portion of their bodies, thus increasing contact with the limited amount of oxygen present.

Given the fact of evolutionary relationships, one would expect that the more closely related the species, the more alike they would be in tolerance to given factors. Although this is the case, we also witness many instances in which a given species shows different levels of tolerance to a given limiting factor over its geographic distribution. These locally adapted populations are called *ecotypes* and may be the result of genetic change resulting in different physiological responses to different environments. Examples of ecotypic variation will be presented later, but for now consider the swimming activity of the jellyfish *Aurelia aurita* in the northern and southern parts of its geographic range.

As shown in Figure 3-3, in the cold waters off Halifax, Nova Scotia (about 45° north latitude), Bullock (1955) found that *Aurelia* maintains active swimming, actually pulsations of the entire body, over a temperature range from about 1° to 25°C. In the tropical waters off Tortuga Island of Haiti (about 20° north latitude), *Aurelia* swims actively over a range from about 13° to 35°C. The northern population reaches its maximal temperature limits well below that of the southern population, and the southern population reaches its minimal temperature limits well above that of the northern population. But both populations swim at about the same rate in their respective environments, and each is eurythermal, maintaining a fairly constant level of pulsations over a broad range of temperature.

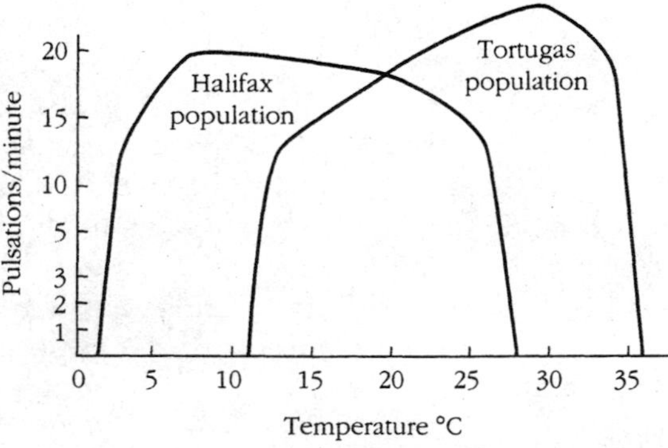

Figure 3-3 Swimming rates of northern and southern populations of the jellyfish *Aurelia aurita* in relation to temperature. (Redrawn with permission from Bullock 1955.)

In addition to ecotypic variation, species populations can also undergo *acclimation,* or short-term changes in response to a changed or changing environment. Often these changes are biochemically based. For example when we ascend a tall mountain, say over a mile high, we will note a higher respiratory rate accompanied by deeper breathing, as well as an accelerated heart rate. These adaptations are a response to the lower barometric pressure and the decreased molecular density of oxygen in the air, and they provide an adequate flow of oxygen to the body tissues. Upon return to lower elevations, both the respiratory and heart rate return to their usual levels. In populations that have inhabited high elevations such as the Andes and Himalayas for many generations, acclimation has given way to permanent genetic changes, as seen in such characteristics as an increased lung capacity and larger number of red blood cells, both adaptations to the low oxygen conditions present.

THE MEDIUM—WATER AND ITS PROPERTIES

The *medium* in which an organism lives is that abiotic component that immediately surrounds it; it is either a liquid or a gas. For aquatic and marine organisms, the medium is freshwater and saltwater respectively; for terrestrial organisms, the medium is air. Air is a physical mixture of gases, including nitrogen, oxygen, carbon dioxide, and various other active and inert gases, whereas water is a compound consisting of hydrogen and oxygen and having unique properties of particular ecological significance. These properties largely relate to water being a dipolar molecule, which results in hydrogen bonds.

In his classic "inquiry into the biological significance of the properties of matter," American biochemist Lawrence Henderson (1913) defended the hypothesis that "the actual environment is the fittest possible abode of life." The properties of water, some of which we shall now explore, provide an excellent defense of Henderson's hypothesis.

Properties of Water

Specific heat. The amount of heat required to raise one gram of water from 0° to 1°C is called *specific heat.* Compared to other liquids, this specific heat is very high, meaning that it requires a considerable amount of heat to alter the temperature of water. Ecologically, this has significance in the tendency of marine and fresh bodies of water to maintain a nearly constant temperature over long periods of time and to be stable over shorter periods, as well. As any early summer swimmer well knows, just because there have been several days of warm (or even hot) weather, the old swimmin' hole can still be icy cold.

Another ecological consequence of the high specific heat of water is the moderation of both summer and winter air temperatures. This is seen in the year-round, consistently moderate, temperatures of coastal areas as contrasted with inland climates. For example, in January the average minimum temperature in Amsterdam, which is essentially a coastal city, is 1°C, whereas in Warsaw, which is at the same latitude, it is −6°C; in July, the average maximum temperature in Amsterdam is 21°C and in Warsaw it is 24°C. Closer to home, in January the average minimum temperature in Boston is −1°C, whereas in Minneapolis, at much the same latitude, it is −12°C. Moderation of temperature is extreme in island climates where the differences between day and night and seasonal temperatures are minimal. In Hilo, Hawaii, for example, the average maximum temperature in January is 27°C and only 28°C in July; likewise, the average minimum temperature in January is 17°C and but 20° in July.

Yet another ecological consequence of the high specific heat of water is its role in the promotion of ocean currents and winds, which will be discussed in Part VII.

Latent heat. There are actually two aspects of latent heat, the latent heat of fusion and the latent heat of evaporation. *The latent heat of fusion* is the amount of energy required to convert one gram of substance from solid to liquid at its melting point; the *latent heat of evaporation* is the amount of energy required to convert one gram of substance from liquid to gas at its boiling point. The latent heat of fusion of water is 80 calories, one of the highest known energy requirements, and the latent heat of evaporation is 536 calories.

A *calorie* (c) is defined as the amount of heat needed to raise the temperature of one gram of water one degree centigrade; it is sometimes referred to as gram calorie or gcal. A "large calorie" or *kilocalorie* (C or kcal) is the amount of heat needed to raise the temperature of one kilogram of water one degree centigrade. Thus 1,000 calories (c) equals one kilocalorie (C). In our discussions of energy we will use c for calories and C for kilocalories.

The ecological consequences of latent heat are several. When a body of water is cooled to its freezing point, further abstraction of heat does not lower its temperature below that point. Heating melts the ice and cooling freezes the water. Secondly, there is a maximal moderating effect on cold climates since the formation of a little ice warms a large amount of colder air. This is why it is always warmer near the edge of a large body of water as it begins to freeze than it is farther on shore. Likewise, since evaporation goes on at all temperatures, it is cooler near a lake in warm weather, since large amounts of heat are consumed in evaporation. It has been estimated that the amount of heat consumed in evaporating water from 100 km^2 is equivalent to 100 million horsepower. The high latent heat of evaporation also explains why transpiration from trees makes them excellent air conditioners, as standing in the shade of one on a hot day readily demonstrates.

Density. One of the unique properties of water is that it reaches its maximum *density* at 4°C, and diminishes above and below that temperature. Ecologically this property is significant because it is largely responsible for the permanent liquid state of many bodies of water in colder climates. If ice were denser than water, colder water would sink to the bottom; ice once formed at the bottom couldn't be melted because warmer water would stay at the surface. Because bodies of water rarely freeze to the bottom, there is a layer of water present in which large numbers of organisms overwinter.

Solvency. No other chemically inert solvent can compare with water in the number and variety of materials it can dissolve nor in the amount of them it can hold in solution. The ecological significance of this *solvency* property is evident in the cycling of nutrients through dissolving soluble materials and thus aiding in their mobilization.

Ionization. *Ionization*, or electrolytic dissociation, accounts for most electrical phenomena and many of the chemical phenomena of solutions. The extent and variety of ionization possible in water far surpasses that in other solvents. This property is ecologically significant due to the environmental richness contributed by large amounts and varieties of ions available.

Surface tension. Next to mercury, water molecules have the greatest tendency to cohere causing its surface to contract to the smallest possible area, a direct result of the hydrogen bonds. In addition to its ecological importance in adsorption and capillary action, water's *surface tension* enables a variety of organisms to inhabit its surface, water striders being but one example.

THE MEDIUM—WATER AND AIR COMPARED

Density and Pressure

The differing properties of water and air pose differing ecological as well as evolutionary challenges to organisms and have resulted in adaptations enabling successful survival in both. The decrease in pressure that accompanies increased altitude, 25 mm mercury/300 m, has significant impact on respiration in warm-blooded animals due to the smaller amount of oxygen that the reduced air pressure is able to sustain. By contrast, the comparatively large increase in pressure that accompanies decreased water depth, namely 760 mm mercury/10 m, has different effects on organisms with air cavities or without air cavities, being potentially adverse for those in the former category. That there are deep as well as shallow ocean communities of organisms indicate that successful evolutionary changes have occurred, adapting these species and thus communities to the challenges of their respective environments.

Density and Buoyancy

The buoyancy effect of water, a property of its density, supports large animals, such as the blue whale, which can attain a weight of 100 tons. By contrast, the largest land plants, the redwoods and sequoias among them, weigh some 6 to 7 tons, and the mighty *Brontosaurus* is estimated to have weighed a mere 40 tons. The buoyancy property enables the existence of plankton communities and such floating species as the Portuguese man-of-war. But this density/buoyancy property has an effect on speed, with speeds attained on land considerably outpacing those attained in water. Duck hawks have been clocked at 290 kph and cheetahs at 112 kph, whereas mackerel may only hit 48 kph.

Waves, Currents, and Wind

Both water and air, through waves and currents in the former and wind in the latter, have abrasive and erosive effects. Pounding waves and currents and strong winds pulverize, erode, and transport substrate, as well as organisms and even whole communities. Both have significant consequences in many ways. For example, on land, most plants are rooted, or sessile, whereas in water, many are free-floating; by contrast, there are no sessile animals on land, but quite a number in water (e.g., sea anemones). Both media transport reproductive materials such as spores and pollen, but the predictability and regularity is greater in water than in air. Finally, both media can effect smoothing and gouging and otherwise mechanically shape the landscape and its inhabitants (Figures 3-4, 3-5, 3-6). Violent waves, such as tsunamis gener-

Figure 3-4 The effects of atmospheric forces on the growth form of Australian pine (*Casuarina equisetifolia*) at South Point on the Island of Hawaii. The flag form results from buds on the windward side of the tree being knocked off by wind and/or destroyed by salt in the air, either or both of which cause excessive evapotranspiration. (Photo by author.)

Figure 3-5 Wind and wind-carried debris have eroded the sandstone creating the sculptured Bryce Canyon of southwest Utah. (Photo by author.)

Figure 3-6 Water erosion has carved the scenic Yellowstone Falls and Canyon of northwest Wyoming. (Photo by author.)

ated by earthquakes (Dudley and Lee 1988), currents exacerbated by flooding, and strong winds such as hurricanes (Boose et al. 1994) can dramatically alter the landscape and devastate ecological communities.

SUMMARY

VOCABULARY

acclimation
calorie (c or gram calorie)
Calorie (C or kilocalorie)
density (water)
ecotypes
eury-
factor interaction
ionization
latent heat of evaporation
latent heat of fusion
law of limiting factors
law of the minimum
law of toleration
limiting factor
maximal conditions
medium
minimal conditions
optimal conditions
solvency
specific heat
steno-
surface tension

KEY POINTS

- An organism's growth and development can be limited if a required factor (e.g., nutrient, sunlight) is present in a minimum amount or exceeds a maximum; distinguishing such limiting factors is not necessarily easy.
- Organisms have a different range of tolerance for different environmental factors, functioning optimally in the central portion of the ranges.
- Environmental factors may act directly or indirectly, as well as independent of or in association with other environmental factors.
- Species with wide ranges of tolerance to a variety of environmental factors are likely to be widely distributed.
- Over its geographic range, a species may show different levels of tolerance to a given environmental factor, resulting in genetic and/or physiological ecotypes or acclimation.
- The physical properties of water are of positive ecological significance, among them moderating ambient temperatures, enabling overwintering under ice, enabling buoyancy, and dissolving a large variety of substances.
- Waves, currents, and winds generally have negative effects on ecosystems even though they function in distributing organisms and their propagation units (e.g., seeds).

CHAPTER 4

Insolation, Precipitation, and Climate

Temperature, precipitation, humidity, wind, and air pressure are the primary determinants of the day to day conditions we call *weather*. Other factors, such as latitude and the presence or absence of mountains also affect weather. On a daily basis, weather can be quite varied—sunny and hot one day, rainy and cool the next. For a still timely, brief introduction to weather see Battan (1974).

Long-term weather, that is, weather over periods of 30 to 40 years or more, constitutes *climate*. Unlike the daily and even seasonal variability of weather, climate tends to be more consistent and hence more predictable because of the long-term averaging of short-term variations. German meteorologist Rudoph Geiger (1941) distinguished between *microclimate*, the more variable climate between the ground surface and 2 meters above it, and *macroclimate*, the less variable climate above 2 meters. It is microclimate that directly impinges on most organisms, being the zone of greatest disturbance and showing the greatest differences within short distances. In general, organisms and ecosystems are influenced broadly by the macroclimate but specifically by microclimate.

This chapter first explores the two major components of climate, temperature and precipitation, and then considers their interaction.

INSOLATION AND TEMPERATURE

Although moonlight, starlight, and luminescent organisms are sources of light, by far the most significant agency is the sun. It is *insolation*, or solar radiation, that produces direct heating of the Earth and is the primary agent in photosynthesis, the starting point of energy flow within ecosystems. And it is the quality (i.e., wavelength, or color), intensity, and duration of insolation that are critical to biological activity. For an in-depth discussion of the quality, intensity, and duration of solar radiation see Clarke (1954) and Gates (1962).

The Sun

The sun is essentially a thermonuclear reactor with a temperature and composition such that hydrogen is transmuted to helium with a release of considerable radiant energy in the form of electromagnetic waves. This radiation extends from high-frequency, shortwave x-rays and gamma rays to low-frequency, longwave radio waves. About 99 percent of the total energy is in the region of wavelengths from 0.136 to 4.0 microns (μ), ranging from the ultraviolet through the near infrared (Figure 4-1). Significantly, about half this energy is in the region that encompasses the visible spectrum (0.38 to 0.77 μ)!

Solar flux. If you pause to consider how small a target the Earth is in the solar system, as well as its distance from the sun, it is not surprising that only about one fifty-millionth of the sun's tremendous energy output reaches the Earth's outer at-

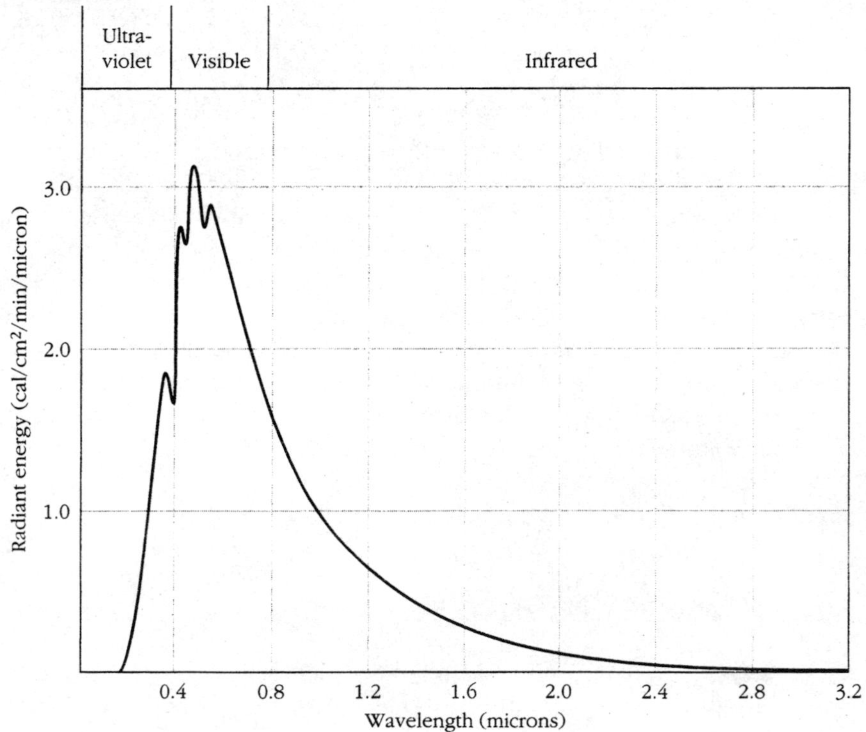

Figure 4-1 Solar radiation outside the atmosphere.

mosphere. Furthermore, it does so at a constant rate referred to as the *solar flux*. This is the amount of radiant energy of all wavelengths that crosses a unit area or surface per unit of time. Using available meteorological data, solar flux is estimated to be 1.94 c per square centimeter per minute (c/cm^2/min), for a total of 13 × 10^{23} c/year. Because of the revolution of the Earth around the sun and the inclination of the Earth's equatorial plane to its orbital plane (Figure 4-2), the flux at any given spot varies seasonally with latitude (Figure 4-3). Because of the Earth's rotation, the flux at a given place varies diurnally.

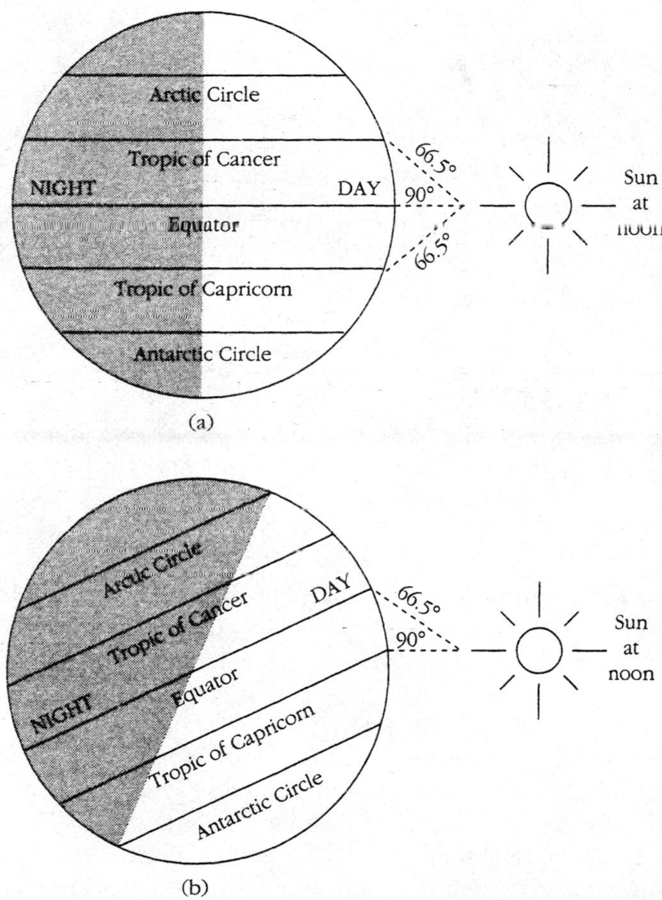

Figure 4-2 (a) Inclination of the sun at fall and spring equinoxes. The sun is directly (90°) over the equator and is at 66.5° at the Tropics of Cancer and Capricorn. (b) Inclination of the sun at the winter solstice. The sun is directly (90°) over the Tropic of Capricorn and is at 66.5° over the equator. At this time, the area above the Arctic Circle is in darkness 24 hours a day, while the area below the Antarctic Circle is in light 24 hours a day; the opposite occurs during the summer solstice when the sun is directly (90°) over the Tropic of Cancer.

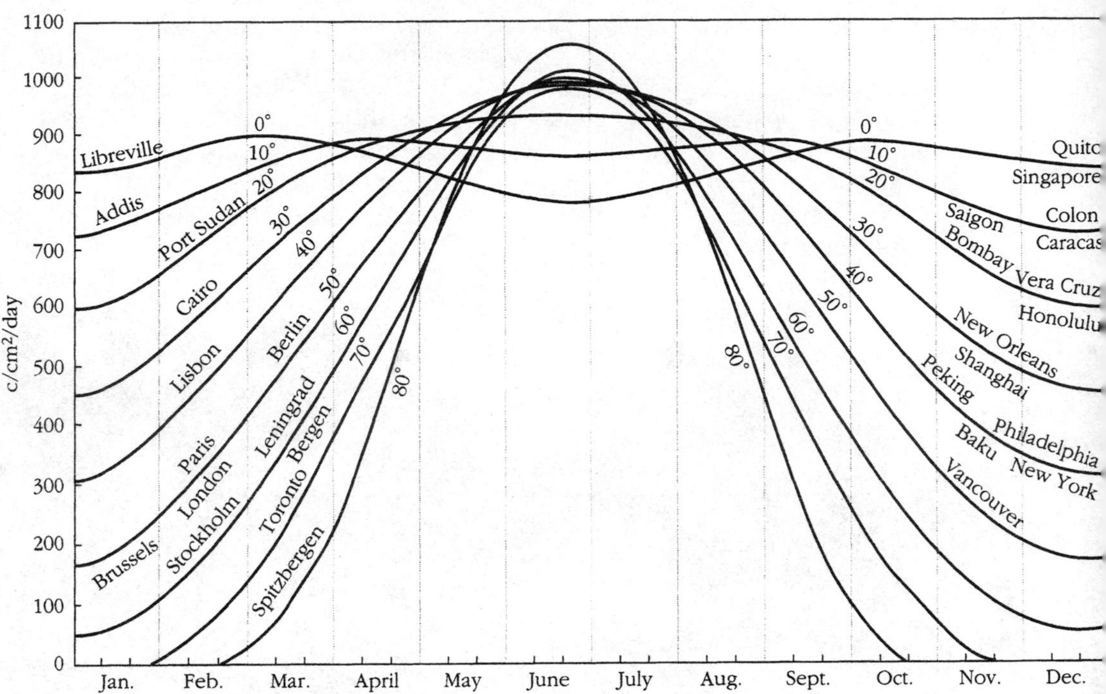

Figure 4-3 Daily totals of solar radiation received on a horizontal surface for different geographical latitudes at different times of the year and based on a solar flux value of 1.94 c/cm²/min. (Redrawn by permission from D. M. Gates. 1962. *Energy Exchange in the Biosphere.* New York: Harper and Row.)

Of significance for biological sytems, however, is the fact that half or more of this flux is depleted as it passes through the troposphere (Figure 4-4). In the Northern Hemisphere about 42 percent of incoming solar radiation is reflected—some 33 percent from clouds and 9 percent from dust (Geiger 1941). This reflecting power, or *albedo*, of the Earth would make our planet about as bright for an inhabitant of space as Venus is for us, according to Geiger. An additional 10 percent of the solar flux is absorbed by ozone, oxygen, water vapor, and carbonic acid or is diffusely scattered by air molecules and small particles. This leaves only about 48 percent of the total solar output that actually reaches the Earth's surface, some of which, in turn, may be reflected back into the atmosphere from light surfaces. As much as 80 percent, for example, is reflected from clean bright sand, as the eyes of beachcombers well recognize. On cloudy days, of course, much less light reaches the Earth's surface, the actual amount being inversely related to the density of the cloud cover. The presence of smoke, soot, and other particulate matter, as well as industrial and automotive gases, also decreases the flow of light to the Earth's surface (Falkowski et al. 1992).

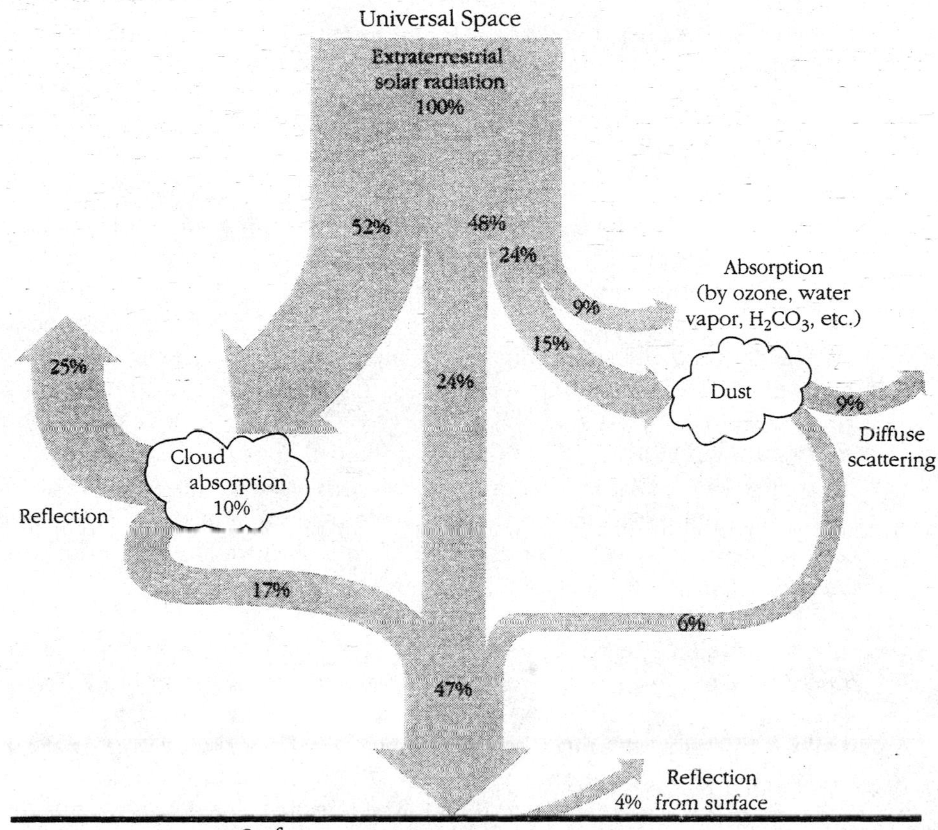

Figure 4-4 Energy intake at the Earth's surface at mid-day.

Infrared radiation. Radiant energy absorbed in the troposphere is also radiated outward in all directions in the far infrared portion of the electromagnetic spectrum. Some of it strikes the Earth and here some of it is, in turn, reradiated. These two energy-income components—direct solar radiation and indirect far *infrared radiation*—heat the lower air, the soil and water at the surface, and the organisms living there, affecting climate, weathering processes, and physiological responses. In addition, the visible light component of direct radiation is needed to initiate the photosynthetic machinery.

Measuring solar radiation. A variety of instruments have been designed to measure the rate and amount of solar radiation, but a discussion of them would take us considerably afield. Pragmatically, ecologists need to know the limitations of particular instruments, but what they really want to know is how much visible light radiation has been or is being received. This factor enables them to calculate the efficiency with which a given ecosystem uses that energy. That is, it is the fate of solar radiation as it becomes incorporated and passes through an ecosystem that is the ecologist's concern.

Changes in Quality and Intensity

Temperature is directly related to the degree, or intensity, of solar radiation. The higher the degree of radiation the higher the temperature; thus, changes in intensity of sunlight directly impact both temperature and biological activity. Likewise, since different portions of the electromagnetic spectrum have different energy levels, changes in the quality of sunlight also impact both temperature and biological activity. Since its output of radiation varies, the sun's warmth and brightness are illusory (Foukal 1990).

As shown in Figure 4-2, there are annual changes in solar intensity caused by the changes in the inclination of the Earth with respect to the Sun. The key factor here is the angle of incidence. The lower the altitude of the Sun, the smaller the angle of incidence and the longer the path of light through the atmosphere; this results in a reduced intensity and consequently temperature.

In water. The most dramatic changes in quality of sunlight occur as light penetrates water where it is reduced in intensity by absorption and by scattering. Thus in pure water the intensity of the red portion of the visible spectrum is reduced to about 1 percent at 4 meters, whereas blue light is reduced only to about 70 percent at a depth of 70 meters (Figure 4-5). Extinction is more rapid at the two ends of the spectrum (violet and red) than in the middle (blue). It is for this reason that deep, clear lakes such as Crater Lake in Oregon (Figure 4-6) and the open ocean ap-

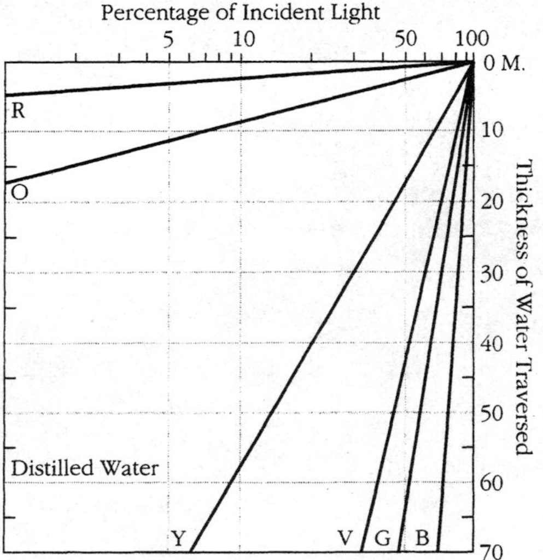

Figure 4-5 Transmission of various wavelengths of light by distilled water showing the percentage of incident light (on a logarithmic scale) remaining after passing through the indicated thickness of water. (Reprinted by permission from G. L. Clarke. 1939. The utilization of solar energy by aquatic organisms. In *Problems of Lake Biology*. Publ. No. 10. American Association for the Advancement of Science.)

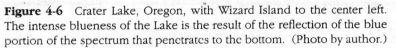

Figure 4-6 Crater Lake, Oregon, with Wizard Island to the center left. The intense blueness of the Lake is the result of the reflection of the blue portion of the spectrum that penetrates to the bottom. (Photo by author.)

pear blue, since blue is the only color reflected, the others having been extinguished. Since natural waters have suspended material, the spectral composition is modified, with the blue component being absorbed or scattered and the green remaining as the most penetrating, resulting in a greenish color to the water. Because photosynthesis is most active in the red-green section of the spectrum, and these wavelengths penetrate water less than blue, it is in the upper portions of deep lakes and the ocean that most photosynthesis occurs and where most aquatic organisms are found.

Light in forests. The light environment in forests is complex, heterogeneous, and subject to change diurnally and seasonally, resulting in effects on communication among animals and between animals and plants; it also has effects on photosynthesis, as well (Endler 1993). Forest color, defined as the spectral composition of light between the near ultraviolet through the visible spectrum, is characterized by yellow-green, blue-gray, reddish, and white depending on the shade or open gap structure of the forest. At dawn and dusk a purplish color is evident. As Endler notes, the perceived colors of animals, flowers, and fruits depends upon the interaction between ambient light color and the reflectance color the animal or plant imparts. As a result, an animal or plant may have a different appearance in different parts of a forest, being relatively cryptic in some light environments and relatively conspicuous in others. In addition, plant growth and form may also be affected by variation in the color of forest light.

Diurnal Changes in Temperature

Because of the high specific heat of water (see Chapter 3), diurnal changes in temperature are minimal in aquatic environments, typically in the order of 1 or 2°C. By contrast, temperature changes from night to day on land can be considerable. The extensive range of diurnal temperature on the desert floor is a case in point (Figure 4-7). In this case, the range at the ground surface is from 18°C at 3:30 A.M. to 65°C at 1:30 P.M., a difference of 47°. Although the air temperature at 120 cm ranges from about 15°C to 38°C over the course of the day, the temperature at 40 cm below ground level remains constant at about 30°C. In such situations, desert shrubs are differentially adapted to withstand daily changes of considerable magnitude at the ground surface while simultaneously experiencing a less marked change a meter above ground and little or no change in their deeper roots. A combination of anatomical, physiological, and behavioral adaptations also enables animals to adapt to the extremes of temperatures in deserts (Figure 4-8).

Although the temperature extremes in this desert situation are quite marked, the temperature pattern at the ground surface interface is typical of terrestrial ecosystems. At the time of the warmest ground surface temperature (1:30 P.M. in Figure 4-7), the

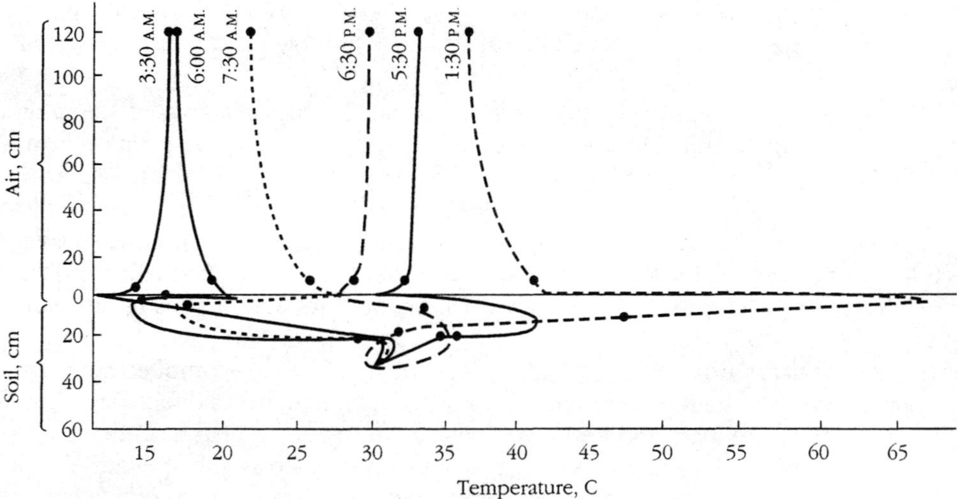

Figure 4-7 The daily cycle of air and soil temperatures from a vegetated sandy area in the Nevada desert on a clear day, July 31, 1953. (Redrawn by permission of the publisher from *Plants and the Ecosystem* by W. D. Billings. Copyright 1964 by Wadsworth Publishing Company, Inc., Belmont, California.)

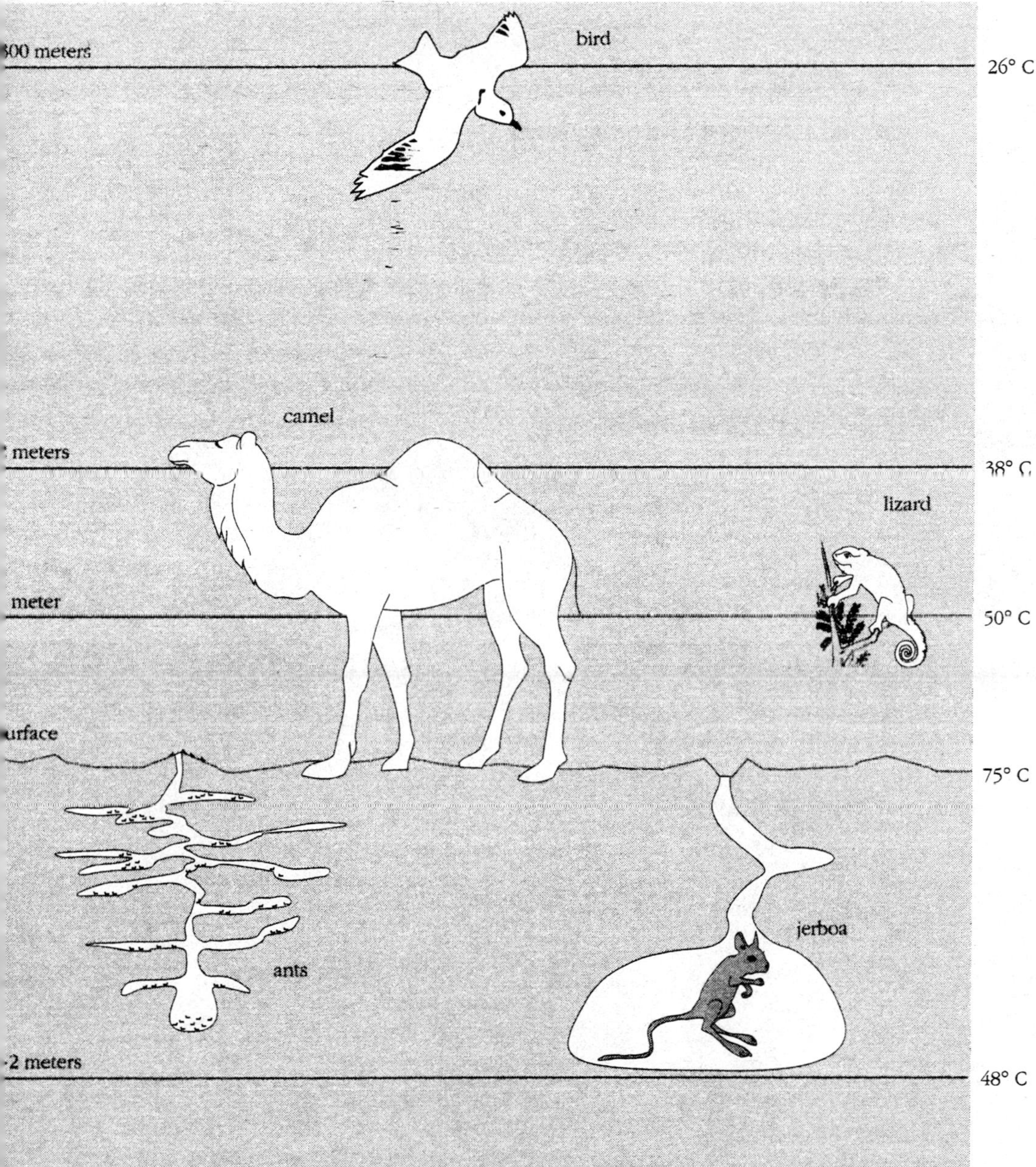

Figure 4-8 Adaptations of animals to temperature in the Negev Desert, Israel. Although the ground surface temperature is 75°C, ants and jerboas 2 meters below the surface are in temperatures that are 27° cooler, birds at 300 meters are in temperatures 49° cooler, and the camel's head is in a temperature 37° cooler than its feet.

temperature above and below the ground decreases; by contrast, at the time of the coolest ground surface temperature (3:30 A.M. in Figure 4-7), the temperature above and below the ground increases.

Annual Changes in Water Temperature

Although daily variations in water temperature are minimal, considerable changes do take place over the course of a year, particularly in the upper waters of temperate aquatic ecosystems (Figure 4-9).

The forces in this series of changes involve the alteration of water density due to temperature cycles and wind action. Because water is most dense at 4°C, warmer or colder water will float on top of a layer of this temperature. So after the ice melts, the surface water warms to 4°C and, as the water becomes more dense, it sinks below the colder, less dense layer immediately beneath it. This overturning of water,

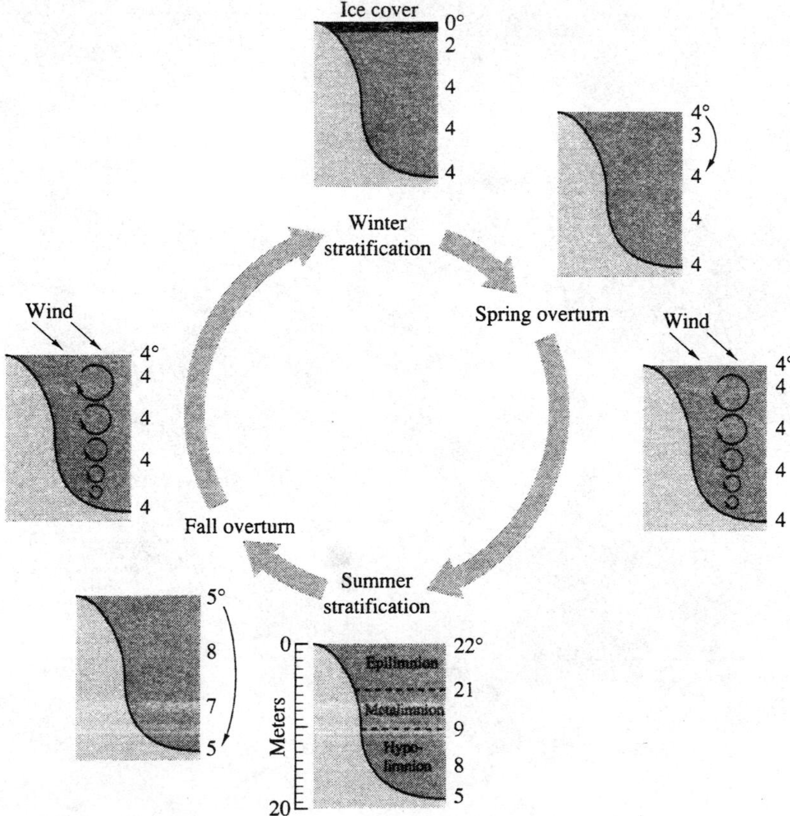

Figure 4-9 The seasonal cycle of temperature (in °C) in a temperate lake. (See text for explanation.)

which, in the absence of wind, will occur at 4°C, ultimately results in a temperature that is uniform from top to bottom. Subsequent heating of surface water, which is accompanied by a corresponding decrease in density, results in warmer water at the surface. Wind action results in stirring of the surface waters, mixing them to depths according to the strength and direction of wind and relative to the exposure or orientation of the system. By midsummer there is a relatively homogeneous surface temperature layer, the *epilimnion*, which is often several meters deep, followed by a zone of rapid temperature change, the *metalimnion*, and then a relatively stable region, the *hypolimnion*, extending to the bottom. As fall progresses, surface temperature changes with corresponding increases in density, eventually resulting in top-heavy conditions. In the absence of wind, turnover will occur at 4°C, but turnover may occur over a longer period of time and at higher temperatures. The end result in either case is the development of a uniform temperature profile from surface to bottom comparable to that of the spring.

Biological significance. The biological significance of this temperature cycle in temperate aquatic ecosystems is indeed broad. Not only are the activities of temperature-dependent organisms directly affected, but the presence of an ice layer, especially when snow covered, severely affects light transmission with a resultant reduction of photosynthesis. Stability of water layers during the summer stratification reduces the amount of mixing so that the hypolimnion tends to become depleted of oxygen even though it may be produced in abundance at the surface. That is, in the stratified condition, there is no atmospheric contact with the hypolimnion. We shall see the untoward effect of this oxygen depletion in a discussion of Lake Erie in Part VIII. On the positive side, however, a significant result of the semiannual mixing of the entire body of water is that nutrients are also mixed and redistributed. Those that became part of the lower water as a result of sinking of dead protoplasm are brought to the surface waters, where they are thus available for photosynthesis and other anabolic activities.

Noncirculating systems. Not all aquatic systems experience such complete mixing. In a very few inland lakes some water, typically in the lower portion, remains unmixed with the main water mass during the circulation periods. Failure to circulate thus largely contributes to the stagnation of nutrient cycling in such systems. In the case of the deep ocean, thermal mixing is restricted to the upper several hundred meters (Figure 4-10). This is largely the explanation of sedimentary stagnation of mineral cycling in the ocean, which will be discussed in Part V. However, on the other side of the ledger, the ocean provides a highly temperature-stable environment. Organisms tolerant of a temperature of only 3 or 4°C have no problem whatsoever below about 1,400 m. George Clarke (1954) has suggested that such a temperature-restricted organism could travel over 60 percent of the globe without exposure to any significantly different temperature.

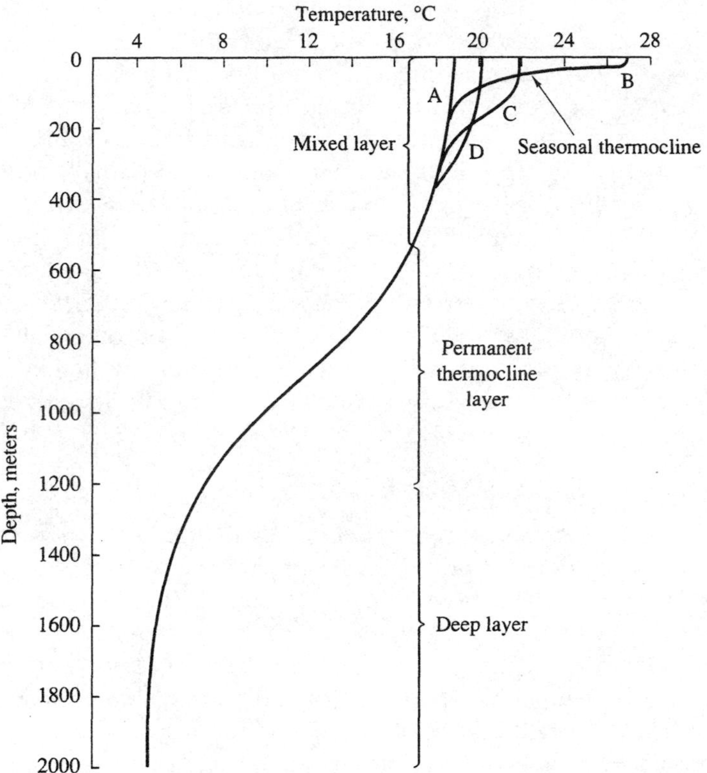

Figure 4-10 The vertical temperature structure in the north central part of the Atlantic Ocean. Seasonal changes in the upper layers are indicated as: A, April; B, August; C, December; D, February. (Redrawn by permission from G. L. Clarke. 1954. *Elements of Ecology.* Copyright © John Wiley and Sons, Inc.)

Temperature and Air Currents

Because solar intensity is greater at the equatorial regions of Earth (Figure 4-2), the lower latitudes are heated more than the polar latitudes. The warmer equatorial air rises to the stratosphere where it moves toward the north and south poles, cooling along the way. The cooler air, being heavier, sinks over the polar regions and flows toward the equator. This circulation pattern is modified because of the spinning of Earth on its axis, which causes the air flow to move toward the right in the northern hemisphere and to the left in the southern hemisphere. The combination of these forces results in several prevailing wind patterns (Figure 4-11), which, in turn, are modified through the annual shift in inclination of the Earth with respect to the Sun (see Figure 4-2).

Temperature is also the major factor in the creation of ocean currents, as will be discussed in Part VII.

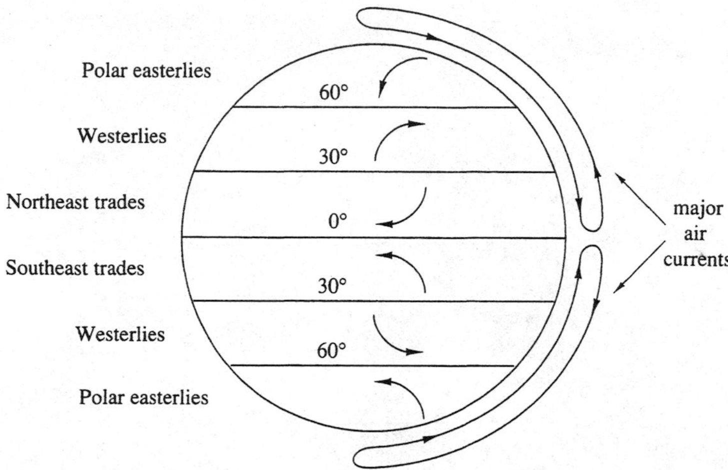

Figure 4-11 The major air currents and prevailing winds on Earth.

PRECIPITATION

As already noted in Chapter 3, the ecological as well as biological significance of water is severalfold and its cyclical movement, globally as well as locally, has considerable influence on the distribution of ecological systems.

The Hydrologic Cycle

The major pathway of the *hydrologic* cycle is an interchange between the Earth's surface and the atmosphere via precipitation and evaporation, the energy for which is derived from the sun (Figure 4-12). The cycle is a steady-state one because the total precipitation is balanced by total evaporation. Although there is a disbalance in favor of the precipitation that occurs over land, there is greater evaporation from the ocean, the compensating factor being runoff from the land. Ecosystems with their biota constitute an accessory whose presence or absence has no significant effect on this major movement, although transpiration does reduce runoff in local environments. However, significant amounts of water are incorporated by ecosystems in protoplasmic synthesis, and a substantial return to the atmosphere occurs by way of transpiration from living plants.

As Figure 4-12 shows, of the total estimated water on the Earth and in its atmosphere, only about 5 percent is actually or potentially free and in circulation, and nearly 99 percent of that is in the ocean. In contrast, 95 percent of the Earth's water is bound in the lithosphere and in sedimentary rocks. Freshwater amounts to only about 0.1 percent of the total supply of water and three quarters of that is bound up in polar ice caps and glaciers. It has been estimated that were this ice-bound water to melt completely, it would equal a water layer 50 m deep over the entire surface of the Earth.

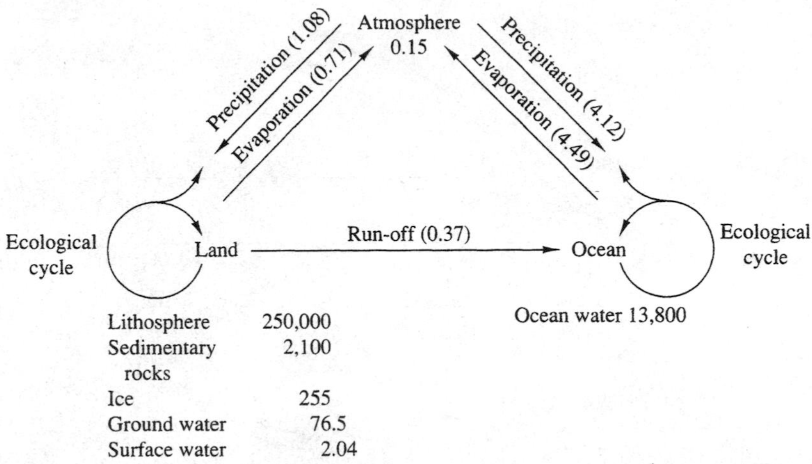

Figure 4-12 The general pattern of the hydrologic cycle and the distribution of water in 10^{17} kg; amounts in parentheses are annual rates. The ecological cycle involves uptake by photosynthesis and imbibition and loss by respiration and transpiration. (Data from Hutchinson 1951, Nace 1967, Kalinin and Bykov 1969, and *International Hydrological Decade* 1970.)

Although total precipitation (5.2×10^{17} kg or 5.2×10^{8} km^3) is an impressive quantity, it is even more significant to note that its source—namely, atmospheric water vapor—constitutes an infinitesimal amount relative to the distribution of water on the Earth. It has been estimated that this amount of water vapor is the equivalent of 2.5 cm of rain covering the entire surface of the Earth. By dividing the total annual precipitation by the amount in the atmosphere, the implication is a turnover about 35 times a year, or approximately every 11 days.

Circulation Patterns

The relative and absolute amounts, as well as the periodicity of precipitation and evaporation, strongly influence the structure and function of ecosystems. Although this topic will be considered in some detail in Parts V and VI, it is relevant here to note some general patterns of precipitation as they relate to the distribution of major types of ecosystems. The circulation pattern of the atmosphere is primarily responsible for the peculiar patterns of precipitation distribution (Figure 4-13). The trade winds, for example, move from cooler latitudes toward the equator, taking up moisture as they move and depositing it in the equatorial region (Figure 4-11). Thus the coastline adjacent to the trade winds north and south of the equator (e.g., southern California, Mexico, northern Chile) is relatively dry, the equatorial region very wet. In contrast, the westerlies, which are north and south of the trade winds, move from warm to cool latitudes, depositing their water load along the coast (Oregon north to Alaska, southern Chile).

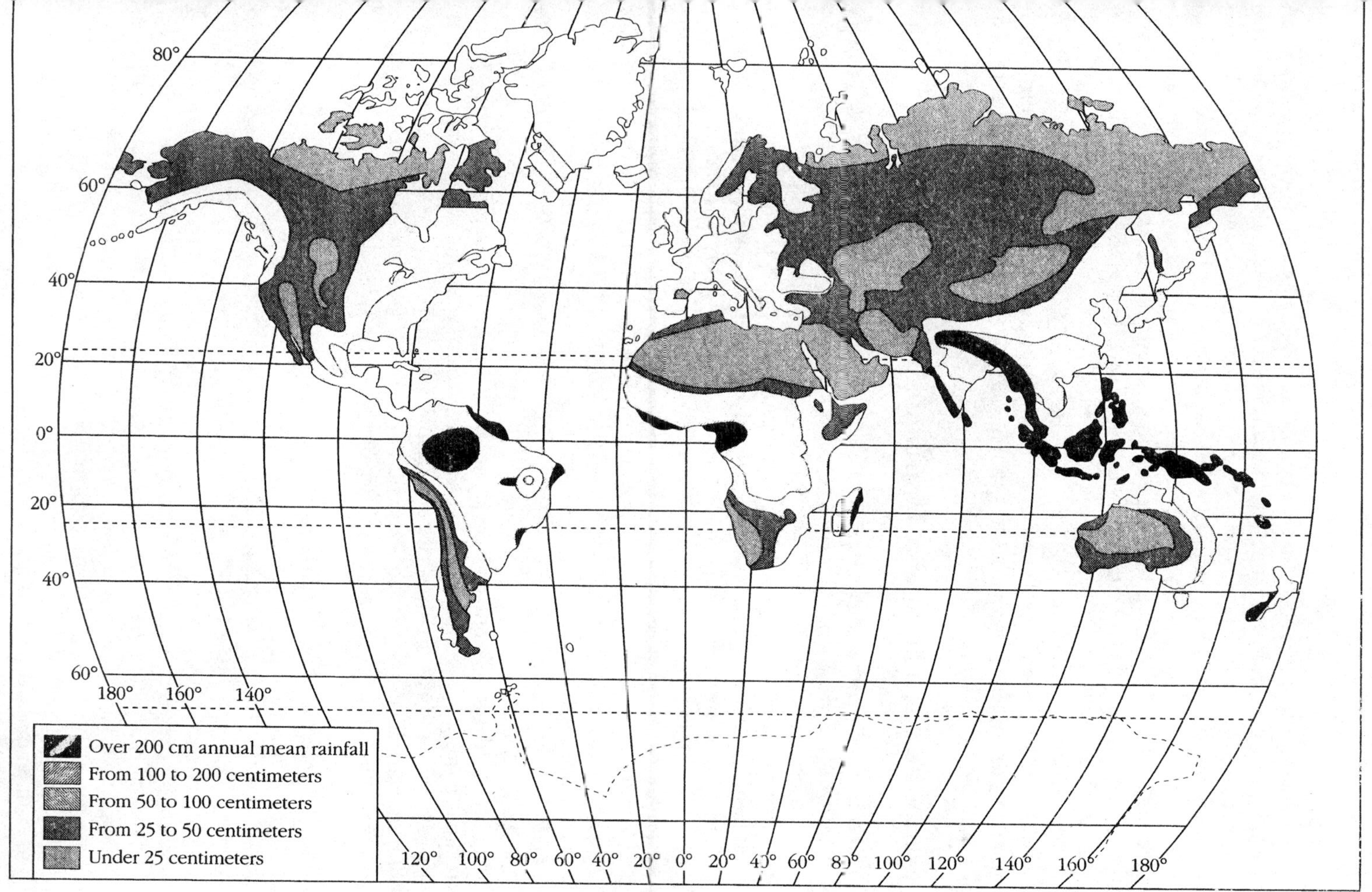

Figure 4-13 The pattern of world precipitation. (Adapted by permission from W. G. Kendrew. 1930. *Climate*. Oxford: Clarendon Press.)

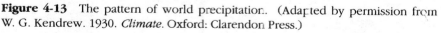

Of considerable importance to continental rainfall distribution are mountain ranges. As moist air moves over a mountain range, it rises and cools to supersaturation, and precipitation results on the windward side of the range (Figure 4-14). As the moisture-depleted air continues to move beyond the crest of the range, it descends, warms, and picks up moisture by evaporation from ground and water surfaces and releases it further to the lee of the mountain. Thus the lee side receives less precipitation than the windward side, an effect often referred to as that of a "rain shadow" analogous to the shadow made by a tree. Examples of the rain-shadow effect can be seen in Figure 4-14 in western North America (Coast Range, Sierra Nevada, Wasatch, and Rocky Mountains), and in Figure 4-13 in western South America (Andes) and southern Asia (Himalayas). More detailed maps of particular areas show this effect even more clearly.

Precipitation and Ecosystem Distribution

The general world precipitation pattern, then, is the result of several forces, primary among which is the interaction between atmospheric circulation and continental or island topography. This interaction has much to do with the distribution of major ecosystems (see Chapter 14). Deserts, in general, receive less than 25 cm of precipitation annually, for example. Even a cursory superpositioning of the geography of annual precipitation (Figure 4-13) and the geography of the major desert areas (Figure 14-1) will show considerable coincidence. As the basic mechanism of the hydrologic cycle suggests, however, the rate and amount of evaporation are as critical as the rate and amount of precipitation. It is the ratio of these two forces that is the crucial factor in determining the distribution of particular types of ecosystems, be they deserts or tropical rainforests. Thus in the western North American desert states (Nevada, eastern Oregon, southern Idaho, western Wyoming, much of Utah and Arizona, and southeastern California) the ratio of precipitation to evaporation is 0.2 or less—that is, potential or actual water loss is greater than water gain. In the relatively arid to moist short-grass prairie states (eastern Montana and western North Dakota, south to eastern New Mexico, and northern Texas), the ratio of precipitation to evaporation is between 0.2 and 0.6. In the more moist eastern half of the country predominated by deciduous forests, the ratio is 0.8 to 1.6.

Precipitation and Nutrients

Another aspect of the hydrologic cycle is worth mentioning briefly. As noted in Figure 4-12, surface runoff to the ocean accounts for 35 percent of the annual intake via precipitation. Because of its capacity for dissolving and carrying nutrients, runoff water is a significant agent in moving nutrients from one ecosystem to another, depleting one, enriching the other. Because water with its nutrient load moves downhill, upland areas tend to be less nutrient rich than lowland areas, such as marshes and continental shelves. This nutrient enrichment is largely responsible for the high productivity of such areas. The importance of surface runoff in nutrient depletion and enrichment is discussed further in Part IV.

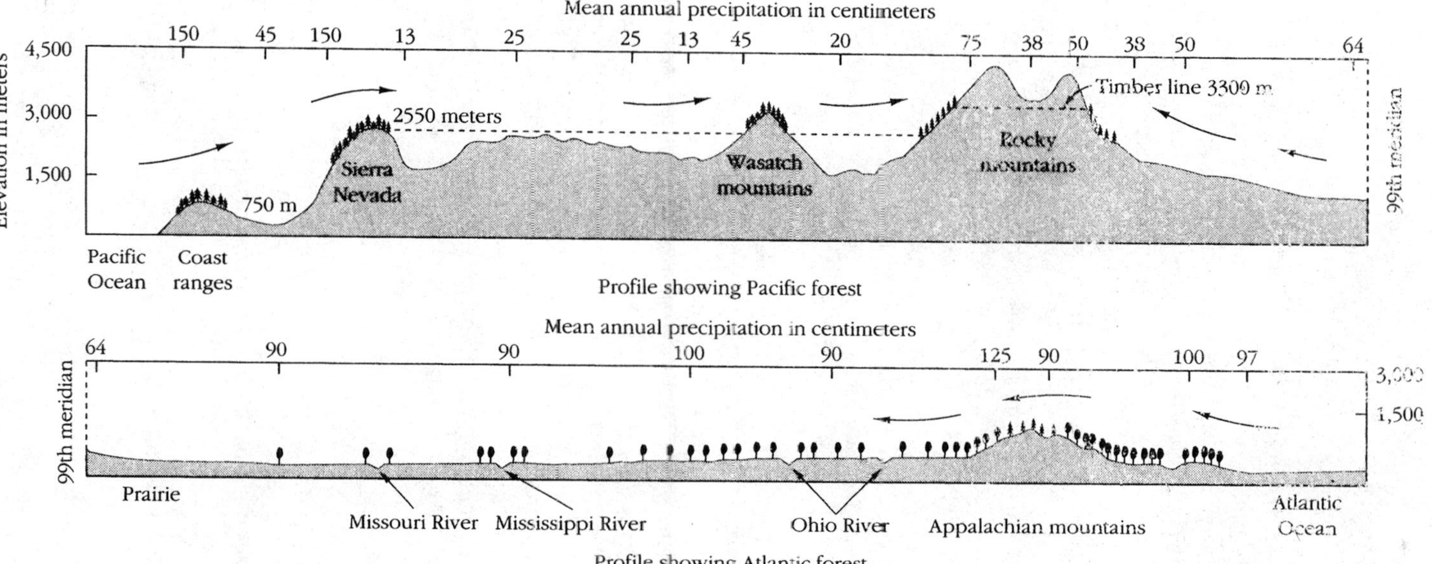

Figure 4-14 The influence of topography on precipitation along 39° north latitude. (Redrawn from R. Zon, 1941. In *Climate and Man.* U.S. Department of Agriculture Yearbook.)

Streams and Groundwater

Streams and groundwater are the greatest sources of water for human use. About 25 percent of the total volume of stream flow is withdrawn annually for human use, nearly 30 percent of which returns to the atmosphere by evaporation. It has been estimated that by the year 2000 some 75 percent of stream flow will be withdrawn for human use for irrigation, cooling power plants, and other industrial uses.

Groundwater is found below the water table, where it is beyond the reach of plant roots or being evaporated. As noted in Figure 4-12, groundwater is some 38 times greater in amount than surface water in streams and lakes. Groundwater sometimes surfaces as springs or may move in rock bodies known as *aquifers*, from which it can be pumped to the surface. Wells that have sufficient hydrostatic pressure to force water above the ground surface to form a fountain are called *artesian wells.*

Groundwater provides about 25 percent of all uses of freshwater in the United States, but meets about 50 percent of drinking water needs. Increasingly, groundwater is used for irrigation purposes, resulting in concern about its availability in the future. Other investigators, however, see the underground reservoirs as a major factor in controlling the water cycle (Ambroggi 1977; Peixoto and Kettani 1973).

TEMPERATURE AND PRECIPITATION

Temperature and moisture constitute excellent examples of factor interaction, a concept discussed in Chapter 3. Whereas temperature has a more limiting effect on organisms when there is either a little or a large amount of moisture, moisture has a more limiting effect when the temperature is very low or very high. Thus some organisms may be very productive or active at high temperatures in the presence of considerable moisture, but the opposite at low temperatures when considerable moisture is present. For example, the American cockroach (*Periplanata americana*) dies at 38°C at high humidities but can survive up to 48°C in dry air; the oriental cockroach (*Blatta orientalis*) can survive 24 hours at 37–39°C if the air is moist but dies at the same temperature in dry air (Chapman 1982).

Kalkat et al. (1961) showed that the insecticide parathion had no effect on mortality of the red floor beetle (*Tribolium castaneum*) at a relative humidity of 90 percent and a temperature of 32°C but a 44 percent mortality at the same relative humidity with a temperature of 43°C. By contrast, the insecticide malathion had a reverse effect with the same combination of humidity and temperature: mortality of 78 percent at a relative humidity of 55% and temperature of 32°C, but 32 percent mortality at the same relative humidity and a temperature of 43°C.

Climate and Treeline

Climate, the long-term interaction of precipitation and temperature, has profound effects not only on the distribution and range of species but, in consequence, on ecosystems (see Chapter 14). One specific case of interaction of climate and species ranges is worthy of brief exploration here; namely the causes of treeline. As Stevens and Fox (1991) note, treelines appear under a variety of circumstances—treelines on tall mountains, along the margins of bogs or swamps, and between forest and grasslands, among others. The correlation of climate and tree distribution is supported by a large body of observations, and the migration of treeline in response to global climate change further supports the role of climate in determining the line of finality in distribution. While the weight of correlative evidence of climate impact on species (and hence ecosystem) distribution is indeed strong, causal analysis and explanation are more difficult. Stevens and Fox (1991) review a number of different hypotheses for the treeline phenomenon, including those related to tree growth (e.g., the proportion of tissues that are photosynthetic decline with growth) and stature (e.g., small plants can exploit the narrow zone of favorable conditions near the ground in cold climates whereas upright trees cannot). Discussion of these hypotheses and the difficulty in testing them would take us too far afield at this juncture, but the point is incontrovertible that climate does exert a profound influence on the distribution of species and ecosystems.

Climographs

One of the effective ways of representing the relationship of temperature and rainfall is the *climograph*, in which temperature is plotted against precipitation (or relative humidity) (Figure 4-15). This kind of graphic representation allows comparison of annual variations and of different major ecosystems. From the applied perspective, climographs have been used to determine the suitability of a particular location for the introduction of new plants and animals.

Classification of Climate

The ancient Greeks recognized a relationship between latitude and temperature, devising a system of climatic zones, namely torrid, temperate, and frigid. In the mid-1850s, French physiologist Alphonse de Candolle advanced a classification based largely on world vegetation. Building on de Candolle's efforts, German biologist Wladimir Koeppen devised his first classification of vegetation zones (1900) and later (1918) gave greater attention to the seasonal characteristics of temperature and rainfall. The major categories in Koeppen's scheme are:

A Tropical forest climates; hot all seasons
B Dry climates
C Warm, temperate, rainy climates; mild winters
D Cold forest climates; severe winters
E Polar climates

To recognize differences within the major climatic types, additional symbols were added: a second letter referring to rainfall and a third to temperature characteristics. For example, BWh refers to tropical desert, arid, hot, whereas BWk refers to mid-latitude desert, arid, cool or cold.

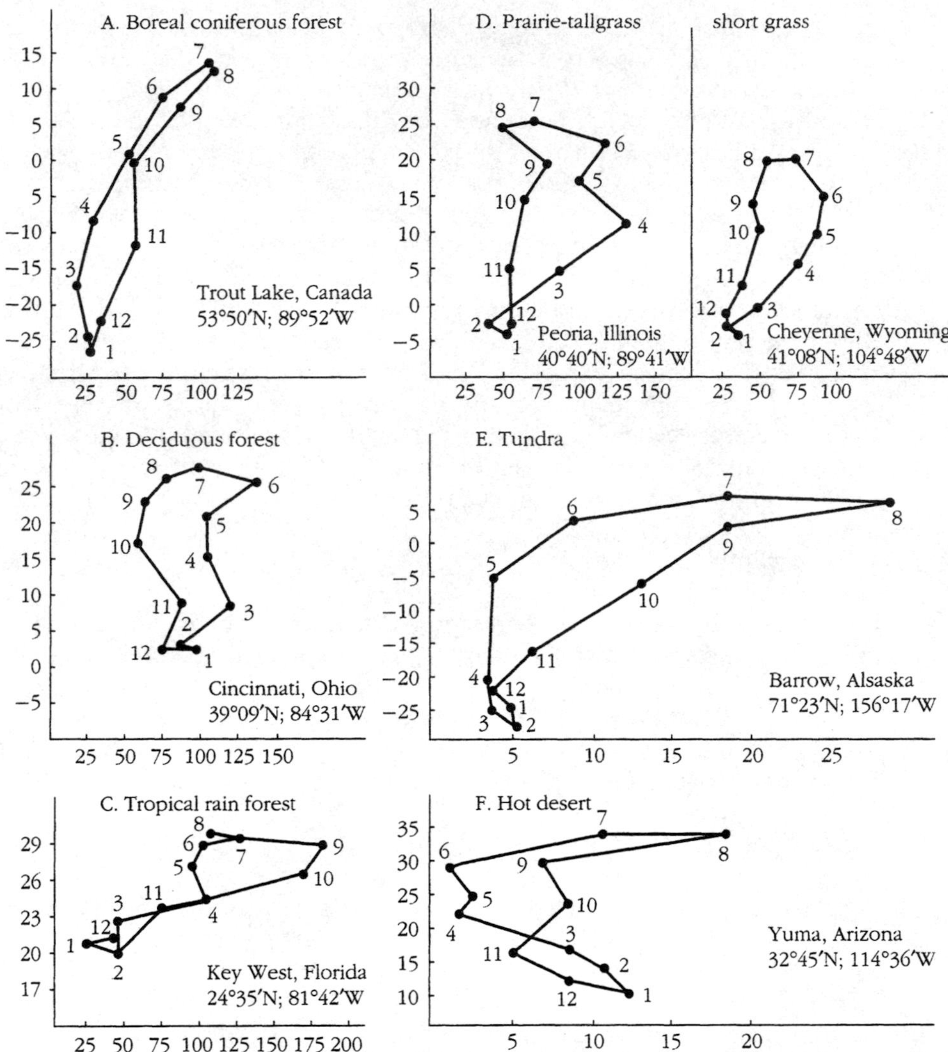

Figure 4-15 Climographs of representative locations in the major biomes of North America. The vertical axis is temperature (°C); the horizontal axis is precipitation (millimeters); numbers refer to months (1 = January, 2 = February, etc.). Data are monthly averages from the period 1941–1960 from *World Weather Records 1941–1950,* U.S. Department of Commerce, 1959.

The Koeppen system is the most widely used, although other climatologists have advanced different classification schemes. American climatologist C. Warren Thornthwaite advanced a scheme for North American climates (1931) and later (1933) extended it to the world. Like Koeppen's classification, Thornthwaite's is based on natural vegetation but, significantly, takes into account the quantity of moisture lost through evaporation from the soil and through evapotranspiration from vegetation.

Further discussion of climatic classification would take us too far afield and is a reminder of Macfadyen's caveat (Chapter 1) that the ecologist must guard against wandering too far from the centrality of ecology. Nonetheless, we will discuss further aspects of temperature and precipitation interaction as they affect geographic distribution of major ecosystems in Part VII.

In departing this brief discussion of the major aspects of climate, it is worth noting that Ruddiman and Kutzbach (1991) believe the formation of giant plateaus in Tibet and the American West may explain why the Earth's climate has grown markedly cooler and more regionally diverse in the past 40 million years. And finally, the role of clouds in global climate changes may be the most elusive factor leading to imprecision in the various climatic models being constructed (La Brecque 1990); the lack of strategic research planning in terrestrial ecology may also be a limitation (Mooney 1991a).

SUMMARY

VOCABULARY

albedo	epilimnion	macroclimate
aquifers	hydrologic	metalimnion
artesian wells	hypolimnion	microclimate
climate	infrared radiation	solar flux
climograph	insolation	weather

KEY POINTS

- The quality, intensity, and duration of solar insolation are critical to biological activity.
- Solar flux at any given location varies seasonally with latitude due to the inclination of the Earth with respect to the sun, as well as diurnally due to the Earth's rotation; nearly half of solar flux is depleted by reflection, absorption, and scattering as it passes through the troposphere.
- Insolation quality changes most dramatically in water as a result of absorption of different wavelenths of light due to depth and the presence of absorbing substances, including the biota.
- Diurnal changes in temperature are most marked in terrestrial as opposed to aquatic situations and are dramatic in desert environments.
- Annual changes in temperature in aquatic environments are more marked in temperate freshwater situations than in other climates or marine environments.

- Temperature is the major factor in the creation of both ocean and air currents.
- The circulation of water between Earth and its atmosphere is essentially in balance; a variety of factors (largely atmospheric circulation and topography) influences the amount and periodicity of precipitation and evaporation in any given location.
- The world precipitation pattern is a substantial factor in the distribution of major ecosystems.
- Temperature and moisture are significant interacting environmental factors affecting species, and hence, community distribution.
- Climographs provide a quick graphic overview of annual temperature-precipitation relationships.
- A number of climate classification schemes have been advanced, the most widely used today being that of Koeppen.

CHAPTER 5

Soils, Nutrients, and Other Factors

SOILS

The study of soils, or pedology, is itself a science, thus our more limited focus is only on some of the basic fundamentals of soil and their effects on and interactions with ecological entities. Historically, soil has been regarded primarily from either a chemical or a geological point of view. The chemical perspective is perhaps best represented by a definition advanced by the early 19th century Swedish chemist Jons Jakob Berzelius, who defined soil as "the chemical laboratory of nature in whose bosom various chemical decomposition and synthesis reactions take place in a hidden manner" (Bridges 1978). The geological view is that soil is pulverized rock containing a certain amount of organic matter, the remains of plants and animals. Subsequently, the view has emerged that soils are related to climate and other physical, chemical, and biological factors, including vegetation, and that they are dynamic systems undergoing constant, perhaps imperceptibly slow, evolutionary changes.

Soil Profile

The mineral and organic components of soil are differentiated into horizons or strata of variable depth. Each of these horizons differs in morphology, physical structure, and chemical and biological characteristics. These horizons are evident when a vertical cut is made through the soil, revealing the *soil profile.* Horizons are categorized by their position in the soil profile as well as by the dynamic processes that have brought them about. The widely accepted structure of the soil profile is as follows (also see Figure 5-1):

- O Organic litter of loose leaves and debris
- A_1 Rich in humus and hence dark in color, the result of decomposed organic material mixed with minerals
- A_2 Zone of maximum leaching of minerals; since the minerals are in solution and hence readily available, plant roots tend to be concentrated in this horizon

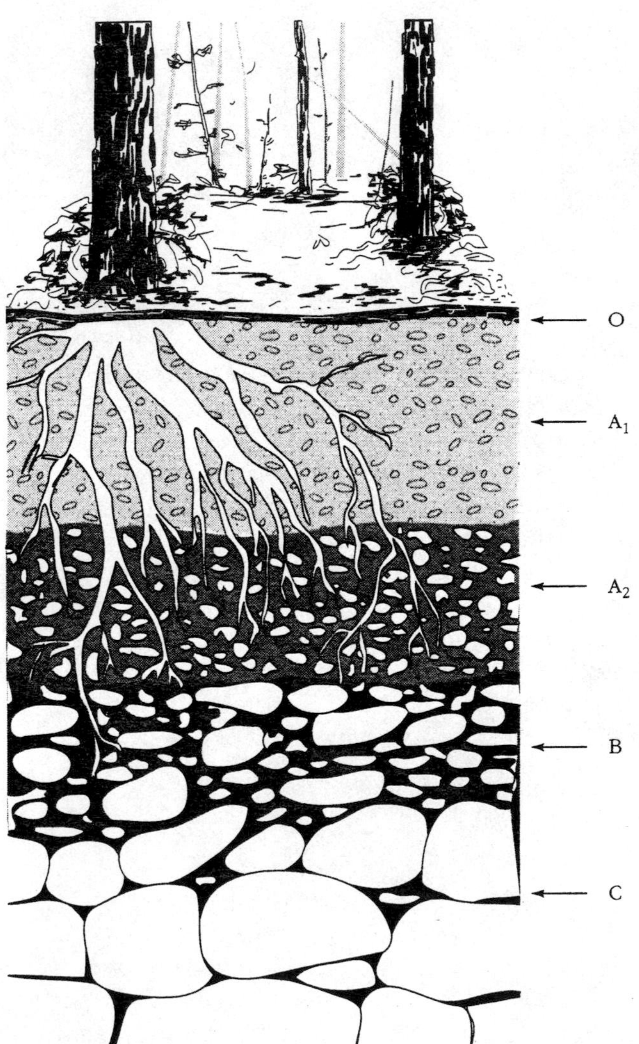

Figure 5-1 Generalized soil profile (see text for details).

B Little organic material and the chemical composition is largely that of the underlying rock; also referred to as the zone of accumulation since minerals from above and below tend to concentrate here
C Parent rock, which is weakly weathered
D Unweathered bedrock

Because of differences in their origin, composition, and formation, the particular soil profile varies in different situations, as we will subsequently see. The horizons are often subdivided for more precise and refined description, but this generalized profile will suffice for our discussion.

Soil Composition

Mineral matter, organic matter, air, and water are the four main constituents of soil. In a typical topsoil, mineral matter constitutes about 45 percent of the soil, organic matter 5 percent, water and air 25 percent each (Bridges 1978).

Mineral matter. Soil mineral matter is derived from the weathering of the parent material (horizon C) and dissolved substances introduced at the soil surface. Its texture is determined by the size and amount of various-sized particles from very small clay particles (less the 0.002 mm diameter) to sand particles (up to 2.0 mm diameter). Although larger particles or stones may occur, for the most part they are not significantly active in soil dynamics.

Organic material. Organic material, as noted in the description of horizon A, is derived from the decay and decomposition of organic matter in the O horizon. Depending on the environmental situation, the litter and debris in the O horizon may completely decay within a year; in other instances they may persist for much longer periods of time. Among the agents active in the degradation process are earthworms, which ingest the organic material and through their movement and egestion mix it with the mineral part of the soil. Dyksterhuis and Schmutz (1947) found that in an undisturbed virgin prairie in Texas, earthworm casts formed a layer 2 to 3 mm thick, which, when air dried, weighed 2,400 g/m^2. In the moderately hot, dry season in Sudan, the earthworm casts weighed (air dried) 475 g/m^2, but 24,000 g/m^2 during the rainy season (Evans and Guild 1947). Other burrowing animals such as prairie dogs, pocket gophers, and ground squirrels move and mix substantial amounts of soil, as do crayfish in poorly drained areas.

In addition to the bacteria and fungi associated with decaying plant and animal material in the organic matter of the O and A_1 horizons, there is often a significant community of living, small photosynthetic organisms, usually referred to as the microflora or microphytes. These can include mosses, lichens, liverworts, and algae. Their role in energy flow and other ecological activities is not, however, well understood (West 1990).

Air and water. Air and water occupy space between the soil particles, the amount of the former depending on the amount of the latter. That is, if the soil is saturated with water, there is little air; but if the soil is depleted of water, air fills the empty spaces.

The composition of soil air is similar to atmospheric air, from which it is derived by diffusion. While the nitrogen content is just about the same as that in the atmosphere, there is typically less oxygen and more carbon dioxide owing to bacterial and fungal decomposition of organic matter in the soil. Anaerobic conditions can arise in soil when the air spaces are filled with water, and oxygen and other gases are unable to diffuse inward from the atmosphere.

The amount of water held by soil depends on its texture and structure; in porous situations, such as sand and volcanic soils, water drains quickly, but in compact situations, such as clay soil, water drains very slowly. In addition to providing water to soil organisms, soil water dissolves soluble entities providing the medium through which plants obtain necessary nutrients. Inorganic salts dissociate into positive and negative ions (cations and anions, respectively) in solution; for example, sodium nitrate ($NaNO_3$) and water (H_2O) dissociate respectively into Na^+ and NO_3^+ and H^+ and OH^-. These cations, which potentially could be lost by water seeping more deeply or evaporating, are, however, retained through the chemical properties of the clay-humus complex.

Colloidal complex. The *colloidal complex*, an intimate association of finely divided humus and mineral soil, especially clay, has been called the "heart and soul" of the soil. It influences the water-holding capacity and rate of circulation through it and is the source of plant nutrients, which are released gradually. Clay-humus particles act like highly charged anions and are referred to chemically as *micelles.* The cations formed when salts dissociate in solution are attracted to and adsorbed by the clay-humus micelle (Figure 5-2). Each of the cations is capable of being exchanged with other cations, thus constituting a dynamic process in nutrient availability and exchange.

Soil Formation

The formation of soil is a dynamic process that depends on climate, organisms, topography or relief, parent material, and, importantly, time (Jenny 1941). But soil formation is also dependent on a number of processes including weathering and leaching, among others. A brief discussion of each of these factors and processes follows.

Climate. The combination of rainfall and temperature plays a significant role in the formation of a soil. Rainfall, for example, promotes movement of minerals and organic matter in soil and is the agent responsible for the solution of soluble salts. However, when rainfall exceeds evaporation *leaching* of soluble salts occurs, that is, they are removed from the soil in the drainage water, resulting in nutrient impoverishment.

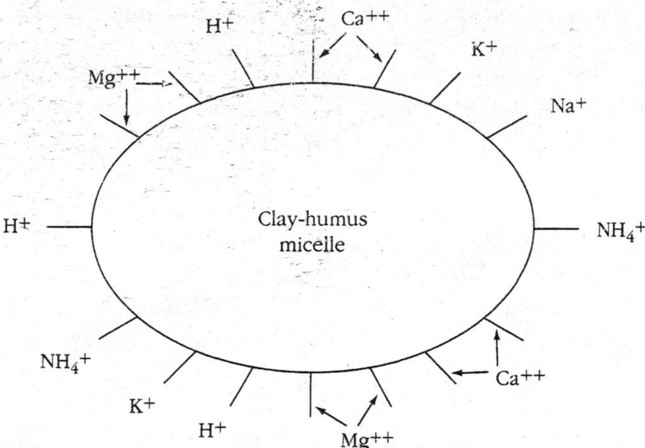

Figure 5-2 Clay-humus micelle showing absorbed ions held by attraction of cation to anion charges.

Rainfall also contributes signficant amounts of minerals to a given area, minerals that are in the atmosphere in the form of dust and other particulates. For example, Carlisle et al. (1966) noted that of the total amount of calcium, magnesium, and sodium added to the soil in an English oak woodland, 17.8, 35.0, and 61.8 percent, respectively, were gained through precipitation.

The expansion and contraction in parent rock caused by shifts in temperature contribute to fragmenting and even pulverizing the rock, a phenomenon referred to as physical *weathering*. This weathering effect is exaggerated when moisture present in rock crevices freezes, causing the formation of ice that expands in the crack, causing further breakage. Chemical weathering occurs when rain containing acids from the breakdown of organic material strikes rock surfaces and results in chemical breakdown of the rock.

Organisms. Incorporation of organic matter into the mineral soil is largely the result of organism activity. Digestion by mites and other arthropods is coupled with decomposition by bacteria and fungi and followed by the mixing activity of earthworms and other burrowing animals. Carlisle et al. (1966) showed that of the total amount of carbon, nitrogen, calcium, and phosphorus that entered into the soil in an English oak woodland, 89.6, 82.3, 62.6, and 58.1 percent, respectively, came from decomposition of the litter.

Organisms like blue-green algae that inhabit the interstices of granite chemically break down parent rock through the release of their metabolic waste products. Shachak et al. (1987) studied two species of snail (*Euchondrus albulus* and *E. desertorum*) that feed on lichens growing at depths of 1 to 3 mm under the surface of limestone rocks in the Negev Desert, Israel. Through their herbivorous activity, the snails are major agents of rock weathering and soil formation in this desert, inasmuch as they physically disrupt and ingest the rock substrate in order to consume the lichens. Rock surface was removed at the rate of 15 mm^3 per snail, per day, or 4 percent of the rock surface area to a depth of 1 mm per year. Based on this rate, the weathering of rock by snails for the whole desert was estimated to be 69.5 to 110.4 g/m^2 (0.695 to 1.104 metric tons/hectare/year).

Topography. The effect of topography on climate, and most particularly rainfall, was described in Chapter 4. It follows that weathering and decomposition processes are strongly influenced by topography. Upland leeward areas are both moister and cooler (see Figure 4-14) thus increasing weathering of parent rock but slowing the decomposition of organic matter and allowing it to become thicker.

Parent material. Since this is the initial source of soil (except in situations where materials are imported through such activities as glaciers, flooding, and wind), parent material is the fundamental basis of the kind of soil existing in any given area. It is this material which is weathered, pulverized, and mixed with organic material.

Time. As with many natural phenomena, time is a critical factor in soild formation. Young soils differ from mature soils because they retain many of the characteristics of their parent material, whereas mature soils have varying amounts of organic material and a clearer definition of their horizons.

Other processes. Depending on environmental and other circumstances a number of additional processes are involved in soil formation. These include:

the movement of colloidal suspensions where soil is subject to desiccation (eluviation);

the development of an extremely acid humus layer in which decomposition is slow, as in coniferous forests (podzolization);

development of a calcium layer in the B or upper C horizon in low rainfall areas (calcification);

the accumulation of oxidized iron and aluminum in humid tropical environments (laterization);

the absence or removal of soluble salts resulting in their accumulation in arid climates (salinization);

the partial removal of salt leaving calcium and magnesium dominant in the clay-humus micelles (calcification);

and the reduction of iron compounds followed by their removal in soils subject to prolonged wetness and consequent anaerobic conditions (gleying).

Classification of Soils

Given scientists' proclivity toward categorization and systematization, it is not surprising that a number of soil classification schemes have been developed (for descriptions, see Bridges 1978, Fitzpatrick 1980, Glaszovskaya 1983, among others). In 1900 the Russian soil scientist V. V. Dikuchaiev appears to have been the first to produce a classification of soils, many terminologies of which have persisted in various forms in subsequent and more sophisticated classifications. Among the most widely used classification, particularly in the United States, is one developed by the U.S. Soil Conservation Service of the Department of Agriculture (1975) (Table 5-1 and Figure 5-3).

TABLE 5-1 U.S. SOIL CONSERVATION SERVICE CLASSIFICATION OF SOILS

Category	Derivation	Description
Alifsols	Aluminum (Al), Iron (Fe)	Soils with aluminum and iron precipitating as silicates in alkaline B horizon; thin humus layer; well-defined horizons (e.g., soils of northcentral U.S.)
Aridisols	Latin, *aridus* = dry	Soils that are arid for long periods, little humus, high alkaline content (e.g., desert soils)
Entisols	Latin, *ent* = recent	Weakly developed, very young soils lacking horizons and dominated by mineral matter (e.g., river banks)
Histosols	Greek, *histos* = tissue	Soils with considerable organic matter (e.g., peats, bogs, muck)
Inceptisols	Latin, *inceptum* = beginning	Moderately developed soils, horizons not well-developed, variable chemical properties, generally humid climates (e.g., brown forest soils)
Mollisols	Latin, *mollis* = soft	Well-developed soils with dark A horizon, high alkaline content, surface layer soft and crumbly (e.g., prairie soils)
Oxisols	French, *oxyde* = oxide	Well-weathered tropical soils rich in iron oxides and typically rich with humus (e.g., lateritic soils of tropics)
Spodosols	Greek, *spodos* = wood ash	A_1 horizon humus-rich, A_2 horizon ash-color with reddish to black B horizon (e.g., the podzol soils of New England)
Ultisols	Latin, *ultimu* = last	Old soils, usually in warm, humid climates, strongly leached, low alkaline content (e.g., soils of southeastern U.S.)
Vertisols	Latin, *verto* = to turn	Clay soils that have deep wide cracks at some time of year, dark in color (e.g., dry river beds)

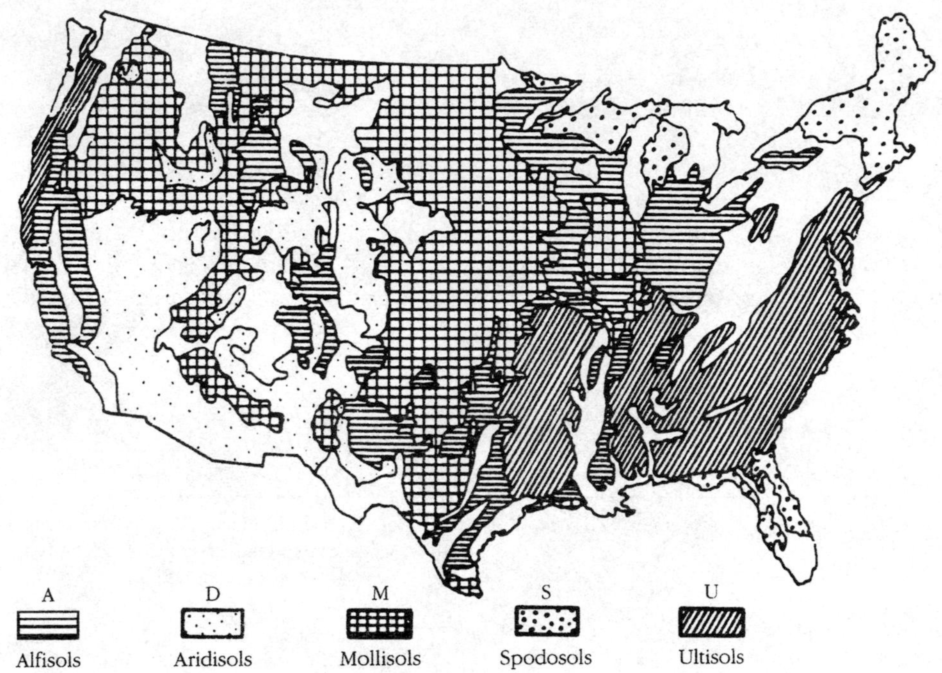

Figure 5-3 The distribution of major soil types in the United States employing the classification of the U.S. Soil Conservation Service. The blank spaces comprise Inceptisols, followed by Entisols and small areas of Histosols and Vertisols; Oxisols are absent. (Redrawn with permission from Jenny 1980.)

A classification of soils advanced by Bridges (1978) (Table 5-2), like earlier Russian traditions, places soils in the different latitudinal zones (Figure 5-4), thereby tying closely with broad climatic regimes and also to the distribution of major biomes, the subject of Part VII.

NUTRIENTS

Life is dependent not only on energy but also on the availability of some 20 chemical elements required in the life processes of most organisms and an additional 10 or so required by different species, usually in trace amounts. Although carbohydrates can be photosynthesized from water and carbon dioxide, the more complex organic substances require additional components either in considerable amounts, as in the case of nitrogen and phosphorus, or in trace amounts, as in the case of zinc and molybdenum. Furthermore, photosynthesis and other metabolic reactions in

TABLE 5-2 ZONAL CLASSIFICATION OF SOILS

Latitude zone	Bio-climatic zone	Zonal soils
Soils of high latitude	Tundra	Arctic brown soils; arctic gley soils
Soils of midlatitudes, cool climates	Northern coniferous forests and heaths	Podzols; brown podzolic soils
	Deciduous woodlands	Brown soils; argillic brown soils
	Wooded steppes	Grey soils
Soils of midlatitudes, warm climates	Mixed moist evergreen woodlands	Brown Mediterranean soils; red Mediterranean soils
	Mixed dry evergreen woodlands	Cinnamon soils
	Mixed oak, pine with summer rainfall	Red podzolic soils; yellow podzolic soils
	Steppe grasslands, dry steppe grasslands	Chernozems, chestnut soils
	Semi-desert	Sierozems
	Desert	Raw mineral soils
Soils of low latitudes	Tropical rainforest and deciduous forests	Ferrallitic soils; podzols
	Tropical grasslands	Ferruginous soils; vertisols, laterite
	Human-modified soils	Acid sulphate soils; paddy soils

Reproduced, with permission, from Bridges 1978.

both plants and animals occur in the presence of enzymes that themselves contain a variety of elements in trace amounts.

Those elements needed in relatively large amounts are generally referred to as *macronutrients*, each constituting 0.2 percent or more of dry organic weight. *Micronutrients* are those elements needed in trace amounts, each constituting less than 0.2 percent of dry organic weight. It is noteworthy that all of the 32 elements that have been identified as nutrients are among the lightest elements. Twenty-eight of the first thirty-five elements in the periodic table are essential for plant and/or animal life; the remaining four are still regarded as light elements, with atomic numbers of 38 (strontium), 42 (molybdenum), 50 (tin), and 53 (iodine).

Macronutrients

Macronutrients may be considered in two groups: (1) those that constitute more than 1 percent each of dry organic weight—carbon, oxygen, hydrogen, nitrogen, and phosphorus; and (2) those that constitute 0.2 to 1 percent of dry organic

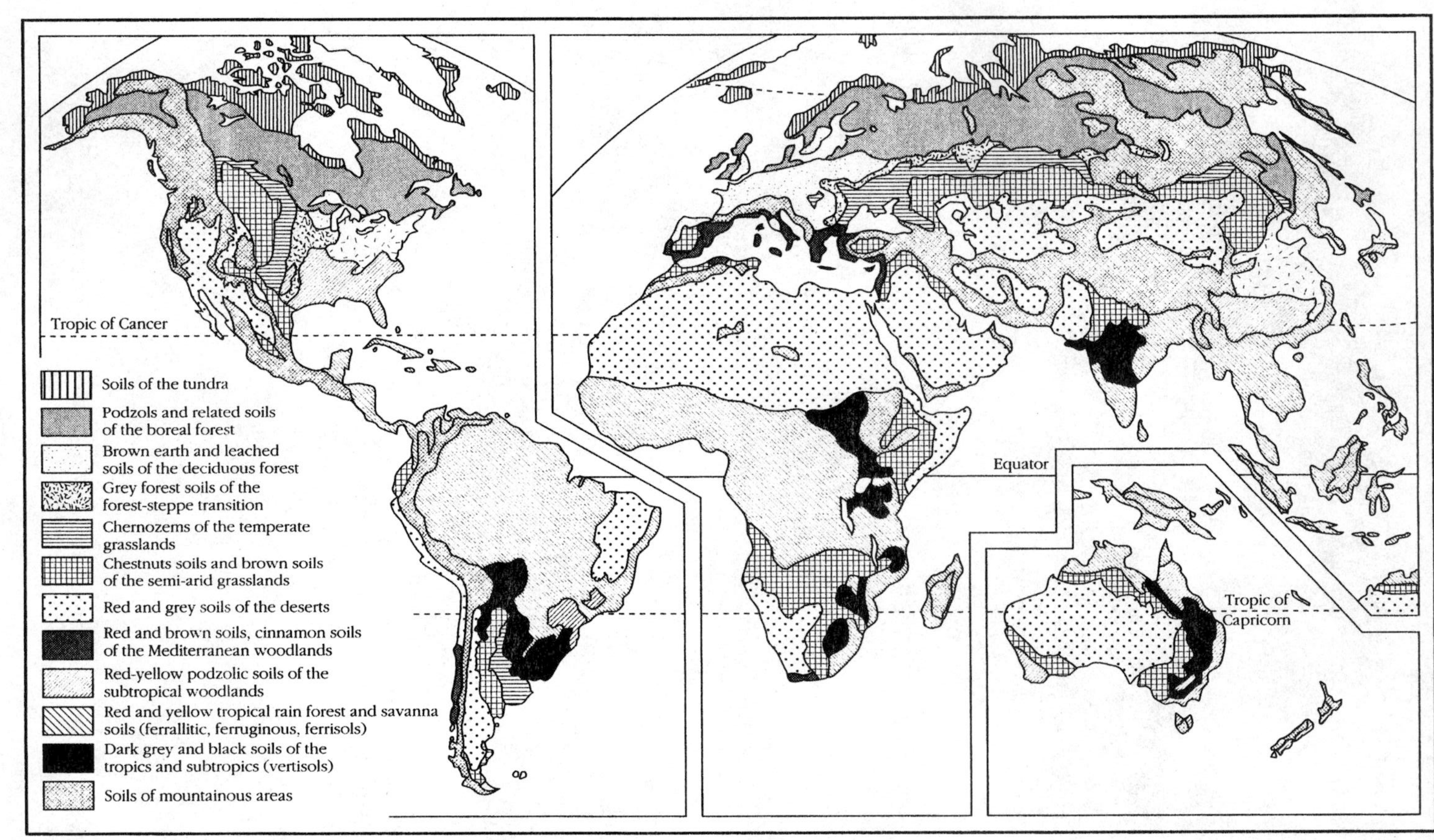

Figure 5-4 Soil map of the world according to the classification by Bridges (1978). (Redrawn by permission.)

weight—sulfur, chlorine, potassium, sodium, calcium, magnesium, iron, and copper. In the human body there are four macronutrients: hydrogen constitutes 63 percent, oxygen 25.5 percent, carbon 9.5 percent, and nitrogen 1.4 percent (Frieden 1972).

Role of macronutrients. The role of the first group of macronutrients is well known: oxygen as a building block of protoplasm and the primary agent in biological oxidation; carbon as the building block of all organic compounds; hydrogen as an acceptor of oxygen; nitrogen as the building block of protein and nucleoprotein, as well as a component of enzymes; and phosphorus as a constituent in photosynthesis, proteins, and nucleoproteins.

The second group of macronutrients plays a variety of roles:

sulfur as a basic constituent of protein;

chlorine, required in minute amounts by plants, but in greater quantities in animals as a key factor in maintaining the acid-base balance in animals;

potassium, involved in the formation of sugars in plants and of proteins in animals;

sodium, an aid in maintaining the acid-base balance in animals;

calcium, a major component of cell walls, needed for normal mitotic divisions and root-tip development and for a variety of functions in animals including maintaining acid-base balance, contraction of muscles, and, in vertebrates, arthropods and mollusks, the key element in providing structural support (bones, exoskeletons, and shells);

magnesium, critical to the structure of chlorophyll and energy transfer from ATP to ADP;

iron, required in electron transfer in photosynthesis and the key element in hemoglobin;

and copper, present in chloroplasts, influencing the rate of photosynthesis.

Micronutrients

The known micronutrients include aluminum, arsenic, boron, bromine, chromium, cobalt, fluorine, gallium, iodine, manganese, molybdenum, nickel, selenium, silicon, strontium, tin, titanium, vanadium, and zinc. Actually, some of these micronutrients may be macronutrients for some species, whereas some of the second group of macronutrients, for example, sodium and chlorine in plants, may be micronutrients. The list of micronutrients continues to expand with four new essential trace elements (arsenic, nickel, silicon, and vanadium) identified in animals in the 1970s (Mertz 1981). At least one, nickel, was the first essential trace element identified in plants in nearly 30 years when its presence was determined to be necessary for development and functioning of root nodules in soybeans (Eskew et al. 1983). As research continues, there is considerable likelihood that additional elements will be identified as essential.

Role of micronutrients. Discussion of all the trace elements would require more space than is appropriate here. In general, however, it can be stated that many of the micronutrients play key roles in enzymatic activity. Some, however, are involved in a wide variety of roles; boron, for example, is involved in cell division, translocation of water in plants, and carbohydrate metabolism, among a number of other functions. Aluminum is necessary for normal development of some ferns, silicon is the structural element of diatom tests, and cobalt is required in root nodules for proper functioning of nitrogen-fixing bacteria. Selenium deficiency, primarily affecting juveniles such as black-tailed deer, results in increased mortality during the neonatal and preweaning period (Flueck 1994).

Nutrient Interaction

As was noted in the earlier discussion of factor interaction, the presence or absence of an essential nutrient may adversely impact the availability or activity of another nutrient. Howarth and Cole (1985) showed that sulfate inhibits the assimilation of molybdenum by phytoplankton, thus making it less available in seawater than it is in freshwater. Molybdenum is required in the conversion of elemental nitrogen to nitrate (nitrogen fixation), the form in which it can be used by plants, as well as in nitrate assimilation. Thus, a low level of molybdenum requires a greater expenditure of energy for these nitrogen processes in seawater than in freshwater and may explain the relatively low levels of nitrogen in coastal marine ecosystems.

RADIATION AND FIRE

Two additional abiotic factors of considerable significance are ionizing radiation and fire. Both are natural phenomena, but both are also affected by human activity. Thus the general ecological impact of each will be discussed here; aspects of their effects on humans will be treated in Part VIII.

Ionizing Radiation

Ionizing radiation. Organisms and ecosystems are and have been exposed from their beginnings to low-level ionizing radiation from cosmic rays, radioactive elements in the Earth's crust (e.g., radium, uranium), and emissions from certain radioactive materials, or *radioisotopes*, that occur naturally in their bodies (e.g., ^{40}K and ^{14}C). *Ionizing radiation* is very high energy that has the capacity to remove electrons from atoms and attach them to other atoms. While the natural sources of ionizing radiation, or *background radiation*, are and have been relatively constant over eons and have resulted in evolutionary adaptation on the part of organisms and ecosystems, human activity has almost doubled the amount of exposure. This doubling comes mostly from medical and dental uses of x-rays, but also from radioisotopes incorporated in building materials, phosphate fertilizers, consumer products such as television sets and smoke detectors, fall-out from atomic weapons, and leakage from nuclear reactors. These aspects of ionizing radiation will be discussed in the section on human ecology (Part VIII).

The ionizing radiations of major concerns are: *alpha*, *beta*, and *gamma radiations*. Alpha and beta radiations are corpuscular, that is they consist of streams of atomic particles. Alpha radiations, actually parts of helium atoms, travel only a few centimeters in air, may be stopped by something as thin as a sheet of paper, but result in a substantial dose of radiation. Beta radiations are high-speed electrons, traveling up to a meter in air and up to several centimeters in organic tissues. Gamma radiations are electromagnetic, with shorter wavelengths than light; they travel considerable distances, readily penetrate matter, and produce ionization over a long pathway.

In addition to ionizing radiations, *neutrons* can induce radioactivity in nonradioactive materials, a phenomenon occurring as a result of atomic explosions and in nuclear reactors. *X-rays* are similar to gamma radiations and are produced by x-ray machines. *Cosmic radiations* consist of both corpuscular and electromagnetic radiations. Under various circumstances, each of these sources of radiation has an ecological impact (e.g., cosmic radiations in space travel and at high altitudes on Earth, exposure to granitic rocks that have considerable levels of radioisotopes).

Sources of exposure. Organisms, and consequently ecosystems, are exposed to ionizing radiation in two ways: externally, from sources outside the body, and internally, from sources inside the body. External radiation sources are cosmic rays, x-rays, gamma rays, and neutrons. Internal radiation sources result from the metabolic incorporation of radioactive materials by plants in photosynthesis and in anabolic construction of their tissues, and by animals through intake of radioactive substances present in food or water. Intake can also occur through inhalation of naturally occurring radioactive gases such as radon and thoron or of radioactive substances occurring as a result of nuclear industrial accidents or atomic explosions.

Fire

As with ionizing radiation, fire is both a natural and human-enhanced factor in ecological systems. Although spontaneous combustion may account for some naturally caused fires, lightning is the primary agent. Climates that consist of hot, dry summers and low humidity are prime situations for the development of thunderstorms with lightning but little or no rain. The vast grasslands of Africa and the United States and the chaparral of southwestern United States and the Mediterranean region are naturally occurring situations conducive to lightning-induced fires. *Chaparral*, vegetational areas characterized by shrubs with leaves that are evergreen, hard, thick, leathery, and small, are in California subject to prolonged summer droughts leading to catastrophic fires that typically recur every 15 to 100 years (Minnich 1983; Riggan et al. 1988). The floral composition of such chaparral can be changed dramatically by these fires, especially if the dominant vegetation must reproduce from seed (Moreno and Oechel 1991). In other situations, a combination of fire, grazing, and the physical environment may produce a static community (Calloway and Davis 1993).

Portions of Yellowstone National Park are another area that is subject to recurring fire, including the great fires of 1988 that consumed 290,000 hectares, an area the size of Rhode Island (Fuller 1991; Jeffrey 1989; Romme and Despain 1989). In a subalpine watershed of the Park, Romme (1982) found evidence for 15 fires since 1600 (of which seven were major ones), consuming more than 4 hectares. Most of the upland area studied was burned by large, destructive fires in the middle and late 1700s; since then fires have been small and have occurred after long intervals. Nonetheless, there appears to be a natural fire cycle of 300–400 years, in which large areas burn during a short period, followed by a long, relatively fire-free period when a highly flammable fuel complex again develops.

The frequency of fires can be affected by changes in climate as well as by human activity and the interaction of the two (Bergeron 1991). In the southern Canadian Rockies, the fire cycle in boreal forests before 1730 was 50 years; since 1970, the cycle has been 90 years. This alteration in frequency is related to a change from a warmer, drier climate to a cooler, moister climate (Johnson and Larsen 1991). Swetman and Betancourt (1990) have determined an association between wildfire occurrences in ponderosa pine forests of the southwestern United States and the El Niño–Southern Oscillation during the past three centuries. Small areas burn after wet springs associated with the low phase of the southern oscillation, whereas large areas burn after dry springs associated with the high phase of this same oscillation. Human activity such as land-use changes can have a dramatic impact on fire frequency. For example, Savage and Swetnam (1990) indicate that there is a coincidence of fire decline in a Navajo ponderosa pine (*Pinus ponderosa*) forest and early sheep herding that developed in the early 1700s. The reduction through this land-use change was compounded further by a climatic change in the 1830s and 1840s toward an extremely wet episode.

Areas that are otherwise not prone to fire, such as the coniferous forests of northwestern United States can be and have been subject to naturally caused fire under conducive weather conditions (Johnson 1992). Over the last 750 years in the boreal forest of northwestern Minnesota, an area dominated by red pine (*Pinus resinosa*), fires were most frequent at about 8-year intervals during the warm and dry 15th and 16th centuries and were at about 13-year intervals from AD 1240 to 1440 and since 1600 (Clark 1990). The longest was an interval of 43 years during the mid-19th century. At the northern limits of the red pine's range, Bergeron and Brisson (1990) found that the interval between fires before 1906 was about 30 years, but fires of sufficient intensity to kill most of the trees occurred every 68 years. The modeling of fire frequency has been reviewed extensively by Johnson and Gutsell (1994).

Types of fire. The two major categories of fire are crown and surface. *Crown fires* are typically very intense, consuming the entire plant community above the ground and often its animal component as well. Combustion of the surface litter, or O horizon, and organic matter in the A_1 horizon not only burns dead leaves and organisms but also seeds and other regenerative structures (e.g., bulbs, tubers). Because of the severity of destruction, such areas present a barren landscape (Figure 5-5) and typically take years to begin even modest recovery. In the interim the area is subject to erosion following even mild levels of precipitation, thus removing the nutrients required for life and regeneration (Kaufman et al. 1993).

Harrison Salisbury (1989) vividly describes the great Heilongjiang, or Black Dragon River, fire of 1987, named for the river that separates the eastern former USSR from China. The Greater Hinggan Forest surrounding the Black Dragon River, known in Russia as the Amur, encompassing an area 500 by 300 miles, contains one-third of China's forest reserves of virgin conifers, and supplies 30 to 40 percent of China's industrial lumber production. The great fire consumed nearly 6 million hectares (equivalent to one-third of the State of Washington), caused at least $5 billion damage, killed 220 people, and injured at least 250 others.

Figure 5-5 Fir tree crown fire during the Gallatin Fire of 1979 in Yellowstone National Park. (Photo courtesy of Tony Caprio.)

Surface fires do not have the same destructive effects as crown fires. In fact, the effect is often beneficial in maintaining the vitality of the biotic component or facilitating the regeneration and even stability of the plant and animal community (Collins and Wallace 1990). *Surface fires*, as the name implies, burn off only the litter layer, or O horizon, the depth burned depending on the intensity and severity of the fire. This surface burning can keep the amount of litter to a minimum, thereby reducing the danger of a crown fire (Figure 5-6).

Beneficial effects of fire. Among the beneficial effects of a surface fire, and even of a crown fire, is the combustion of dead litter into ash, resulting in the release of mineral nutrients for recycling. This may be coupled with the increased growth of nitrogen-fixing legumes that often occurs after a moderate surface fire.

In some situations fire is necessary for regeneration of fire-resistant, or fire-adapted, species. The seeds of a number of species, among them jack pine, white pine, paper birch, and eucalyptus, require fire to release seeds from cones. Long-

Figure 5-6 Surface fire near Crown King, Arizona. (Photo courtesy of John H. Dieterich.)

leaf pine of the southeastern United States requires fire to compete with the scrub hardwoods that otherwise choke them out. In a crown or severe surface fire, the scrub hardwoods are subject to combustion, but the terminal bud of long-leaf pine is well protected by long, fire-resistant needles. Subsequent to the fire, the long-leaf pine is freed from competition and able to mature. In grassland situations, the combustion of trees and shrubs frees the grass from competition for light, moisture, and nutrients.

Clark (1986) has shown that in the history of coastal forest trees in southeastern Long Island, fire was a significant factor in insuring opportunities for seedling establishment at higher elevations as conditions became unsuitable for regeneration on low-elevation sites because of a change in the physical environment, namely a rise in sea level.

Fire-tolerant trees such as oaks (*Quercus alba, Q. borealis, G. ellipsoidalis, Q. macrocarpa,* among others) and aspen (*Populus tremuloides*), along with physical firebreaks such as bodies of water and physiographic discontinuities, played a substantial role in preventing prairie fires from affecting the Big Woods vegetation of Minnesota in the mid-nineteenth century (Grimm 1984). The Big Woods was a large area (circa 16,900 km^2) of deciduous forest, the most common member of which was elm (*Ulmus americana*), composing 27 percent of the trees. Today, the area is almost entirely urban or agricultural, with only scattered woodlots.

Adverse effects of fire. The major impact of a severe crown or surface fire is disturbance of the ecological community, specifically the destruction of the organisms and their intertwined relationships. Crown fires and severe surface fires leave the ground surface subject to erosion, the severity of which depends on the nature of the soil surface and the amount and intensity of precipitation. For the human eye, the aesthetic pleasure derived from the sight of broad expanses of forest and prairie is fractured and the mind-refreshing experience of a walk in the woods is shattered. Salisbury (1989) put it dramatically: "There is no lens wide enough to picture a forest that has just died, black trees extending to the horizon."

Although fire has a beneficial effect in mobilizing nutrients in ash, there are substantial losses of nutrients as particulate matter in smoke and, especially for nitrogen, by volatization (Hobbs and Gimingham 1987). Losses of nitrogen and sulfur are especially large at higher temperatures, reaching 57 percent and 36 percent, respectively, at 750°C; losses of nitrogen and phosphorus were four to five times greater in a fire at 800°C than in one at 600°C. Substantial amounts of sodium, potassium iron, zinc, and copper were also volatilized in both field and laboratory experiments.

These several effects of fire on ecosystems suggest it is more beneficial in some circumstances than others. However, this aspect of fire has not always held popular sway and thus well-intended efforts to avoid fire have led to even worse crown and surface fires. Intelligent forest management practices based on sound ecological principles and data are imperative and seem to be gaining ground (Mooney et al. 1981).

SUMMARY

VOCABULARY

alpha radiation
background radiation
beta radiation
chaparral
colloidal complex
cosmic radiation
crown fire
gamma radiation
ionizing radiation
leaching
macronutrients
micelles
micronutrients
neutrons
radioisotopes
soil profile
surface fire
weathering
x-rays

KEY POINTS

- The differentiation of mineral and organic components into horizons gives rise to a soil profile, organic matter generally being more abundant at the upper levels and inorganic matter more abundant at lower levels; the amount of air and water varies in different kinds of soils.
- Mineral matter in the soil is derived from the parent material and organic matter from the decay and decomposition of organisms or their byproducts.
- Soil is formed through the interaction of climate and organisms on parent material, and is influenced by topography, time, and other factors.
- Several major soil classification schemes have been proposed based on chemical composition and latitude.
- Macronutrients and micronutrients play a variety of key roles in the functioning of organisms and ecosystems; while many may act independently, most interact directly or indirectly.
- Organisms and ecosystems have been exposed to background ionizing radiation since their beginnings, a probable factor contributing to their evolution.
- Fire in some ecosystems is a naturally occurring factor necessary for maintenance of the system; in other situations, fire can be a negative factor, destroying ecosystems.

PART III

Energy Flow in Ecosystems

CHAPTER 6

Energy Fixation by Autotrophs

The conversion of solar radiation to chemical energy by photosynthesis is the starting point of energy flow within ecosystems. As was noted in Chapter 4, the quality and intensity of insolation varies daily and seasonally with latitude, the inclination of the Earth with respect to the Sun (as well as with altitude), and the medium through which sunlight travels. Whatever its quality and intensity, solar energy is transformed from the radiant to the chemical form in photosynthesis and from the chemical to mechanical and heat forms in cellular metabolism. These conversions and sequences are fundamental to the energetics of organisms and ecosystems as energy passes into and through producers, to and through consumers and decomposers.

Sun —radiant energy→ Producers —chemical energy→ Consumers and decomposers

Producers —heat energy→

Consumers and decomposers —heat energy→

Consumers and decomposers —mechanical energy→

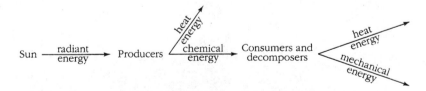

RADIANT ENERGY AND PHOTOSYNTHESIS

Because radiant energy is converted by autotrophs in photosynthesis, measurement of the amount and rate of energy fixationis based on the photosynthetic equation

$$6\,CO_2 + 12\,H_2O \xrightarrow[\text{chlorophyll (or other pigment)}]{673\ C} C_6H_{12}O_6 + 6\,O_2 + 6\,H_2O$$

The stoichiometry of the equation is such that if the amount of one component is known, other components of the equation can be computed. In some situations, for example, it may be more appropriate and convenient to measure the amount of carbon dioxide used in a given period of time than it would be to measure the amount of oxygen or carbohydrate produced. Having done so, the investigator could then compute the amount of carbohydrate produced because one unit of carbohydrate is produced for each six units of carbon dioxide assimilated. Precise calculations involve certain qualifications of these equivalences; the basic equation is adequate, however, for developing the general principle.

The various methods developed to date are beset with technical and/or interpretational limitations. Nonetheless, they do provide approximations of energy fixation, a process that is difficult to measure under both field and laboratory conditions. It is important to recognize these difficulties as well as to gain an understanding of the methodology of ecological study. Doing so will help in recognizing the limitations inherent in the data subsequently discussed. Thus a brief consideration of the major techniques of measuring primary production—that is, the amount and rate of energy fixation—is in order.

MEASURING PRIMARY PRODUCTION

Harvest Methods

One of the most common methods of measuring autotroph or *primary production*, and one of the oldest, is that of harvesting. For many years farmers have used the harvest method in reporting their yields as so many bushels of wheat, so many pounds of beef, or so many gallons of milk. Ecologists also use the harvest method by removing vegetation at periodic intervals and weighing the material. The part of the plant harvested varies with the intent of the study; in harvesting grasses, for example, plants are typically clipped at ground level. Aquatic vegetation has been sampled in the same way. However, as will be shown later in this chapter, a considerable error is introduced by failing to harvest the roots, particularly the large rootstocks common to many aquatic plants and the roots of trees. Because translocation of organic matter occurs between above- and belowground parts of plants, substantial error can be introduced if only certain parts of the plant are sampled. For comparative studies, the frequency of sampling, as well as the time of year of the sampling, can be additional sources of error.

After drying to constant weight, the harvest can be expressed in terms of the biomass (the total mass or weight) per unit area per unit of time—for example, as grams per square meter per year ($g/m^2/yr$ or $gm^{-2}y^{-1}$). If the exact caloric content of the material is known, biomass can be converted and expressed in energy terms. Although the caloric content of many species is known, a harvest more often than not is a mixture of many different plants. For most purposes, a gross estimate can be reached on the basis that each gram of dried terrestrial plant material has an energy content of about 4.5 C. A gram of dried algae, for example, has an energy content of 4.9 C.

One major limitation of the harvest method is that it fails to account for the amount of material (or energy) that may have been consumed by herbivores. No less crucial is its failure to account for the energy used by the autotroph in its own metabolism, growth, and development. What is actually measured, then, is the *standing crop*, the amount of autotroph at a given time, or at given time intervals, that is in excess of the demands of its own metabolism or of the herbivores that depend on it. If this harvest is corrected for the amounts lost by respiration (metabolism), as well as any other avenues of loss (e.g., herbivory), an estimate of the total or *gross production* will have been made. In some instances, the ecologist is more interested in *net production*—that is, gross production minus plant respiration. The significance of these two ways of considering production will be apparent later in this chapter.

Although the harvest or standing crop approach is widely used in ecological studies, particularly in terrestrial situations, other methods are more sophisticated. Often they are no more precise and certainly no less beset with methodological limitations. These methods are directed at other components of the photosynthetic equation, primarily at measurements of CO_2 assimilation, O_2 release, and chlorophyll production.

Carbon Dioxide Assimilation

The uptake of CO_2 in photosynthesis or its loss in respiration under natural terrestrial conditions can be determined by use of infrared gas analysis. Such analyzers measure the amount of CO_2 entering and leaving an airtight enclosure of a known area or volume that is constructed of a light-transmitting substance, such as plexiglass, glass, or plastic, in which a plant leaf or branch is placed. It is assumed that any CO_2 removed from the incoming air has been incorporated into organic matter during the period of observation, thereby indicating both the amount and the rate of photosynthesis. Yet while photosynthesis has been going on, so has metabolic or respiratory activity. Consequently, what has again been measured is a short-term aspect of net production. If a comparable study area is established, using a light-tight container, no photosynthesis will take place, but respiration will. Therefore the amount of CO_2 released from the chamber in a given period, or at periodic intervals, will be a measure of the amount and rate of respiratory activity. This value, added to that obtained in the light-exposed chamber, can be used to approximate the total or gross production of the system. The major limitation of the gas analysis method undoubtedly is that it requires the use of sophisticated and expensive equipment in the field. Further, extrapolating from individual leaves of plants to whole communities is beset with the possibility of significant errors.

An advance over the enclosure method is the aerodynamic method in which periodic measurements are made, using CO_2 sensors placed at intervals along a vertical pole. By measuring the difference in CO_2 above and in the autotrophic layer, a measure of net production can be achieved. This method has been used successfully to measure photosynthesis rates in crops, grasslands, and forests, sometimes by combining it with the enclosure method.

Oxygen Production

Infrared gas analysis of carbon dioxide is not feasible in an aquatic ecosystem and so alternatives and other carbon dioxide-based methods have been developed. The measurement of oxygen evolution has been one of the most widely used procedures, partly because of its relative simplicity. Samples of water containing autotrophs, such as algae (and almost invariably some heterotrophs such as bacteria and zooplankton), are taken from a given depth of a pond, lake, or ocean and are distributed into pairs of smaller sample bottles, typically of 125 to 300 ml capacity. One of the paired bottles is of light-transmitting material, typically glass, permitting photosynthesis to take place; the other bottle is darkened to preclude photosynthesis but not respiratory activity. The setup is then comparable to that used in the study of CO_2 assimilation. The bottles are resuspended, usually at the level from which the samples were taken, and allowed to incubate for a period of time (up to 24 hours but usually on the order of 4 to 8 hours). At the end of this time the oxygen content is determined either by standard chemical titration or with an electronic sensor. This oxygen change in the light bottle can be substituted in the photosynthetic equation to determine the amount and rate of net photosynthesis. The normal respiratory consumption of oxygen can be accounted for by using the dark bottle data. Total or gross production can be approximated by combining this amount with the value obtained in the light bottle.

A basic assumption in the light-dark bottle method is that plant respiration is the same under the two conditions. For very short-term experiments and for some species, this assumption appears to be true, but many plants show different respiratory rates under dark conditions. Yet a further limitation of the method is the very enclosing of a sample of an entire community. The sample can fail to reflect the actual population being sampled (corrected to some extent by replicating the samples a number of times). Also, variation in the number of heterotrophs will skew the respiratory consumption. The absence of water movement with its effect on moving nutrients to and from sites of photosynthesis also must be acknowledged. The chemical composition of the sampling device and of the containers themselves can alter the rate of photosynthesis. Finally, if the water sample, particularly one taken from considerably below the surface, is left exposed to light too long, photoinhibition may occur. In spite of these limitations, the light-dark bottle method enjoys wide use.

In another variant of the oxygen-evolution method, the oxygen content of an aquatic ecosystem is determined every 2 to 3 hours for a period of 24 hours. Using an electronic sensor attached to a recording device, it is also possible to sample the system continuously for a 24-hour period. The natural photoperiod in both these methods thus substitutes for the nighttime simulation achieved in the dark-light bottle technique, and the total ecosystem instead of only subsamples is evaluated. Using various calculations, it is possible to determine the net production of oxygen (i.e., the excess over that consumed in respiration) and in turn, to determine by the photosynthetic equation, the gross production in the system.

In some situations (e.g., in the deep parts of lakes where no photosynthesis occurs), oxygen disappearance can be used to measure the net production of an entire community. Such techniques have been particularly significant in studies of lake pollution.

Radioisotope Methods

Considerable precision and sensitivity can be gained in measuring primary production by using a radioisotope tracer in the photosynthetic process. Although other isotopes, notably phosphorus-32, have been used, the radioactive form of carbon known as carbon-14 (^{14}C) has been the most fruitful. It can be used in a method similar to the light-dark bottle technique for measuring oxygen. Unlike the O_2 method, which measures gross production, however, it is now recognized that the ^{14}C method measures net production (or a quantity closer to net than to gross production). Water samples are taken from a desired level and distributed in paired light and dark bottles. A known quantity of ^{14}C, usually in the form of a bicarbonate such as $NaH^{14}CO_3$, is then introduced into the bottles, and they are resuspended to incubate, usually for about 6 hours. During the time of exposure both the stable carbon, which is present as CO_2 or HCO_3, and the unstable ^{14}C are assimilated into carbohydrate and incorporated in the protoplasm of the autotrophs. At the conclusion of the incubation period the samples are filtered, and the stable and radioactive carbon that has been incorporated biologically is captured on the filter. The filters are dried and placed in a counting chamber to determine their level of radioactivity and hence the amount of radioactive carbohydrate produced. Calculations are then made from the photosynthetic formula, based on the assumption that the uptake of radioactive carbon is proportional to the stable carbon (actually, it is not exact and is corrected by multiplying the amount of radioactive carbohydrate by an appropriate factor):

$$\frac{6\ ^{14}CO_2}{6\ CO_2} = \frac{^{14}C_6H_{12}O_6}{C_6H_{12}O_6}$$

The effects of the chemical composition of the sampling and incubation chambers, as well as possible photoinhibition from prolonged exposure of the sample to light, can also contribute to error in the ^{14}C method. In addition, the relative acidity-alkalinity of the sample, the possible presence of inhibiting ions in the ^{14}C solution, the nature of the filter, and the efficiency of the counting chamber can also affect the uptake of the ^{14}C or the accuracy of its determination. Comparative studies using alternate methods, such as changes in adenosine triphosphate (ATP), indicate that the ^{14}C method often underestimates carbon uptake and hence the amount and rate of production (Peterson 1980). Nonetheless, the method has been used for over 40 years and remains the most feasible and practical in aquatic environments.

Other Methods

Among other methods is one based on a determination of chlorophyll, the premise being that there is a close correlation between the amount of chlorophyll and the amount and rate of photosynthesis. The procedure involves periodic sampling of plant material and the subsequent extraction of chlorophyll in a suitable organic solvent, the concentration being determined by spectrophotometric means. Based on the assumption of given rates of photosynthesis per unit of chlorophyll, it is then possible to estimate the production of the area sampled. As an advantage over some of the other methods, samples need not be confined in dark or light containers. This method is well adapted to and has been used quite extensively in the study of aquatic systems, but only recently has it received much application in terrestrial situations. It has considerable promise, especially in the comparison of chlorophyll to other pigments (notably the yellow carotenoids) as an index to heterotrophic-autotrophic conditions, the former being more characterized by carotenoids and the latter by chlorophylls.

In another aquatic method the pH is measured. This process is based on its relationship to dissolved carbon dioxide, the pH changing as the CO_2 is taken up in photosynthesis or released in respiration. Although this method involves specific calibration of the relationship between pH and CO_2 for each system in which it is used, its advantage is that it does not disrupt the system because the pH electrode can be left in place.

Other methods try to measure the rate of disappearance of other raw materials of significance in the photosynthesis process (e.g., nitrogen and phosphorus). Because such disappearance can also result from the action of nonliving processes, considerable caution is required in their use.

Satellite remote sensing of marine primary production shows promise for ocean-scale studies (Platt and Sathyendranath 1988; Orrigo and McClain 1994). Estimations of primary production have also been based on the age of ocean basins (MacDonald and Carmack 1991) and by simulation models (Reich et al. 1991, among others).

EARLY ESTIMATES OF PRIMARY PRODUCTION

Terrestrial Studies

Cornfield in Illinois. In a study that proved considerably in advance of its time, Edgar Transeau (1926) estimated the accumulation of energy by a midwestern cornfield in a single growing season (Table 6.1). His calculations were based on an estimated harvest of 10,000 plants per 0.405 *hectare** (one acre) weighing

*Hectare (ha) = 10,000 m^2 or 2.471 acres.

TABLE 6-1 ENERGY BUDGET OF AN ACRE (0.405 HECTARE) OF CORN DURING ONE GROWING SEASON OF 100 DAYS

	Glucose (kg)	Calories (C) (millions)	Solar energy utilized (%)
Incident solar radiation		2043	100.0
Biological utilization			
Energy incorporated			
Net production (NP)	6687	25.3	1.2
Respiration (R)	2045	7.7	0.4
Gross production (GP) (= NP + R)	8732	33.0	1.6
Energy used in transpiration	—	910	44.4
Energy not used	—	1100	54.0

Data of E. N. Transeau, 1926. *Ohio Journal of Science* 26:1–10.

6,000 kilograms (kg) and on available information on the chemical composition of the plants. Based on these assumptions, he calculated that there were 2,675 kg of carbon in the 10,000 corn plants. Because this carbon entered the plants only through photosynthesis, it would be equivalent to a net production of 6,687 kg of glucose.

Molecular composition	Molecular weight	Metric weight

$$\frac{C_6}{C_6H_{12}O_6} = \frac{72}{180} = \frac{2675\ \text{kg}}{X\ \text{kg}}$$

$$X = 6687\ \text{kg glucose}$$

To this net production, Transeau added the equivalent amount of glucose that he estimated had been metabolized in cellular respiration during the growing season, 2,045 kg, to obtain the total or gross production of 8,732 kg of glucose.

He used 3,760 kilocalories as an estimate of the amount of energy required to produce one kilogram of glucose, obtaining 33 million C (8,732 kg × 3,760 C/kg) that would have been incorporated in gross production. Of this amount, 7.7 million C were used in metabolic activities.

He also estimated that 1.5 million kg of water had transpired from the area—about 1,545 m^3, sufficient to cover the area to a depth of 38 cm. This transpiration would have required an expenditure of 910 million C of energy, none of which was used in photosynthesis.

Because the total solar energy available to the cornfield ($2{,}043 \times 10^6$ C) was known from other studies, it was possible for Transeau to calculate the efficiency of energy use. It may be surprising to learn that the field of corn incorporated only

1.6 percent of the total energy available:

$$\frac{\text{Gross production}}{\text{Solar radiation}} \times 100 = \frac{33 \times 10^6 \text{ C}}{2043 \times 10^6 \text{ C}} \times 100 = 1.6 \text{ percent}$$

Transeau noted, however, that because only about 20 percent of the solar radiation measured by the instrument used was actually effective in photosynthesis, the actual efficiency would be somewhat higher—namely, about 8 percent.

Another point of interest is the amount of energy used in growth, development, and maintenance. It is reflected in cellular respiration and is measured as the loss via respiratory activity. In the case of the cornfield, the estimated energy loss by respiratory activity is 23.4 percent:

$$\frac{\text{Energy of respiration}}{\text{Energy of gross production}} \times 100 = \frac{7.7 \times 10^6 \text{ C}}{33.0 \times 10^6 \text{ C}} \times 100$$
$$= 23.4 \text{ percent}$$

Thus although relatively little of the total available energy was used in photosynthesis, the plants were quite efficient (76.6 percent) in converting the captured energy to biomass.

This study shows, in some ways, how information on primary energy fixation can be interpreted to direct attention to at least two major aspects of primary production—amount (i.e., biomass) and rate (i.e., production). The amount for the growing season is what has been discussed so far. Because the growing season for corn is assumed to be 100 days in the region that Transeau considered, a rough approximation of a daily rate of energy fixation might be made. Because there are differences in rate of photosynthesis per unit of leaf area by age and because the total amount of photosynthetic surface in a plant varies with age, only the crudest approximations to a daily rate can be made. Transeau himself estimated the average daily increment to be around 8 percent. Periodic measurement, using one of the nonharvest methods described above, would have been needed for a more accurate measurement of daily or hourly rates of energy fixation.

Old field in Michigan. Transeau's study was based on several different sources of data on managed agricultural land in northern Illinois. In the same general temperate region, Frank Golley (1960) investigated an old-field ecosystem in southern Michigan some 30 years later. The harvest and respiratory data for the perennial grass-herb vegetation of the field are shown in Table 6-2. By calculation, the efficiency of energy utilization by the grass-herb component is seen to be 1.2 percent (i.e., $5.83 \times 10^6 \div 471.0 \times 10^6$) and the percentage of energy lost in plant respiration is 15.1 percent ($0.88 \times 10^6 \div 5.83 \times 10^6$).

Although the efficiency of energy capture in this terrestrial ecosystem is about three-fourths that of the cornfield (1.2 versus 1.6 percent), the respiratory loss is only about two-thirds (15.1 versus 23.4 percent). So although the efficiency of energy capture is less, the use of that energy (i.e., the amount conserved in plant tissue) is

TABLE 6-2 ANNUAL ENERGY BUDGET FOR PERENNIAL GRASS-HERB VEGETATION IN SOUTHERN MICHIGAN

	$c/m^2/yr$
Incident solar radiation	471.0×10^6
Plant utilization	
Net production (NP)	4.95×10^6
Respiration (R)	0.88×10^6
Gross production (GP)	5.83×10^6

Data of F. B. Golley, 1960. *Ecological Monographs* 30:187–206.

greater in the natural old field than in the managed cornfield. In a comparable old-field study in South Carolina, Golley and Gentry (1965) found that during the 1960 growing season 48 percent of gross production was lost in respiration by the dominant plant, broom sedge. And in a wet marine grassland, a salt marsh, John Teal (1957) found that 77 percent of gross production was consumed in respiration by plants. In these last two instances, then, a substantial portion of the initial energy captured was dissipated out of the ecosystem at the first trophic level.

Freshwater Studies

Lake Mendota, Wisconsin. The data from Lake Mendota, Wisconsin, shown in Table 6-3 were collected by Chancey Juday (1940), one of the significant figures in the early development of scientific investigation of aquatic ecosystems. According to these data, only 0.35 percent of the solar flux is incorporated in gross production at the autotroph level (428 ÷ 118,872). This is about one-fourth that found by

TABLE 6-3 ANNUAL ENERGY BUDGET OF LAKE MENDOTA, WISCONSIN

		$c/cm^2/yr$
Incident solar radiation		118,872
Plant utilization		
Phytoplankton		
Net production (NP)	299	
Respiration (R)	100	
Gross production (GP)		399
Bottom flora		
Net production	22	
Respiration	7	
Gross production		29
Gross production by autotrophs		428

Data of C. Juday, 1940. *Ecology* 21:438–50.

Transeau and one-third that found by Golley. For maintenance and growth, Lake Mendota phytoplankton used 25 percent of their energy intake in metabolism and the bottom flora used 24 percent—results virtually identical to those obtained for corn by Transeau and higher than that found by Golley for the grass-herb community of the old field in Michigan.

Juday, however, failed to account for energy lost to herbivores as well as to decomposers; thus these estimates of efficiencies are in error.

Cedar Bog Lake, Minnesota. In a study of a bog lake several years after Juday's study, Raymond Lindeman (1942) computed the loss due to herbivory as 42 $c/cm^2/yr$ and to decomposition as 10 $c/cm^2/yr$. By adding the components, the gross autotroph production rises from 428 to 470 $c/cm^2/yr$ with a consequent increase in energy fixation efficiency, from 0.35 to 0.39 percent. The decrease in loss due to respiration is from 25 to 22.3 percent.

With the same incident solar radiation as Lake Mendota, Cedar Bog Lake in Minnesota is only one-fourth as efficient according to the data assembled by Lindeman (Table 6-4). Gross production, including losses due to herbivory and decomposition, was 111 $c/cm^2/yr$ for an efficiency of energy capture of 0.10 percent. Energy lost in respiration at the autotroph level amounted to 23.4 $c/cm^2/yr$, or 21 percent.

Comparison of Classic Studies on Primary Production

In Lake Mendota and Cedar Bog Lake, energy capture is considerably less than in the two terrestrial systems discussed (0.10 and 0.39 percent versus 1.2 and 1.6 percent). This difference in primary energy capture is chiefly related to diminished light penetration in water. That is, what was measured and used for comparison was the incident solar radiation at the surface rather than the amount that was transmitted through the water to the actual sites of photosynthesis. Thus actual efficiency is somewhat higher than that indicated by Lindeman and Juday, perhaps on the order of 1 to 3 percent, respectively. On the other hand, respiratory losses at the autotroph level in the two aquatic ecosystems (21.0 and 22.3 percent) are about the same as in the cornfield (23.4 percent) but are more than in the old-field grass-herb community (15.1 percent).

TABLE 6-4 ANNUAL ENERGY BUDGET OF CEDAR BOG LAKE, MINNESOTA

	$c/cm^2/yr$
Incident solar radiation	118,872.0
Plant utilization	
Net production (NP)	87.9
Respiration (R)	23.4
Gross production (GP)	111.3

Data of R. Lindeman, 1942. *Ecology* 23: 399–418.

COMPARISONS OF PRIMARY PRODUCTIVITY

Efficiency of Energy Capture

The data on primary energy capture in these two terrestrial and two aquatic ecosystems indicate that only 0.1 to 1.6 percent of the incident solar radiation is initially incorporated into protoplasm. According to a great many studies, largely since the 1940s, it appears that the efficiency of energy capture in gross production under natural conditions is seldom more than 3 percent, although efficiencies of 6 to 8 percent have been recorded under intensive agriculture. Gross production efficiencies in fertile regions are approximately 1 to 2 percent and 0.1 percent in infertile regions; the average for the biosphere, as a whole, is about 0.2 to 0.5 percent.

It is important to note in these calculations investigators have generally used total incident radiation and have not discriminated for the fact that only about half the wavelengths of visible light, mostly in the blue and red ranges, are captured in photosynthesis (Figure 6-1). Making this correction would double efficiency. In

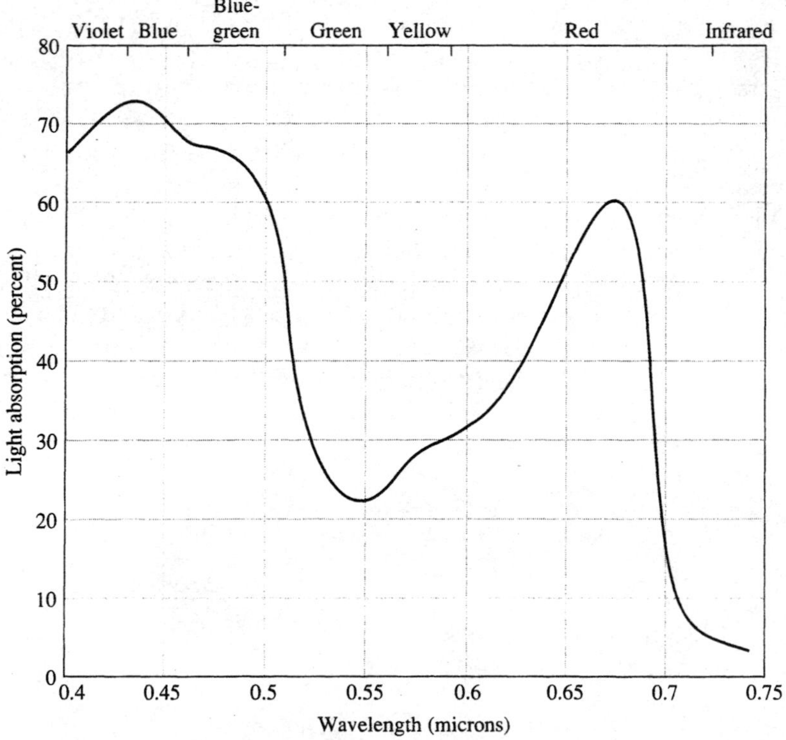

Figure 6-1 Absorption by photosynthetic pigments of solar energy in the visible spectrum is greatest in the violet, blue, blue-green, and red portions, constituting about one-half the total.

addition, most investigators tend to use yearly solar radiation values, but such usage fails to consider the seasonal differences in photosynthesis. In Northern Hemisphere temperate regions, for example, photosynthetic activity is chiefly concentrated in the period from April to October, peaking in June, July, and August. So under these conditions the effective solar flux would be about half the yearly total and efficiency would be doubled. Also, investigators have not always compensated for reduced light penetration in aquatic ecosystems. Had these several kinds of adjustments been made, gross production efficiency would be increased to a range of 2 to 6 percent.

In addition to these factors, the unit of measurement itself can greatly distort actual production efficiencies. For example, based on data collected from 30 sites (15 forests, 11 grasslands, and 4 deserts) in the continental United States, photosynthetic efficiency was highest for forests, being more than 100 times greater than for deserts (Webb et al. 1983). However, when this efficiency was evaluated per unit of foliage, the differences were much less, with a hot-desert site having the highest efficiency.

Finally, photosynthetic efficiency is typically based on the organs responsible for photosynthesis, and in the case of terrestrial ecosystems, on aboveground parts. Stanton (1988) has shown that 60 to 90 percent of the net primary productivity in grasslands occurs in the soil as roots. In tallgrass prairie, belowground production is 48–64 percent of the total net primary production, mixed grass prairie is 61–80 percent, and shortgrass prairie is 70–78 percent.

Gross and Net Productivity

Energy expended in self-maintenance, growth, and respiration also varies considerably, from 15 to 24 percent, as we have already seen in the old field and the cornfield. In temperate forests this expenditure reaches 50 to 60 percent and in tropical forests 70 to 75 percent. Based on a large number of studies, it appears that, in general, autotroph respiration accounts for 30 to 40 percent of gross production. Thus only about 60 to 70 percent (and often less) of the initial energy captured generally results in net production. As a rough rule of thumb, we can estimate net primary production to be about half of gross production. It must be remembered, however, that any "rule of thumb" has exceptions.

Factors Influencing Productivity

It is well recognized that various factors influence the rate and amount of photosynthesis. In addition to the availability of the basic chemical components of photosynthesis, changes in physical and biological factors influence the rate and amount of photosynthesis, as do diurnal and seasonal changes in such factors as light and temperature. A brief consideration of recent studies demonstrating some of the factors affecting productivity follows.

Moisture. The central grassland region of the United States encompasses precipitation ranges from 260 to 1200 mm and temperature ranges from 3° to 22°C. The importance of precipitation is seen in Figure 6-2, which shows that the lowest values of aboveground net primary production occur in the west and the highest values occur in the east in normal, favorable and unfavorable years (Sala et al. 1988). In unfavorable years (Figure 6-2b), the normal aboveground net primary productivity is reduced 17 percent in the eastern (wetter) region and 33 percent in the western (drier) region. In favorable years (Figure 6-2c), production is increased by the same respective percentages in both regions.

Moisture and soil. Soil characeristics also impact the level of productivity (as described above) in these central United States grasslands. When precipitation is more than 370 mm/yr, sandy soils with low water-holding capacity are more productive than loamy soils with high water-holding capacity; the opposite pattern occurs when precipitation is less than 370 mm/yr (Sala et al. 1988).

Nutrients. Given the operation of the law of the minimum and law of tolerance, it is not surprising that nutrients exert considerable influence on the rate and amount of productivity in ecosystems. For example, aboveground net primary production was measured in a lodgepole pine (*Pinus contorta latifolia*) forest, a meadow dominated by western wheatgrass (*Agrophron smithii*) in southwestern Wyoming, and in shortgrass (*Bouteloua gracilis*) prairie in northeastern Colorado (Hunt et al. 1988). In response to the addition of nitrogen in the form of ammonium nitrate (NH_4NO_3), production increased in the forest by 52 percent, in the meadow by 102 percent, and in the prairie by 81 percent over control plots in which no nitrogen was added.

Estuaries and coastal marine ecosystems are more likely to be nitrogen-limited than are lakes, at least for northern temperate-zone estuaries and lakes (Howarth et al. 1988). However, it seems likely that some estuaries are phosphorus limited or that they switch seasonally between phosphorus limitation and nitrogen limitation. Since estuaries receive far more nutrient inputs than any other type of ecosystem, many receiving 1,000-fold greater quantities than heavily fertilized agricultural fields, control of input of nutrients containing nitrogen and phosphorus is important. As we will discuss in Part VIII, nutrient enrichment of aquatic ecosystems, including estuaries, results in a phenomenon known as *eutrophication* in which anoxic bottom waters are the consequence of the decomposition of the increased biomass produced in the upper waters because of the enrichment of the nutrients.

Studies on desert antarctic lakes suggest that nutrient supply, rather than light or temperature, determines the variations in primary productivity and that low light rather than low temperature dampens algal photosynthesis (Vincent 1981). Lake Fryxell in the Dry Valley region of South Victoria Land in Antarctica, with an average depth of 8 m and a maximum depth of 19 m, is permanently covered by about 4 m of ice. Carbon-14 measurements of primary production showed that the phyto-

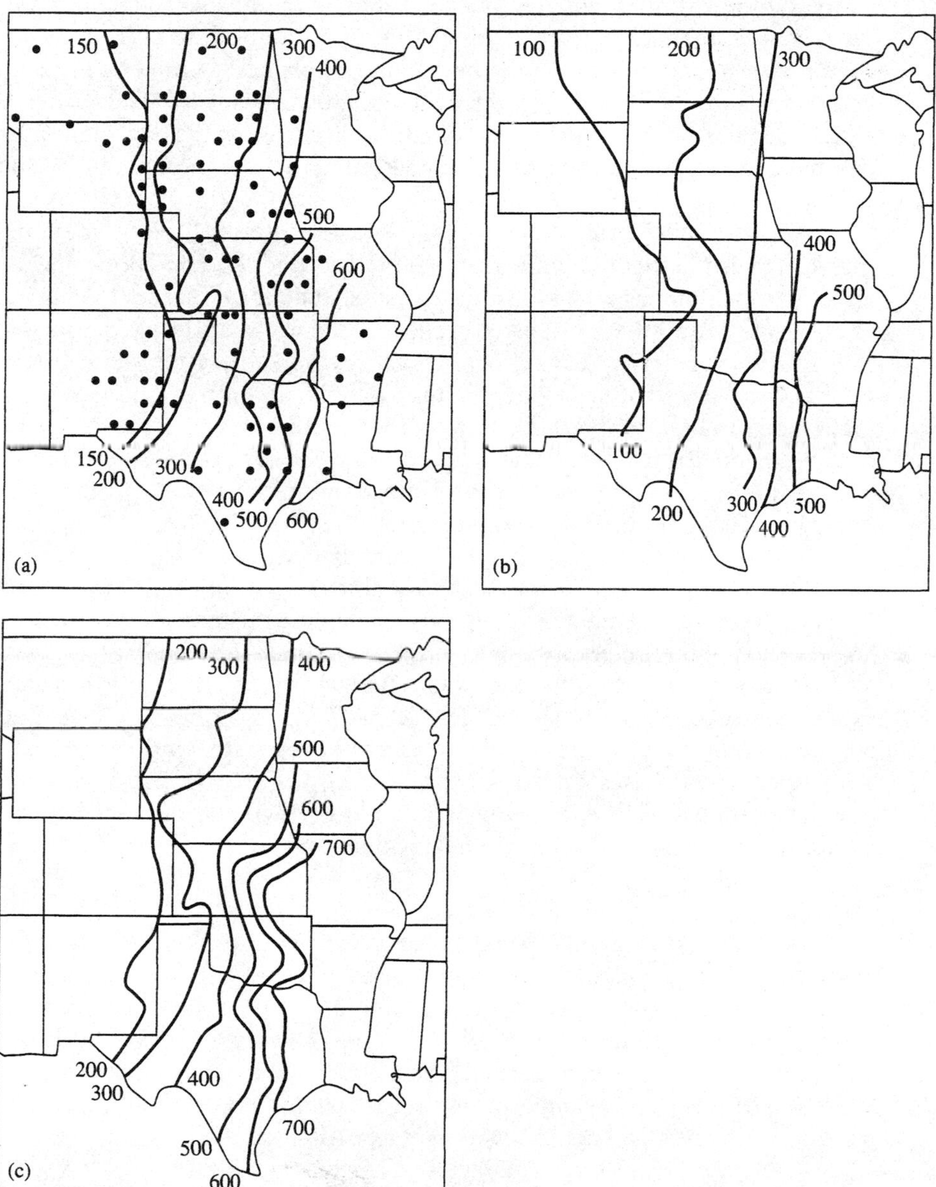

Figure 6-2 Aboveground net primary production (g/m²) for the central grassland region: (a) During years of average precipitation; (b) during unfavorable years; and (c) during favorable years. (Reproduced with permission from Sala et al. 1988. *Ecology* 69:40-5.)

plankton (*Ochromonas* and *Chlamydomonas*) immediately under the ice were adapted to relatively bright light but were limited by the availability of nitrogen; by contrast, the shade-adapted euphotic zone phytoplankton (*Chroomonas* and *Pyramimonas*) did not respond to nitrogen or phosphorus enrichment. Both populations showed net increases very early in the growing season. The flagellated algae in the middle zone swam up to areas of greater light during the day, returned to lower depths of nutrient supply at night, and continued to grow throughout the summer.

Moisture and nutrients. Wetland ecosystems are subject to periodic flooding, and their plant communities are adapted to flooded conditions, including the anaerobic conditions in the soils and sediments that occur during flooding. Brinson et al. (1981) found that in forested wetlands, the lowest primary productivity rates (5.8 g organic matter/m^2/day) occur in dwarf cypress ecosystems that grow on sterile shallow soils and on soils where rainfall is the main source of nutrients and water; higher rates (52.1 g/m^2/day) occur in floodplain forests, which receive nutrient-rich flowing water. Both forested and nonforested wetlands exhibit similar rates of gross primary productivity, but forested wetlands exhibit lower rates of net biomass production than their nonforested counterparts.

Organisms. In the Serengeti National Park, Tanzania, and Masai Mara Game Reserve, Kenya, McNaughton (1985), using harvest techniques, found that aboveground productivity averaged 664 g/m^2/yr and 3.78 g/m^2/day, the latter peaking at more than 30 g/m^2/day. This level was stimulated by the grazing of ungulates such as wildebeests, zebras, Thomson's gazelle, buffalo, and topi. The ungulates maintain the vegetation in an immature, rapidly growing state similar to that at the beginning of the rainy season. It is possible that this level of production may have been overestimated since, as shown by Stanton (1988) and discussed above, a significant proportion of the aboveground production may have been translocated from below ground.

Density of organisms. Productivity in individual red alder trees (*Alnus rubra*) as well as forests, as might be expected, varies with density. Low, intermediate, and high density plots (respectively 1,240, 4,068, and 10,091 trees/ha) showed a total aboveground net production of 8.83, 13.30 and 12.48 mg/ha, respectively (Bormann and Gordon 1984). When measured on an individual tree basis, a reverse trend was found, namely 7.07, 3.27, and 1.25 mg/tree. In other words, numbers of trees compensate for the reduced net production per tree in high density plots.

Dissolved substances. Using ^{14}C and chlorophyll methods on the Orinoco River Systems in Venezuela, Lewis (1988) found that annual mean gross primary production was highest in a whitewater tributary (Apure River) draining the plains or llanos north of the Orinoco River and lower in the blackwater tributaries to the south (Caura and Caroni Rivers), while in the Orinoco proper, production was intermediate to these levels and above that of the Apure. The presence of dissolved and suspended substances, however, suppressed potential production by 85 percent in whitewaters, 40 percent in blackwaters, and intermediate levels in the Orinoco main stream.

Growth and development. In a study of balsam fir (*Abies balsamea*) development on Whiteface Mountain, the highest peak in the Adirondack Mountains in northeastern New York, Sprugel (1984) determined the aboveground productivity for a mature stand to be 960 g/m^2/year distributed as follows: foliage, 380 g/m^2/yr; twigs, 150 g/m^2/yr; branch wood and bark, 170 g/m^2/yr; bole bark, 25 g/m^2/yr; and bole wood, 235 g/m^2/yr. As a stand develops over approximately a 60-year period, productivity remains generally constant. However, during the early stages, about 45 percent of the total aboveground production produces bole wood and bark in contrast to about 25 percent in a mature stand.

The influence of age is also evident in the annual net production in a plantation of Scots pine (*Pinus sylvestris*) in England (Figure 6-3). According to the investigator, J. D. Ovington (1962), net production increases to a maximum of about 52×10^6 C/ha (i.e., 5200 C/m^2) at about 30 years of age and levels off thereafter. Eugene Odum found that production in an old field in Georgia was highest in the first year after the field was allowed to go fallow and reached a plateau in the third year.

Physical and biological interactions. In marine environments the interaction of physical and biological processes is extremely important in structuring biological communities and influencing production (Daly and Smith 1993). Phytoplankton production accounts for over 95 percent of marine photosynthesis and occurs in only about 5 percent of the ocean by total volume of water. In restricted situations, behavior such as vertical migration and predation may control plankton production. On the larger scale, physical factors such as movement and thermal properties of ocean currents, coupled with different amounts of critical nutrients, let alone sunlight, directly affects plankton production. Thus while some ecologists have tended to view the ocean as a homogeneous environment of physical and chemical factors, or at least less heterogeneous than terrestrial ones, this homogeneity may be true only in small scale situations rather than throughout the ocean environment.

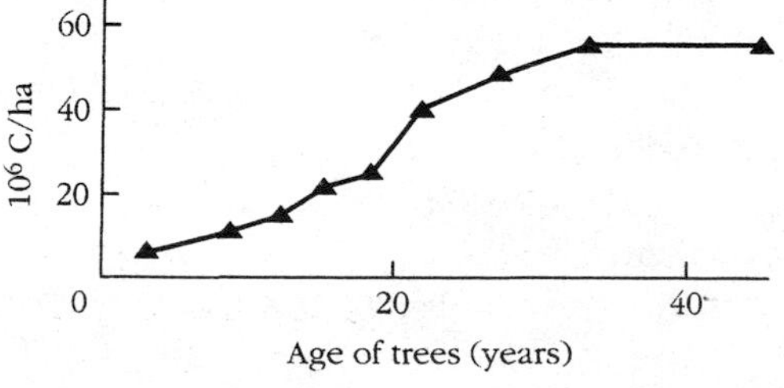

Figure 6-3 Annual mean productivity of trees and understory vegetation in a Scots pine (*Pinus sylvestris*) plantation in England. (Adapted by permission from J. D. Ovington. 1962. *Advances in Ecological Research* Vol. 1. New York: Academic Press.

WORLD PRIMARY PRODUCTIVITY

Perhaps the earliest estimate of world primary productivity was that of Justus von Liebig (1862), whose law of the minimum was discussed in Chapter 3. His estimate of world productivity at 500 g (dry matter)/m^2/yr was based on the assumption of the world's surface being covered by a moderately productive meadow. Whittaker and Likens (1973a) noted that this estimate was not too far from current estimates of the weighted mean production of the world, namely 320 g/m^2/yr.

Major summaries of estimates of world primary productivity in terrestrial and aquatic ecosystems have been made by a number of investigators. Among them, those by Whittaker and Likens (1973b) and Leith and Whittaker (1975) are the most comprehensive and detailed. Figure 6-4 depicts the average annual rate of production in kilocalories per square meter in the world's major ecosystems. The original data were expressed in grams of dry matter, but have been converted to energy units to be consistent with the usage throughout this chapter.

As mentioned, as a rule of thumb, one gram of dry protoplasm has a caloric value of about 4.5 C; this is the conversion value used for Figure 6-4. Actually, however, wide variation in caloric values exists. For example, protoplasm that is largely carbohydrate, like sugar beets, will be of lower caloric content (roughly 4.0 C/g) than that rich in oils, like peanuts (roughly 9.0 C/g). Also, if the material is dried and the amount of mineral ash is determined and subtracted to give the true organic weight, the caloric equivalent would be higher than 4.5 C/g. A number of determinations suggest a normal range of 4.1 to 5.2 C/g of dry organic weight, but some algae have values of up to 8.7! Any selected conversion is thus arbitrary and we have chosen here to err on the low side by using the value of 4.5 C/g. Caloric values in Figure 6-4 were originally expressed as grams of carbon/m^2/yr and were converted to energy units according to the following assumptions: (1) that dry organic matter is about 49 percent carbon, and (2) that dry organic matter has a caloric value of 4.5 C/g. Arithmetically, then, we multiply the gram carbon value by 9.2 (i.e., $4.5 \div 0.49$).

Examination of Figures 6-4 and 6-5 and Table 6-5 shows that the terrestrial ecosystems with the highest productivity rates are in the tropics: tropical forests are nearly twice as productive as temperate forests and nearly three times as productive as boreal coniferous ones. Data examined by Westlake (1963) extend this generalization to annual and perennial agricultural plants as well as to submerged and emergent aquatic plants in tropical and temperate conditions. In both instances, the tropical counterparts are about twice as productive. Annual production (rate of production $\times$ area of production) shows a corresponding pattern: tropical forests—20×10^3 C/m^2; temperate forests—11.2×10^3 C/m^2; coniferous forests—4.8×10^3 C/m^2 Westlake 1963. The effect of latitude on incident solar radiation is significant here (see Figure 4-3).

Although net annual productivity is higher in tropical ecosystems, the efficiency of conversion of gross to net productivity is lower. Box (1978) has shown

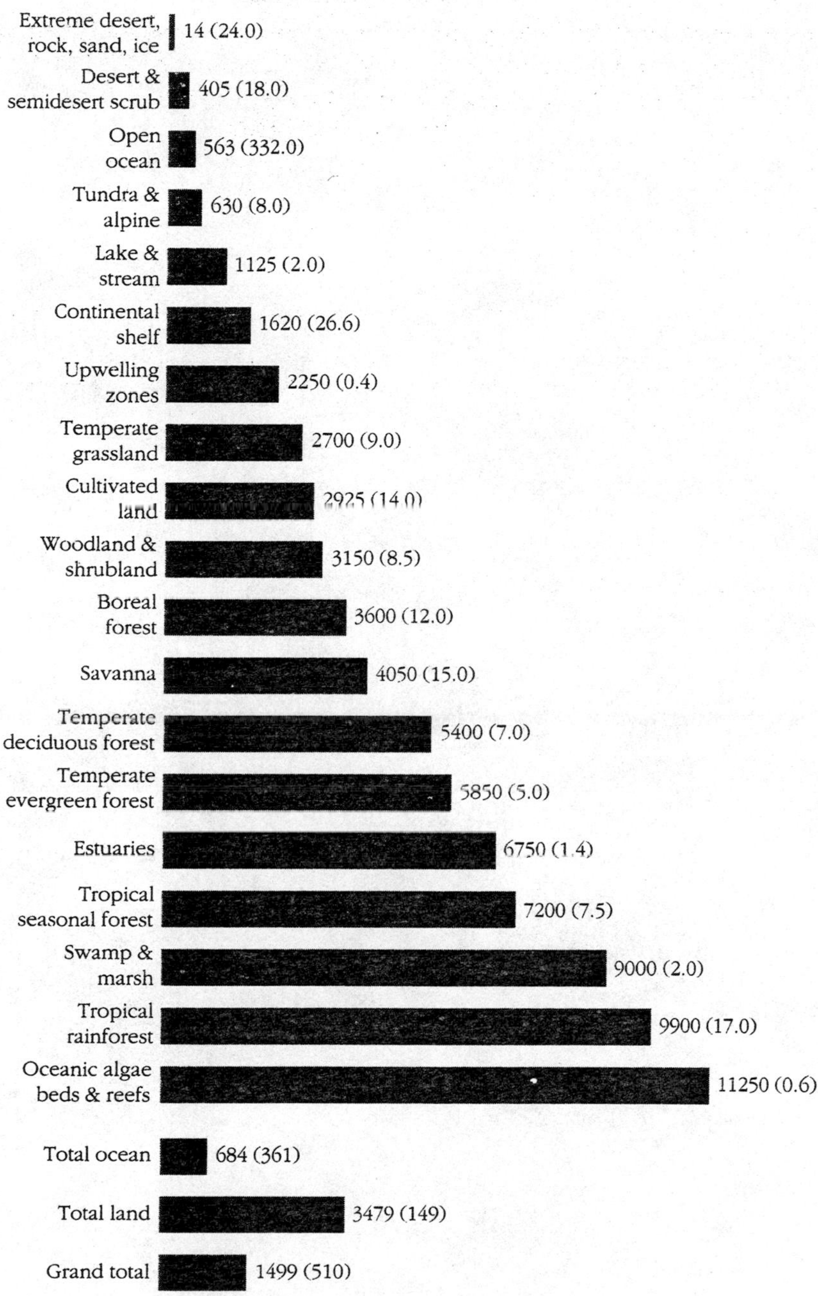

Figure 6-4 Annual average rate of net plant production. The number after the bar is C/m^2/yr; the number within the parentheses is area in 10^6km^2. (Derived from data in R. J. Whittaker, 1975. *Communities and Ecosystems,* 2d ed. New York: Macmillan Co.

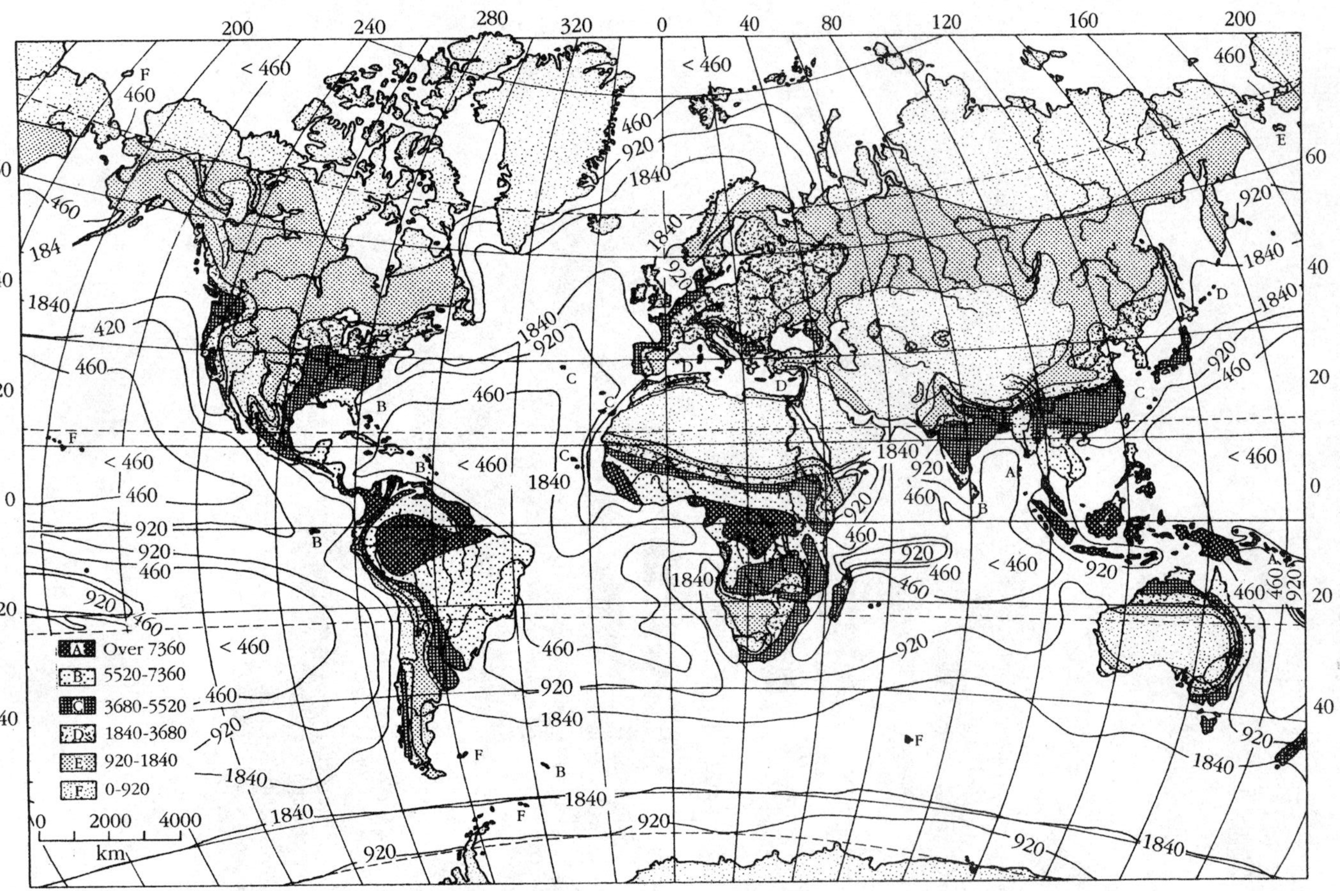

Figure 6-5 Annual net primary production in $C/m^2/yr$. Mapped values are approximate and are incompletely adjusted for losses to consumers, decomposers, and substrate. (Redrawn by permission from a map in H. Lieth and D. E. Reichle (eds.), 1970. *Analysis of Temperate Forest Ecosystems.* Ecological Studies 1. New York: Springer-Verlag.)

TABLE 6-5 RANGES OF NET PRIMARY PRODUCTIVITY (C/m²/yr)

Ecosystem	Ranges: Low	Medium	High
Forest	$10.0–1.3 \times 10^3$ Boreal forest (China) woodland (Kenya); chaparral (Mexico)	$4.0–6.0 \times 10^3$ Mangroves (Burma); temperate deciduous forest (Mexico); broadleaf forest (China)	$7.9–9.0 \times 10^3$ Tropical seasonal forest (Brazil); tropical rain forest (Brazil, Zaire, Indonesia)
Grassland	$0.5–1.5 \times 10^3$ Stressed subtropical and montane formations (China, Sahel, Andes)	$1.5–2.5 \times 10^3$ Temperate grassland (China, Turkey, Argentina)	$2.5–3.5 \times 10^3$ Tropical grassland (Brazil, Sudan, India)
Cereals	$0.15–0.3 \times 10^3$ Upper Volta (now called Burkina Faso), Nigeria, Zaire	$0.45–0.6 \times 10^3$ Pakistan, Brazil, Mexico	$0.75–0.9 \times 10^3$ Vietnam, China, Indonesia

Data from V. Smil, 1979. *American Scientist* 64: 522–31.

that the percent of gross primary productivity that results in net production is in the range of 40 to 50 percent whereas in temperate regions the percentage is in the range of 60 to 70 percent and reaches 80 percent in the Antarctic. The high temperatures of tropical environments result in plants expending more of their gross production energy in respiration, thus reducing the amount of energy converted to net production.

Figures 6-4 and 6-5 also show that the open oceans and deserts are much less productive when compared to the ocean-land surface regions, such as the continental shelf, coral reefs, upwelling areas, and estuaries. Because the oceans constitute 70 percent of the Earth's surface and deserts constitute about 28 percent of the land area, about 80 percent of the Earth's surface consists of the least productive ecosystems. Generally, low production in the sea is a function of nutrient limitation, as we shall see in Chapter 8. The ocean-land interface, in contrast, is rich in nutrients derived from runoff from the land and/or from upwellings of nutrient-rich sediments. In the desert, lack of moisture is the primary factor accounting for low productivity. Periodically the availability of nutrients in the ocean or rain in the desert results in daily rates of production comparable to those of the higher productive regions. They are transitory, however; as a result, the daily rate based on annual data is much lower.

Finally, although the large surface area of the sea compensates considerably for the low rate of production on a unit area basis (Figure 6-4), the total production of the sea (rate × area) of 2.5×10^{17} C/yr is only half that of the land, 5.2×10^{17} C/yr.

SUMMARY

VOCABULARY

eutrophication
gross production
hectare
net production
primary production
productivity
standing crop

KEY POINTS

- The stoichiometry of the photosynthetic equation enables a variety of methods for measuring primary production including harvesting, carbon dioxide assimilation, oxygen production, radioisotopes, chlorophyll, and pH. Satellite sensing and computer simulations are more recent approaches.
- Studies of primary production have been conducted in both terrestrial and aquatic ecosystems, the significant early ones, now recognized as classic, are a cornfield in Illinois, an old field in Michigan, and two lakes, one in Wisconsin and the other in Minnesota, the latter study having had profound impact on subsequent ecological studies.
- Data on gross production efficiency (production relative to total incident energy) often are in error because they usually do not account for the actual wavelengths of energy captured in photosynthesis, the seasonal variation in photosynthesis, nor the fact that considerable primary production occurs below ground.
- In general, autotroph respiration accounts for 30 to 40 percent of gross production, or as a "rule of thumb," about one half of the gross production.
- Net autotroph production (gross production less respiration) is available for the next trophic level and/or decomposition.
- A number of factors influence productivity including moisture, soil, nutrients, and the type, density, and developmental stage of organisms.
- The highest productivity rates are in the tropics, tropical forests being nearly twice as productive as temperate forests and three times as productive as boreal forests. However, the higher tropical temperatures reduce the percentage of gross primary productivity that results in net productivity to 40 to 50 percent, as compared to the 60 to 70 percent in temperate regions.
- The open ocean and deserts are less productive than ocean-land interface regions; about 80 percent of the Earth's surface consists of the least productive ecosystems.
- Although the oceans constitute some 70 percent of the Earth's surface, their total production (rate × area) is about one-half that of the land.

CHAPTER 7

Energy Flow Beyond the Producers

In Chapter 6 we considered the initial incorporation and utilization of energy by autotrophs and primary productivity in selected ecosystems as well as on a global level. We now turn to the fate of energy as it passes beyond this initial trophic level in several aquatic and terrestrial ecosystems to provide the background for developing a general model of energy flow.

AUTOTROPH-BASED ECOSYSTEMS

Cedar Bog Lake, Minnesota

Cedar Bog Lake is a good place to begin a discussion of energy flow beyond the producers (autotrophs) since Raymond Lindeman (1942), who studied the Lake, was perhaps the most significant person in developing the concept of the energy dynamics or, as he termed it, the "*trophic dynamics,*" of ecosystems. Also, since we discussed the incorporation of solar energy in Cedar Bog Lake in the preceding chapter, the next steps of energy flow in that ecosystem will be easier to develop.

Fate of energy incorporated by autotrophs. Of the gross production of 111 c/cm^2/yr in Cedar Bog Lake, 20.7 percent, or 23 c/cm^2/yr, is transformed in biological reactions occurring at the cellular level, reactions that underwrite the growth, development, maintenance, and reproduction of the autotrophs (see Table 6-4). This results in a net production of 88 c/cm^2/yr. Figure 7-1 shows the fate of this net production.

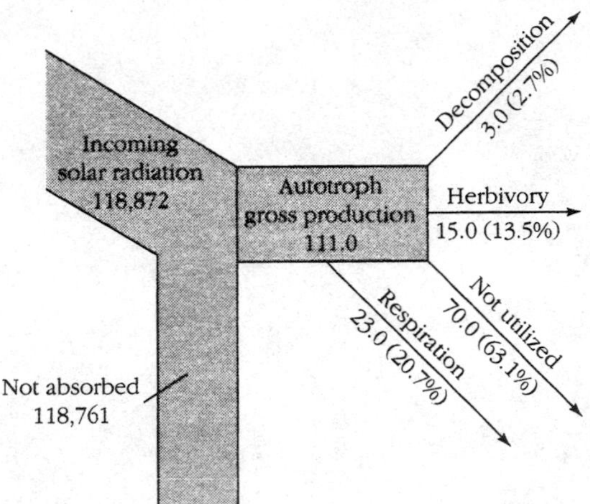

Figure 7-1 Fate of energy incorporated by autotrophs in Cedar Bog Lake, Minnesota, in c/cm²/yr. (Data of R. Lindeman. 1942. *Ecology* 23:399–418.)

Herbivores consume 15 c/cm²/yr as they graze or feed on autotrophs—this amounts to 13.5 percent of net autotroph production (see Figure 7-1). Decomposition accounts for about 2.7 percent of net production (3 c/cm²/yr). The remainder of the plant material, 70 c/cm²/yr, or 63.1 percent of net production, is not used at all but becomes part of the accumulating sediments. It is obvious, then, that much more energy is available for herbivory than is consumed and that this ecosystem is especially prodigal in energy use beyond the autotroph level.

In a further observation about Figure 7-1, note that the various pathways of loss are equivalent to and account for the total energy capture of the autotrophs—that is, gross production. Also, collectively the three "fates" (decomposition, herbivory, and not utilized) are equivalent to net production. Because similar diagrams will be used in developing the fate of energy at subsequent levels, recognition of this basic budget pattern is important to understanding what follows.

Of the total energy incorporated at the level of the herbivore—namely, 15 c/cm²/yr—30 percent or 4.5 c/cm²/yr is used in metabolic activity (Figure 7-2). There is, then, a considerably higher proportion of energy lost via respiration by herbivores (30 percent) than by autotrophs (21 percent). There is also considerable energy available for the carnivores—namely, 10.5 c/cm²/yr or 70 percent—that is not entirely utilized. In fact, only 3.0 c/cm²/yr or 20 percent of net production passes to the carnivores. This is a more efficient use of resources than occurs at the autotroph → herbivore transfer level, but still it is quite profligate in this particular ecosystem. As we will see, other systems use available energy more efficiently.

At the level of the carnivore (Figure 7-3), about 60 percent of the carnivore's energy intake is consumed in metabolic activity and the remainder becomes part of the nonutilized sediments. Only an insignificant amount is subject to decomposition yearly. This high respiratory loss compares with 30 percent by herbivores and 21 percent by autotrophs in this ecosystem. A number of factors are involved in this greater energy utilization. Among the more significant is the generally greater locomotor activity of carnivores—in effect, they must expend considerable energy to get their energy.

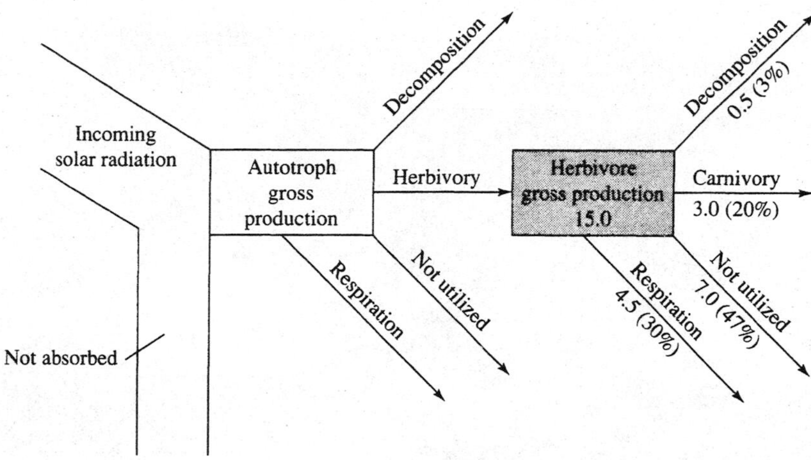

Figure 7-2 Fate of energy incorporated by herbivores in Cedar Bog Lake, Minnesota, in c/cm²/yr. (Data of R. Lindeman. 1942. *Ecology* 23:399–418.)

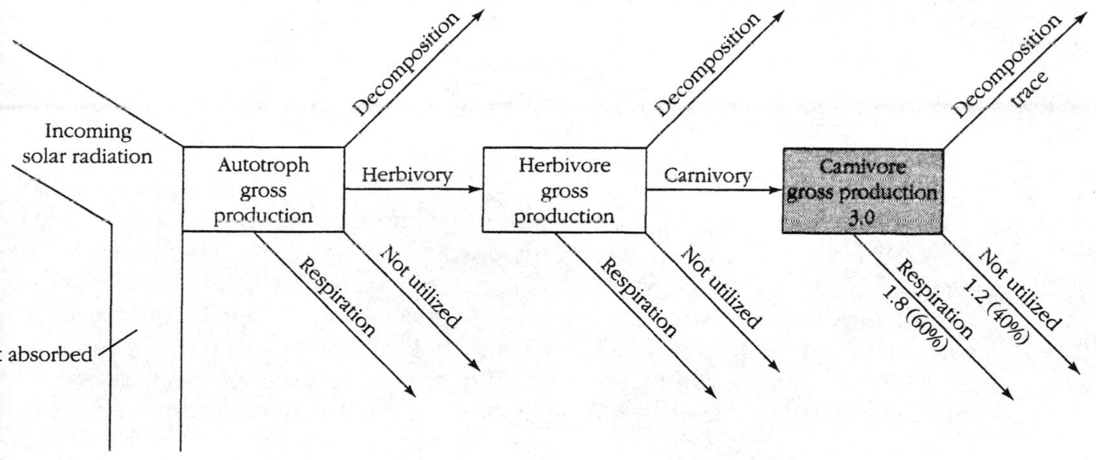

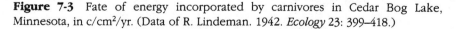

Figure 7-3 Fate of energy incorporated by carnivores in Cedar Bog Lake, Minnesota, in c/cm²/yr. (Data of R. Lindeman. 1942. *Ecology* 23: 399–418.)

Having developed in stepwise fashion the fate of energy in this ecosystem, we can now look at it in a composite way (Figure 7-4), slightly modified from the previous diagram. Using it as a model, we find that several points are immediately evident. Most important is the one-way street along which energy moves. The energy that is captured by the autotrophs does not revert back to solar input; what passes to the herbivore does not pass back to the autotrophs, and so on. As it moves progressively through the various trophic levels, it is no longer available to the previous level. The immediate implication of this unidirectional energy flow in an

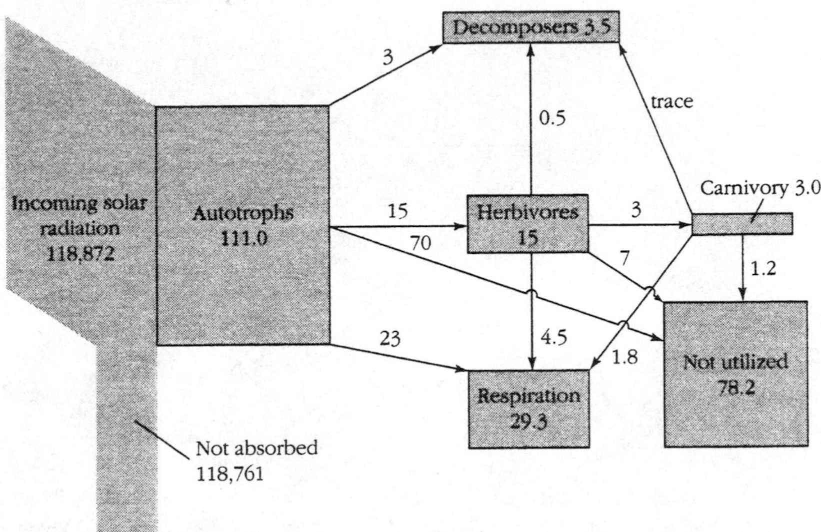

Figure 7-4 Energy flow diagram for Cedar Bog Lake, Minnesota, in c/cm²/yr. (Data of R. Lindeman. 1942. *Ecology* 23: 399–418.)

ecosystem is that the system would collapse if the primary energy source, the Sun, were cut off.

The next major point to be noted is the progressive decrease in energy at each trophic level. This fact is largely explained by the energy dissipated as heat in metabolic activity and measured here as respiration. This particular ecosystem also has a large amount of energy that is not used. Even if more of the nonutilized energy were used in a more efficient system, there would still be a considerable loss in respiration. If all the nonutilized plant material were consumed by herbivores, for example, there would be 85 $c/cm^2/yr$ gross production at the herbivore level. However, about 30 percent, or 25 $c/cm^2/yr$, would still be lost via respiration at this trophic level. Similarly, if there were 100 percent efficiency of energy transfer to the carnivore level, of the approximately 60 $c/cm^2/yr$ gross production at that level, some 36 $c/cm^2/yr$ gross production would be lost via respiration. Thus even with a higher efficiency of energy use, considerable energy would still be required to maintain the system.

Laws of Thermodynamics

These factors—unidirectional flow and efficiency of energy utilization—account for the requirement of a continual source of energy to preclude collapse of an ecosystem. An ecosystem simply cannot maintain itself if deprived of energy income for any extended period of time.

The operation of an ecosystem, then, is consistent with the laws of thermodynamics that deal with the relationships between energy and matter in a system. First, we can account for the total energy input to an ecosystem in budgetary fashion. This is consistent with the *first law of thermodynamics*, which says, in various ways, that the sum total of energy in a system is a constant or, more familiarly, that energy can neither be created nor destroyed. Energy may change form (e.g., from radiant to chemical) but not amount. But as energy moves through an ecosystem, it does change form, ultimately to heat, which is not directly usable by the system. Thus the system tends to run down, dissipating its energy and losing its organized structure. This is the nature of the *second law of thermodynamics*, a tendency toward entropy or maximum disorganization of structure and maximum dissipation of usable energy. This trend is countered by the continual input of energy from the outside.

Silver Springs, Florida

The model of energy flow in Cedar Bog Lake is, of course, highly diagrammatic and greatly oversimplified. It is hoped that the Cedar Bog Lake study will help in understanding a more sophisticated representation of energy flow, one developed by Howard T. Odum (1957) from a study on a river system, the famous tourist attraction of Silver Springs, Florida.

There are several elegant refinements in Figure 7-5 over the Cedar Bog Lake diagram. For example, the light actually absorbed by the plants is the basis for considering autotroph efficiency, not the total incident solar radiation. Also, there is an accounting of material (energy) entering the system from contributory streams or the surrounding terrain (labeled "import"). And, most significantly, the large energy loss by way of decomposer respiration is considered.

A comparison of these two systems (Table 7-1) is useful for several reasons: both systems are classic studies in modern ecology; they represent two quite different ecosystems, one a bog-water lake, the other a clear-water river; and the amounts, rates, and efficiency of energy movement are quite different. (*Note:* To facilitate comparison, Cedar Bog Lake data have been converted from c to C and from cm^2 to m^2.)

Among the differences that can be noted, the autotrophs in Silver Springs are at least ten times more efficient in capturing solar energy than their counterparts in Cedar Bog Lake. But subsequent to initial capture, about two and one-half times more energy is lost in respiration in Silver Springs, the higher loss being observed at all trophic levels. Although the percentage of energy lost to decomposition and nonutilization combined appears more or less comparable at each trophic level, the actual percentage loss to each is quite different (see totals). Of the total, only about one-third is decomposed yearly in Cedar Bog. This accounts for the progressive accumulation of sediments typical of boreal bog lakes, sediments known as peat. In contrast, most of the nonutilized material in Silver Springs is exported downstream.

Old Field in Michigan

For consideration of energy flow in a terrestrial ecosystem, we will build on the discussion of primary energy capture in an old field in Michigan (Golley 1960), which

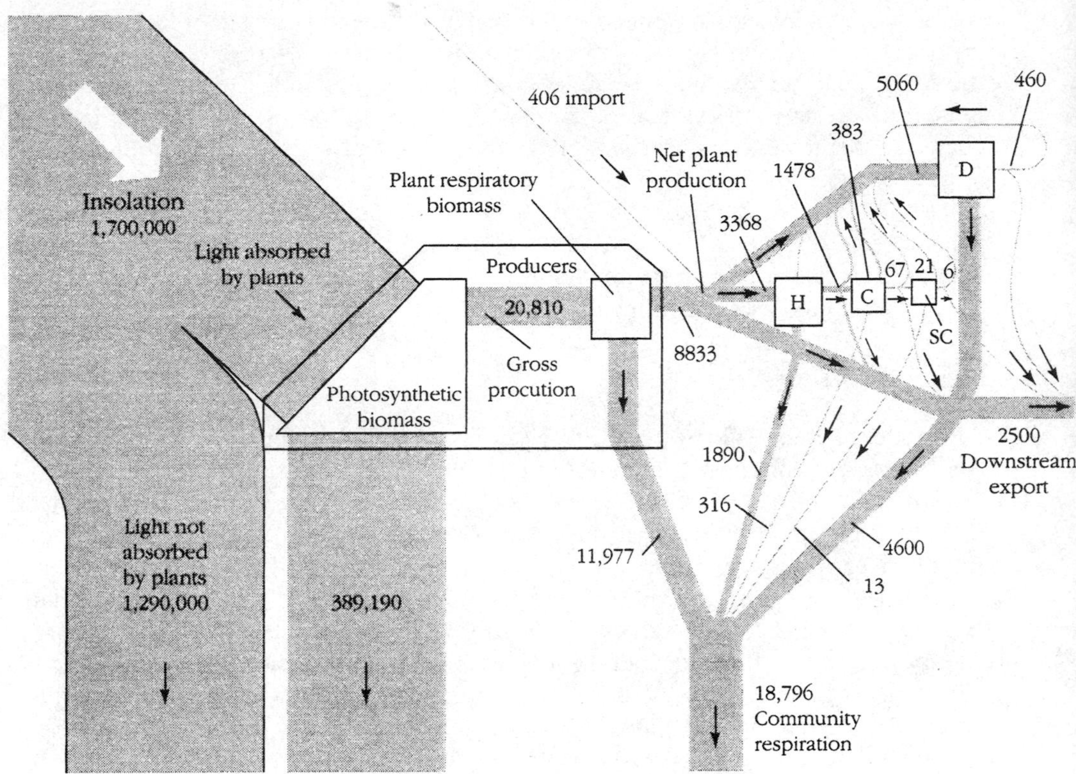

Figure 7-5 Energy flow diagram for Silver Springs, Florida, in C/m²/yr. (C=primary carnivores; D=decomposers; H=herbivores; SC=secondary carnivores.) (Redrawn by permission from H. T. Odum. 1957. *Ecological Monographs* 27: 55–112.)

was introduced in Chapter 6 (see Table 6-2). Golley's study, like those of Lindeman and Odum, was a pace-setting attempt to account for the energy flow in a terrestrial ecosystem, thus constitutes a good place to expand on the topic of autotroph-based ecosystems.

According to Golley's results, the total energy in the vegetation that was available to mice (*Microtus pennsylvanicus*) and other herbivores amounted to 15.8×10^3 C/hectare/year; of this amount, only 1.6 percent (250×10^3 C) was consumed by mice and other herbivores (Figure 7-6). Fifteen percent of the gross production in the vegetation was expended in respiration; fifty-eight percent was unused. Owing to the many sampling difficulties involved, Golley did not follow the flow of energy that was not consumed by mice.

Given the large amount of available energy left ungrazed by the mice and other herbivores (the unused compartment), it is evident that grazing accounted for a very small portion of energy flow in this terrestrial ecosystem. As we shall see in the dis-

TABLE 7-1 COMPARISON OF ENERGY FLOW IN A BOG LAKE[a] AND A RIVER[b]

	Cedar Bog Lake, Minnesota		Silver Springs, Florida	
	($C/m^2/yr$)	(%)	($C/m^2/yr$)	(%)
Incoming solar radiation (S)	1,188,720		1,700,000	
Effective solar radiation (ES)	?		410,000	
Producers (autotrophs)				
Gross production (GP_A)	1,113		20,810	
Efficiency $\left(\frac{GP_A}{S \text{ or } ES}\right)$		0.10		1.2 or 5.1
Respiration (R)	234		11,977	
Respiratory loss $\left(\frac{R}{GP_A}\right)$		21.0		57.6
Net production (NP_A)	879		8,833	
Decomposed or not utilized		83.1		61.9
Herbivores				
Gross production	148		3,368	
Efficiency $\left(\frac{GP_H}{NP_A}\right)$		16.8		38.1
Respiration	44		1,890	
Respiratory loss		29.7		56.1
Net production (NP_H)	104		1,478	
Decomposed or not utilized		70.2		72.7
Carnivores				
Gross production	31		404	
Efficiency $\left(\frac{GP_C}{NP_H}\right)$		29.8		27.3
Respiration	18		329	
Respiratory loss		58.1		81.4
Net production	13		73	
Decomposed or not utilized		100.0		100.0
Total loss via respiration	296	26.6	14,196	68.2
Total loss via decomposition	310	27.9	5,060	24.3
Total loss via nonutilization	507	45.5	1,554	7.5

[a] Data of R. Lindeman. 1942. *Ecology* 23:399-418.
[b] Data of H. Odum. 1957. *Ecological Monographs* 27:55–112.

cussion of detritus-based ecosystems, this is a common pattern; most of the energy follows a decomposition pathway on the "unused" component.

In addition to the energy derived by mice from vegetation, there was a net import of energy (13.5 C/hectare/year) resulting from immigration of mice from adjacent ecosystems. Of the total amount of energy in the mouse population, more than 60 percent was consumed in metabolic activity and at the next trophic level, the least weasels (*Mustela rixosa allegheniensis*) expended 93 percent of their energy in life-sustaining activities. Unlike plants, animals move about and expend energy

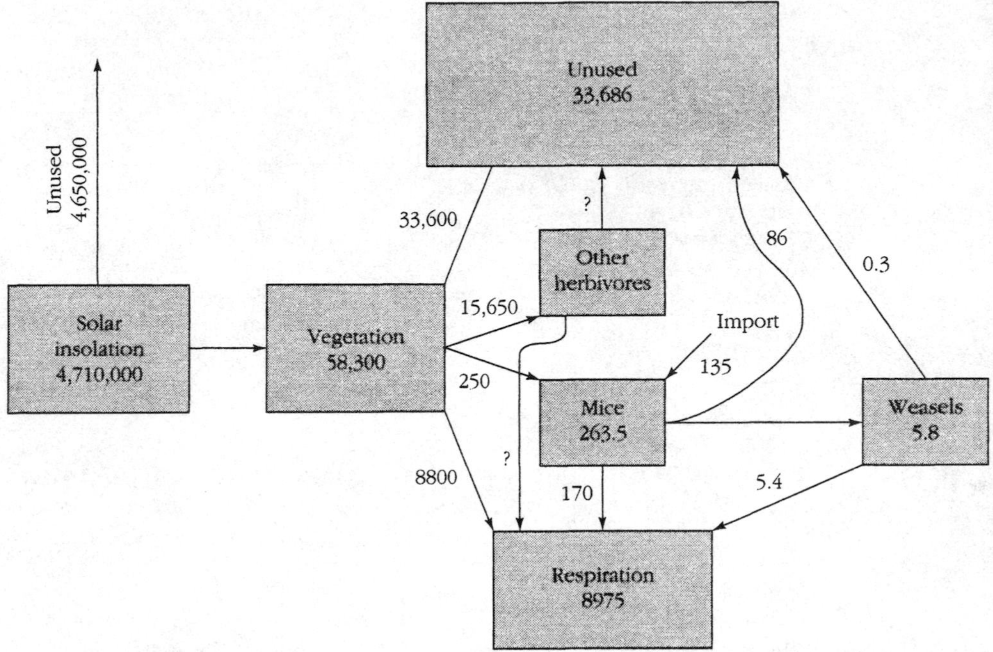

Figure 7-6 Energy flow diagram for an old field in Southern Michigan in 10^3C/hectare/year. (Data of F. Golley. 1960. *Ecological Monographs* 30:187–206.)

in maintaining a more or less constant body temperature; these activities account for the greater respiratory losses in heterotrophs as contrasted to autotrophs.

In summary, of the total amount of energy incorporated in this terrestrial ecosystem (58,300 C/hectare/year):

transmitted to herbivores	27.00%
transmitted to carnivores	0.01%
lost via respiration	15.00%
unused	58.00%

After consideration of detritus-based ecosystems in the next section, we will attempt to generalize the pattern of energy flow in different ecosystems.

DETRITUS AND DETRITUS-BASED ECOSYSTEMS

The ecosystems that have just been discussed depend directly on the influx of solar radiation. They are characterized by a dependence on energy capture by photosynthetic autotrophs and secondarily by movement of that captured energy through the system via herbivory and carnivory. A large number of ecosystems function in this way, and numerous herbivores, carnivores, and omnivores, including humans, are completely dependent on such autotrophic ecosystems.

Other ecosystems depend less on direct solar energy incorporation and more on the influx of dead organic material, or *detritus*, produced in another ecosystem. Indeed, some ecosystems, such as caves, are completely independent of direct solar energy and are instead completely energy dependent on the influx of detritus. These entities can be regarded as detritus-based ecosystems. In other instances, subcomponents of an ecosystem derive their energy entirely from that system's detritus through decomposition, as was noted in the foregoing discussions of autotroph-based ecosystems. Because of the significance of decomposition in autotrophic- and detritus-based ecosystems, as well as compartments of autotrophic ecosystems, a brief discussion of this important ecological process is in order before delving into its impact on various ecosystems.

Decomposition

Decomposition of organic material occurs in a variety of ways, among them leaching and fragmentation, but primarily by the activity of organisms that may, in turn, facilitate both leaching and fragmentation. The primary agents of the final stages of decomposition are microbes that act through the process of metabolism.

Leaching. *Leaching* results from water percolating through an organic structure, such as a fallen tree, and dissolving some of its material content. Not only is there loss of mass from the structure, the minerals released are then available for transport as well as recycling. Leaching is a very important process in the loss of mass and materials from decomposing leaf litter in terrestrial and aquatic ecosystems, as well as from living plants (Harmon et al. 1986).

Fragmentation. Fragmentation, or communition, is the process of pulverizing or reducing material into small particles. *Physical fragmentation* involves physical activity as when a dead tree, or snag, falls to the ground and shatters into smaller pieces. In streams and rivers, abrasion against rocks or other structures also results in fragmentation. In a dry environment the seasoning of wood that results from the sun's bleaching and drying of the outer portion, followed by loss of moisture and formation of cracks that increase access to microbes, may also be considered to be a kind of physical fragmentation.

Biological fragmentation can be caused by both plants and animals. Invertebrates chew, ingest, excavate, and tunnel woody products resulting not only in pulverization but often in providing access, if not actual transport, to microbes that continue the decomposition by metabolizing the fragmented material. Biological fragmentation by plants occurs when, for example, another plant grows on a fallen tree, causing the latter to split and otherwise fragment.

Metabolism. Metabolism of organic material is primarily a microbial activity, although, to a lesser degree, invertebrates play a role. In terrestrial systems, basidiomycetes (fungi) are the primary agents of decomposition, whereas in aquatic systems, bacteria, including actinomycetes, are most important.

Not all plant or animal parts are decomposed at the same rate; for example, lignin decays more slowly than cellulose. Also, as decay proceeds and carbon is lost via metabolism (respiration), the concentration of nutrients may increase (Harmon et al. 1986).

Decomposition of Detritus in Forest Ecosystems

Accumulation of detritus. Forest floor detritus includes recent litterfall as well as decomposing organic matter in the A horizon above the mineral soil. It includes leaves and the coarser woody materials, such as downed boles and large branches, and is not only a habitat for organisms but is significant in influencing soil and sediment transport and storage. Even though coarse woody material is low in nutrient concentration and is slow to decompose, as we shall see, it plays an important role in energy flow and nutrient cycling (Harmon et al. 1986). Coarse woody material contains a large amount of organic matter in the form of celluoses, lignins, and other organic materials and thus constitutes a major long-term source of energy. Although the slow rate of decomposition of coarse woody material results in its being a nutrient sink in the short-term, it is a long-term source of these nutrients.

Ovington (1965) proposed that: (1) greater litterfall accumulation occurs with increasing distance from the equator, that is less organic matter accumulates in tropical than temperate forests; and (2) temperate softwood forests contain about four times the amount of litterfall present in termperate hardwood forests. However, based on much more extensive data, Vogt et al. (1986) have shown that this generalization does not hold true on a global scale. As Table 7-2 demonstrates, patterns of forest floor accumulations are separated by the evergreenness or deciduousness of the foliage. Tropical, subtropical, and warm temperate forests located at latitudes less than 40° had similar mean accumulations of forest floor mass under deciduous (8,789, 8,145, and 11,480 kg/ha) or evergreen (22,456, 22,185, 19,148, and 20,026 kg/ha) foliage. As Table 7-2 demonstrates, deciduous forests generally have only one-half the forest floor accumulation compared to that of the evergreen forests, regardless of the form of the leaf (broadleaf, needleleaf). Further, broadleaf evergreen forests have similar accumulations of forest floor mass, as do needleleaf forests. At latitudes greater than 40°, deciduous forests accumulate an average of three-fourths of the forest floor mass of evergreen forests.

The general pattern, which is demonstrated by the data in Table 7-2, is that of greater average forest floor mass accumulations occurring at latitudes above 40° while lower average accumulations occurred at latitudes less than 40°.

Detritus energy flow in a temperate deciduous forest. Gene Likens and F. Herbert Bormann have carried out extensive and long-term studies on the Hubbard

TABLE 7-2 FOREST FLOOR MASS AND LITTERFALL MASS IN FOREST ECOSYSTEMS

Forest type	Temperature (annual°C)	Precipitation (annual mm)	Forest floor mass (kg/ha)	Litterfall (kg/ha/yr)	Woody Litterfall (kg/ha/yr)
Tropical broadleaf deciduous	23.0	2,147	8,789	9,438	—
Tropical broadleaf evergreen	26.1	2,504	22,547	9,369	3,114
Tropical broadleaf semideciduous	22.5	1,431	2,170	5,890	3,477
Subtropical broadleaf deciduous	12.5	738	8,145	3,333	637
Subtropical broadleaf evergreen	12.8	1,705	22,185	5,098	2,902
Mediterranean broadleaf evergreen	12.4	987	11,400	3,042	800
Warm temperate broadleaf deciduous	13.9	1,391	11,480	4,236	891
Warm temperate broadleaf evergreen	12.8	1,409	19,148	6,484	—
Warm temperate needleleaf evergreen	13.9	1,374	20,026	4,432	1,107
Cold temperate broadleaf deciduous	5.4	875	32,207	3,854	1,046
Cold temperate needleleaf deciduous	10.2	1,806	13,900	3,590	—
Cold temperate needleleaf evergreen	8.1	1,278	44,574	3,144	602
Boreal needleleaf evergreen	2.1	694	44,693	2,428	991

Data extracted by permission from Vogt et al. 1986. *Advances in Ecological Research* 15:303-409.

Brook Experimental forest, a sugar maple, beech, and yellow birch forest in New Hampshire. The cycling of several major nutrients will be discussed in Part V; here we will consider the flow of energy through detritus (Gosz et al. 1978).

As Figure 7-7 shows, annual net primary production in the Hubbard Brook Forest is 45 percent of gross production, 55 percent being lost in the forest's metabolism. Of the annual net production (4,680 C/m^2/yr), 75 percent (3,481 C/m^2/yr) enters the grazing and detritus pathways; grazing, however, is insignificant, amounting to about 1 percent a year. The remaining net annual production (1,199 C/m^2/yr) is stored above and below ground as accumulated litter and unused detritus.

Because additional organic material comes from sources other than the forest plants, the actual amount of energy flowing through the detritus chain is greater (3,505 C/m^2/yr) than that derived from the annual net production (3,481 C/m^2/yr)

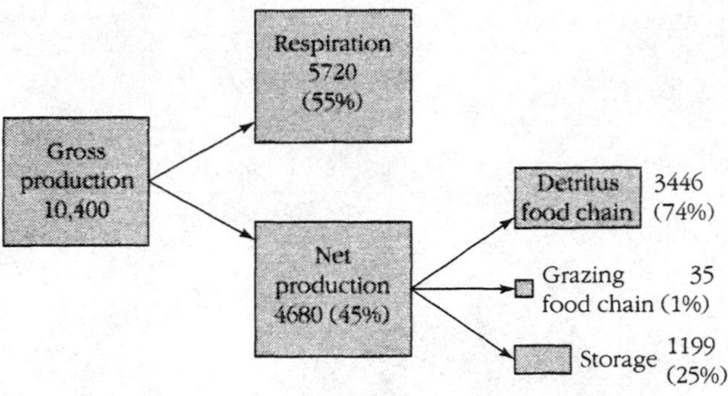

Figure 7-7 Fate of energy in C/m²/yr in the Hubbard Brook Experimental Forest. (Data from Gosz et al. 1978. *Scientific American* 238: 93–102).

of a given year. The sources of all of the energy flowing through the detritus pathway is:

Leaves	83%
Root death	12
Nonleaf litter fall	2
Organic matter via precipitation	2
Fecal material	0.9
Exuded by roots	0.1
Animal death	trace
Total	100%

In this 60-year-old forest about 150 C (4 percent) of this detritus material is not consumed by detritus feeders (bacteria, fungi, many invertebrates) or transferred to carnivores (beetles, centipedes) or omnivores (salamanders, rodents, birds) and thus it accumulates annually on the forest floor. Total annual net production of this ecosystem therefore consists of 1,199 C/m^2 of living material and 150 C/m^2 of organic material for a total of 1,349 C/m^2 or 29 percent of the total net primary production.

Detritus energy flow in a temperate evergreen forest. Energy flow over an 18-year-period in a mature, managed Scots pine plantation in England between 17 and 35 years after planting was studied by J. D. Ovington (1962). The data in Figure 7.8 show that: (1) a substantial proportion of the energy initially incorporated in the system passes through a detritus food chain (38 percent of net production); and (2) relatively little of the living material is grazed by herbivores but rather is harvested (24 percent of net production). Whereas such harvesting is certainly performed by a consumer (humans), it is scarcely grazing in an energy-flow sense. Another significant factor to be recognized is that 30 percent of the harvested tree production is not actually harvested but remains in roots. In Chapter 6 we noted that harvest measurements of plant production have typically bypassed analysis of roots—for the very good reason that they are so very difficult to remove. Numerous data indicate, however, that this is a component that must be considered in assessing the energy budget of an ecosystem.

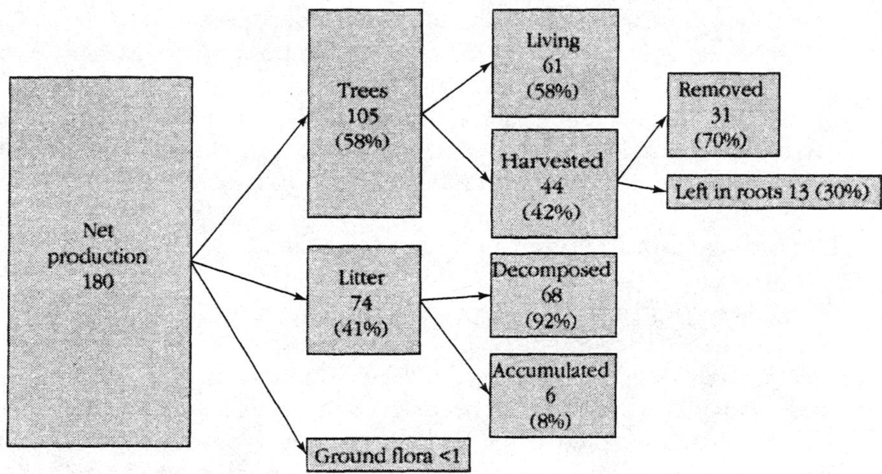

Figure 7-8 Fate of energy in a Scots pine (*Pinus sylvestris*) plantation over 18 years, in 10^7C/ha. (Based on data of J. D. Ovington. 1962. *Advances in Ecological Research* 1:103–92.)

Effects of clearcutting. Based on studies of northern hardwood stands in the White Mountains of New Hampshire, Covington (1981) found dramatic changes in the availability of organic matter because of *clearcutting*, that is, the complete removal of trees. During the first 15 years following clearcutting, forest floor organic matter decreased by over 50 percent, or 30.7 mg/ha. This decrease is the result of lower leaf and wood litter fall and of more rapid decay resulting from the higher temperatures due to the greater exposure of the soil to insolation, among other factors. During the next 50 years after clearcutting, forest floor organic matter increased by 28.1 mg/ha and by year 65 was close to 56.0 mg/ha.

Decomposition in subarctic woodlands. Decomposition in subarctic woodland soils is slow largely because of the low soil temperatures, low soil pH, and the paucity of readily available nitrogen (Moore 1981). For example, over a 2-year period loss in mass was between 10 and 60 percent, with 60 to 90 percent of the loss occurring in the first year (Moore 1984).

Invertebrate decomposition. Because of the relatively slow decomposition of arthropod chitinous exoskeletons, a reservoir of nutrients remains unavailable in forest litter. Seastedt and Tate (1981) found that 30 percent of the original mass of dead millipedes and 14 percent of the original mass of dead crickets was recovered after 1 year in the litter. Since the standing crop of mass of remains was greater than the standing crop of living forest floor arthropods, the untappable nutrient pool is considerable.

Nitrogen limitation in forest detritus decomposition. In a study of several ecosystems, Hunt et al. (1988) found decomposition to be the fastest in a lodgepole pine forest (*Pinus contorta*), intermediate in a blue grama (*Bouteloua gracilis*) prairie, and slowest in a western wheatgrass (*Agropyron smithii*) meadow. In all cases, an increase in the nitrogen supply in the form of ammonium nitrate resulted in a small but significant effect on decomposition rate.

Decomposition of Detritus in Grasslands

In prairie ecosystems, 60 to 90 percent of the net primary productivity and 90 percent of the secondary productivity occur in the soil (Stanton 1988). Herbivory and decomposition are the two major trophic pathways in these belowground prairie ecosystems. With respect to the decomposer component, the standing microbial biomass has been estimated as second only to primary producers, with a caloric value of this biomass at about one-half that of the net primary production of the system; for this reason, the microbial community has been called "the invisible prairie" (Paul et al. 1979). Fungi constitute the dominant organisms in the microbial component, and the bacterial biomass, including such genera as *Pseudomonas*, *Flavobacterium*, and *Alcaligenes*, constitutes one-fourth to one-half that of fungi (Coupland 1979).

In a study of Wisconsin prairie soil subjected to experimental burning and raking as compared to undisturbed soil, Lussenhop (1981) found that the physically disturbed soil significantly increased the processing of plant material and fungal hyphae by bacteria, as well as by several groups of mites (oribatid, prostigmatid, and astigmatid).

Most of the annual grass (and forest) production is never eaten by grazers; it accumulates and what dies becomes available to the detritus feeders. In a study of the broom sedge field in South Carolina, for example, Golley and Gentry (1965) reported that of the 1960 net production of 2,692 C/m^2/yr, about 53 percent (1,419 C) remained unconsumed as standing dead vegetation and about 9 percent (237 C) became part of that year's litter. Most significantly, less than 2 percent of the year's net plant production passed through the grazing herbivores. This figure is comparable to the 1.6 percent Golley (1960) obtained in his old field study in Michigan (see Table 6-2).

Decomposition of Detritus in Chaparral

Decomposition of leaf litter in coastal mountain chaparral (see Chapter 14) in southern California during the interval between natural fires may be a substantial source of plant nutrients for chaparral growth. Schlesinger and Hasey (1981) showed that over a 1-year period, the predominant evergreen sclerophyll shrub (*Ceanothus megacarpus*) lost 15 and 19 percent of its original ash-free dry mass at sites at 910 and 350 m elevation, respectively; a deciduous shrub (*Salvia mellifera*) lost 20 and 24 percent, respectively.

In a later study, Schlesinger (1985) followed the decomposition of these same species at the same elevations over a 3-year period. He found similar decomposition rates between sites and between species, with a mean residence time for the organic matter in the litter of 5.3 to 7.7 years. This further supports the contention that the decomposition of the litter layer is important to the supply of available nutrients in mature chaparral.

Decomposition of Detritus in Deserts

Microarthropods (mites, *Collembola* and *Psocoptera*, among others) affect litter decomposition in desert ecosystems by inoculating litter with fungal spores, grazing on fungi, and by preying on free-living nematodes (Santos and Whitford 1981). In an experimental treatment of desert shrub litter with insecticides and fungicides in New Mexico, they found that suppression of fungi reduced decomposition by 29 percent, but exclusion of microarthropods reduced decomposition by 53 percent. In untreated litter, 55 percent decomposed during the growing season, and 23 to 29 percent disappeared in the winter months.

In an associated study, Santos et al. (1981) found that elimination of microarthropods (primarily tydeid mites) resulted in increased numbers of bacteriophagic nematodes and reduction in numbers of bacteria; elimination of both nematodes and microarthropods resulted in increased numbers of bacteria. Elimination of mites resulted in a 40 percent reduction in decomposition, suggesting that in a desert, tydeid mites affect decomposition of buried litter by regulating the population size of the bacterial grazers, cephalobid nematodes.

Decomposition and Detritus Energy Flow in Aquatic Ecosystems

Lentic ecosystems. (See Chapter 15). Microbial decomposition is a dominant process in litter decomposition in aquatic situations and is influenced by environmental factors such as temperature, oxygen, and nutrient availability and the quality of the material being decomposed (e.g., stems versus leaves, inhibiting chemicals, particle size) (Brinson et al. 1981; Polunin 1984). Unfortunately, there is little specific information about the microorganisms that carry out this activity. Several exemplary studies are nonetheless worthy of consideration with regard to the flow of energy through the detritus pathway in aquatic ecosystems. Aquatic ecosystems consist of several categories: *lentic*, or standing water (e.g., ponds, lakes); *lotic*, or running water (e.g., streams, rivers); *estuaries*, or coastal waters where fresh water and ocean water mix; and *oceans*, characterized by a high salt content.

Root Spring, 2 m in diameter and 10 to 20 cm deep, is located in Concord, Massachusetts, not far from another aquatic locale of significance to naturalists, Thoreau's Walden Pond. Teal (1957) determined that of the total energy income of the pond, photosynthesis accounted for 710 C/m^2/yr. In contrast, terrestrial plant debris falling on the spring accounted for 2,350 C/m^2/yr. More than three times as much energy entered the system in the form of dead organic material or detritus than was fixed through photosynthesis. Of this total income (3,060 C/m^2/yr), the detritus-feeding herbivores consumed about 75 percent (2,300 C/m^2/yr), the remainder being deposited in the system. Thus in Root Spring most of the energy supporting the higher trophic levels is neither produced within the system nor derived from living material. The energy is largely of external origin in the form of detritus.

In a study of Cone Spring, Iowa, Tilly (1968) found that energy income to the spring was autotrophic (by three species of flowering plants). No grazing herbivores were present, the plant material being consumed by a variety of detritus-feeding primary consumers. Of the net production of 9,508 C/m^2/yr, 25 percent (2,384 C/m^2/yr) passed through the detritus-feeding herbivores and 36 percent (3,400 C/m^2/yr) through the decomposer route. The remaining 39 percent was exported to the marsh surrounding the spring—that is, to storage. Odum and Heald (1972) described a similar situation for brackish-water mangrove swamps in South Florida. Here there is an interaction of microbial decomposers and detritus-feeding herbivores on leaf fragments. This material is successively eaten (herbivores), absorbed (microbes), egested and re-eaten (coprophagy), or absorbed by yet other herbivores or microbes until the plant material is exhausted. In both Cone Spring and the mangrove swamp the detritus-feeding herbivores serve as an energy link supporting a carnivorous consumer population. In the case of the mangrove detritus chain, the linkage is significant in supporting the commercial fishing industry of southern Florida.

Lotic ecosystems. (See Chapter 15.) In a woodland stream, Petersen et al. (1989) found that the standing crop of detritus ranged from 87.8 to 970.9 g/m^2 over the course of 1 year (1972–1973). The average standing crop of detritus was calculated at 426.4 g/m^2. Microbial metabolism removed 150 percent of the average standing crop annually, mostly the smallest particles. Macroinvertebrates (those greater than 75 mm long) removed 11.6 percent of the standing crop by ingestion and assimilation. Of the macroinvertebrates, shredders (e.g., caddisflies, stoneflies, isopods, and amphipods) accounted for 20 percent, and collectors and grazers (e.g., midges, mayflies, beetles, oligochaetes, and a pelecypod) accounted for the remaining 80 percent.

Estuaries. In their analysis of the Chesapeake Bay estuary, Baird and Ulanowicz (1989) demonstrated that detritivory (by pelagic microbes and at the bottom by polychaetes, amphipods, and blue crabs) is about 10 times greater than herbivorous grazing. They showed further that 70 percent of the detritus is produced within the system, the remainder being imported from lotic systems feeding the Bay.

Ocean ecosystems. In a study in the Aleutian Islands, Duggins et al. (1989) showed that particulate and dissolved organic detritus originating from kelp, predominately *Laminaria* and *Alaria*, greatly increased the growth rate of benthic suspension feeders, the mussel *Mytilus edulis*, and the barnacle *Balanus glandula*. Mussels in the kelp-dominated habitats grew approximately up to four times as fast as mussels in urchin-dominated habitats; barnacles grew up to five times as fast in kelp-dominated environments.

In the oceans, photosynthetic plankton are the autotrophic producers and are at the base of the grazing and detritus food pathways. Numerous studies have shown that a large fraction of the primary production is not consumed directly by grazing herbivores, but is subject to detrital decomposition, largely by bacteria. Thus, the structure of these plankton-based ecosystems resemble benthic and terrestrial systems (Fenchel 1988).

Concluding Comment

It is evident from the foregoing account of decomposition and detritus-based ecosystems that the detritus chain cannot be disregarded in energy flow studies. The considerable amount of evidence that has been accumulated over recent years strongly indicates that the detritus chain is of greater importance in energy flow in many ecosystems than are herbivorous/carnivorous/omnivorous grazing chains. As we have seen, as much as 90 percent of the energy flow in various communities may be through detritus feeders. In addition, the decomposition of detritus liberates stored nutrients, thereby making them available for the feeding of autotrophic productivity and, in turn, initiating the cyclic turnover of these nutrients. For example, Waring and Schlesinger (1985) have shown that decomposition of detritus provides 69 to 87 percent of the nutrients needed annually for forest growth.

AN ENERGY FLOW MODEL

It should be obvious by now that the various ecosystem models used to far (Figures 2-2, 2-3, 7-4, 7-5, 7-6, 7-7, and 7-8) need to be modified further. The model shown in Figure 7-9 attempts to recognize the various inputs and fates of energy. An equal flow of energy from the two primary sources, imported living and dead organic matter and from photosynthesis, is assumed. Moreover, the contribution from each in the rest of the chain is assumed to be about equal. To apply the model to any given system, we need merely expand or contract the amount of shading to correspond to the energy recruitment by organic matter or by photosynthesis, the width of the energy channels, and the dimensions of the trophic level boxes. A model of Cedar Bog Lake and the Scots pine plantation, for example, would have a large "storage" box, whereas Root Spring would have a width for the organic matter import channel 3.3 times greater than the photosynthetic channel.

It should be obvious that the type of information required to construct an energy flow diagram for any particular ecosystem is extremely difficult to obtain. For this reason, relatively few systems have been studied in detail. In such studies as have been done, the critical reader, as well as the investigator, is well aware of the limitations of the data. Not only are there many parameters to be assessed simultaneously, but many variables are also inherent in the material being studied. Young trees undergo photosynthesis at a rate different from old trees (see Figure 6-3), and younger, smaller animals metabolize at higher rates than older and larger ones, and so on.

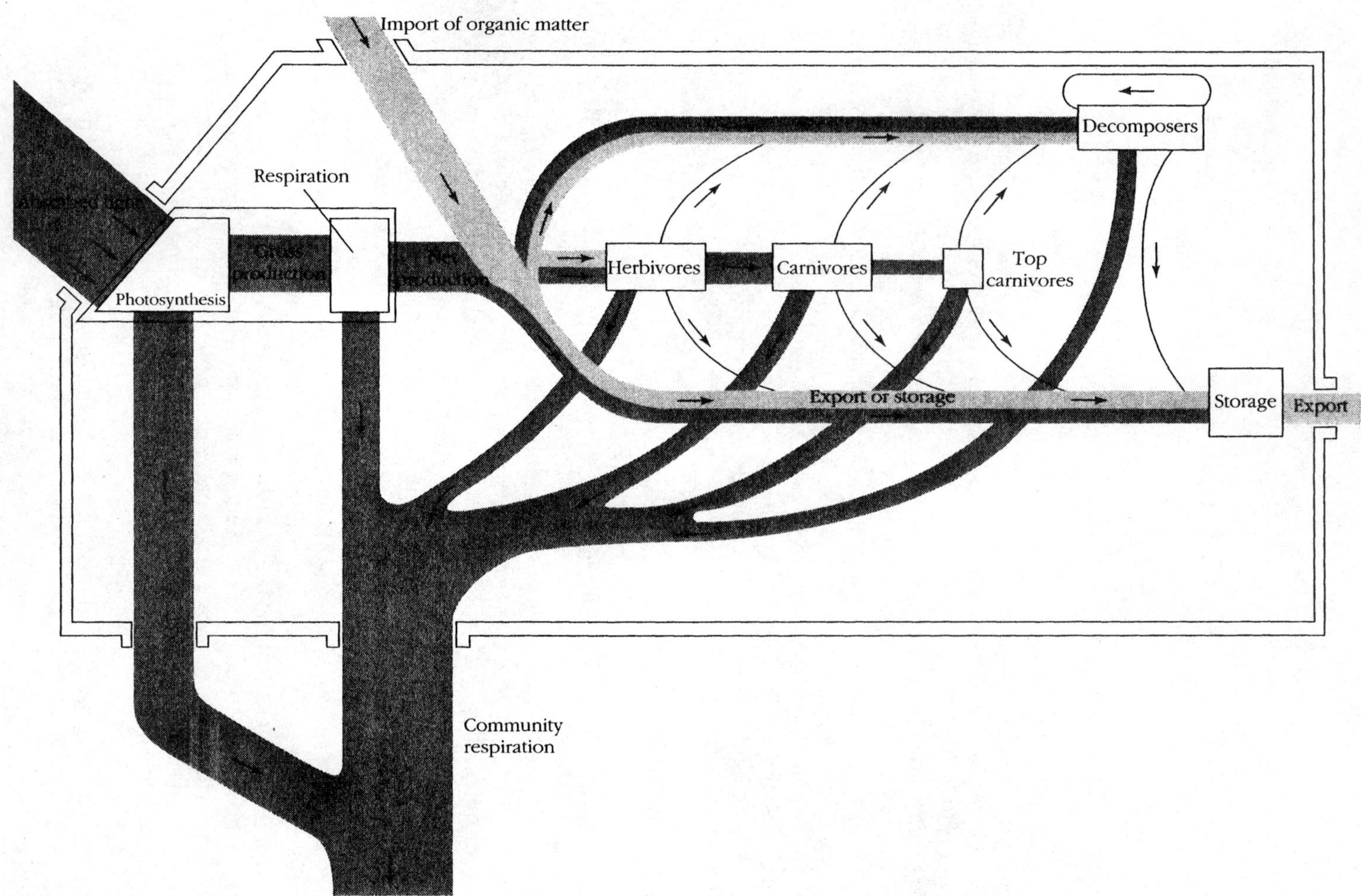

Figure 7-9 Energy flow diagram of a generalized ecosystem. (Adapted by permission from E. P. Odum. 1959. *Fundamentals of Ecology*, 2nd ed. Philadelphia: W. B. Saunders Co.)

Seasonal shifts in quality and amount of sunlight must be explained (see Figure 4-3), and, of no small importance, the inevitable question of adequate sampling to ensure replicability is a continual thorn in the ecologist's (and any scientist's) side.

Because of these limitations some ecologists have taken to the laboratory or to computer simulation, where some variables can be more limited, or at least regulated. From such studies, often done concurrently with field studies, have come several significant generalizations about energy flow that appear to agree fairly well with available field data.

From studies based on laboratory populations of algae, microcrustacea, and hydra, for example, Lawrence Slobodkin (1960) suggested that in the transfer of energy from one trophic level to the next the *gross ecological efficiency*, as he termed it, is on the order of 10 percent. Thus if there are 100 calories of net plant production, only about 10 calories net production would be expected at the herbivore level and only 1 calorie at the carnivore level. Various studies to date, however, show a range in gross ecological efficiency from 5 to about 30 percent, averaging about 10 percent from producer to herbivore, and 15 percent from herbivore to carnivore.

Generalizations about efficiency are still uncertain, however. Besides the sampling limitations previously discussed, very few ecosystems have been studied and those that have are not fully comprehensive. Furthermore, it seems questionable to apply the same definitions of efficiency to the different trophic levels because the biological processes in each are or may be different. Both warm- and cold-blooded consumers may be grouped in the same trophic level, for instance, and a general efficiency may be recorded. It has been shown, however, that efficiency of the warm-blooded forms is but about one-fourth that generally proposed for cold-blooded forms.

Regardless of the precision of any efficiency "constant," the point inherent in this discussion is clear— namely, that unless an ecosystem is supplied with energy either in the form of sunlight or organic matter from another photosynthesis-dominated ecosystem, it will collapse. The substantial loss of energy as unusable heat via the respiratory process must continually be replaced. This is, then, the elementary principle to be derived from energy flow studies.

OOD CHAINS AND FOOD WEBS

Implicit in the diagrams and discussions of energy flow, either through herbivory/carnivory or through detrital decomposition, is a connectedness or linkage of organisms dependent for their very existence on other organisms in the next lower trophic level. Such linkages are usually referred to as a *food chain*, a simple version of which was introduced in Chapter 2 in the model of energy flow from reindeer moss to reindeer to humans (see Figure 2-1). A food chain is an energy sequence of links or trophic levels that starts with a species that eats no other species and ends with a species that is eaten by no other species.

More often than not, such simple food chains are oversimplified versions of the reality of feeding relationships. Instead, there are often multiple and interconnecting pathways, as well as numbers of different species involved at each trophic level. These complex pathways resemble a web rather than a simple chain and are referred to as *food webs.* Since several species may occur in each of the trophic levels, the collection of organisms that feed on a common set of organisms and are fed on by another common set of organisms are referred to as a *trophic species* (Briand and Cohen 1984).

Also implicit in the flow of energy through an ecosystem is that the number of trophic levels is limited because of the decreasing availability of energy resulting from the inefficiencies in energy transfer from one trophic level to the next. If this is the case, then one seemingly intuitive hypothesis concerning food chain length that has been advanced is that chains should be longer in ecosystems with higher primary productivity. This hypothesis is referred to as the "energetic hypothesis" (Hutchinson 1959). Based on a review of studies on impoverished to highly productive systems, Pimm (1982) concluded that there is no evidence to support the energetic hypothesis.

A second hypothesis concerning food chain length is known as the dynamical stability hypothesis and was advanced from theoretical or mathematical models by Pimm and Lawton (1978). Although the technical aspects of this hypothesis are beyond our consideration, the gist is that food chains should be longer in ecosystems that are not subjected to disturbances by catastrophic events such as fire, flooding, widespread disease, etc. The idea is developed from some mathematical models that ecosystems with longer chains take longer to return to equilibrium once disturbed, so that webs with longer chains may be less likely to persist in nature. According to Briand and Cohen (1987), there is no reported evidence to support or negate this hypothesis.

In their review of 113 community food webs from natural communities, Briand and Cohen (1987) found no evidence to support either the energetic or the dynamic stability hypotheses. They did conclude, however, that the spatial dimensions of an ecosystem are related to food chain length, with three-dimensional environments having longer food chains than two-dimensional ones. Two-dimensional environments are regarded as essentially flat, like a grassland, the tundra, a sea or lake bottom, a stream bed, or the rocky intertidal zone; three-dimensional environments were regarded as solid, like a forest canopy or the water column of the open ocean.

Of the 113 natural food webs reviewed by Briand and Cohen (1987), it can be noted that although the smallest food chain length was 2 species and the maximum 10, the more general trend is 3 to 5 species. With respect to the number of trophic species, the numbers ranged from 3 to 48, the more general being about 15 to 20. The actual number of links of all the chains in the webs ranged from 2 to 138, with the general range being 20 to 30. As only one of many possible examples, the food web in St. Martin in the northern Lesser Antilles (Figure 7-10) is a

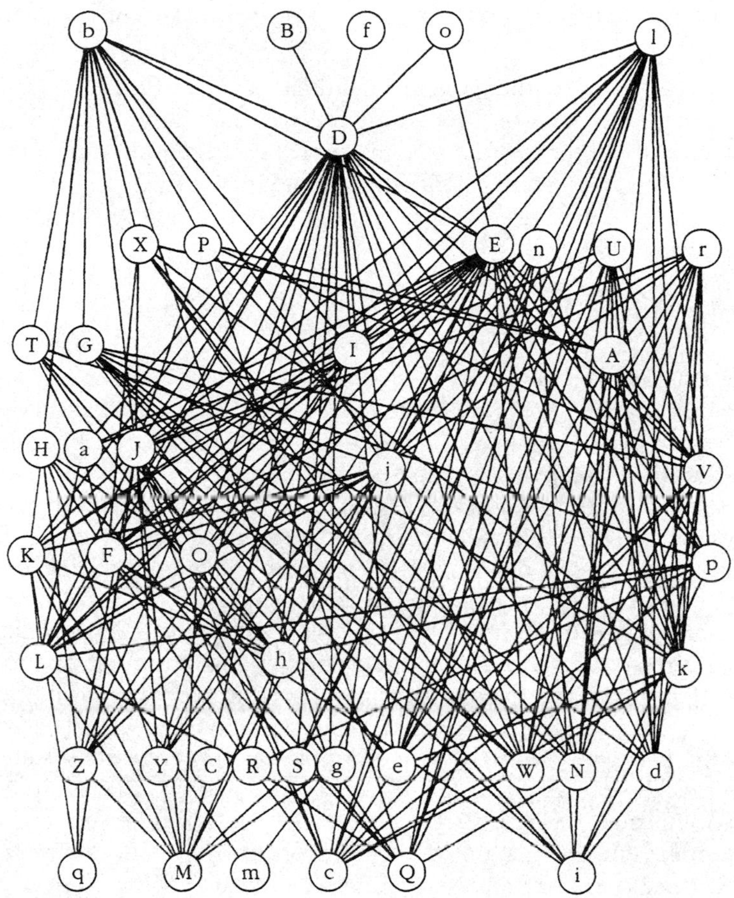

Figure 7-10 The food web from the island of St. Martin in the northern Lesser Antilles consisting of 44 species, eight trophic levels, and a multitude of linkages with detritus (M) and fruit and seeds (Q) at the bottom and kestrels (b), nematodes (B, f, and o), and thrashers (l) at the top. (Reproduced with permission from L. Goldwasser and J. Roughgarden. 1993. *Ecology* 74: 1216–33).

complex one in which there are 44 species involved in eight trophic levels, forming an intricate web of linkages between and within trophic levels (Goldwasser and Roughgarden 1993).

A collection of papers on food web theory was presented as a Special Feature in the December, 1988, issue of *Ecology* (Vol. 69: 1647–76); this source is recommended for further discussion of this intriguing topic. Also of importance is the overview of food webs that appears in Lawton (1989), in Hall and Raffaelli (1993), and in the critiques offered by Martinez (1992) and Paine (1988).

PYRAMIDS OF ENERGY, BIOMASS, AND NUMBERS

In the 1920s, one of the pioneers of animal ecology, Britisher Charles Elton, used an old proverb to describe another aspect of energy flow in ecosystems, namely, "one hill cannot shelter two tigers." Although he was alluding directly to territorial behavior, Elton also established the principle that at the "top" of the energy hill—that is, at the carnivore level—there is not enough energy to support more than a very few carnivores.

Energy Pyramids

The trophic or food chain, together with its concomitant energy losses at each step, can be visualized not only as an elegant flowing model (Figure 7-9) but also as a static *energy pyramid* (Figure 7-11) and can readily convey the image of the lone tiger on the hill. This pyramid is based on the energy data from Silver Springs already considered (Table 7-1 and Figure 7-5). As a quick reference, an energy pyramid conveys the general trophic relations in the system and readily reconfirms the major principle of progressive diminution of energy in the higher trophic levels. A less steep-sided pyramid might suggest reduced primary production, less efficient energy transfer, or less efficient use of available energy. If a phytoplankton-based system were sampled immediately after a period of maximum grazing that resulted in exhaustion of producer populations, an inverted pyramid would be obtained. Such patterns must be temporary; if not, their top-heaviness would be an indication of imminent collapse. Similarly, if one samples a temperate ecosystem in winter and in summer, the energy pyramid will likely be quite different (Figure 7- 11) and likewise for ocean studies (Figure 7-12) (Ravera 1969). For these reasons, it is important to determine energy flow not on a short-run or instantaneous basis, but over a relatively long period of time of at least a full year. Doing so permits making adjustments for such components as the phytoplankton, which turn over energy rather rapidly. The concept of turnover of energy and matter is discussed further in Part V.

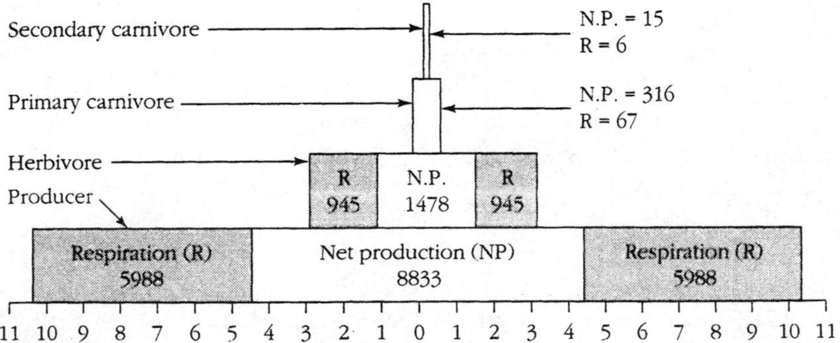

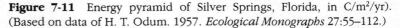

Figure 7-11 Energy pyramid of Silver Springs, Florida, in C/m²/yr). (Based on data of H. T. Odum. 1957. *Ecological Monographs* 27:55–112.)

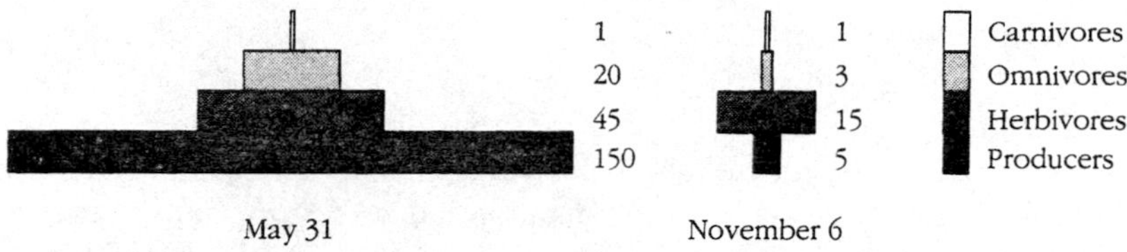

Figure 7-12 Comparison of winter and summer 1967 biomass pyramids of net plankton in mg (dry weight/m^3) in Lake Maggiore, Italy. (Redrawn by permission from O. Ravera. 1969. *Verhandlungen der Internationalen Vereinigung für Theoretische und Angewandte Limnologie* 17:237–54.)

Biomass Pyramids

Pyramids of biomass can also be constructed either by direct measurement or by appropriate conversion of energy units to biomass. Assuming, as we did earlier, that each gram of dry organic matter was equivalent to 4.5 C, we could express the energy pyramid for Silver Springs as a biomass pyramid by changing the scale along the bottom such that each kilocalorie unit would be equivalent to 222 g. Within the limits of this assumption, the shape of the pyramid would be unchanged. The actual biomass data from Silver Springs (in grams per square meter) are: producers, 809; herbivores, 37; primary carnivores, 11; secondary carnivores, 1.5. The reader may wish to construct a biomass pyramid using these data to compare it with the assumed one and with the energy pyramid shown in Figure 7-11.

Numbers Pyramids

It is also possible to move yet another step away from the structure of an ecosystem in energy terms by conducting a census of each of the major categories to produce a *pyramid of numbers.* An assessment of "who and how many live there" is often of critical signficance, as we shall see in Parts V, VI, and VIII. As would be expected, numerous producers are requisite to fewer herbivores and the latter to still fewer carnivores.

Figure 7-13 presents a comparison of the energy, biomass, and numbers pyramids at the consumer level in the same ecosystem, an old field in Michigan.

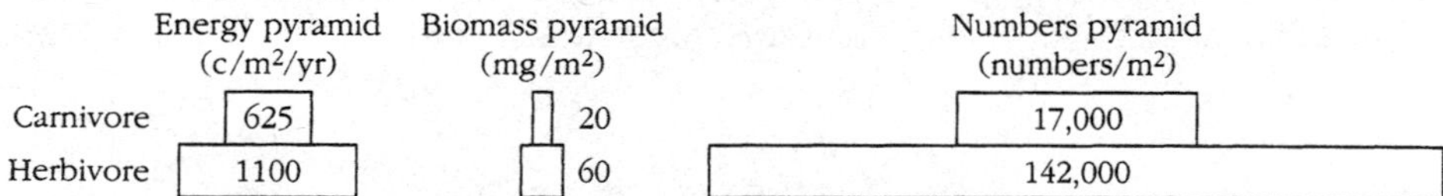

Figure 7-13 Comparison of pyramids of energy, biomass, and numbers of herbivores and carnivores in an old field in Michigan. (Redrawn by permission from Odum 1971 and based on data in M. D. Engelmann. 1968. *American Zoologist* 8:61–9.)

Critique of Pyramids

Although pyramids of energy, biomass, or number are valuable in gaining an assessment of an ecosystem, their limitations are substantial in gaining a comprehensive picture of energy flow. The energy pyramid has no appropriate or easy place for locating the decomposers and it does not allow us to represent the loss due to storage easily. The numbers pyramid, in addition to these limitations, equates all organisms as identical units. Thus one clump of grass at the producer level would have the same value as the one tiger at the top of the hill. A comparable difficulty inherent in the biomass pyramid is in assuming comparable ecological (energetic) significance to unit weights of widely disparate groups. British ecologist John Phillipson (1966), for instance, has noted that 1 g of the mollusk *Ensis* has a caloric equivalent of 3.5 C whereas 1 g of the microcrustacean *Calanus* is equivalent to 7.4 C. It is for these reasons that the static pyramid model of an ecosystem is not as functional in describing the energetics of an ecosystem as the energy flow model (e.g., Figure 7-9). Neither is it as significant as a mathematical formulation or a computer simulation.

ECOSYSTEMS AND THERMODYNAMICS

So far we have emphasized the complexity of energy flow and its progressive diminution through ecosystems. The critical reader, however, will have already recognized that ecosystems do not controvert but conform precisely with basic thermodynamic principles. In ecosystems, consistent with the first law of thermodynamics, energy is neither created nor destroyed, but it is transformed and the sum total entered can be accounted for on a budget- balance sheet (Table 7-1). And consonant with the second law of thermodynamics, energy is transformed ultimately into a nonusable form—heat. But a system that continually transfers its chemical energy to heat tends toward a state of thermodynamic equilibrium, a state of maximum entropy, of increased randomness, and hence of disorganization. We have observed, however, that ecosystems are ordered, and later we shall see that they are relatively stable but dynamic steady-state systems. Thus according to physical laws of the universe, they must have a continual source of energy to survive. The physicist Erwin Schroedinger (1955) offered a resolution of this paradox in a thermodynamic context by stating that a biological system delays its decay to thermodynamical equilibrium by "feeding on negative entropy" (i.e., on a more ordered, less random system—in this case, the Sun) and that it maintains its high level of orderliness by "sucking orderliness from its environment."

SUMMARY

VOCABULARY

biological fragmentation	clearcutting	energy pyramid
	detritus	estuaries

first law of thermodynamics
food chain
food web
gross ecological efficiency
leaching
lentic ecosystems
lotic ecosystems
metabolism
oceans
physical fragmentation
pyramid of biomass
pyramid of numbers
second law of thermodynamics
trophic dynamics
trophic species

KEY POINTS

- In autotrophic ecosystems, the energy accumulated through net primary production is used to support higher trophic levels, moves unidirectionally through those levels, and decreases in amount at each level.
- All the energy incorporated into and moving through an ecosystem can be accounted for consistent with the first law of thermodynamics; consistent with the second law of thermodynamics, energy is eventually lost irretrievably as heat and must be continually supplied to an ecosystem lest it run down.
- In detritus-based ecosystems, energy also moves unidirectionally, with losses at each trophic level. The detritus chain is of greater importance in energy flow in many ecosystems than is that of the producer/consumer chain.
- Decomposition of organic material in detritus ecosystems (and elsewhere) occurs through leaching, fragmentation, and by the activity of organisms, the rate and amount varying in different kinds of ecosystems.
- In general, the transfer of energy from one trophic level to the next is in the order of 10 percent.
- In general, food chains have 3 to 5 trophic links with 15 to 20 species; within food webs, the number of links, in general, ranges from 20 to 30.
- Three-dimensional environments such as a forest canopy or a water column in the open ocean have longer food chains than do two-dimensional environments such as a grassland or lake bottom.
- Relations within an ecosystem can be represented graphically by a pyramid of energy, biomass, and numbers, each of which has limitations regarding the flow of energy.

PART IV

Biogeochemical Cycles and Ecosystems

CHAPTER 8

Gaseous and Sedimentary Nutrient Cycles

The importance of nutrients and their cycling through ecosystems has been touched upon a number of times thus far. It is now appropriate to recognize more fully the cyclical patterns of these nutrients, to seek out the general patterns that may exist among them, and subsequently to consider the effects resulting from inadvertent or purposeful human interaction with them.

We have already described the *hydrological cycle*, the movement of water from the atmosphere to the Earth's surface and through a variety of pathways back to the atmosphere. Now we will consider the movement of elements such as carbon and nitrogen. The movement of *chemical* elements involves *bio*logical organisms and their *geo*logical (atmosphere or lithosphere) environment. Collectively, these movements are referred to as *biogeochemical* cycles.

In one group of biogeochemical cycles, the atmosphere constitutes a major reservoir of the element that exists there in a *gaseous* phase. Such cycles, called *gaseous nutrient cycles*, show little or no permanent change in the distribution and abundance of the element, though over shorter periods of geological time there may be greater accumulations in a given compartment of the cycle. Carbon and nitrogen are prime representatives of biogeochemical cycles with a prominent gaseous phase.

In a *sedimentary cycle* the major reservoir is the lithosphere from which the elements are released largely by weathering. The sedimentary cycles, exemplified by phosphorus, sulfur, iodine, and most of the other biologically important elements, have a tendency to stagnate. In such cycles, a portion of the supply may accumulate in large quantities, as in the deep ocean sediments, and thereby become inaccessible to organisms and to continual cycling. Some of the elements that are characterized by sedimentary cycles do have a gaseous phase, sulfur and iodine being among them, but these phases are insignificant in that there is no large gaseous reservoir.

As these biogeochemical cycles are discussed, note their common, fundamental characteristic: large available pools are often coupled with small available pools that are extremely labile and highly dependent on continual input. Further, it will become evident that although there is interaction among biological, chemical, and geological components of the cycles, there is significant biological control of the chemical factors in the environment (Redfield 1958; Lovelock and Margulis 1974).

GASEOUS NUTRIENT CYCLES

Carbon Cycle

Perhaps the simplest of the nutrient cycles is one whose general components are quite well recognized—the carbon cycle (Figure 8-1). Next to water, carbon is the most significant element in organisms, constituting 49 percent of their dry weight. The carbon cycle is essentially a perfect one in that carbon is returned to the environment about as fast as it is removed. Also, it is one that involves a gaseous phase,

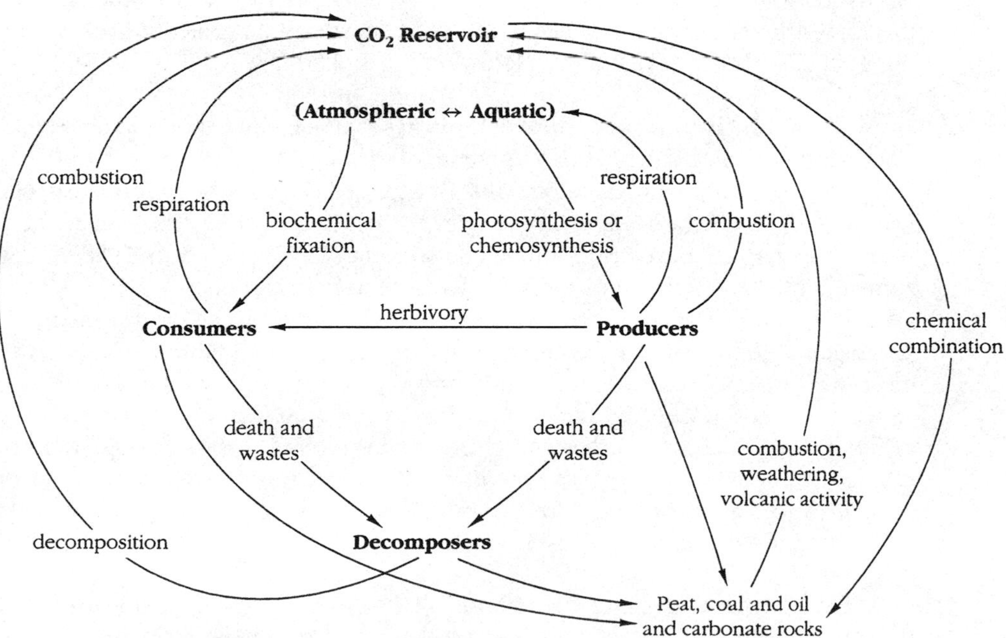

Figure 8-1 The carbon cycle.

atmopheric carbon dioxide. The basic movement of carbon is from the atmospheric reservoir to producers to consumers, and from both these groups to the decomposers, and then back to the reservoir. In this cycle the gaseous reservoir is the atmosphere, which has an average concentration of about 0.032 percent [or 320 parts per million (ppm)] of carbon dioxide.

Daily and seasonal variation. There is considerable daily as well as seasonal variation in the amount of atmospheric carbon dioxide because of geographical and other factors that influence removal through photosynthesis and return by respiration. As a result of respiration, for instance, CO_2 levels near the ground may rise to 0.05 percent at night and drop to well below the average concentration of 0.032 percent during the day as CO_2 is taken up in photosynthesis. Seasonally photosynthetic uptake of CO_2 exceeds return to the atmosphere by respiration and decomposition during the summer; the reverse is true during the winter. As a result, in northern latitudes CO_2 content reaches a maximum in April and September and the winter-to-summer differences may vary by as much as 0.002 percent, or 20 ppm.

In a study of a mixed deciduous forest in Harvard Forest, Massachusetts, Bazzaz and Williams (1991) made continuous measurements of carbon dioxide concentrations (by infrared gas analysis) at 0.05, 0.2, 3, and 12 m above the soil surface from March through November 1985. Concentrations generally decreased with height throughout the season and were highest near the soil surface in the summer (Figure 8-2). A diurnal pattern also is noted, with increased carbon dioxide concentration during the night and decreased concentration during the day, hitting the lowest level shortly after noon. Marked seasonal patterns are also evident in Figure 8-2, with low levels during late winter and late fall.

Biological cycling of carbon. Carbon dioxide is the major state in which carbon is involved in biological or trophic level cycling. Between 4 to 9×10^{13} kg of carbon are used annually in autotrophic photosynthesis. For some organisms in the vicinity of hydrocarbon seeps on the ocean floor, notably mussels, clams, and tube worms, methane is the carbon source (Brooks et al. 1987; Cary et al. 1988) and is assimilated by the collaboration of symbiotic methane bacteria.

Respiratory activity in the producers and consumers accounts for the return of a considerable amount of the biologically fixed carbon as gaseous CO_2 to the atmosphere. The most substantial return, however, is accomplished by the respiratory activity of decomposers in their processing of the waste materials and dead remains of other trophic levels. Additional return from the biota occurs through the nonbiological process of combusion, both through the purposive use of wood in a fireplace and accidental fire in a forest or building. Such combustion can and does involve consumers and decomposers as well as producers.

Geochemical cycling of carbon. The distribution of carbon on Earth is shown in Table 8-1. Earlier estimates by Bolin (1970), from whom the data in the table are derived, placed the amount of carbon in fossil fuels and in the sediments at twice the amounts shown in the table. Estimates by Berner and Lasaga (1989) give

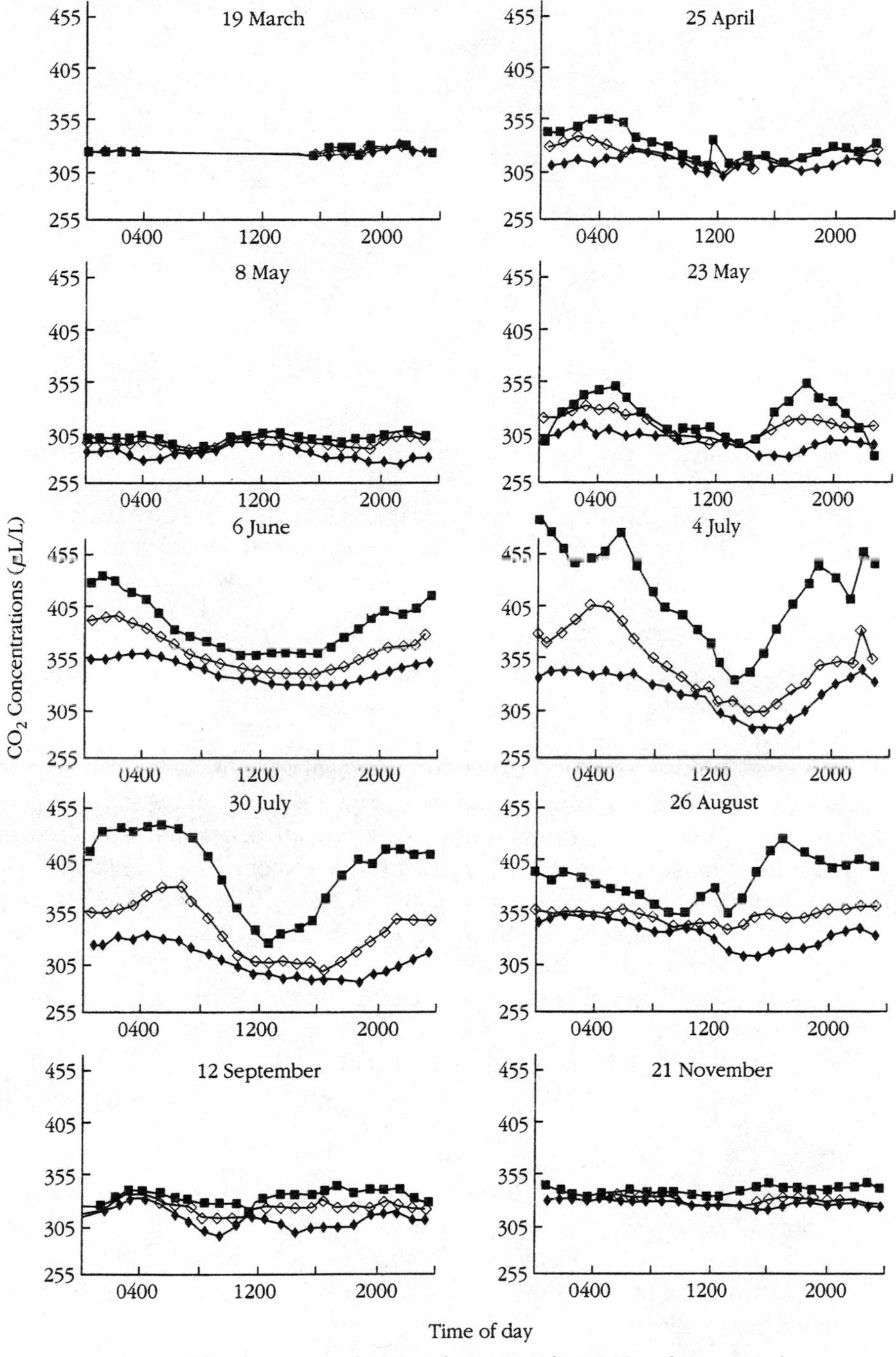

Figure 8-2 Daily and seasonal variations in carbon dioxide concentrations in a deciduous forest in New England. (Redrawn by permission from Bazzaz and Williams. 1991. *Ecology* 72:12–6.)

TABLE 8-1 CARBON IN THE BIOSPHERE IN 10^{9*} METRIC TONS

Atmosphere	692	Organisms	3,432
Ocean water	35,000	Living	592
Sediments	>10,000,000	Dead	2,840
		Fossil fuels	>5,000
Total organic	8,432		
Total inorganic	10,035,692		

*One metric ton = 2204.6 pounds
Data derived from B. Bolin et al. 1979.

the amount in the sediments as seven times larger and that in dead organisms as five times larger than those shown in Table 8-1. Whichever of these estimates is taken, the amount of carbon in organic form is but a fraction (less than 0.1 percent) of that occurring in the geological component of the environment; further, the major reservoir of carbon is inorganic, with some 99 percent being in sedimentary form. This means that the major supply of carbon to the Earth's surface, where it becomes biologically involved, is by erosion and metamorphism of sediments. Another way of conceptualizing the distribution of carbon is to note that at any given time, one part in a thousand is in the ocean-atmosphere system, and the rest is in sedimentary rocks (Broecker 1973).

Sedimentary carbon includes such animal remains as protozoan tests, coral, mollusk shells, and echinoderm and vertebrate skeletal material, as well as carbonate rocks. In addition, a number of aquatic plants occurring in alkaline waters release calcium carbonate as a byproduct of photosynthetic assimilation. For example, 100 kg of *Elodea canadensis* can precipitate 2 kg of $CaCO_3$ in 10 hours of sunlight under natural conditions. This pure calcium carbonate precipitate mixes with clay to form marl, which over a long period of time can be compacted as limestone. Most of the limestone deposits in the world are presumed to have this kind of biological origin. On weathering and dissolution of carbonate rocks, the combustion of the fossil fuels (peat, coal, oil), and volcanic activity involving deposits of both fossil fuel and carbonate rocks, the bound carbon is returned to the atmospheric-aquatic reservoir.

Aquatic-atmospheric interchange of carbon. Because of its significance, the interplay between atmospheric and aquatic carbon dioxide must be considered briefly. Interchange between the two phases occurs through diffusion, the direction of which, among other factors, depends on relative concentrations. Passage into the aquatic phase also takes place through precipitation. A liter of rainwater, for example, contains about 0.3 cc of gaseous CO_2. The dissolved CO_2 combines with water in the soil or in aquatic ecosystem to form carbonic acid (H_2CO_3) in a reversible reaction. In turn, carbonic acid dissociates in a reversible reaction into hydrogen and bicarbonate ions (HCO_3^-). The latter ion, in turn, dissociates in another reversible reaction into hydrogen and carbonate ions. The various reactions are summarized as follows:

$$\begin{array}{l} \text{Atmospheric } CO_2 \\ \quad \downarrow \uparrow \\ \text{Dissolved } CO_2 \quad + H_2O \leftrightarrows + H_2CO_3 \leftrightarrows H^+ + HCO_3^- \leftrightarrows H^+ + CO_3^= \end{array}$$

Because all these reactions are reversible, the direction of the reaction depends on the concentration of critical components. Thus a local depletion of atmospheric CO_2 would result in a net movement of CO_2 into the atmosphere from the dissolved phase, triggering a set of compensating reactions. Similarly, the assimilation of the bicarbonate ion (HCO_3^-) in photosynthesis by acquatic plants would tend to shift the equilibrium the other way. Actually, however, the equilibrium system is much more complicated. The amount of carbon present as bicarbonate and carbonate, for example, is also dependent on the pH of the water. At high pH values (i.e., alkaline conditions) more carbon is present as carbonate; at lower values (i.e., acid conditions) more occurs in the dissolved phase. Thus manipulation of either pH or CO_2 concentration affects the operation of the system. A further complication to the stability of the system is that days or weeks may be required to achieve equilibrium with the atmosphere across a water surface.

The Global Cycling of Carbon

Carbon regulation in the ocean and atmosphere. According to Broecker (1973), the *residence time* (the length of time spent in a given compartment of the cycle) of carbon in the sediments is in the order of 100 million years (10^8 years), whereas residence time in the ocean-atmosphere system is 100,000 years (10^5). Because of both the greater content of carbon in the ocean than in the atmosphere (Table 8-1) and the chemistry of the aquatic-atmosphere inter-change of carbon, the ocean, in effect, dictates to the atmosphere what its CO_2 content will be. The ocean and its sediments thus take on a singular importance in the global cycling of carbon. Given the large size of the Pacific Ocean, it is not surprising that portions of it (central and eastern equatorial) are the largest natural ocean source of carbon dioxide (Murray et al. 1994). Lakes are a small but potentially important conduit for carbon from terrestrial sources (Cole et al. 1994), and very significant carbon sinks are forest ecosystems (Dixon et al. 1994). Forests cover more than 4.1×10^9 hectares of Earth's land area (37 percent), with approximately 37 percent of the carbon in low latitude forests, 14 percent in midlatitudes and 49 percent at high latitudes.

Broecker (1973) has further noted that organisms produce about five times as much calcium carbonate ($CaCO_3$) as that brought into the oceans from land drainage. If all this carbon remained preserved in the sediments, the ocean would run a very large carbon deficit and that, in turn, would pull carbon from the atmosphere, creating a deficit there. Such a situation is avoided by the feedback mechanism described above for the reversible reaction of the carbonate ion. This feedback adjusts the ocean's chemistry such that roughly 80 percent of the ocean floor is bathed in waters corrosive to $CaCO_3$, leaving only 20 percent of the ocean floor supersaturated with carbonate. Likewise, if too much of the calcium carbonate were to be discarded, the carbonate-ion content would be lowered, and the compensating chemical reaction would restore the balance. Thus, the carbon balance in the ocean is controlled by keeping the carbonate-ion concentration in balance, and this balancing, in turn, keeps the atmospheric carbon content under regulation.

Modeling the global carbon cycle. Houghton et al. (1983) present a straightforward equation to express the annual changes in the components of the carbon cycle, as follows:

$A = F \pm S - L$, where

$A =$ the annual increase in atmospheric CO_2

$F =$ the annual release of CO_2 from combustion of fossil fuels

$S =$ the average annual net movement of carbon as CO_2 into the oceans

$L =$ the net exchange of carbon between the atmosphere and the land surface

In this equation, A and F are quantities that have been estimated with some considerable precision; S is a less precise quantity as is L.

Using data from 69 regional ecosystems, Houghton et al. (1983) have developed a model to attain a more precise assessment of L and have concluded that changes in land use over the past two centuries have caused a significant release of CO_2 to the atmosphere from the terrestrial biota and soils. These changes include the harvest of forests and the conversion of forests to pasture and agricultural areas, resulting in a net release of carbon to the atmosphere; afforestation, the growth of harvested forests, and build-up of soil organic matter result in storage of carbon. Earlier, Bolin et al. (1979) estimated that between 40×10^{15} and 120×10^{15} grams of carbon have been added to the atmosphere by deforestation and the expanding use of agriculture.

Bramryd (1979) has noted that in addition to the increases in the outflow of carbon to the atmosphere from exploitation of terrestrial ecosystems through clearcutting and burning, agricultural practices such as deeper ploughing, which increases the mineralization rate of accumulated organic matter in agricultural lands, aggravate the situation. Further, plans to drain and plow large areas of peatland will create aerobic conditions in what had been an anaerobic environment and thereby increase the mineralization of the long-stored carbon. Bramryd also notes that the impact of solid waste disposal needs to be considered. In well-compacted landfills, anaerobic conditions exist and mineralization of organic matter is strongly reduced;

however, incineration will rapidly oxidize the organic matter, releasing the stored carbon to the atmosphere.

According to the model developed by Houghton et al. (1983), there has been a net release of carbon from terrestrial ecosystems since at least 1860, and until about 1960 the annual release was greater than the release of carbon from fossil fuels. They conclude, however, that given the uncertainties in estimating the other components of the formula given above, the combined errors are large enough that the degree to which the global carbon budget is balanced depends on which combination of estimates is used. If the low estimate for the biotic release of carbon is combined with the higher estimate of oceanic uptake, the budget is balanced; if the higher estimates for biotic release are used, the budget would not be balanced.

An earlier mathematical model developed by Bjorkstrom (1979) attempted to analyze the distribution of excess carbon released in the period 1860–1970. The model indicated a variation between 25 and 70 percent for the amount of remaining airborne carbon released by human activity. The major causes of these variations are the uncertainty regarding the atmospheric CO_2 concentration prevailing before 1860 and the potential for the terrestrial biomass to accumulate carbon in organic material on land. Similarly, Bolin et al. (1979) noted that based on the assumption that the preindustrial atmospheric CO_2 concentration was about 290 ppm, at least 70×10^{15} g/C of the sum total released by human activity has remained in the atmosphere; if, however, a preindustrial atmospheric CO_2 base line of 260 ppm is assumed, the amount of the remaining CO_2 would be twice that value.

Long-term implications in global carbon cycling. Although the introductory explanation of the carbon cycle conveyed a rather simple situation, it should be obvious that the cycle is, in fact, relatively complex in the number of ways, as well as duration of time, in which carbon is used, stored, and restored as it circulates. It is also obvious that there are many unknowns and uncertainties as to quantities and rates involved in the many transfers that take place. In a very general way the various pathways do, however, constitute self-regulating feedback mechanisms resulting in a relatively homeostatic system (Garrels et al. 1976). Over time, additions and deletions can be eliminated and/or compensated. However, the degree to which the Earth as a total ecosystem can withstand or adapt to a long-term disturbance of the existing equilibrium is at this time both uncertain and a matter of contention among ecologists and other scientists (Holmen 1992).

Nonetheless, most authorities agree that the more than 100 years of increased fossil fuel use, deforestation, and various agricultural practices have resulted in increases of atmospheric carbon, the precise amount of which is, as just noted, debated (Solomon et al. 1985; Trabalka et al. 1985). Assuming a concentration of about 260 to 275 ppm of CO_2 in the mid-nineteenth century, there has been about a 20 percent increase by the 1990s. Based on measurements made in the period 1959–69, when a 3 percent increase (9 ppm) in atmospheric CO_2 was recorded, coupled with the accelerated rate in the use of fossil fuels, increased deforestation (especially of tropical rain forests) and increasingly intensive and expanding agricultural practices, the prediction is for an increase in atmospheric CO_2 to between 375 and 400 ppm (an increase of 14 to 21 percent) by the end of the century. The unknown factor, however, is the extent to which the ocean can continue to act as a sink for the extra carbon inserted into the atmosphere. Present combustion rates of 5 to 6 $\times$ 10^9 metric tons per year, for example, should be increasing atmospheric carbon dioxide by about 2 ppm per year. The increase, however, is only about one-third that amount. The other two-thirds is absorbed largely by the ocean and, to some extent, perhaps by increased plant production. The long-term effects of such increases is a matter of considerable concern and will be considered in Part VIII.

In an extensive review of the response of natural ecosystems to rising levels of carbon dioxide, Bazzaz (1990) indicates that on the individual, physiological level of plants there is strong documentation for enhanced photosynthesis and growth, increased allocation to underground parts, and increased water use efficiency. In some species, increased photosynthesis and growth can be of limited duration, however. At the ecosystem level, increased productivity that has been observed may result mainly from changes in species composition brought about by differential species responses to increased carbon dioxide. Because of the complexity of interactions involved, and limited knowledge of them, it is not possible to predict with certainty the direction of change in response to increasing carbon dioxide (Bazzaz 1990; Post et al. 1990).

Nitrogen Cycle

Because of its role in the construction of proteins and nucleic acids and its importance as a potential limiting factor in many biological phenomena, nitrogen is a significant element biochemically and thereby akin to oxygen, carbon, and hydrogen. Since nitrogen makes up a large portion of the Earth's atmosphere (about 78 percent), the problem with its availability is not an absolute shortage, but a relative one; that is, scarcity of usable or metabolizable nitrogen during critical growth periods. For example, plants encounter shortages of inorganic nitrogen, such as nitrate and/or ammonium ions, and animals experience shortages of organic nitrogen for specific proteins and/or amino acids (Mattson 1980).

The main steps of the nitrogen cycle were developed in the nineteenth century by the French agricultural chemist Jean-Baptiste Boussingault who published some 150 scientific papers, primarily on the role of nitrogen in animal nutrition and plant physiology (Aulie 1971). The biogeochemical cycle of nitrogen is complex, but is essentially a complete or perfect cycle (Figure 8-3).

The nitrogen cycle bears general similarities to the carbon cycle (compare with Figure 8-1), but there are a number of marked differences. Although organisms live in a nitrogen-rich atmosphere (79 percent) in contrast to that of carbon (0.03 to 0.04 percent), the gaseous form of the nitrogen (N_2), unlike CO_2, can be used by very few organisms. And whereas organisms dispense both carbon and nitrogen as metabolic waste products, little or none of this nitrogen is lost in gaseous form. Finally,

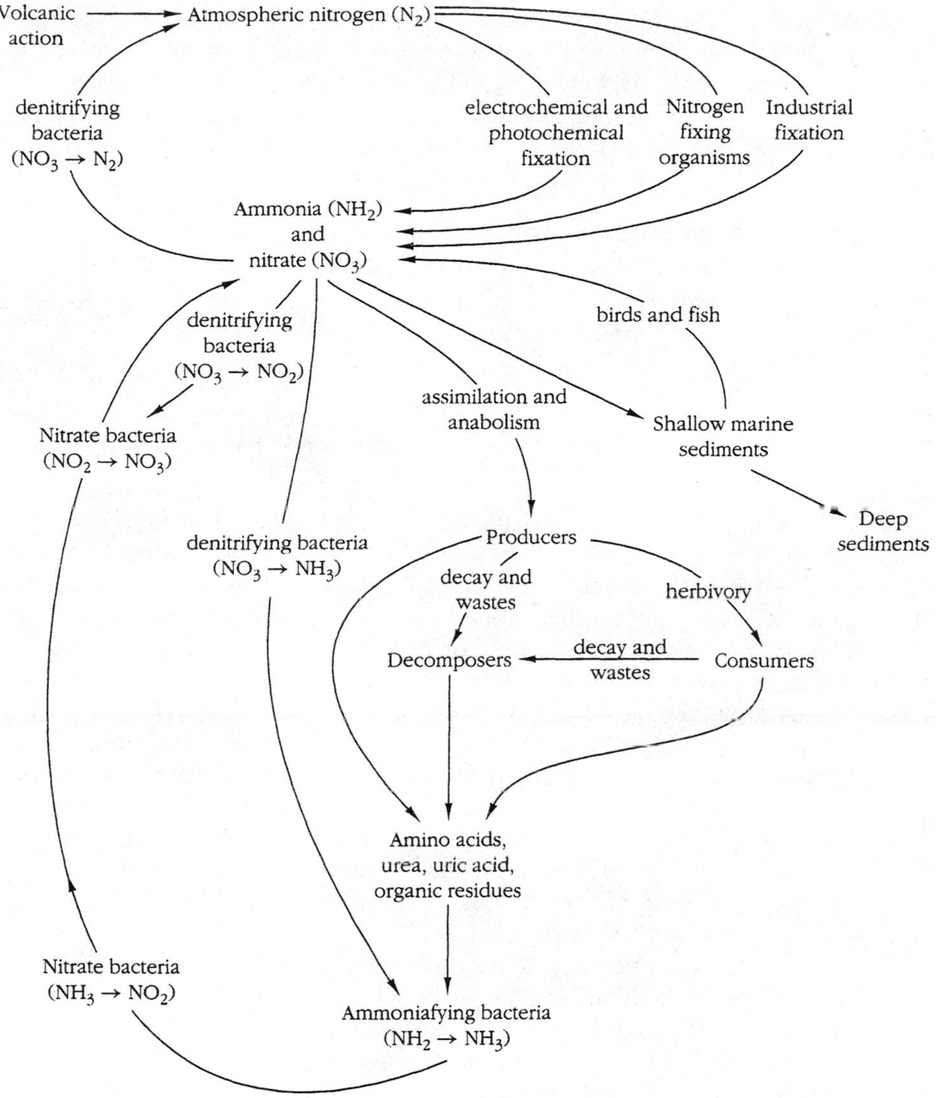

Figure 8-3 The nitrogen cycle.

biological involvement in the nitrogen cycle is far more extensive, complicated, and ordered. It is also highly specific in that certain organisms are able to act only in certain phases of the cycle.

With an atmospheric concentration of 78 percent, it would seem that the nitrogen reservoir is the atmosphere. Because most organisms are unable to use elemental nitrogen, however, the crucial reservoir is the store of nitrogen occurring in both inorganic (ammonia, nitrite, and nitrate) and organic (urea, protein, nucleic acids) form. Unlike carbon, which is readily available in reservoir quantities both in the air and water, atmospheric nitrogen must be fixed into an inorganic form, largely nitrate and ammonia, before it can be tapped for biological processes.

Because of its complexity, we will discuss in detail each of the major steps in the nitrogen cycle, as follows (see Figure 8-3):

nitrogen fixation, the conversion of dinitrogen gas (N_2) to form any nitrogen compound; biological nitrogen fixation is the enzyme-catalyzed reduction of N to ammonia (NH_3), nitrate (NO_3), or any organic compound;

ammoniafication, the breaking down or conversion of organic nitrogen (e.g., compounds containing the amine group ($-NH_2$)) to ammonia;

nitrification, the conversion (by oxidation) of ammonia to nitrite (NO_2) or nitrate;

denitrification, the reduction of nitrate to any gaseous nitrogen species, generally N_2 or nitrous oxide (N_2O).

For comprehensive reviews of the nitrogen cycle, see Sprent (1987) and Jaffe (1992).

Nitrogen fixation. Fixation of nitrogen first requires the activation of molecular nitrogen by splitting it into two atoms of free nitrogen ($N_2 \rightarrow 2N$). It is an energy-requiring step that, in biological nitrogen fixation, consumes 160 C/mole.* The combination of free nitrogen with hydrogen to form ammonia ($N \rightarrow NH_3$) releases 13 C/mole, resulting in a net energy input for nitrogen fixation of 147 C/mole. Except for photosynthetic types, all nitrogen-fixing organisms require an external supply of carbon compounds as an energy source to effect this endothermic reaction. The significance of this fixation, which is regulated by two enzymes, nitrogenase (Moffat 1990a) and hydrogenase, is its low-energy requirement (Shanmugam and Valentine 1975). This situation is in contrast to the extremely high temperatures and pressures required for industrial fixation (i.e., a temperature of 400°C and a pressure of 200 atmospheres).

Although fixation can occur by both physicochemical and biological means, the latter is by far the more significant. It has been estimated that electrochemical and photochemical fixation results in an average amount of nitrate in the order of 7.6×10^6 metric tons/yr. In contrast, biological fixation is estimated at 54×10^6 metric tons/yr. Of no small significance is the amount fixed annually in the manufacture of synthetic fertilizers, some 30×10^6 metric tons. This amount has been increasing at such a rate that it is estimated production will reach 100×10^6 metric tons by the year 2000. C. C. Delwiche (1970) has suggested that the current amount of industrially fixed nitrogen is equal to the amount fixed biologically prior to the advent of modern agriculture. And, Yung and McElroy (1979) have shown that lightning and subsequent atmospheric chemistry could have provided a source of nitrate for the primitive, that is prebiotic, ocean as large as 10^6 tons of nitrogen a year, sufficient to fill the ocean to its present level of nitrate in less than 10^6 years.

*A mole is the quantity of substance in grams numerically equal to its molecular weight. Thus, because the molecular weight of N_2 is 28, a mole of N_2 is 28 grams.

The commercial production of nitrate, initially for weapons and today for agricultural fertilizer, is energy-expensive. In what is known as the Haber process in honor of Fritz Haber, who discovered it in 1909 (and who, along with Carl Bosch, began the first commercial plant in 1913), nitrogen is combined with hydrogen to form ammonia. The reaction requires, as noted above, temperatures of 400°C and 200 atmospheres of pressure, as well as a source of hydrogen atoms, requirements that are met by the use of fossil fuels such as oil and natural gas. But, with the increasing cost of fossil fuel, investigations are underway to model commercial production on the natural biological nitrogen fixation process, including adaptations involving genetic engineering of nitrogen-fixing bacteria, especially *Klebsiella pneumoniae* (Brill 1979; Marx 1977). However, biological nitrogen fixation also uses considerable energy, namely between 12 and 36 adenosine triphosphate (ATP) molecules for each N_2 molecule fixed. Nitrogen-fixing bacteria use a variety of substrates, including sugars and fatty acids, as sources of both ATP and electrons. The calculation of the energy absorbed by the reactants in nitrogen-fixation, or the thermodynamics of the reaction, imposes a constraint on all possible methods of fixation, namely that energy must be supplied (Safrany 1974).

The list of bacteria and algae known to be capable of fixing free nitrogen is substantial. For convenience, however, they can be considered in two major groups: symbiotic nitrogen fixers (largely bacteria but also including fungi and algae) and free-living nitrogen fixers, including both bacteria and algae and perhaps other microorganisms. Symbiotic nitrogen fixation largely occurs in terrestrial situations, whereas fixation by free-living forms occurs in both terrestrial and aquatic situations. Nonetheless, symbiotic nitrogen fixation is quantitatively much more significant than that by free-living forms, exceeding the latter at least a hundredfold.

Nitrogen fixation by symbiotic and free-living (nonsymbiotic) bacteria accounts for approximately 60 percent of the 2×10^8 metric tons of nitrogen fixed annually by biological and nonbiological processes (Leschine et al. 1988).

It is important to note again that all symbiotic and nonsymbiotic bacteria require an external supply of carbon compounds as an energy source to effect the endothermic reaction of nitrogen fixation (approximately 147 C/mole), for very few of these bacteria are capable of photosynthesis. Among the exceptions is the purple bacteria *Rhodopseudomonas capsulata*, which can grow with dinitrogen gas as the sole nitrogen source under anaerobic conditions with light as the energy source (Madigan et al. 1979). But this bacterium can also fix nitrogen in darkness by depending on anaerobic sugar fermentation for energy. Although growth on dinitrogen is optimal under photosynthetic conditions, reduction of dinitrogen is not necessarily coupled to the photosynthetic process.

For an extensive review of nitrogen fixation, see Broughton (1981–86).

Symbiotic nitrogen fixation. Although not the only bacterial group of symbiotic nitrogen fixers, species of the root-nodule bacteria *Rhizobium* are both the most important and the best understood. Species of *Rhizobium* are highly host-specific to particular species of leguminous plants (peas, clover, beans, etc.); that is, particular strains of *Rhizobium* are associated only with particular species or strains of legumes. These bacteria penetrate root hairs of the legume, presumably by lysing the cell wall and membrane, and then reproduce. The root hair responds to this invasion by a differential growth, resulting in an enlargement, the nodule, in which nitrogen fixation occurs. Rhizobial bacteria have recently been induced to form nodules in the roots of important crop plants, including rice, wheat, and oilseed rape (a source of low-cholesterol oil), although none of the groups has yet demonstrated the production of useful amounts of fixed nitrogen (Moffat 1990b).

Symbiotic nitrogen fixation also occurs in nonleguminous plants such as alder (*Alnus*), bayberry and wax-myrtles (*Myrica*), and sweet-fern (*Comptonia peregrina*), and buckthorn (*Rhamnus*). These species are often found in nutrient-poor soils that harbor actinomycete bacteria of the genus *Frankia* in nodules that develop on their roots, similar to the situation for *Rhizobium* (Van Raalte 1982). It is noteworthy that alders are the first trees to reforest coniferous areas of the Northwest, doubtless playing a significant role in re-establishing the nitrogen balance (see later discussion on effects of deforestation on nitrogen balance). Bormann and Gordon (1984) have shown that stand density of red alder (*Alnus rubra*) in northwest Oregon affects the amount of nitrogen fixation. As Figure 8-4 showns, fixation is greatest at an intermediate density; this is also the case for nitrogen fixation per individual tree.

Diurnal and seasonal changes in nitrogen fixation by lupines (*Lupinus lepidus* and *L. latifolius*) on pyroclastic volcanic sites on Mt. St. Helens, Washington, show the important role of nitrogen fixation in re-establishing vegetation in highly dis-

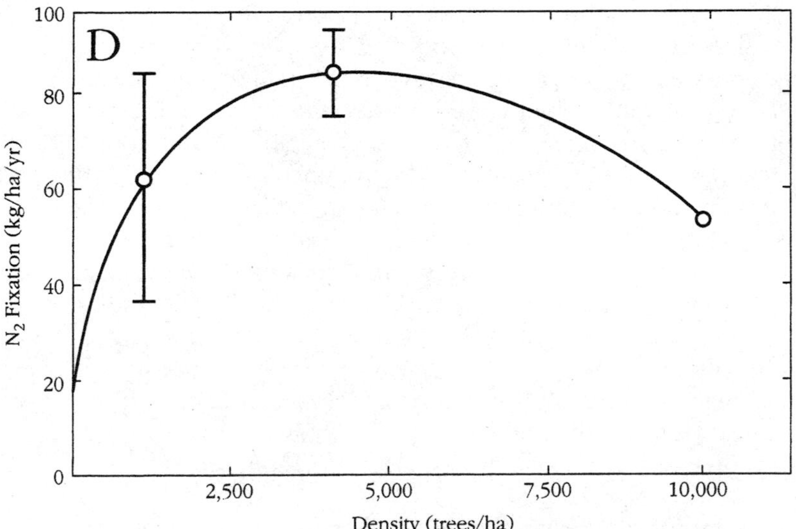

Figure 8-4 The effect of density of young alder (*Alder*) on nitrogen fixation in northwest Oregon. (Redrawn by permission from Bormann and Gordon. 1984. *Ecology* 65:394–402.)

turbed situations (Halvorson et al. 1992). Species differences in the ability to fix nitrogen symbiotically under comparable conditions is evident in black locust and black alder trees, which fixed two and five times more nitrogen respectively than did pines in a "sandbox" ecosystem (Bormann et al. 1993).

As is typical of most mutualistic symbiotic relationships, legumes and the several nonlegumes cannot fix nitrogen in the absence of their nodule bacteria nor can the latter do so when cultured outside the plant.

Symbiotic nitrogen fixation also occurs in animals that feed on plant products. For example, intestinal bacterial fixation of nitrogen is a significant source of nitrogen in the termite *Kalotermes minor* and perhaps other termite species (Benemann 1973).

As noted above, the bulk of fixation in aquatic systems occurs in nonsymbiotic associations. Two exceptions are worthy of note, one being of considerable importance to human food production. This is the water fern *Azolla* and its blue-green algal symbiont *Anabaena*, which is grown extensively in subtropical and temperate rice fields in China as a fertilizer crop. The other instance is the first-known example of nitrogen fixation involving a symbiosis of diatoms (*Rhizosolenia castracanei* and *R. imbricata* var. *shrubsolei*) and bacteria (*Richelia intracellularis*) in the pelagic zone in the eastern Pacific Ocean (Martinez et al. 1983). Such symbiotic systems may contribute a significant amount of nitrogen to nutrient-poor waters.

Although the manner in which plants regulate dinitrogen-fixing symbiotic associations is an unresolved question, there is evidence to suggest that the glutamine synthetase-catalyzed step in ammonia assimilation is a strong influence. Knight and Langston-Unkefer (1988) showed that a toxin released by the bacteria *Pseudomonas syrgingae* pv. *tabaci*, a tobacco leaf pathogen, led to doubling nitrogen-fixation in alfalfa (*Medicago sative*) by inhibiting glutamine synthetase by which nitrogen is reduced to ammonia.

Free-living nitrogen fixation. Among the nonsymbiotic nitrogen fixers are free-living bacteria of both aerobic and anaerobic types and blue-green algae. Aerobic nitrogen-fixing bacteria, such as *Azotobacter*, are widely distributed in soils as well as in fresh- and marine waters; the same is true for anaerobic forms, such as *Clostridium*. In fact, accumulating evidence indicates that many soil and water bacteria are capable of *free-living nitrogen fixation*; and because they often occur in abundance, the total amount of nitrogen fixed may be considerable. Groffman et al. (1993) showed that the amount of nitrogen increased 10-fold during the spring period of most active nitrification, the microbial biomass appearing to be the key regulator of the fate of the added nitrogen. In salt-marsh sediments, for example, bacterial nitrogen fixation has been shown to be 10 times that of algae. Also, four strains of bacteria from fresh water and mud are the first anaerobic bacteria known to use cellulose as an energy source for nitrogen fixation (Leschine et al. 1988). Because cellulose abounds in nature, such nitrogen-fixing, cellulose-fermenting bacteria, if widespread, could play a major role in nitrogen (as well as carbon) cycling.

Cyanobacteria of the genus *Oscillatoria* (= *Trichodesmium*) account for annual inputs of nitrogen to the world's oceans of about 4.8×10^{12} grams, while benthic environments contribute 15×10^{12} grams (Capone and Carpenter 1982; Carpenter and Romans 1991). This amount is one-fifth of current estimates of nitrogen fixation in terrestrial environments and one-half of the present rate of industrial synthesis of ammonia. When the total of all nitrogen inputs to the sea is compared with estimated losses through denitrification, the marine nitrogen cycle approximates a steady state.

Because of nitrogen's significance in soil fertility, data on nitrogen fixation in the soil are extensive. Generally, however, fewer quantitative studies have been made on the contribution that these nonsymbiotic organisms make to the pool of inorganic nitrogen in natural waters.

Studying aquatic nitrogen fixation. To study aquatic nitrogen fixation, the dissolved atmospheric nitrogen is removed (usually by purging with helium) and replaced with a tracer, the stable isotope ^{15}N. The fate of the isotope is then detected by mass spectrometry in methods roughly analogous to those described for using ^{14}C to measure productivity. In a study on Sanctuary Lake in northwest Pennsylvania, Richard and Vera Dugdale (1962) showed a positive correlation between high rates of nitrogen fixation and the presence of large populations of three species of the blue-green algae *Anabaena* (Figure 8-5). Similarly, high fixation rates have been associated with large populations of other blue-green algae, including *Gleotrichia echinulata* in Lakes Mendota and Wingra in Wisconsin and *Trichodesmium* in the Sargasso Sea. The energy required for the process is at the expense of the photosynthetic process of which these algae are capable; that is, the organic compounds that they produce supply the energy needed (147 C/mole) for the fixation process.

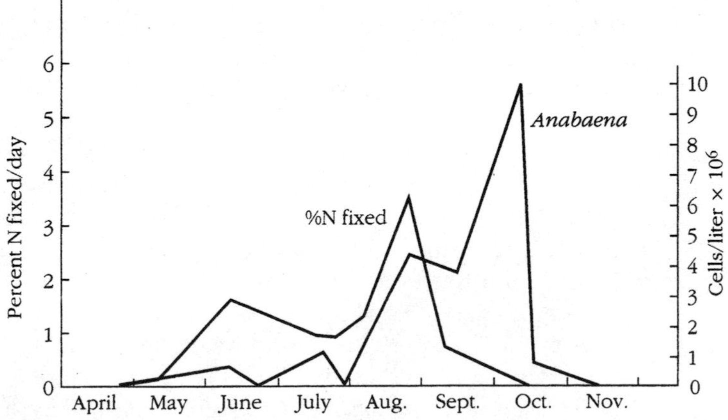

Figure 8-5 The rate of nitrogen fixation (%/day) and the abundance of blue-green algae (*Anabaena*) in Sanctuary Lake, Pennsylvania. (Adapted by permission from V. A. and R. C. Dugdale. 1962. *Limnology and Oceanography* 7:170–77.)

The absence of nitrogen fixation in October in Sanctuary Lake while large populations of *Anabaena* were still present (Figure 8-5) is probably the result of other factors extrinsic to the algae, factors that regulate expression of the capacity to fix the nitrogen. Paerl and Keller (1979) describe a number of physiological adaptations by which *Anabaena* dominate surface lake waters in spite of the fact that high light intensity and oxygen concentration can lead to inactivation of photosynthetic pigments. Adaptations include temporal separation of fixation of CO_2 and N_2 and alteration of pigments. It has also been shown that the presence of nitrogen combined as ammonia affects fixation and that, within limits, there is a direct relationship to temperature and light intensity. Other investigators have shown that various nutrients, such as phosphorus, iron, calcium, molybdenum, and cobalt, are important in controlling growth and nitrogen fixation in nitrogen-fixing organisms. As little as 150 g/ha of molybdenum is required for nitrogen fixation, for example. Thus it is apparent that the production of nitrate, the all-important regulative nutrient for non-nitrogen fixers, is, in turn, regulated by other gradient and nutrient factors. The regulator is itself regulated.

That nutrient cycles are inter-related and even interdependent has been discussed by Likens (1981) and Butcher et al. (1992), and relations of carbon and nitrogen cycles have been demonstrated by Hurd et al. (1994), among others.

Ammoniafication or mineralization. After the incorporation of inorganic nitrogen (NO_3) into an organic form (largely as the amine group, $-NH_2$) via protein and nucleic acid synthesis, it is metabolized and returned to the major part of the cycle as waste products of that metabolism (e.g., urea or uric acid) or as organized protoplasm in dead organisms. Many heterotrophic bacteria, actinomycetes, and fungi in soil and water use this organic nitrogen-rich substrate. In metabolizing organic nitrogen, they convert it to and release it in an inorganic form, ammonia. The process is referred to as *ammoniafication* or *mineralization.* Actually, what these microbes are doing is excreting what for them is excess nitrogen. Some of the ammoniafying microorganisms are subtrate-specific, using only peptone but not simple amino acids, or using urea but not uric acid. In contrast, other species appear to be able to use a wide variety of organic nitrogen sources. The ammoniafication process is an energy-releasing, or exothermic, reaction. If, for example, the protein is glycine-based, ammoniafication results in the release of 176 C/mole. This energy is used by the bacteria in maintaining their life processes.

Ammoniafication in forests. The major source of available nitrogen in forest soils is by the release of ammonium from decomposing organic material. Leaching, denitrification, and atmospheric inputs are generally a small fraction of annual mineralization and uptake in intact forests. Because the uptake of available nitrogen is relatively rapid in intact forests, the pool sizes of ammonium and nitrate are small, and turnovers are rapid. Disturbance, such as clearcutting or dieback, causes increased nitrogen mineralization and, if converted to the mobile nitrate form, leads to elevated losses of nitrogen to stream water (Vitousek and Matson 1985, among others).

Leaching of litter may result in an increase of nitrogen (immobilization) in the soil before mineralization begins. Berg and Ekbohm (1983) showed that both immobilization and mineralization are affected by the ratio of carbon to nitrogen. In a mature forest in central Sweden, net immobilization occurs if the C:N ratio was above 109, net mineralization occurring below that level; by contrast in a clearcut area in the same region, net immobilization occurred if the C:N ratio was above about 63, net mineralization occurring below that level.

In the Oregon Cascades, relatively pure stands of mountain hemlock (*Tsuga mertensiana*) die in distinct waves after their roots become infected by laminated root rot, *Phellinus (Poria) weirii*, the dieback occurring in a radial pattern as root infection and subsequent tree death spread outward from a central infection point (Matson and Boone 1984). Following this pathogen-induced disturbance, nitrogen mineralization doubles but declines as the regenerating stands develop. The availability of nitrogen appears to influence tree resistance to the root rot thus suggesting that the pattern of nitrogen availability is both a consequence and cause of natural disturbance.

Whether natural vegetation acts as a sink for or source of ammonia appears to depend on atmospheric concentrations of the compound. The canopy of a montane-subalpine forest in the Colorado mountains would seem to be a source for ammonia when exposed to air containing low concentrations but be a sink when exposed to air enriched by nearby agricultural sources (Langford and Fehsenfeld 1992). Since ammonia is the primary neutralizing agent for atmospheric acids, the role of the forest canopy in offsetting the effects of acid rain (to be discussed in Part VIII) can be of significance.

Ammoniafication in grasslands and steppes. In a sagebrush (*Artemesia tridentata* spp. *vaseyana*) steppe community, Burke (1989) found that both soil microclimate and organic matter quality control nitrogen turnover, with soil microclimate limiting mineralization to a short season in early spring and summer; only during this time did soil organic matter exert control.

Experimental treatment of blue grama grass (*Bouteloua gracilis*) showed that uptake of nitrogen was greater and plant growth faster in soils containing bacteria (*Pseudomonas paucimobilis* or *P. stutzeri*) and bacterial-feeding nematodes (*Pelodera* sp. or *Acrobeloides* sp.) than plants in soil with only bacteria, because of increased nitrogen mineralization by bacteria and NH^4+–N excretion by nematodes (Ingham et al. 1985). While these nematode microbial-grazers appear to provide a short-term increase in mineralization, they may perform important regulatory functions at critical times in the growth of plants.

Ammoniafication in subalpine lake. In a nutrient-poor subalpine lake (Castle Lake in northern California), Axler et al. (1981) found that nitrogen regeneration through zooplankton excretion of urea and microbial mineralization enabled high rates of ammonium assimilation and consequent phytoplankton production in spite of low levels of inorganic nitrogen.

Nitrification. Although some autotrophic and many heterotrophic marine bacteria can use ammonia or ammonium salts to synthesize their own protoplasm, it is not generally accessible in this form and must be converted to nitrite or nitrate in a process termed *nitrification.* This process, which is pH dependent and which occurs slowly, if at all, in acid conditions, takes place in two steps: first, the conversion of ammonia or ammonium salts to nitrite ($NH_4^+ \rightarrow NO_2^-$); and second, the conversion of nitrite to nitrate ($NO_2^- \rightarrow NO_3^-$).

Nitrosomonas can convert ammonia to nitrite, a toxic form in even small concentrations. Others, such as *Nitrobacter,* act on nitrite, completing the conversion to nitrate. These nitrifying bacteria are all chemosynthetic autotrophs, obtaining their energy from this oxidation process (65 C/mole in the case of *Nitrosomonas* and 17 C/mole in the case of *Nitrobacter*). They then use some of the energy to obtain their needed carbon by reduction from the dioxide or bicarbonate form. In doing so, they produce large quantities of nitrite or nitrate relative to their own growth gain. In their review of nitrogen metabolism in the soil Jackson and Raw (1966), of the Rothamsted Experimental Station in England, report that for each unit of carbon dioxide assimilated, *Nitrosomonas europaea* oxidized 35 units of ammonia to nitrite and *Nitrobacter agilis* oxidized between 76 and 135 units of nitrite to nitrate.

Nitrate (as well as nitrite) is easily leached out of soil, especially in acidic conditions. Unless it is assimilated by plants, it may become "lost" to that ecosystem but become available through groundwater circulation to another one elsewhere. The deserts of Chile, for instance, are rich in nitrates washed from the Andes Mountains. They are so rich, in fact, that they provided a sufficient supply of nitrates to meet the need for explosives by the Western powers in World War I.

Nitrification in soils. The concentration of nitrous oxide dissolved in soil water increased by two orders of magnitude after whole-tree harvesting, as compared with an intact, second-growth forest at the Hubbard Brook Experimental Forest in New Hampshire (Bowden and Bormann 1986). The nitrous oxide, produced in the soil by nitrification, dissolves in soil water and is transported to streams where it rapidly degasses from the solution and is released into the atmosphere. Such losses may be important to the global atmospheric budget of nitrous oxides.

Robertson and Vitousek (1981) found that nitrification is closely correlated with nitrogen mineralization that increased in soils in various stages of primary succession (see Chapter 13 for discussion on succession) and remained more or less constant in soils in various stages of secondary succession, with the exception being the oldest soils where the highest rates were observed.

Nitrification in the ocean. Nitrification in the open ocean and its sediments is not entirely understood. Stanley Watson (1965), of the Woods Hole Oceanographic Institute, described the first nitrifying bacterium ever isolated from open ocean waters, *Nitrosocystis oceanus.* In subsequent studies Watson has shown that it is an obligate autotroph, using only ammonia as an energy source and carbon dioxide as a carbon source. Watson's estimates of total nitrification by *Nitrosocystis* would hardly account for the relatively high nitrate levels in the sea. It is probable, then, that populations of nitrifying bacteria and/or their levels of activity are much higher in the open ocean than studies to date suggest. Even more likely has been the failure to adequately measure the production of dissolved organic nitrogen (Bronk et al. 1994). In addition, however, depending on latitude, considerable nitrate moves into the open ocean from salt marshes, which abound on the world's seacoasts, and from coral reefs. Nonetheless, fixed nitrogen is regarded by some scientists as the major limiting nutrient for phytoplankton production in the ocean, particularly in tropical waters.

Nitrous oxide and ozone destruction. Recognition of the role of nitrous oxide (N_2O) in the destruction of stratospheric *ozone* (O_3) has led to increased interest in the sources, sinks, and biogeochemical cycles of this trace gas. It has been determined, for example, that creating the ubiquitous material nylon produces nitrous oxide as a byproduct (through synthesis of adipic acid) and may account for 10 percent of its increase in the atmosphere (Thiemens and Trogler 1991).

Increases in the global production of N_2O have been predicted as a consequence of increasing use of nitrogen fertilizer and the combustion of fossil fuel. The marine environment plays a major role in the global cycling of N_2O with sources in surface ocean waters and sinks in low-oxygen environments. Coastal marine sediments are shown to be a net source of nitrous oxide, and polluted sediments to have higher rates of N_2O production than relatively unpolluted sites (Seitzinger et al. 1983).

Denitrification. The route of nitrate through an assimilatory-anabolism circuit and its return via ammoniafication and nitrification is but one possible alternative. *Denitrification* to molecular or gaseous nitrogen (N_2), as well as to nitric oxide (NO) and nitrous oxide (N_2O), is effected by bacteria, such as *Pseudomonas*, and fungi. They use the nitrate as an oxygen source in the presence of glucose and phosphate. Most denitrifiers reduce nitrate only to nitrite, however, and others reduce it to ammonia. Under anaerobic conditions denitrification of nitrate to nitrous oxide in the presence of glucose is an exothermic reaction yielding 545 C/mole. Denitrification to molecular nitrogen yields 570 C/mole. These yields are in contrast to aerobic oxidation of glucose, which yields 686 C/mole.

Unless recaptured in nitrogen fixation, molecular nitrogen released in denitrification can be returned to the atmospheric pool. But whether or not molecular nitrogen or one of its oxides will result from denitrification is known to be pH dependent. Thus nitric oxide evolution becomes significant at pH values below 7.0; above a pH value of 7.3, nitrous oxide tends to be reabsorbed and further denitrified to molecular nitrogen.

Because denitrification to molecular nitrogen is known to occur under anaerobic or partially anaerobic conditions, it would be expected to occur in soils that are poorly aerated. It would also be expected in those soils with considerable organic matter, for the latter have a high oxygen demand and so are typically anaerobic.

Denitrification in tallgrass prairie. Denitrification was found to be higher in unburned tallgrass prairie in central Kansas than in annually burned, annually burned and grazed, and cultivated sites (Groffman et al. 1993). Further, the cultivated sites had consistently lower rates of denitrification than the native prairie sites even when water and nitrate were added. In all sites, although denitrification was generally related to patterns of soil moisture, water addition did not stimulate the process in ungrazed prairie soils; water plus nitrate additions did stimulate it, however. Unburned prairie is wetter and has higher concentrations of nitrate in soil solution than burned sites and thus plays a significant role in site fertility, landscape water quality, and atmospheric chemistry in tallgrass prairie.

Denitrification in aquatic ecosystems. Little denitrification would be expected in the typically oxygen-rich surface waters of lakes and seas, where theoretically the process is otherwise possible. During anoxic periods, however, molecular nitrogen can be formed in aquatic ecosystems, as shown by John Goering and Vera Dugdale (1966) in an Alaskan lake. Winter water samples taken both at the bottom and at 1 m below the ice surface were inoculated with nitrogen-labeled nitrate ($K^{15}NO_3$) and then incubated at lake temperature. Analysis by mass spectrometry showed denitrification in the bottom samples to be about six times faster than that near the surface, but the only significant end product was molecular nitrogen. There was a small reduction to ammonia, but no nitrous or nitric oxide was detected.

Global Nitrogen Budget

The global distribution of nitrogen is shown in Table 8-2. The global circulation of this nitrogen is, as has been noted several times in the discussion of the nitrogen cycle, subject to disruption and warrants further consideration.

From the foregoing discussion, it is clear that the movement of nitrogen is by no means unidirectional, unregulated, or energy independent. Numerous routes are available at virtually every major way station, each route is biologically and/or nonbiologically regulated, and energy is consumed or released in each process. These numerous self-regulating, energy-dependent feedback mechanisms lead to the assumption that the global nitrogen cycle is balanced, nitrogen fixation being balanced

TABLE 8-2 NITROGEN IN THE BIOSPHERE IN 10^6 METRIC TONS

Atmosphere	3,800,000	Ocean water (dissolved)	20,000
Land organisms	772	Ocean organisms	901
Living	12	Living	1
Dead	760	Dead	900
Inorganic N (land)	140	Inorganic N (ocean)	100
Earth's crust	14,000,000	Sediments	4,000,000
	Total organic	1,673	
	Total inorganic	21,820,240	

Data derived from C. C. Delwiche, *Scientific American*, September 1970.

by denitrification. This assumption is in question, however. Any possible balance is increasingly stressed by the accelerating rate of industrial nitrogen fixation without being compensated by an increased rate of denitrification. A further disturbance of the system is created by the burning of fossil fuels in automobiles. This source of NO_2 release to the atmosphere constitutes a major source of air pollution and is a pollutant especially detrimental to the respiratory system and to the maintenance of the ozone layer in the atmosphere.

Numerous studies have indicated that large amounts of nitrogen (and other nutrients) are lost from surface soils after clearcutting, especially if followed by burning. According to Matson et al. (1987), if overall losses of nitrogen average 1000 kg/ha, then 20 to 25 $\times$ 10^9 kg of nitrogen are mobilized annually from the 20 to 25 $\times$ 10^6 hectares cleared in forest conversion. This amount is equivalent to more than half the industrial nitrogen fixation globally and is greater than the total amount of nitrogen delivered by rivers to the ocean. If this mobilized nitrogen is lost to the atmosphere or to aquatic systems, there would be untoward consequences on the productivity of tropical ecosystems.

Within some ecosystems, however, the major processes of nitrification and denitrification are well attuned to the productivity demands of the ecosystems. These processes are most rapid in winter in temperate zones, for example, resulting in maximal amounts of nitrate in the spring and early summer, the time when nitrate demand for plant growth and reproduction is highest. In many ecosystems, however, no denitrification occurs. Thus although nitrogen fixation enables ecosystems to stay ahead of nitrogen deficiency, denitrification, where it occurs, ensures that the margin is always slim.

A major contribution to understanding the global nitrogen budget is found in Delwiche (1981); also see Butcher et al. (1992). Earlier, Delwiche (1970) estimated the total amount of nitrogen in the biosphere as nearly 22 $\times$ 10^2 metric tons with most of it in the inorganic state (Table 8-2). He also estimated that the total annual nitrogen fixation of annual nitrogen fixation of 92 $\times$ 10^6 metric tons (54 biological, 30 industrial, 7.6 photochemical, and 0.2 volcanic) is replaced through denitrification by only 83 $\times$ 10^6 metric tons (43 terrestrial, 40 marine, and 0.2 in sediments). The disbalance (9 $\times$ 10^6 metric tons), which is largely exacerbated by the increased rate of industrial fixation, is being built up annually in groundwater, rivers, lakes, and the ocean. As shall be seen later, this fixed nitrogen is a significant factor in pollution of aquatic ecosystems. Recent studies in the marine environment indicate an additional 20 $\times$ 10^6 metric tons are added to the nitrogen pool by nitrification. Marine nitrogen input from all sources is nearly balanced by denitrification, thereby approximating a steady state.

SEDIMENTARY NUTRIENT CYCLES

Although some essential nutrients such as sulfur may have a gaseous phase (e.g., sulfur dioxide), such phases are relatively insignificant, for there is no large gaseous reservoir. Furthermore, none of these nutrients cycles so readily as carbon and nitrogen. In their pathways there are fewer self-correcting, homeostatic mechanisms and more stages in which short- or long-term stagnation can occur. Most significant of the stagnation stages is sedimentation in the oceans and deep continental lakes.

Having considered the complexities of the nitrogen cycle in detail, it will be prudent to examine briefly the essentials of sedimentary cycles—of sulfur and phosphorus. Then we shall consider the phosphorus cycle in given ecosystems in some detail. These two elements have been selected because of their critical significance in growth and metabolism (e.g., sulfhydryl bonding in proteins, phosphate-energy transfer compounds) and because they have relatively well-understood biogeochemical cycles.

Sulfur Cycle

Sulfur, the 14th most abundant element in the Earth's crust, is not only a critical component of essential biological compounds such as the amino acids cysteine and methionine, its cycle, like that of nitrogen, plays a substantial role in the regulation of other nutrients, including oxygen and phosphorus. The heart of the sulfur cycle involves the uptake of sulfate ($SO_4^=$) by producers, largely through their roots, and the release and transformation of the sulfur in a number of different steps and variety of forms, including sulfhydryl (–SH), hydrogen sulfide (H_2S), sulfites ($SO_3^=$), and molecular sulfur (S). Like the nitrogen cycle, the sulfur cycle is complex (Figure 8-6); unlike the nitrogen cycle, it doesn't fall into discretely packaged steps such as nitrogen fixation, ammoniafication, etc. Hence where we begin to describe the cycle is somewhat arbitrary, but we shall start with the assimilation of sulfur into plants (more or less at the center of Figure 8-6).

Assimilation and release of sulfur by plants. Sulfur enters the trophic cycle in terrestrial plants via root adsorption in the form of inorganic sulfates (e.g., calcium sulfate, sodium sulfate) or by direct assimilation of amino acids released in the decomposition of dead or excreted organic matter. Bacterial and fungal (*Aspergillus* and *Neurospora*) mineralization of the organic sulfhydryl in amino acids followed by oxidation results in sulfate; this adds to the sulfate pool for root adsorption.

Under anaerobic conditions, sulfuric acid may be reduced directly to sulfides, including hydrogen sulfide, by such bacteria as *Escherichia* and *Proteus* ($SO_4^= + 2H^= = H_2S + 2\,O_2$). Sulfate is also reduced under anaerobic conditions to elemental sulfur or to sulfides, including hydrogen sulfide, by such heterotrophic bacteria as *Desulfavibrio*, as well as by *Escherichia* and *Aerobacter*. These sulfate-reducing anaerobic bacteria are heterotrophic, using the sulfate as a hydrogen acceptor in metabolic oxidation in a manner comparable to the use of nitrite and nitrate by denitrifying bacteria.

Until recently, sulfate reduction had been thought to occur only under anaerobic conditions since in the presence of even traces of oxygen, nitrates, or other electron acceptors, sulfate reduction was inhibited. However, Canfield and Des Marais (1991) found that sulfate reduction occurs in the upper, well-oxygenated,

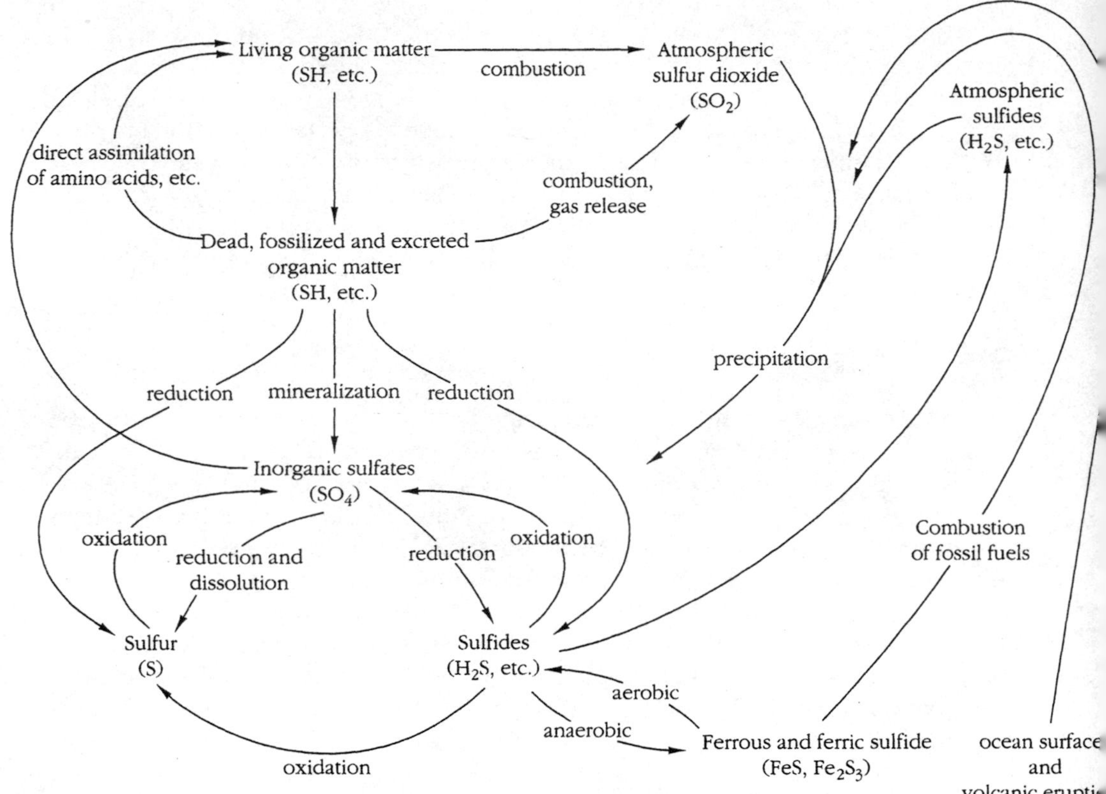

Figure 8-6 The sulfur cycle.

photosynthetic zone of hypersaline microbial mats from Baja California, Mexico. Sulfate reduction occurs at the same rate at noon, when photosynthetic bacteria (the cynaobacterium *Microcoleus chthonoplastes*) were releasing oxygen, and at night under anaerobic conditions. Thus, sulfate reduction is not strictly an anaerobic process, but the degree to which aerobic reduction contributes on the broad scale is yet to be determined.

It is important to note that although photosynthesis accounts for the bulk of oxygen in the atmosphere, without denitrifying and sulfate-reducing bacteria, the oxygen cycle would be out of balance. There are no mechanisms for releasing the considerable amount of oxygen bound up in nitrates and sulfates other than the reduction processes that occur in the nitrogen and sulfur cycles.

The presence of large quantities of hydrogen sulfide in the anaerobic, and usually deeper, portions of aquatic ecosystems is inimical to most life. For example, hydrogen sulfide inhibits oxygen release from and nutrient uptake by rice plants; but, when present, the filamentous bacterium *Beggiatoa* significantly reduces hydrogen sulfide levels and increases oxygen release from the rice plants (Joshi and Hollis 1977). The presence of hydrogen sulfide probably accounts for the absence of higher animals below 200 m in the Black Sea and has been considered responsible for kills of fish in valley impoundments polluted by pulp-mill effluents, which are rich in sulfates.

The existence of archaebacterial sulfate reducers (*Methanobacterium thermautotrophicum* and *Methanococcus thermolithotrophicus*) at extremely high temperatures (70°C to 100°C) can explain the formation of hydrogen sulfide observed in submarine hydrothermal systems, deep oil wells, and fumaroles (steam vents) (Stetter et al. 1987). A more recent study has demonstrated sulfate reaction near hydrothermal vents at temperatures up to 100°C with an optimum rate between 103° to 106°C by hyperthermophilic bacteria (Jörgensen, Isaksen, and Jannasch 1992).

As is true of ecological balances, one organism's poison is another's treat, so to speak. Colorless sulfur bacteria, such as species of *Beggiatoa*, oxidize hydrogen sulfide to elemental sulfur, and species of *Thiobacillus* oxidize it to sulfate; other species of *Thiobacillus* oxidize sulfide to sulfur, and still others oxidize sulfur to sulfate. For some species, even those of the same genus, the oxidation processes can occur only in the presence of oxygen; for others, oxygen availability is irrelevant. These latter bacteria are chemosynthetic autotrophs, using the energy liberated in the oxidation to obtain their needed carbon by reduction of carbon dioxide:

$$6\,CO2 + 12\,H_2S \rightarrow C_6\,H_{12}O_6 + 6\,H_2O + 12\,S$$

These bacteria are thus comparable to the chemosynthetic autotrophic nitrifying bacteria that oxidize ammonia to nitrite and nitrite to nitrate. In addition, they also include the green and purple photosynthetic bacteria that use the hydrogen of hydrogen sulfide as the oxygen acceptor in reducing carbon dioxide. The green bacteria apparently are able to oxidize sulfide only to elemental sulfur, whereas the purple bacteria can carry the oxidation to the sulfate stage:

$$6\,CO_2 + 12\,H_2O + 3\,H_2S \rightarrow C_6H_{12}O_6 + 6\,H_2O + 3\,SO_4^{=} + 6\,H^+$$

Sulfur in the atmosphere. Sulfur in the atmosphere comes from several different sources: decomposition and/or combustion of organic matter, combustion of fossil fuels, and ocean surfaces and volcanic eruptions. The most prevalent form of sulfur entering the atmosphere is sulfur dioxide (SO_2); it, along with other atmospheric forms such as elemental sulfur and hydrogen sulfide, is oxidized to sulfur trioxide (SO_3), which combines with water to form sulfuric acid (H_2SO_4).

Atmospheric sulfur, largely in the form of sulfuric acid, is removed by two general processes: *rainout,* which includes all processes within clouds that result in removal; and *washout,* which is the removal by precipitation below the clouds (Kellogg et al. 1972). Depending on the amount of the various sulfur compounds available to form the sulfuric acid, the degree of acidity can be strong enough to approximate that of battery acid. Atmospheric inputs of sulfuric acid provide the dominant source of both hydrogen ions (H^+) for cation replacement and mobile anions for cation transport in subalpine soils of northeastern United States that are affected by acid precipitation (Cronan et al. 1978).

Sulfur in the sediments. The sedimentary aspect of the cycle involves the precipitation of sulfur in the presence of such cations as iron (Fe) and calcium (Ca) as highly insoluble ferrous sulfide (FeS) and ferric sulfide (Fe_2S_3, pyrite) or relatively insoluble calcium sulfate ($CaSO_4$). Of considerable ecological significance is ferrous sulfide, which is formed under anaerobic conditions. It is insoluble in neutral or alkaline water; consequently, sulfur has the potential for being bound up under these conditions to the limits of the amount of iron present. Once buried, pyrite appears to be geologically stable and is a primary reservoir of both iron and sulfur in salt-marsh as well as other sediments (Luther et al. 1986).

The oxidation of sulfides in marine sediments is a key process, though poorly understood. It has been shown, however, that thiosulfate (S_2O_3) acts as a shunt in this process constituting the bulk of the hydrogen sulfide oxidation products; subsequent to this process it was reduced back to hydrogen sulfide and oxidized to sulfate (Jörgensen 1990).

Because of the thermodynamics of this ferrous sulfide system, other nutrients important to biological systems can also get trapped for varying periods of time. Among them are copper, cadmium, zinc, and cobalt. On the contrary side, however, the very binding of these iron compounds allows the conversion of phosphorus from insoluble to soluble form and thereby makes it accessible. There is little question, then, but that sulfate and sulfide reduction in the anaerobic sulfur-containing muds controls to a significant degree the biological chemistry of the ecosystem. This portion of the sulfur cycle provides an excellent example of the interaction and regulation that exists between different mineral cycles, besides demonstrating again the complex biological and chemical regulation within such cycles.

Global cycle of sulfur. Unlike the situations for carbon and nitrogen, the total amount of sulfur in various compartments of the biosphere and the rate at which the various forms of sulfur turnover are not as well understood. Kellogg et al. (1972) have approximated the annual turnover of sulfur as shown in Figure 8-7, but the amount of the more or less long-term losses to the sediments are not acknowledged. Given the holding-action of ferris and ferric sulfide in the sediments, this part of the cycle suggests that the overall sulfur cycle is not balanced, although the ebb and flow from ocean and land to atmosphere and back would appear to be so. For example, according to Figure 8-7, the annual total output of sulfur to the atmosphere of 550×10^6 tons is balanced by the return of the same amount to the Earth's surface. Although the total amount of sulfur in the biosphere is a constant, the ever increasing release of sulfur by the combustion of fossil fuels will continue to increase the amount annually cycled (Charlson et al. 1992). The resulting increase in the amount

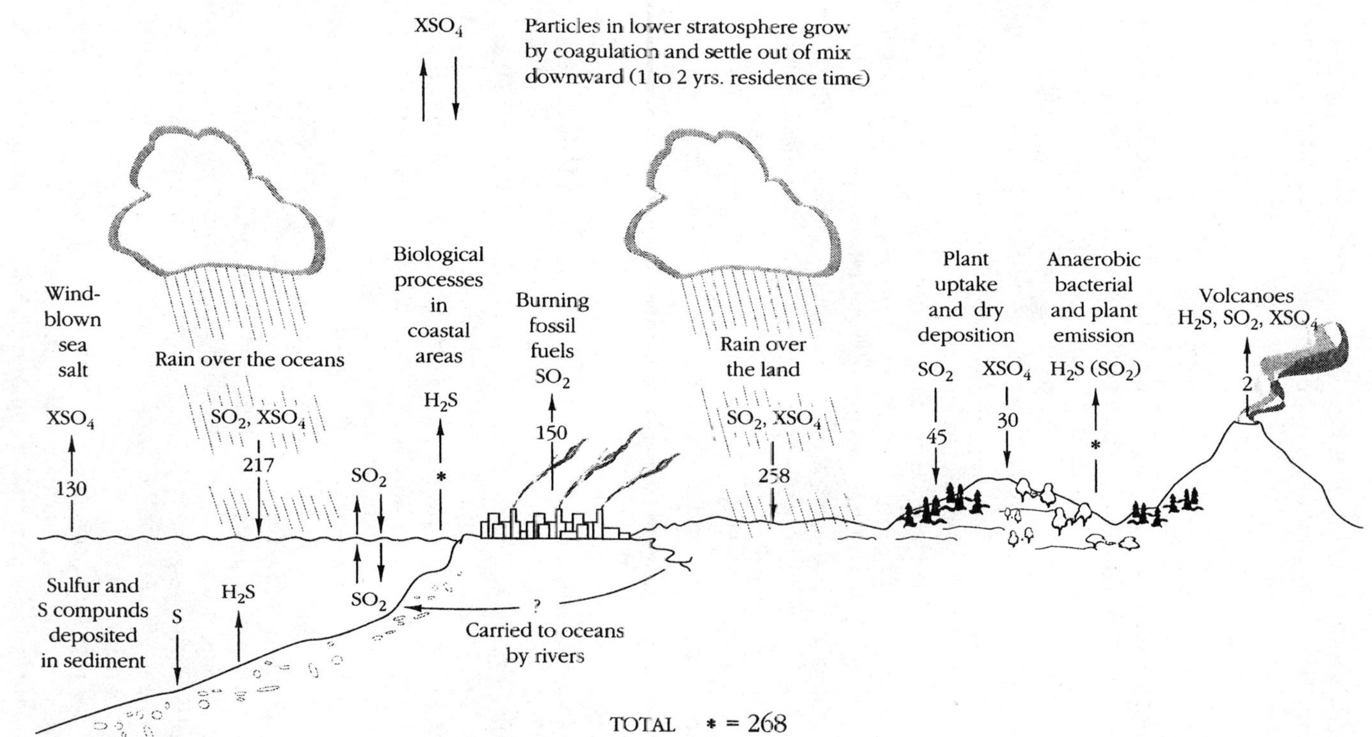

Figure 8-7 The annual cycle of atmospheric sulfur compounds calculated as 10^6 tons of sulfate per year. (Redrawn with permission from Kellogg et al. 1972. *Science* 175:587–96.)

of sulfuric acid (and other acids) in precipitation carries the consequence of correlative increases in the damage to ecosystems in particular parts of the world (see discussion of acid precipitation in Part VIII).

Phosphorus Cycle

As a constituent of nucleic acids, phospholipids, and numerous phosphorylated compounds, phosphorus is one of the nutrients of major importance to biological systems. The ratio of phosphorus to other elements in organisms tends to be considerably greater than the ratio of phosphorus in the available and primary sources; thus, phosphorus becomes ecologically significant as the most likely limiting or regulating element in productivity. For example, there is now general agreement that phytoplankton growth will be limited by the amount and availability of phosphorus in most lakes because of the precise relationship between the concentration of total phosphorus and the standing crop of phytoplankton in a wide variety of lakes. This includes lakes in which low nitrogen-to-phosphorus ratios should favor limitation by nitrogen (Schindler 1977). Schindler further suggests that the "evolution" of appropriate nutrient ratios for phytoplankton growth involves a complex series of interrelated biological, geological, and physical processes, including photosyntheses, the selection for species of algae that can use atmospheric nitrogen, alkalinity, nutrient supplies and concentrations, and rates of water renewal and turbulence. For a timely review of the phosphorus cycle, see Jahnke (1992).

Plants require inorganic phosphate for their nutrition typically as orthophosphate ions. In typical mineral-cycling fashion, this phosphate is transferred to consumers and decomposers as organic phosphate (Figure 8-8). Subsequently, it is made available as inorganic phosphate for recycling via mineralizing decomposition. Lean (1973) has found that the excretion of organic phosphorus by plankton results in the extracellular formation of a colloidal substance that accounts for most of the nonparticulate phosphorus in lake water. In marine situations, bacterial degradation is too slow to account for the rapid turnover of phosphorus; most regeneration is done by protozoans and small metazoans.

Most phosphate becomes lost to this central cycle by two physical processes, one short term, the other long term. Physical adsorption by sediments and soils is of great importance in controlling the concentration of dissolved phosphorus in soils and lakes. Sedimentation, in contrast, involves binding the phosphorus with such cations as aluminum, calcium, iron, and magnesium, thus forming insoluble compounds that precipitate out. In deeper bodies of water, this process takes the phosphorus out of the reach of upwelling and major water circulation, and consequently, only extremely small portions are returned by fish and seabirds. In shallower circumstances, however, the absorbed and/or sedimented phosphorus is the source for aquatic plants. For example, Carnigan and Kalff (1980) showed that nine common species of aquatic macrophytes (including *Myriophyllum*, *Potamogeton*, *Callitriche*, *Elodea*, and *Najas*) obtained all their phosphorus from sediments and thereby serve as potential pumps of the phosphorus to the open water.

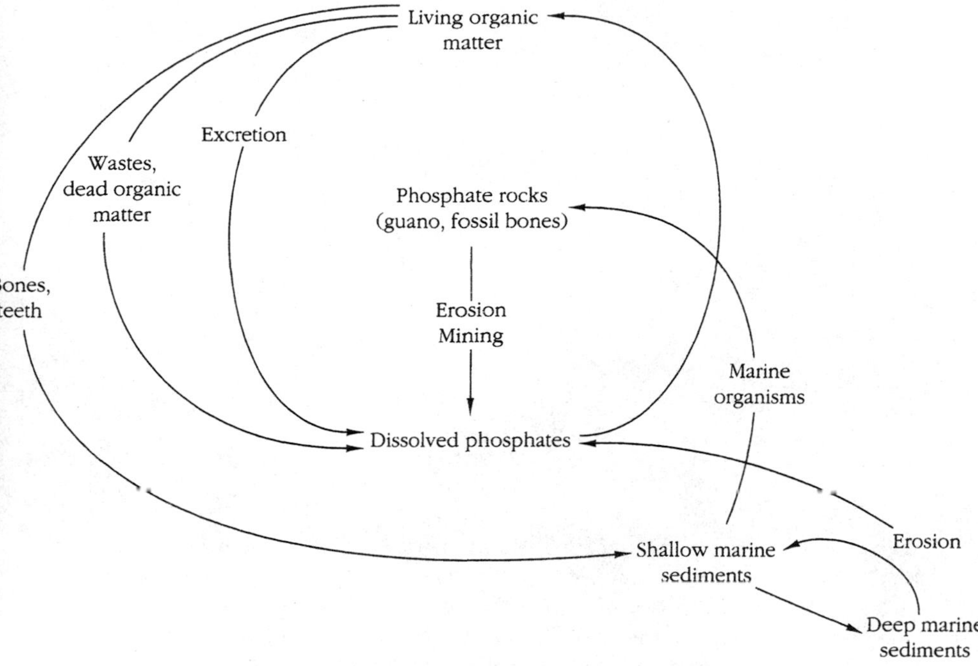

Figure 8-8 The phosphorus cycle.

Sedimentation of phosphorus, as introduced above in the discussion of the sedimentary phase of the sulfur cycle, is tied to the ferrous-ferric sulfide system, but denitrification is also involved. For example, phosphate deposits on the ocean floor are particularly common in regions of coastal upwelling; for the most part these deposits are of ancient origin with modern formation occurring primarily in areas of intense coastal upwelling, as off the coasts of Peru and Namibia. Piper and Codispoti (1975) attribute these ancient deposits to high denitrification rates during only a few intervals of geological time, intervals that coincided with episodes of animal extinctions (Figure 8-9). On the other hand, the new phosphate deposits, which grow downward into soft sediment, appear to originate from regeneration in the sediments rather than from direct precipitation from bottom waters (Burnett et al. 1982).

Biological processes, such as the formation of teeth and bone—both of which are more or less resistant to weather—and excretion, also account for considerable losses from the biotic portion of the phosphorus cycle. The tremendous deposits of guano (seabird wastes) that have accumulated over hundreds of years along the western coast of South America and are "mined" for fertilizer are a case in point in such removal.

The cycling of phosphorus thus bears many characteristics of nutrient cycling in general and of sedimentary cycles in particular. However, unlike carbon, nitrogen,

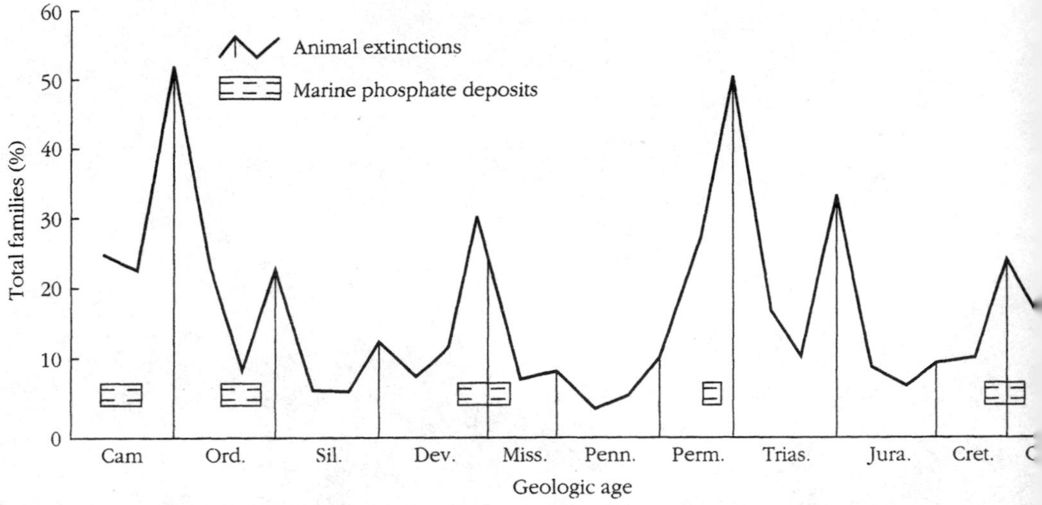

Figure 8-9 Episodes of extinction in the animal kingdom and the approximate ages of extensive marine phosphate deposits that are tied to periods of extensive denitrification. The main episodes of extinction occurred near the close of the Cambrian, Devonian, Permian, Triassic, and Cretaceous periods. The last appearances of animal families are given in percentages. (Redrawn with permission from D. Z. Piper and L. A. Codispoti. 1975. *Science* 188:15–8.)

and sulfur, there is no atmospheric phase, and the sedimentary phase constitutes a more dominant role in the phosphorus cycle.

Before discussing in-depth phosphorus movement in a particular set of aquatic ecosystems, examples of several other ecosystems is in order.

Phosphorus cycling in a northern hardwood forest. Phosphorous is conserved biologically within the Hubbard Brook northern hardwood forest in New Hampshire by close coupling of biological decomposition and uptake processes in the surface soils (Wood et al. 1984). The underlying iron- and aluminum-rich B horizons function as a geological buffer that regulates the constant and low-level losses of dissolved phosphorus in stream water. It appears that after clearcutting, the phosphorus in the A horizon migrates into the B horizon from which it may represent an important source of the nutrient for forest regrowth.

Phosphorus cycling in desert soils. The total phosphorus in desert soils in southern New Mexico decreased with the increase of the soil age and was removed from the ecosystem as readily as the most easily leachable base cations (Lajtha and Schlesinger 1988). Calcium forms of phosphorus (e.g., calcium phosphate) constituted the single largest amount, and iron and aluminum forms constituted very small percentages. Thus, unlike forested soils where iron and aluminum oxides in the B horizon fix the phosphorus and serve as a phosphorus sink after clearcutting, the small percentage of iron- and aluminum-bound phosphorus in desert soils precludes their playing a significant role in phosphorus conservation.

Phosphorus cycling in Alaskan tundra. Channels of subsurface water draining, known as water tracks, sampled in Alaska tundra had more rapid cycling of nitrogen and phosphorus than adjacent nontrack areas due to slightly warmer soil temperature, deeper thaw, higher soil phosphatase and protease activities, and more rapid nitrogen mineralization (Chapin et al. 1988). This latter condition indicates again the interrelationships of nutrient cycles, the regulation of the regulated.

Phosphorus cycling in a woodland stream. Nutrient cycles in streams may be viewed as spirals because of the continuous downstream displacement of nutrients as they cycle (Webster and Patten 1979). In a small woodland stream in eastern Tennessee, Mulholland et al. (1985) showed that the total spiraling length of phosphorus varied from 23 meters in November, just after peak autumn leaf fall, to 99 meters in August. This primarily reflected the travel of phosphorus in the dissolved form. The movement of phosphorus is affected directly by the amount of both coarse and fine organic matter in the stream, the levels of which are reduced considerably after late autumn and winter storms literally flush the stream. In an earlier study of this same stream, Newbold et al. (1983) showed that phosphorus moved downstream at an average velocity of 10.4 m/day, cycling once every 18.4 days.

Phosphorus cycling in a fen peatland. In a peatland in central Michigan, Richardson and Marshall (1986) demonstrated that adsorption by the soil and the accumulation of dead organic matter as peat (with phosphorus stored therein) control the long-term storage of phosphorus. Initially, microorganisms and fine sediments take up most of the phosphorus that becomes available, doing so within an hour, as demonstrated by introducing radioactive phosphorus (^{32}P); by contrast, uptake by plants (sedges belonging to the genus *Carex*) was slow even after 45 days. After the introduction of sewage effluent, algal populations quickly absorbed significant amounts of the phosphorus contained in the effluent.

Phosphorus Cycling in Ontario Lakes

Although we have discussed the phosphorus cycle in general as well as aspects of it in several different ecosystems, a detailed and comparative consideration is in order because of the significance of phosphorus in productivity and other metabolic and ecological processes. This in-depth treatment will also demonstrate the complexities of and differences in the cycle in one category of ecosystems, namely aquatic ecosystems. For this purpose, we will concentrate on a series of nine lakes in Ontario studied by F. H. Rigler, of the University of Toronto (1964). The lakes were quite varied, including hard-water lakes with pH values between 7.7 and 8.7 and soft-water lakes with pH values between 6.7 and 7.2. Maximum depth in the lakes varied between 6 and 53 meters and surface area varied between 0.5 and 2,180 hectares.

Phosphorus in these lakes, as in any aquatic system, occurs in three forms and cycles as follows:

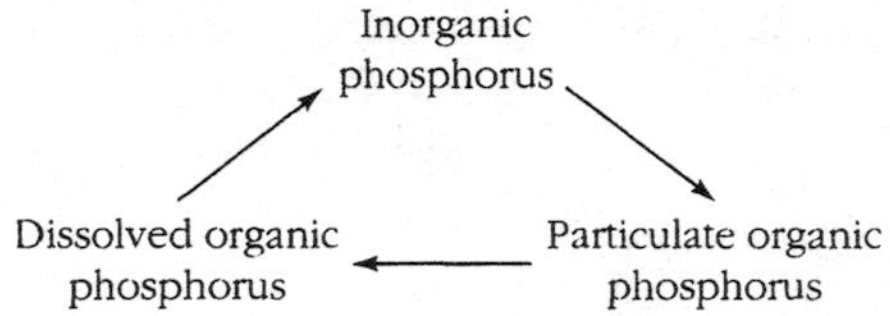

The inorganic phosphorus is typically orthophosphate and is readily soluble. Particulate organic phosphate is that which occurs in suspension in living and dead protoplasm and is insoluble and in suspension. Dissolved or soluble organic phosphorus is derived from the particulate matter by excretion and decomposition. The total phosphorus in a system consists of that in each of the three compartments, plus that occurring as mineral phosphate in sediments or rock substrates. For discussion purposes, only the motile phosphate will be considered, the total of which in the nine lakes of Rigler's study ranged between 5 and 133 mg/l.

According to Rigler's findings, Teapot Lake and Heart Lake can be used to represent the basic patterns and conditions of the entire series. The annual distribution of the three forms of phosphorus in these two lakes appears in the lower part of Figure 8-10. Particulate organic phosphorus showed no consistent seasonal pattern, but if there was an increase, it tended to be in the winter, as shown for Teapot Lake. The dissolved organic phosphorus showed no seasonal maximum. Inorganic phosphorus was consistently low during most of the year, but increased between December and April.

Seasonal patterns. This seasonal pattern makes phosphorus available in the inorganic form, the only form that can be used biologically at a time of the year when conditions approach optimum for productivity. Here there is a parallel to nitrification, which is also largely a winter event and which provides a maximum of nitrate in the spring. Phosphorus tends to become bound in the sediments under oxidized conditions such as occur in the fall and to be released under anoxic conditions, a situation that develops in many temperate lakes during winter stagnation. Thus the very environment that leads to a binding of sulfur in the presence of iron—namely, a reducing environment—is the situation under which, as already noted, phosphorus is released. In contrast, as Pomeroy (1960) has noted, in shallow estuaries inorganic phosphate stock is highest in late summer or early fall. This situation results from the interaction of river flow and rainfall with the metabolic processes of estuarine organisms and is compounded by the chemistry of the sediments.

The proportion of phosphorus in the three major compartments was strikingly similar during the summer period in eight of the lakes--inorganic phosphorus ranged from 4.8 to 7.8 percent, dissolved organic phosphorus ranged from 25.0 to 31.7 percent, and particulate organic phosphorus ranged from 61.8 to 68.2 percent. The one

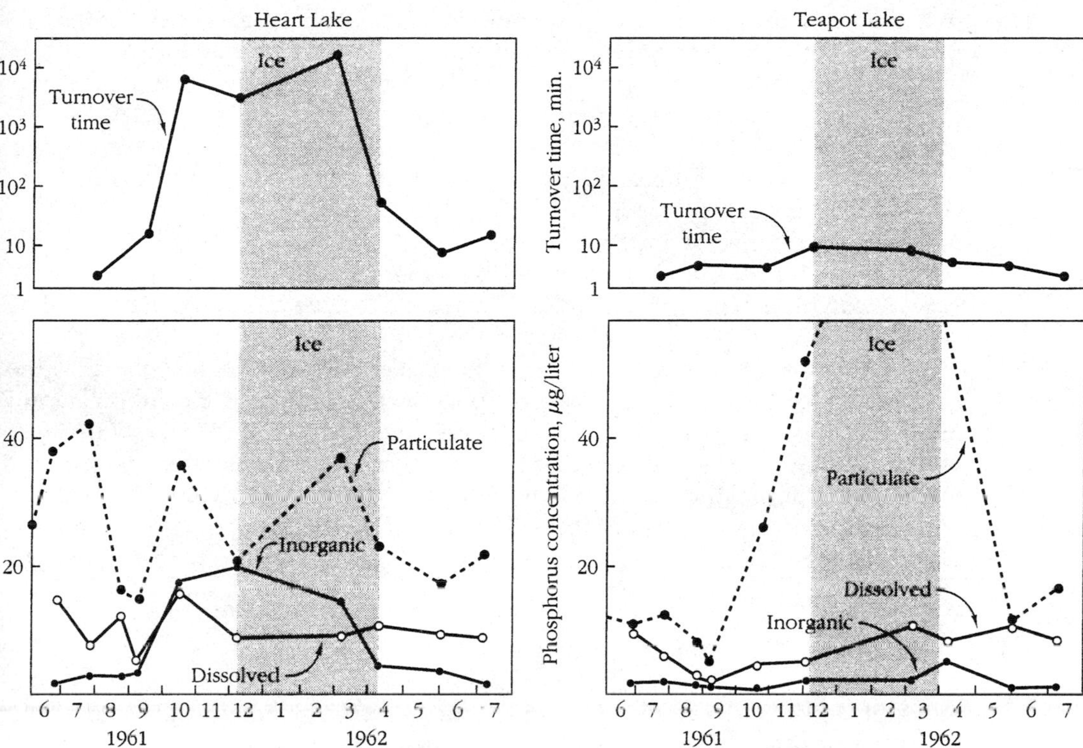

Figure 8-10 Seasonal changes of turnover of inorganic phosphorus and amounts of the three forms of phosphorus in two lakes in Ontario. (Adapted by permission from F. H. Rigler. 1964. *Limnology and Oceanography* 9: 511–8.)

exceptional lake, Grenadier Pond, had 83.7 percent of its phosphorus in particulate form and 12.5 percent in dissolved organic from. The inorganic component was, however, comparable to that of the other lakes, 4.8 percent. Grenadier Pond is very productive, partly because it is subject to urban drainage known to be rich in particulate phosphorus.

Turnover time. The summer distribution of phosphorus just described for these lakes is a static rather than a dynamic situation. The shifts in distribution shown in Figure 8-10 suggest movement of phosphorus from one compartment to another but give no indication of the rate at which it occurs. By tagging water samples at different times of the year with radioactive phosphorus, ^{32}P, in the form of the inorganic orthophosphate, Rigler was able to get a preliminary estimate of the rate at which phosphorus is used or turned over. Thus if 100 units of radioactive phosphorus were introduced as inorganic phosphate and only 90 units were left in the water an hour later, the amount of turnover would be 10/100 units, or 10 percent per hour. If this rate of turnover is constant, it would then take 10 hours to turn over the phosphorus completely, a period referred to as *turnover time.* The actual computation of turnover time is a bit more complicated, for we must account simultaneously for the presence of stable phosphorus and the shift in the ratio of radioactive to stable forms.

The general annual pattern of turnover time in the lakes that Rigler studied was much the same: during the summer average turnover time ranged from 0.9 to 7.5 minutes. As progression through the fall and winter took place, there was a lengthening of turnover time to the maximum, which generally occurred under ice and snow cover. This pattern is shown in the upper part of Figure 8-10. These two lakes also represent the extremes in magnitude of turnover time, the winter average being only 7 minutes in Teapot Lake, a 3.5-fold increase from the summer. This winter rate is, however, atypical of the other lakes. In Heart Lake the average winter turnover time was 10,000 minutes (i.e., 7 days) versus a summer average of 7.3 minutes.

A number of studies like Rigler's have led to the generalization that the turnover time of dissolved and inorganic phosphate is characteristically a matter of minutes in freshwater lakes. Not so in marine water, however. Work by Pomeroy (1960), among others, indicates that summer turnover time is a matter of hours (1 to as many as 56) in marine stiuations (Table 8-3). Interestingly, the winter range of values (30–196 hours) is not especially different from the range observed in most of the lakes that Rigler studied (13–166 hours). Turnover time in terrestrial ecosystems may range up to 200 years.

Phosphorus: Nutrient at Large

Although little of the phosphorus in an ecosystem may be readily available in the inorganic state, the turnover of phosphorus may be—and usually appears to be—rapid

TABLE 8-3 TURNOVER TIME OF DISSOLVED PHOSPHATE IN MARINE WATERS AT ABOUT THE SAME LATITUDE AND LONGITUDE

Location	Time of year	Turnover time (hours)
Estuaries		
Salt-marsh creek, Georgia	July	1
Altamaha River, Georgia	May	1
Altamaha River, Georgia	November	13
Open water		
Continental shelf	July	5
Gulf Stream, surface	July	4
Gulf Stream, 50 m	July	12
Coastal seawater	April	4
Coastal seawater	October and November	46–155
Doboy Sound, Georgia	February	50
Doboy Sound, Georgia	June and July	37–56
Marsh		
Salt marsh, low tide	January and October	40–49
Salt marsh, high tide	November	169

Adapted by permission from L. Pomeroy. 1960. *Science* 161:1731-2. Copyright 1960 by the American Association for the Advancement of Science.

enough to maintain a constant supply sufficient to turn the ecosystem dynamo. Yet ecosystems have been subject increasingly to heavier loads of phosphates because of their use in fertilizers and, until recently, in laundry detergents. In the early 1970s in the United States alone, for example, 10^9 kg of phosphorus entered the nation's waters annually from phosphate-based detergents. This amount accounted for 30 to 40 percent of all phosphorus entering aquatic environments. Because of its usually limiting role in primary production, the overload of phosphorus results in algal-clogged ponds and streams. This situation is followed by severe oxygen depletion, the result of oxygen demands by decomposers acting on the large amount of dead algae. Oxygen depletion leads, in turn, to suffocation of organisms, including fish. Legislation restricting the amount of phosphorus in detergents, coupled with the rapidity with which phosphorus is partioned to sediments, has tended to stablize the amount of inorganic phosphate in aquatic habitats. This relatively rapid stabilization via sedimentation is of singular importance in maintaining homeostasis in ecosystems. As a case in point, Alfred Redfield, of the Woods Hole Oceanographic Institute, maintained (1958) that the nitrate of the sea and the oxygen in the atmosphere are controlled by the overall biogeochemical cycle. This cycle is, in turn, ultimately regulated by phosphorus, particularly under anaerobic conditions.

SUMMARY

VOCABULARY

ammoniafication
biogeochemical cycle
denitrification
free-living nitrogen fixation
gaseous nutrient cycle
mineralization
nitrification
nitrogen fixation
ozone
rainout
residence time
sedimentary cycle
symbiotic nitrogen fixation
turnover time
washout

KEY POINTS

- Biogeochemical cycles consist of those demonstrating a dominant atmospheric gaseous phase (e.g., carbon and nitrogen) and those having a dominant sedimentary phase (e.g., sulfur and phosphorus); most nutrients have sedimentary cycles.
- All biogeochemical cycles have large available pools, coupled with small, extremely labile pools, that are highly dependent on continual input.
- The carbon cycle is deceptively simple and would seem to be "perfect" in the sense that the incorporation of carbon dioxide from the atmosphere into the energetics of ecosystems is more or less balanced by its return in the form of respiration. This "perfect" state is, however, subject to considerable question.

- The amount of carbon in organic form is less than 0.1 percent of that occurring in the geological component from which it is eroded/leached; its residence time in the sediments is in the order of 10^8 years.
- The interchange of carbon dioxide between the atmosphere and water is largely regulated by the dynamics of the dissociation characteristics of carbonic acid.
- The level of carbon dioxide in the atmosphere has increased by some 20 percent since the beginning of the industrial revolution through the burning of fossil fuels; this has been augmented by changes in land use, including extensive clearcutting and conversion of forest to pasture and agricultural land. The impact of this increase is highly debated, some reporting enhanced photosynthesis and growth, others concerned about overheating of the Earth's atmosphere.
- In contrast to the relative simplicity of the carbon cycle, the nitrogen cycle involves four major steps (nitrogen fixation, ammoniafication, nitrification, and denitrification), each involving a number of different organisms, some free-living, some in symbiotic relationships, and each conditioned by other factors such as relative acidity and dependence on energy.
- Although there are numerous self-regulating, energy-dependent, feedback mechanisms in the nitrogen cycle, there are questions as to whether the global nitrogen cycle is balanced because of the excess load of nitrogen-compounds released to the air from the burning of fossil fuels since the beginning of the industrial revolution.
- In their cycles considerable amounts of both sulfur and phosphorus are lost to the sediments for extended periods of time owing to their being precipitated through binding with various cations such as iron, calcium, magnesium and aluminum. The global cycles of sulfur and phosphorus are not well understood.

CHAPTER 9

Nutrient Budgets and Ecosystems

For clarity, we have been concerned up to this point with the movement of individual nutrients, with major emphasis on the more global biological and chemical aspects of their biogeochemistry. Except in passing, no particular attention has been given to such geological processes in ecosystems as the input of nutrients by way of precipitation, dust, and weathering or output by way of runoff and erosion. Moreover, although a given nutrient may be limiting in a given case, a large number of chemicals is requisite to growth. So there is a need to broaden the scope from individual to aggregate cycles. Finally, because nutrients may get bound up in the biomass of an ecosystem for long periods of time, as in the trunks of standing trees in a forest, consideration must be given to the availability and source of nutrients in an ecosystem—that is, to the budget of nutrients.

In a sense, there are two nutrient budgets in an ecosystem (Figure 9-1). The internal nutrient budget is concerned with the circulation of nutrients through the biotic and abiotic compartments of a given ecosystem. It involves the input and output that occurs along the producer → consumer → decomposer food chain and the exchanges between reservoirs and sediments of that ecosystem. In contrast, the external nutrient budget pertains to the intake and output of the entire ecosystem in relation to other ecosystems. Geological actions, such as volcanic eruptions, spew materials into the atmosphere or spread lava over the terrain, thereby transferring nutrients from one place to another, sometimes thousands of miles distant. Meteorological actions, such as rock weathering or wind that carries nutrients whipped up into dust or evaporated into the atmosphere, cause exchanges of nutrients between ecosystems. Animals that feed in one

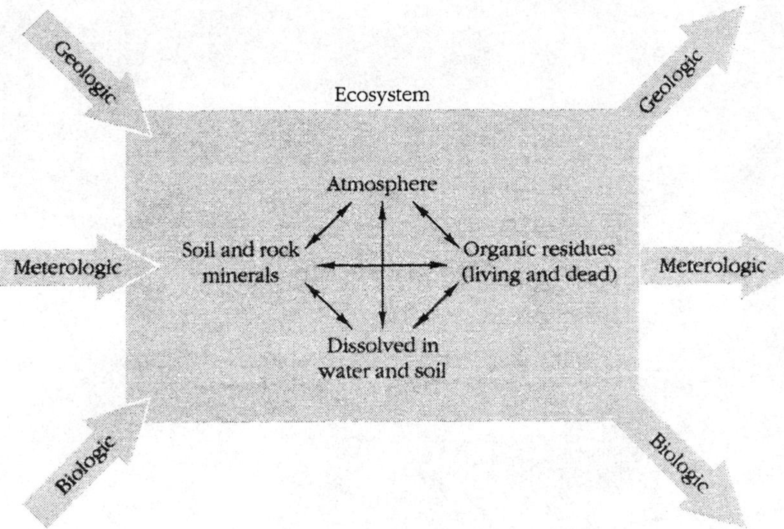

Figure 9-1 The internal and external nutrient budgets of ecosystems. In addition to the various internal nutrient interchanges among biotic and abiotic components of an ecosystem, exchanges occur externally among ecosystems through geological, meteorological, and biological processes. (Reproduced from E. J. Kormondy. 1974. In F. Sargent II (ed.) *Human Ecology*. Amsterdam: North Holland Publishing Co.)

ecosystem and defecate or die in another or trees grown in one ecosystem and burned in another are obvious examples of external nutrient exchange resulting from biological activity. With their tremendous capacity for movement of self and organic (e.g., food) and inorganic (e.g., fertilizers) nutrients, humans are without question the most powerful biological agents affecting internal and external nutrient budgets.

Obviously internal and external budgets are interrelated: internal cycles are dependent on inputs; outputs are mainly a function of internal processes. Well-worked-out nutrient budgets, either internal or external, are still relatively few in number, chiefly because of the technical and interpretive difficulties such processes entail.

INTERNAL NUTRIENT BUDGETS

When considering nutrient budgets, it is a major task to determine the mineral content of the biotic components of an ecosystem and to assess shifts in this content with time, much less trace the flow of that content through the food chain. Consequently, most investigators have chosen to work with the cycle of one or two nutrients. Even this kind of study is extremely demanding and fraught with technological as well as interpretational difficulties.

Forest Ecosystems

Cesium and white oak trees. Using the radioisotope (^{134}Cs), Witherspoon (1964) traced the movement of cesium in white oak *(Quercus alba)* trees at the Oak Ridge National Laboratory (Figure 9-2). Following the injection of 2 millicuries of cesium into the vascular tissues, Witherspoon found that, over two growing seasons, the maximum concentration in the leaves occurred in early June and amounted to about 40 percent of the total input. The remainder had spread to the woody tissues in the roots, stem, and branches. Of the total leaf content, 33 percent was lost through leaf fall, 15 percent was leached out by rain, and the remainder was incorporated in woody tissues. By November about 70 percent of the rain-leached cesium was in the top 10 cm of the soil, and 17 percent was added to the leaf litter, supplementing that which had come through leaf fall, for a grand total of about 19 percent of the original input.

Turnover and cycling time of calcium. The movement of cesium in white oak suggests that the turnover time differs for different parts of the tree. This is a generally recognized phenomenon that warrants additional discussion. Table 9-1

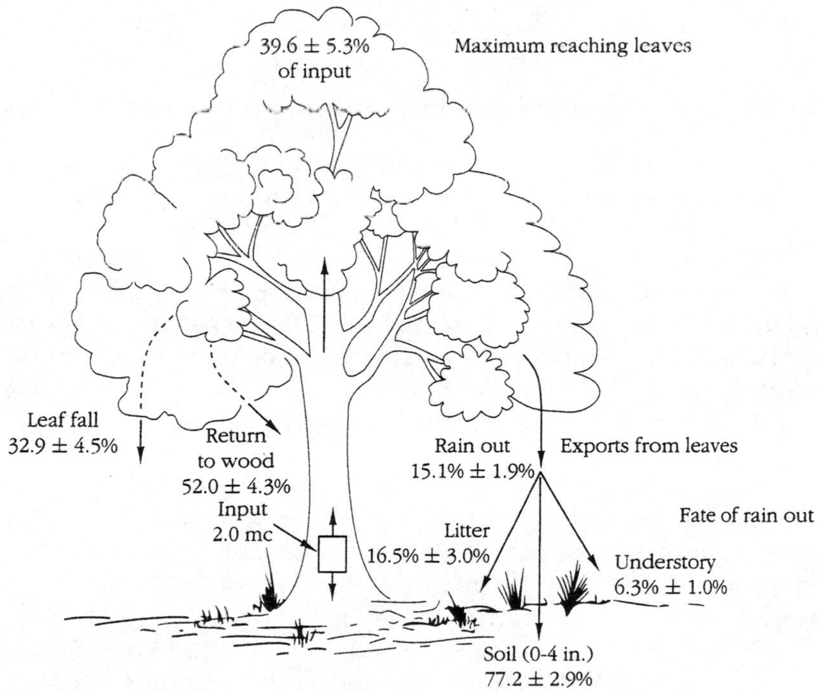

Figure 9-2 The cycle of ^{132}Cesium in white oak, an average based on 12 trees observed for two growing seasons. (Redrawn by permission from J. P. Witherspoon. 1964. *Ecological Monographs* 34:403–20.)

TABLE 9-1 TURNOVER AND CYCLING TIMES FOR CALCIUM IN DIFFERENT ECOSYSTEMS

	Latitude degrees north	Growing season (days)	Turnover time in years					
			Soil	Wood	Canopy	Litter	Total	Total less soil
Moist tropical forest (Ghana)	7	240	8.2	6.8	1.5	0.2	16.7	8.5
Montane tropical rainforest (Puerto Rico)	18	365	3.0	6.4	0.9	0.2	10.5	7.5
Northern hardwoods (New Hampshire)	44	120	14.0	10.8	0.8	34.8	60.4	46.4
Douglas fir (Washington State)	48	200	57.4	20.2	5.5	10.2	93.3	35.9
Mixed oak (Belgium)	51	200	108.8	12.6	0.4	0.9	122.7	13.9
Oak-ash (Belgium)	51	200	184.9	21.5	0.4	0.6	207.4	22.5
Scotch pine (England)	53	200	11.2	6.1	0.8	3.4	21.5	10.3
Spruce (Russia)	55	120	5.1	22.2	1.3	9.0	37.6	32.5
Spruce (Russia)	65	105	7.6	18.3	3.2	13.6	42.7	35.1

Reproduced with permission from "Mineral Cycling: Some Basic Concepts and Their Application in a Tropical Rain Forest" by C. F. Jordan and J. R. Kline. *Annual Review of Ecology and Systematics*, vol. 3.

shows the turnover and cycling time of calcium in four compartments of a series of ecosystems. The table is based on a simulation study, using actual data from other studies. With the exception of the two tropical forests, turnover time in the trees is shortest in the canopy (primarily leaves but also including flowers and fruits), next shortest in the litter, and in all cases is longest in wood. Cycle time tends to expand with increasing latitude. A trend to increased length of intratree cycling time with increasing latitude would be expected because uptake (and release) is directly related to the rate of primary production and that rate, as we have seen, decreases with increasing latitude (see Figure 6-5). Element availability patterns in the soil have a considerable effect on turnover time in the soil, as can be seen by the lack of a trend in soil turnover time.

Carbon and yellow poplar trees. The annual internal budget of carbon in a yellow poplar *(Liriodendron tulipfera)* forest, also in the Oak Ridge National Laboratory, is shown in Figure 9.3. According to W. F. Harris and his associates (1975) net photosynthesis involving the overstory (1,205) and understory (496) canopy and the herbaceous layer (55) amounts to 1,756 grams of carbon/m^2/yr. Annually 1,570 g/m^2 of carbon in the products of photosynthesis are transferred to branches-bole-stump (826) and to lateral roots (744). Total movement of carbon to the atmosphere in respiration from autotrophs is 1,436 g/m^2. Carbon in the litterfall amounts to 229 g/m^2. On the decomposer level, carbon moves from litter (11) and

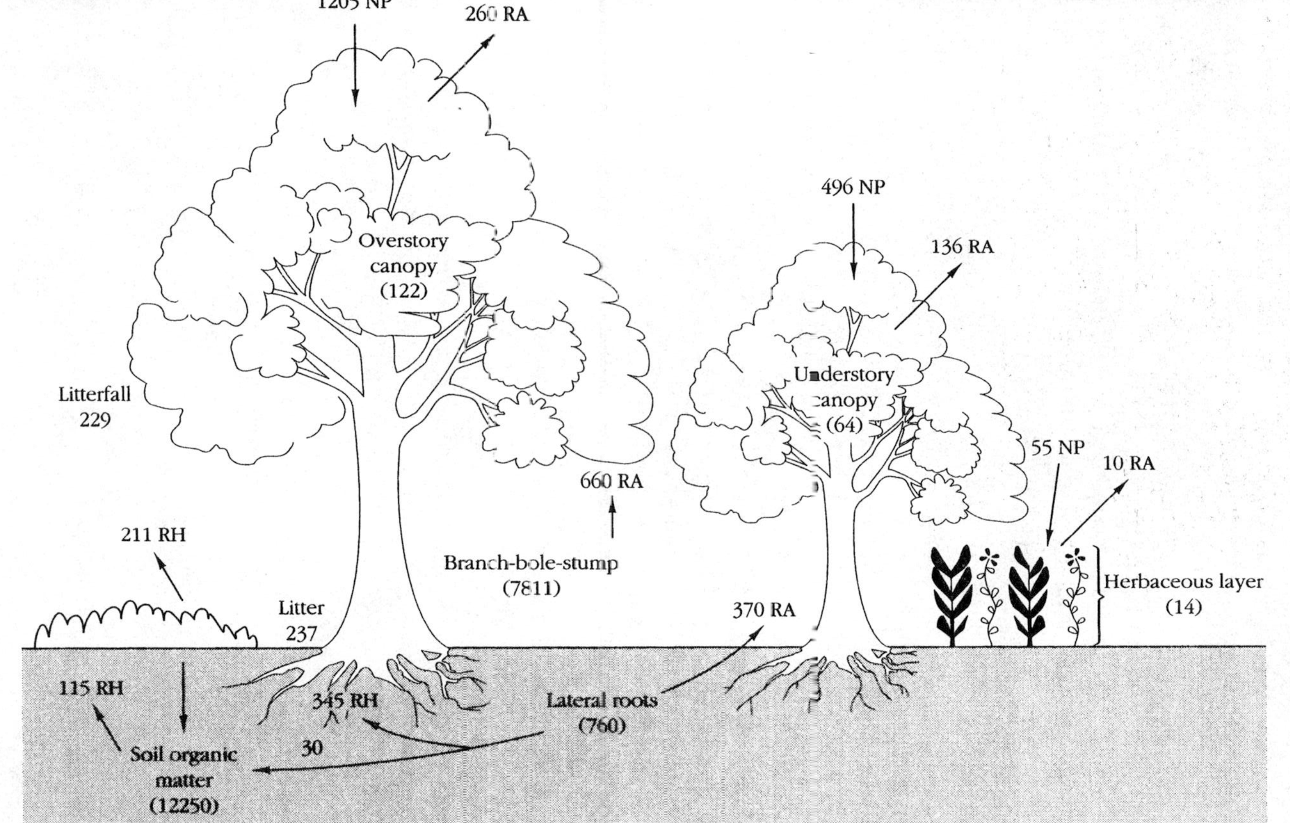

Figure 9-3 Internal budget of carbon in a temperate deciduous forest. The figures in parentheses represent the mean annual standing crop in grams of carbon/m². NP = net photosynthesis; RA = autotroph respiration; RH = heterotroph respiration. NP, RA, and RH are in grams of carbon/m². (Adapted from W. F. Harris et al. 1975. In *Productivity of World Ecosystems.* Washington, D.C.: National Academy of Sciences.)

lateral roots (30) into the soil; by way of respiration, carbon moves back to the atmosphere from the litter (211), lateral roots (345), and soil (115). Arthropods feeding on the canopy and preying on such herbivores constitute a relatively insignificant component of the carbon budget and are not indicated in the figure. The increase in stored carbon each year amounts to 191 g/m^2, 166 g/m^2 in the branch-bole-stump component, and 25 g/m^2 in the lateral roots.

As in the case of cesium and calcium, there are differences in the turnover rate and residence time of carbon in different parts of this yellow poplar forest ecosystem (Table 9-2). The short turnover time in the litter and lateral roots (the latter due to sloughing of the roots) contrasts with the lengthy time carbon remains in the aboveground wood (156 years) and in the soil (107 years). This comparatively long residence time in wood and soil indicates that forests may act as carbon sinks, thus buffering increments to atmospheric CO_2, a point that will be discussed further in Part VIII. The soil and litter residence time compares favorably with that of calcium in mixed oak and oak-ash forests in Belgium (see Table 9-5) but not so in the case of the woody component.

Six elements in Douglas fir forest. The annual internal cycles of six elements were studied in an old-growth Douglas fir *(Pseudotsuga menziesii)* forest in western Oregon by Phillip Sollins and his associates (1980), (Table 9-3). The dominant Douglas firs ranged in age from 350 to 550 years. Weathering of bedrock was the major source of all elements except phosphorus and nitrogen. Precipitation accounted for more of the input of phosphorus and biological fixation accounted for 40 percent (to as much as 70 percent) of the total input of nitrogen. Although accumulation in the biomass of the live vegetation was decreasing in this old-growth forest, there was a net accumulation in the detrital component (fallen logs and foliage, fine woody litter, and soil organic matter). Over 99 percent of the input of phosphorus, potassium, calcium, magnesium, and sodium was lost annually; there is also a 20 percent loss of nitrogen from the rooting zone.

TABLE 9-2 TURNOVER TIME FOR CARBON IN A YELLOW POPLAR FOREST

Compartment	Turnover time in years
Litterfall	1.1
Lateral roots	2.0
Wood, aboveground	156.0
Total wood	20.0
Soil organic matter	107.0
Forest ecosystem	10.0

Data from W. F. Harris et al. 1975. In *Productivity of World Ecosystems.* Washington, D.C.: National Academy of Sciences.

TABLE 9-3 INPUT, ACCUMULATION, LOSS, AND RATIOS OF TRANSFERS OF ELEMENTS IN AN OLD-GROWTH DOUGLAS FIR FOREST

	Nitrogen	Phosphorus	Potassium	Calcium	Magnesium	Sodium
Input						
Precipitation	2.0	0.3	0.9	3.6	1.2	5.7
Nitrogen fixation	2.8	—	—	—	—	—
Weathering	0.0	0.2	4.7	118.9	7.2	47.0
Unknown sources	1.7	0.0	3.4	0.0	0.0	0.0
Total input	6.5	0.5	9.0	122.5	8.4	52.7
Accumulation						
In vegetation	−2.8	−0.4	−1.0	−4.5	−0.6	−0.2
In detritus	7.8	0.2	0.5	3.9	0.4	0.1
Total loss	1.5	0.8	9.5	123.1	8.6	52.8
Ratios of transfers						
Uptake by vegetation/ total input	6.5	18.2	2.99	0.55	1.19	0.03
Uptake by vegetation/ total loss	28.0	13.0	2.83	0.54	1.16	0.03
Uptake from litter solution by vegetation/total uptake by vegetation	0.46	0.48	0.0	0.34	0.14	0.12
Redistribution to new foliage/total requirement for new foliage production	0.56	0.31	0.30	−1.8	−0.38	−2.2

Adapted by permission from P. Sollins et al. 1980. *Ecological Monographs* 50:261–85.

The cycling patterns of these elements, obtained by dividing uptake by the vegetation by the total input, and similarly dividing by total loss (Table 9-3), are of two types: a tightly closed cycle, indicated by large ratios, for nitrogen, phosphorus, and potassium; and open cycles for calcium, magnesium, and sodium, as indicated by small ratios. By dividing uptake from litter solution by total uptake of vegetation, it is evident that much nitrogen, phosphorus, and calcium are absorbed from litter solution, whereas calcium, magnesium, and sodium are obtained from soil solutions. Nitrogen and phosphorus are present in a form that makes them readily available for translocation. In contrast, the low calcium translocation is due to the calcium being locked into calcium pectate, the substance that glues cells together in plants. Finally, redistribution of elements from old foliage to new foliage is of importance for nitrogen (56 percent), phosphorus (31 percent), and potassium (30 percent) but not for calcium, magnesium, and sodium. These various sets of comparisons seem directly related to the fact that soils are rarely deficient in calcium, magnesium, and sodium whereas deficiencies of nitrogen, potassium, and phosphorus are common.

Six elements in a Scots pine forest. For another consideration of the internal nutrient budget of several minerals, Ovington's (1962) data on Scots pine plantation in England over a 55-year period are instructive (Table 9-4). In each instance, about half or more of the nutrient was taken up by trees rather than by ground flora. Also, there was a very substantial amount of nutrient in the litter and in unharvested tree roots and crowns. Significantly, most of this (95 to 100 percent) becomes available for recycling through decomposition. (Compare entries on total return to litter and total released by decomposition.) Nonetheless, uptake by the plants and release by decomposition are not in balance. More is required for growth and is tied up in organized form than is available through decomposition. This growth has occurred by a shift of part of the nutrient capital of the entire ecosystem from the abiotic to

TABLE 9-4 BUDGET OF SELECTED NUTRIENTS (in kg/ha) FOR PLANTATIONS OF *Pinus sylvestris* FOR 55 YEARS FROM THE TIME OF PLANTING

	Nitrogen	Calcium	Potassium	Magnesium	Phosphorus	Sodium
Trees						
Total uptake	4,817	2,272	1,933	431	413	132
Distribution						
In living trees	453	272	150	64	41	36
Harvested	161	210	98	42	14	5
In crowns and roots of harvested trees	704	386	279	80	75	48
Litterfall (leaves, branches, cones)	3,499	1,404	1,406	245	283	43
Ground flora						
Total uptake	2,058	771	876	169	182	72
Distribution						
Present in ground	40	19	22	5	3	4
Litterfall from ground flora	2,131	775	877	170	187	70
Total uptake (trees and ground flora)	6,875	3,043	2,809	600	595	204
Total return to litter and soil	6,334	2,565	2,562	495	545	161
Total released by decomposition	6,084	2,450	2,538	473	526	161
Net uptake from soil	791	593	271	127	73	43

Adapted by permission from J. D. Ovington. 1962. *Advances in Ecological Research*, vol. 1. New York: Academic Press.

the biotic component. (See entries on net uptake from soil.) Relative to total uptake, the net uptake from the soil in this ecosystem ranges from 9.6 percent for potassium to 21 percent for sodium.

Nutrient residence time in lodgepole pine ecosystems. In many cold-temperate and montane pine forests, slow decomposition and mineralization in the forest floor lead to a high accumulation of organic matter and nutrients whereas in warmer regions, forest floor accumulation is much slower. Lodgepole pine *(Pinus contorta* spp. *latifolia)* forests dominate the mountain landscape over much of the central and northern Rocky Mountains, a region characterized by long, cold winters and short, dry summers, conditions that lead to slow mineralization of organic matter and high forest floor accumulations. Fahey (1983) found that accumulation of biomass and nutrients in six stands of lodgepole pine in southeastern Wyoming was very high and that, in consequence, residence times for organic matter and nutrients were considerably longer than those that have been calculated for pine forests from warmer and/or moister climates (Table 9-5). The exception to this pattern, a pine forest in New Brunswick, Canada, is the shorter nutrient residence times and faster nutrient release from decomposing leaf litter. Those forests that exhibit high accumulation and slow release of nutrients such as nitrogen and phosphorus probably play a significant role in limiting primary productivity.

Nutrient budgets in two deciduous forests. A difference between uptake by plants and release by decomposition is to be anticipated, for the former occurs because of the growth process, which, in turn, results in the accumulation of matter as biomass. This growth capacity is sustained by internal relocation of nutrients within the ecosystem. A comparison of this difference on an annual basis in two deciduous forests growing on different soil types is shown in Table 9-6. The different

TABLE 9-5 FOREST FLOOR BIOMASS AND ELEMENT RESIDENCE TIMES IN PINE- DOMINATED ECOSYSTEMS OF THE WORLD

			Residence time (yr)					
Location	Forest type	Stand age (yr)	Dry mass	N	P	Ca	Mg	K
South Carolina, USA	*Pinus taeda*	17	4.3	9.9	9.5	3.3	2.4	3.3
North Carolina, USA	*Pinus taeda*	32	5.4	11.2	7.0	5.9	4.3	2.3
England	*Pinus sylvestris*	45	5.2	3.0	1.2	2.8	3.1	0.6
Minnesota, USA	*Pinus resinosa*	60	5.0	ND[a]	7.8	4.9	3.1	3.9
Northern Ontario, Canada	*Pinus banksiana*	30	ND	13.0	12.4	5.9	8.6	6.1
New Brunswick, Canada	*Pinus banksiana*	57	23.8	38.0	20.6	12.2	12.7	3.0
Wyoming, USA	*Pinus contorta*[b]		18.0	53.5	20.3	21.8	21.4	17.7

[a]ND = not determined.
[b]Mean of six stands ranging in age from 80 to 240 yr.
Modified with permission from T. J. Fahey. 1983. *Ecological Monographs* 53:51–72.

TABLE 9-6 ANNUAL NUTRIENT BUDGET (in kg/ha) IN TWO DECIDUOUS FORESTS IN BELGIUM

	Oak-Ash forest (115–160 years)					Mixed Oak-Beech-Hornbeam forest (30–75 years)				
	N	Ca	K	Mg	P	N	Ca	K	Mg	P
Total nutrient in vegetation	1,260	1,648	624	156	95	533	1,248	342	102	44
Total annual uptake	123	129	99	24	9.4	92	201	67	19	6.9
Total annual return to litter and soil	79	87	78	19	5.4	62	127	53	13	4.7
Net annual uptake from soil	44	42	21	5	4.0	30	74	14	6	2.2

Derived from data in P. Dugivneaud and S. Denaeyer-DeSmet. 1969. In *Analysis of Temperate Forest Ecosystems*, ed. D. E. Reichle. New York: Springer Verlag.

flux rates in the two forests is a function of age, species composition, and nutrient availability and requirements, in spite of the fact that the annual production in the two forests was the same. That different ecosystems accumulate different amounts of nutrients would be expected because of some of the same phenomena. Although they are not precisely the same age, total nutrient content of the mixed oak-beech-hornbeam forest (first horizontal entry of Table 9-6) and that of the 55-year-old mature pine forest (last horizontal entry of Table 9-7) show that the general proportion of the different nutrients is similar except for calcium, which is concentrated to a greater degree in the deciduous forest.

TABLE 9-7 NUTRIENT CONTENTS AND ACCUMULATION RATES FOR BALSAM FIR (*Abies balsamea*) FORESTS OF VARIOUS AGES

	Age of stand (years)						
	0	10	20	30	40	50	60
Nutrient content (kg/ha)							
Nitrogen	0	28	369	393	427	460	494
Potassium	0	88	121	134	158	174	188
Magnesium	0	18	30	33	40	44	47
Calcium	0	117	174	210	27	294	313

Nutrient accumulation rate (kg/ha/yr)						
Years	0–10	10–20	20–30	30–40	40–50	50–60
Nitrogen	28.8	8.1	2.4	3.4	3.3	3.3
Potassium	8.8	3.3	1.3	2.4	1.6	1.5
Magnesium	1.8	1.2	0.4	0.6	0.4	0.4
Calcium	12.0	6.0	4.0	7.0	2.0	2.0

Adapted with permission from D. G. Sprugel. 1984. *Ecological Monographs* 54:165–86.

Nutrient cycling changes in balsam fir forests during succession. Owing to the phenomenon of wave regeneration in high-altitude balsam fir (*Abies balsamea*) forests of the northeastern United States, a gradient of fir stands of different ages is created, making an ideal situation for the study of successional changes in a variety of ecological events. Sprugel (1984) examined the nutrient accumulation and storage of nitrogen, potassium, magnesium, and calcium in such fir stands ranging in age from 10 to 60 years of age on Whiteface Mountain, New York. As shown in Table 9-7, all four of these nutrients accumulate rapidly in the early stages of development, although the greatest difference between uptake and release is greatest later in stand development. Thus, as Sprugle notes, if ecosystem nutrient losses are controlled by nutrient accumulation in living biomass, the period of maximum nutrient conservation occurs in the years immediately after disturbance, when the stand is accumulating nutrients most rapidly.

Nutrient cycling changes in slash pine plantations during succession. Observations comparable to those seen in balsam fir forests undergoing succession were made by Gholz et al. (1985) in plantations of slash pine (*Pinus elliottii* var. *elliottii*) of different ages in northern Florida. Plant nutrient uptake was already high in 2-year-old plantations because of rapid colonization by grasses. Thereafter, uptake of nitrogen, phosphorus, potassium, and magnesium increased about 50 percent to a maximum at 8 years; calcium uptake increased 130 percent during the same time period. After 8 years, uptake declined for all elements with nutrient return in litter-fall remaining relatively constant after 5 years.

Nutrient cycling in coniferous forests. In comparing the internal cycling of nitrogen in a 10-year-old conifer plantation and a mature conifer forest, Davidson et al. (1992) found that although net ammoniafication (mineralization) rates were higher in the young forest than in the mature one, the reverse was the case for gross ammoniafication, which was two to three times higher. This indicates a more rapid turnover of nitrogen in the mature forest. The mature forest had a smaller nitrate pool and a lower net nitrification rate suggesting that it is a non-nitrifying system and that nitrate is important only in the nitrogen cycle of the young plantation; however, gross nitrification rates were similar in both forests. Finally, microbial assimilation of nitrate was significant in both forests indicating a rapid turnover of the small but significantly important nitrate pool, serving as an important pathway to its retention.

Nutrient cycling in moist tropical forests. In their review of moist tropical forests, Vitousek and Sanford (1986) note that patterns of nutrient cycling, and hence budgets, are diverse, differing in forests on different soils. Nonetheless, these variations follow coherent, explicable patterns and are influenced by climate, species composition, successional status (time since disturbance), and soil fertility. Their review was confined to forests between the Tropics of Cancer and Capricorn with an average annual precipitation in excess of 1500 mm and a dry season (less than 100 mm/month) of 4 months or less.

Soils of moderate to very low fertility (oxisols and ultisols) constitute 63 percent of the 106 hectares occupied by moist tropical forests; moderately fertile soils (alfisols, mollisols, vertisols, and others) account for 15 percent of the area; soils of low to very low fertility (spodosols, histosols, and others) constitute the remainder. In their analysis, Vitousek and Sanford (1986) clustered the soils into four major categories: infertile oxisols/ultisols, alfisols and other moderately fertile soils, spodosols and other very low fertile soils, and montane.

Analysis of data from a large number of different investigations showed that biomass does not differ strongly among forests with different soils except that it is lower in the more infertile white-sand soils (Table 9-8). They concluded that an association between soil fertility and aboveground biomass was unlikely in any but the most extreme cases.

Nutrient concentrations in leaves can serve as an index of the influence of soil fertility since all leaves have the same basic function and all use the same nutrients in photosynthesis and construction of organic materials. Table 9-9 shows that concentrations of all the major nutrients in leaves are elevated on the more fertile tropical soils and are intermediate on the infertile oxisols/ultisols.

As in other ecosystems, nutrients in tropical forests are added by precipitation, mineral weathering, and gas absorption (including nitrogen fixation). They are trans-

TABLE 9-8 BIOMASS AND NUTRIENT CONTENT IN MOIST TROPICAL FORESTS

Soil type	Biomass (T/ha)	Nutrient (kg/ha)				
		N	P	K	Ca	Mg
Moderately fertile soils	317	1,833	1,87	1,864	3,217	344
Infertile oxisols/ultisols	381	1,331	59	393	845	362
Spodosols	102	300	31	294	277	76
Montane soils	263	671	39	664	828	193

Derived from data in P. M. Vitousek and R. L. Sanford, Jr. 1986. *Annual Review of Ecology and Systematics* 17:137–67.

TABLE 9-9 NUTRIENT CONCENTRATIONS IN LEAVES IN MOIST TROPICAL FORESTS

Soil type	Nutrients (%)				
	N	P	K	Ca	Mg
Moderately fertile soils	2.40	0.14	1.50	1.61	0.48
Infertile oxisols/ultisols	1.71	0.06	0.47	0.31	0.18
Spodosols	1.08	0.06	0.60	0.60	0.23
Montane soils	1.25	0.07	0.63	0.82	0.26

Derived from data in P. M. Vitousek and R. L. Sanford, Jr. 1986. *Annual Review of Ecology and Systematics* 17:137–67.

ferred from plants to soil through litterfall and *throughfall* (nutrients that are added to precipitation as it passes through the canopy) and become subject to recycling through decomposition of the surface litter and organic matter in the soil. Nutrient losses occur through leaching and volatization.

Because of both the small numbers of studies to date and the use of different techniques, Vitousek and Sanford were unable to detect any useful patterns of nutrient output budgets. In general, however, nutrient input through precipitation is higher in more seasonal tropical forest sites. Also, chemical weathering is rapid in humid tropical climates, and weatherable materials are lost early during soil development. Since many tropical soils are old, it is not surprising that many are now poor in nutrients.

This review by Vitousek and Sanford, as well as other studies, indicates that moderately fertile tropical soils support productive forests that cycle large quantities of nutrient elements. The widespread infertile oxisols/ultisols cycle smaller quantities of phosphorus and calcium although they are quite rich in nitrogen. Forests on sandy spodosol soils cycle small quantities of nitrogen, and montane tropical forests are, in general, low in nitrogen, the upper montane forests cycling substantially less nitrogen than lower montane forests.

Nutrient cycles and soil age. Taking advantage of a well-established chronological sequence of soils of the progressively older islands of Hawaii, the youngest being the farthest east (Hawaii) and the oldest the farthest west (Kauai), Riley and Vitousek (1995) have provided a dramatic example of the increase in nitrogen pools and nitrogen cycling with increasing soil age. The age range of the basaltic soils was from 200 years on the Kilauea site on the island of Hawaii to 4.5×10^6 years at the Kauai site, with soils on the other island sites being 6,000, 185,000, and 1.65×10^6 years old. At the beginning of soil development nitrogen is virtually absent from these basaltic, and most other, parent soil materials; subsequent nitrogen fixation leads to an accumulation in terrestrial ecosystems, and its biological availability increases concurrently.

Grasslands and Soils

Ecosystem stability in relation to nutrient cycling involves different sorts of pools: those that are small, turning over rapidly, and principally microbe-fauna mediated; and others that are very large, nonlabile, and turning over slowly, in a few to several centuries. In soils, the bulk of the nutrients are present in large inactive pools as compared to the small pools of very biologically active components. Small pools of rapidly cycling components may account for a major proportion of the total available nitrogen and phosphorus taken up by plants in terrestrial ecosystems (Coleman et al. 1983).

The principal biological activities in nutrient cycling in the soil are immobilization (e.g., storage in the flora and fauna) and mineralization by microbes, as we have seen in Chapter 8. However, each nutrient has specific flows and storage compartments; what applies to one nutrient, carbon for example, is not directly transferable to nitrogen or to sulfur or to phosphorus. Quantification of these activities in terrestrial ecosystems is more difficult than in aquatic systems because of greater heterogeneity and because relatively lower proportions of nitrogen, phosphorus, and sulfur are biologically active. Further, the ratios of biological activity of elements in terrestrial systems may or may not approximate the elemental ratios established in aquatic systems described by Redfield (1958). Finally, seasonal measurements of levels of nutrients in various soil pools are not reliable indices of biological activity; changes in levels of labile inorganic phosphorus, usually used to assess biological availability, seriously underestimates amounts of phosphorus biological active in grassland soils (Cole et al. 1977).

Nitrogen cycle in tallgrass prairie. The nitrogen cycle in tallgrass prairie is impacted by the seasonal dynamics of nitrogen in the various ecosystem components and must be evaluated as the grassland responds to manmade or natural events, such as grazing, drought, fertilization, and fire. Based on studies, including a simulation model, on a tallgrass prairie in Osage County in northeastern Oklahoma, Risser and Parton (1982) showed that grazing, fertilization, and irrigation increased nitrogen-cycling rates. By contrast, annual spring fires reduced plant uptake and mineralization of nitrogen from plant parts, while mineralization of nitrogen from organic matter increased.

The increase in nitrogen cycling during grazing is the result of increased mineralization of nitrogen from dead plant material, which, in turn, is the result of higher decomposition rates and an increased mineralization of nitrogen in the plant material consumed by cattle and returned as urine and feces. The addition of water resulted in increased nitrogen uptake by roots and transfer to litter as well as shoot death. And, as might be expected, the addition of nitrogen-containing fertilizers made more nitrogen available for uptake. The reduction in nitrogen cycling following annual fires is tied to a reduction by both root and shoot uptake flows; also, nitrogen transfer from the plant canopy to the surface litter, root death, mineralization of nitrogen from the plant material, and the formation of soil-organic nitrogen were all factors in decreasing the nitrogen flow. By contrast, fire hastened mineralization of nitrogen from the soil-organic pool and volatization of nitrogen was also increased.

On a seasonal basis, Risser and Parton (1982) showed that the total nitrogen content in live shoots increased to a peak in the middle of the growing season (July and August), but nitrogen concentration in these live shoots was highest in May, decreasing throughout the season to a minimum in September. This decrease in concentration is caused by dilution of nitrogen in plants as a result of plant growth. By contrast, the nitrogen concentration of standing dead material showed no seasonal patterns and the higher nitrogen content was the result of a greater amount of standing dead biomass. Although nitrogen content and concentration were variable in litter and root, sharp seasonal patterns were not observed.

Aquatic Ecosystems

Thoreau's Pond. Well known in literary history based on the writings of New England's curmudgeon, Henry David Thoreau (1906), Thoreau's Pond (actually a bog), or its less well-known name Gowing's Swamp, occupies an equally intriguing place in ecology because of its unusual nutrient cycling properties. The "pond," located in Concord, Massachusetts, is 40 meters above mean sea level and has a total area of 0.38 hectares. Structurally, it is a classic floating-mass *Sphagnum* bog with a restricted flora of plant species normally found far to the north. Typical of peat bogs, the substrate is highly acidic and nutrient poor. It is also characterized by receiving its water from precipitation rather than groundwater and by a very high rate of evapotranspiration, the latter playing a dominant role in determining concentrations of relatively inert chemicals.

According to Hemond (1980), the bog accumulates metal from precipitation primarily by ion exchange and, in the case of potassium, by active uptake by *Sphagnum*. Ion exchange increases the mineral acidity of the water as does acid rain, but these potentials are counteracted by the reduction of sulfate and the biological uptake of nitrate. The affinity of peat and *Sphagnum* for metals varies widely from metal to metal. Lead, for example, is relatively immobile and thus results in the peat profile containing an historical record of metal inputs and in its isotope form (^{210}Pb) serving as a basis for estimating the age of the bog, namely about 500 years.

Based on limited seasonal sampling with estimated corrections for the same, the input of potassium, magnesium, and lead are respectively 146, 132, and 54 mg/m^2/year; the outputs are respectively 110, 72, and 1 mg/m^2/year, resulting in net accumulations, respectively, of 36, 60, and 51 mg/m^2/year. In contrast to many other ecosystems discussed in this chapter, wherein either larger net outputs or essentially balanced input/outputs are characteristic, this *Sphagnum* bog retains and stores many elements. Further, its homeostatic mechanisms that counter acid rain also set it apart from many other ecosystems.

In a later study on Thoreau's Pond, Hemond (1983) demonstrated that it also stores nearly 80 percent of the annual nitrogen input, again setting this kind of ecosystem in a category by itself.

Carbon in a swamp-stream. Creeping Swamp is an undisturbed swamp-stream in the Coastal Plain of North Carolina in which J. P. Mulholland (1981) followed the cycling of organic carbon. In contrast to upland streams and typical of lowland streams, Creeping Swamp has poorly defined channels with low water velocities and, consequently, little erosive power. As discharge rates increase, these kinds of stream waters expand laterally, maintaining shallow depths and low velocities. Further, flow pathways are often tortuous and beset with trees, logs, and other obstructions that result in greater water retention times.

According to Mulholland, nearly all of the organic carbon in this 8 km section of Creeping Swamp is *allochthonous*, that is, external in origin (this in contrast to *autochthonous*, or internal in origin). Nearly 98 percent of the allochthonous organic carbon is dominated by leaf litter inputs (60.2 percent) and dissolved organic carbon from upstream, including inflow from a tributary (33.1 percent) (Figure 9-4). During February and March, filamentous diatoms and green algae, along with diatoms, mosses, liverworts, and submerged vascular plants, made a small contribution to the annual carbon input (2.4 percent). The annual output through fluvial processes, 95 percent of the carbon being in the form of dissolved organic carbon, constituted 37 percent of the total output; the remaining 63 percent was retained within the ecosystem, 56.5 percent being consumed in respiration, and storage accounting for the remaining 6.5 percent.

Nitrogen dynamics during succession in a desert stream. In a study of seven successional sites in Sycamore Creek, Arizona, a lowland Sonoran Desert stream, Grimm (1987) showed that total nitrogen storage (3–9 gm/m^2) was lower than that in forest streams of Oregon (12 g/m^2) and Quebec (22 g/m^2). Further, unlike the Oregon and Quebec streams in which nearly all of the nitrogen is in the form of allochthonous detritus entering the stream, benthic algae and autochthonous detritus account for about 90 percent of the nitrogen in this desert stream. The total input of nitrogen from internal and external sources exceeded the total output except on one study date, the retention largely being inorganic nitrogen due to autotrophic assimilation.

Grimm used the nitrogen data to evaluate a model of nitrogen retention in succession advanced earlier by Vitousek and Reiners (1975), namely that patterns of nitrogen retention during succession reflect patterns of net ecosystem production and biomass accumulation. Specifically, the model proposes that ecosystem nitrogen retention will increase as rates of net ecosystem production and standing stocks increase, then decline as biomass approaches a steady state. As implied by the notion of a steady state, nutrient input would be balanced by output, although changes in the form of the nutrient may occur. Grimm concluded that the desert stream data supported this nitrogen retention model.

General comments on nutrient cycling in stream ecosystems. Prior to discussing internal budgets of other types of ecosystems, it will be helpful to generalize the salient aspects of nutrient cycling in stream ecosystems. Meyer et al. (1988) provided an excellent summary in noting the following:

1. as in all ecosystems, nutrients regulate ecological processes such as decomposition and primary production, particularly if the nutrient is limiting;
2. nutrients link terrestrial to aquatic ecosystems through runoff and other processes;
3. stream processes such as nutrient uptake and release by biota and denitrification can alter the timing, magnitude and form of element fluxes to downstream ecosystems and thus determine nutrient availability to biological communities downstream.

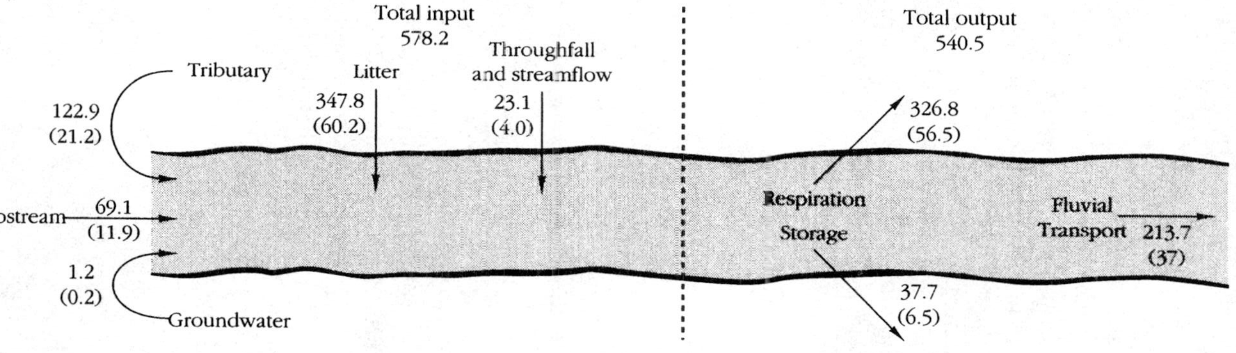

Figure 9-4 Annual organic carbon flow in a 8 km section of Creeping Swamp, North Carolina, in g/m^2 with percentages with parentheses. (Based on data in P. J. Mullholland. 1981. *Ecological Monographs* 51:307–22.)

Nitrogen cycle in tidal freshwater marsh. Plant production is high in tidal freshwater wetlands, as it is in inland freshwater marshes and salt marshes at similar latitudes. This high plant production promotes the formation of nitrogen-rich, sedimented organic soils in marshes and causes a high demand for nitrogen for plant tissues. In a study of tidal fresh-to-brackish water marshes on the North River in Massachusetts, Bowden (1984) demonstrated that gross production of ammonium-nitrogen in the sediments could supply all of the nitrogen required to support annual plant production (22.3 g/m^2/yr) and microbial decomposition (29.5 g/m^2/yr). In a follow-up study, Bowden (1986) described how the microbial transformations of ammonium and nitrate in the sediments involving nitrification and nitrate reduction, and nitrogen immobilization in sediments and litter, regulate the nitrogen cycle. Under favorable conditions, mineralization of sedimented organic nitrogen may supply a major portion of the annual plant requirements for nitrogen and thus render the plant community less dependent on nitrogen imported from the river into the marsh.

Computer modeling in salt marsh sediments. As might be expected, computer simulations of internal (as well as external) nutrient budgets are of increasing importance in contemporary ecological research. Gardner (1990) developed a computer model for simulating vertical profiles of the concentration of organic matter, pyrite (FeS_2), dissolved oxygen, and carbon in the sediments of a *Spartina* marsh. The model was developed from and tested against known values based on a number of different studies.

In the model, organic matter enters the sediment in three ways: by sedimentation, belowground production of roots, and chemoautotrophic fixation of CO_2 associated with the oxidation of pyrite. Perturbation of the sediment by the burrowing activity of fiddler crabs results in exchanges of organic matter, carbon, pyrite, and dissolved oxygen between the sediments and the surface environment. The model suggests further that accumulation of pyrite and organic matter in the marsh sediments is regulated by the turnover time of roots and by the mean diameter of roots. Intense aeration of the sediment, caused by diffusion of oxygen assumed to be present in roots, prevents the buildup of pyrite. The model suggests a complex series of interacting regulatory processes that are made even more complex by variations in tides, as well as diurnal and seasonal environmental perturbations, none of which are accounted for in the model.

Agroecosystems

Agroecosystems are ideal for nutrient budget studies because they have a simple vegetational structure, well-defined boundaries, and the inputs and outputs of nutrients can be easily manipulated (Odum 1984).

Nitrogen cycling in an agroecosystem. In a study of nitrogen cycling in a Georgia piedmont agroecosystem near Athens, Georgia, Groffman et al. (1986) found that the addition of nitrogen fertilizer adversely affected the production of grain sorghum (*Sorghum bicolor*) by promoting the production of competing weeds, most notably *Amaranthus*, which accumulates large amounts of nitrate and, in these fertilized plots, exploited the available nitrate. At the end of the four-month growing season (June–September), weed biomass in the fertlized plots was in excess of 9,000 kg/ha and sorghum biomass about half that. By contrast, in unfertilized plots sorghum outcompeted *Cassia obtusifolia*, the dominant weed, and the resulting biomass was essentially the reverse found in fertilized plots: sorghum biomass was more than 9,000 kg/ha and weed biomass about half that. *Cassia obtusifolia* is a common pest in soybean agroecosystems, which generally receive no nitrogen fertilizer in southereastern USA.

The results of this study suggest that when nitrogen fertilizer inputs are used without other inputs (i.e., weed control) they are likely to be poorly used by the agroecosystem.

EXTERNAL NUTRIENT BUDGETS

Because of the magnitude and technical difficulties involved in their study, comprehensive external nutrient budgets are yet relatively uncommon. Early work tended to emphasize nutrient inputs and outputs or process mechanisms in which a particular nutrient flux such as throughfall or litterfall was characterized in detail. Currently, measurement of water and nutrient fluxes is a major focus of ecosystem ecology. Synthesis of input-output studies with measurement of internal fluxes and cycling mechanisms provides the opportunity to examine fundamental processes that control nutrient fluxes through an ecosystem or an aggregate of ecosystems. For a comprehensive review of fluxes and cycling mechanisms in different ecosystems, see Thurman (1985).

Importance of Throughfall in Forest Systems

As might be expected, precipitation accounts for the largest nutrient input to most ecosystems. *Incipient precipitation* includes water falling vertically on a forest as well as mist, fog, and cloud water, which is laterally intercepted by the forest canopy. As noted earlier, precipitation that passes through the canopy and falls to the ground is referred to as throughfall; that which runs down the branches and trunk to the base of the tree is called *stemflow.* Some precipitation does not reach the forest floor because some stemflow never reaches the ground, some water is evaporated from the canopy during precipitation, and some is held by the canopy and evaporated afterwards.

In an extensive review, Parker (1983) concluded that throughfall and stemflow are major pathways in nutrient recycling in forests. Annual nutrient return to the forest soil for potassium, sodium and sulfur is predominately through these processes, and little is due to litterfall. For example, the throughfall flux of potassium ranges as much as 11 times higher than that in incident precipitation. For other critical nutrients, the flux, based on the average of a large number of studies, exceeds that in incident precipitation as follows: potassium, 7.2 times; chlorine, 4.1; sodium, 3.9; magnesium, 3.5; calcium, 3.2; sulfate sulfur, 2.7; phosphorus, 2.6; ammonium nitrogen, 2.0; and nitrate nitrogen, 1.5.

Importantly, as Parker notes, these nutrients are immediately available for reabsorption unlike those in the litter that must undergo decomposition. In throughfall, leaching makes the major contribution to nutrient enhancement accompanied by dry deposition of nutrients (e.g., in dust) between storms, exudates from leaf surfaces, and decomposition of dead leaves before they fall. Stemflow contributes 5 to 20 percent of the nutrient deposition and is the major mineral flux to a narrow area about the fole.

Importance of Allochthonous Inputs

Small streams draining heavily forested watersheds depend on allochthonous organic inputs from the adjacent terrestrial environment as a source of carbon and nutrients for their biological processes. A study by Triska et al. (1984) is instructive on this point. The study pertained to the nitrogen budget of a small stream in a 10-hectare watershed in a 450-year-old Douglas fir (*Pseudotsuga menziesii*) and western hemlock (*Tsuga heterophylla*) forest in the western Cascades of Oregon in 1974–75.

As shown in Figure 9-5, of the total annual input of nitrogen in the stream (15.25 $g/m^2/yr$), only 5 percent (0.76 $g/m^2/yr$) was due to biological fixation within the stream. Seventy-three percent of the nitrogen (11.06 $g/m^2/yr$) entered in the form of dissolved organic nitrogen and nitrate and twenty-two percent in the form of litterfall, lateral movement, and throughfall (3.43 $g/m^2/yr$). The total annual nitrogen input exceeded the output (11.36 $g/m^2/yr$) by 3.89 $g/m^2/yr$, indicating the stream was not operating on a steady state, at least for that year. Although dissolved organic nitrogen constituted the greatest form of input (10.56 $g/m^2/yr$) and output (8.38 $g/m^2/yr$), the largest pool of nitrogen within the stream was in the form of particulate organic matter (11.93 $g/m^2/yr$).

Estuaries and coastal marine ecosystems receive large allochthonous inputs of nutrients, organic carbon as well as sediments from non-point-source runoff from terrestrial ecosystems. Such runoff accounts for more than one-half the total nitrogen inputs to many major estuaries such as Delaware Bay, Narragansett Bay, and Chesapeake Bay (Nixon and Pilson 1983). In the tidal, freshwater Hudson River estuary (Figure 9-6), such inputs are the major sources of organic carbon, driving ecosystem metabolism, and thus strongly influencing dissolved oxygen concentrations (Howarth et al. 1991). Although the Hudson River estuary is dominated by forests, sediment and organic carbon come overwhelmingly from urban and suburban areas and from agricultural fields. Changes in land use within the basin will doubtless change the allochthonous inputs into the estuary, thereby altering its metabolism.

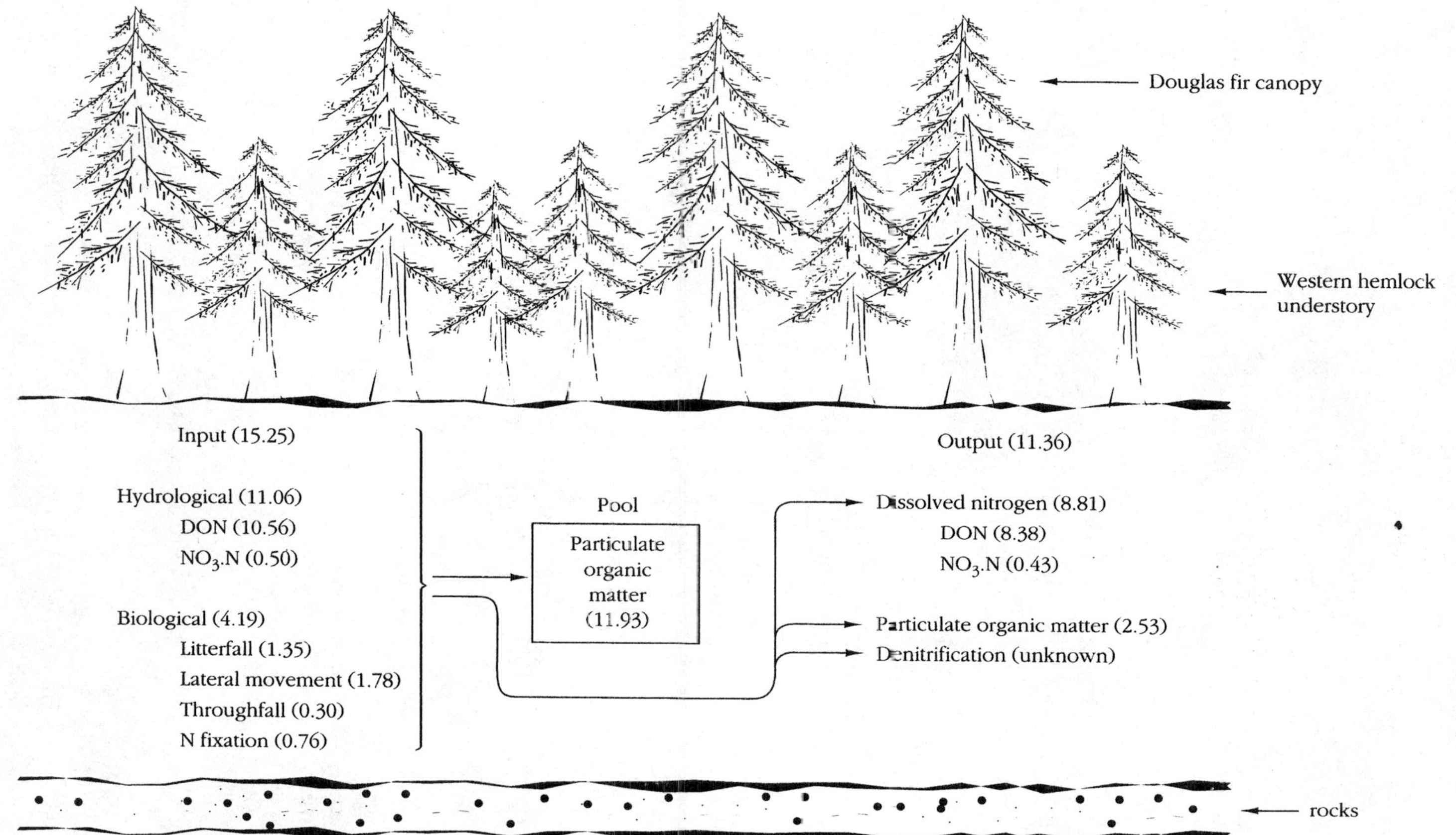

Figure 9-5 Annual nitrogen budget (in g/m²/yr) in a small stream in the western Cascades in Oregon. DON = dissolved organic nitrogen. (Based on data in Triska et al. 1984. *Ecological Monographs* 54:119–40.)

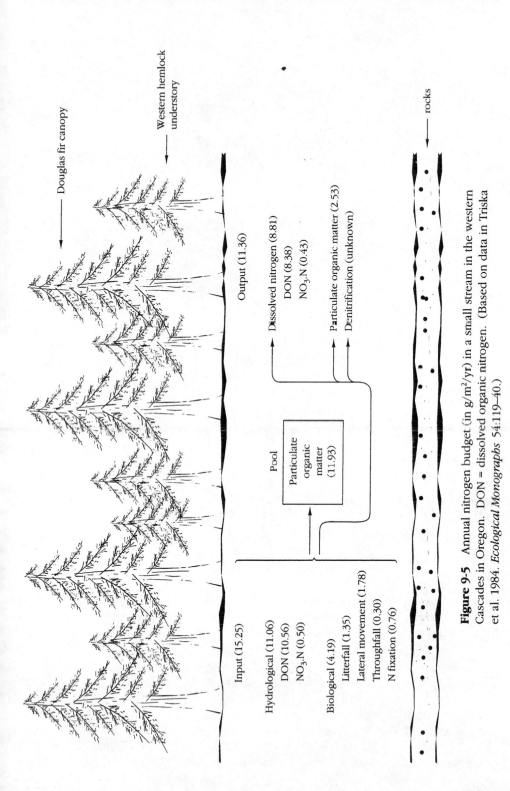

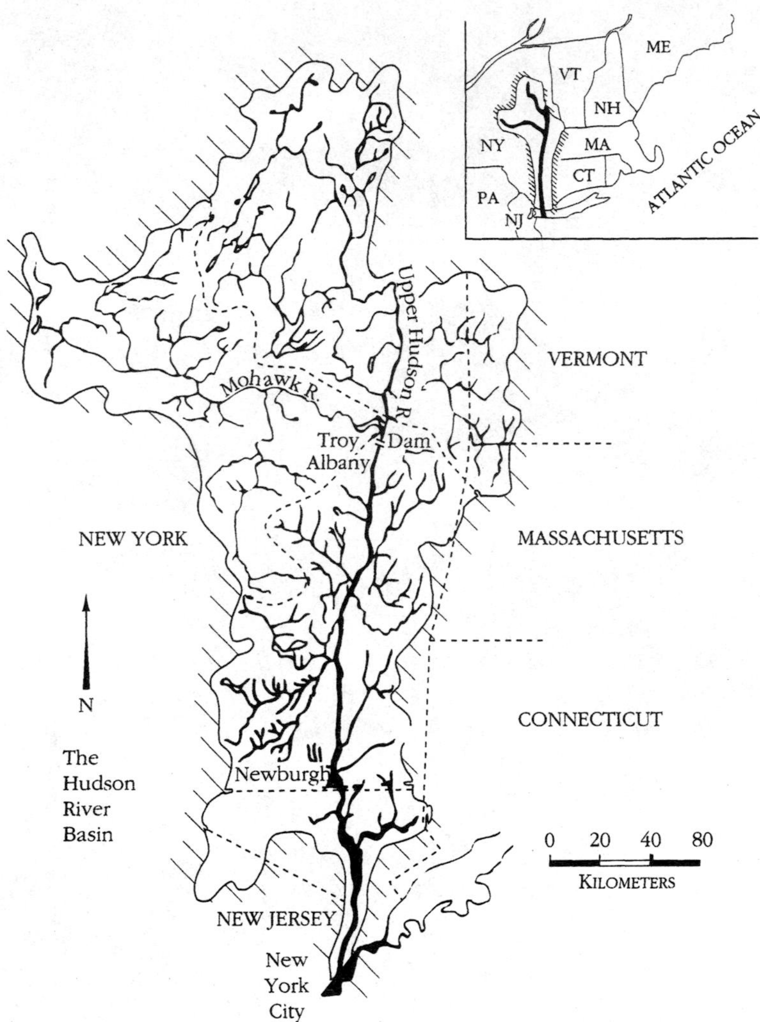

Figure 9-6 The Hudson River watershed, composed of the upper Hudson River basin, the Mohawk River basin, and the Lower Hudson River basin. (Redrawn by permission from R. W. Howarth et al. 1991. *Ecological Applications* 1:27–39.)

Standing Crop and Flux of Dissolved Organic Carbon

The concentration, composition, and flux of dissolved organic carbon (DOC) were measured by McDowell and Likens (1988) in Bear Brook, a stream within the Hubbard Brook Valley, New Hampshire (Figure 9-7), over a period of 7 years. Although the data were not all collected simultaneously nor with the same methodology, the synthesis of the available data does provide some valuable perspectives on the changes in DOC flux and concentration and the mechanisms involved in regulating them.

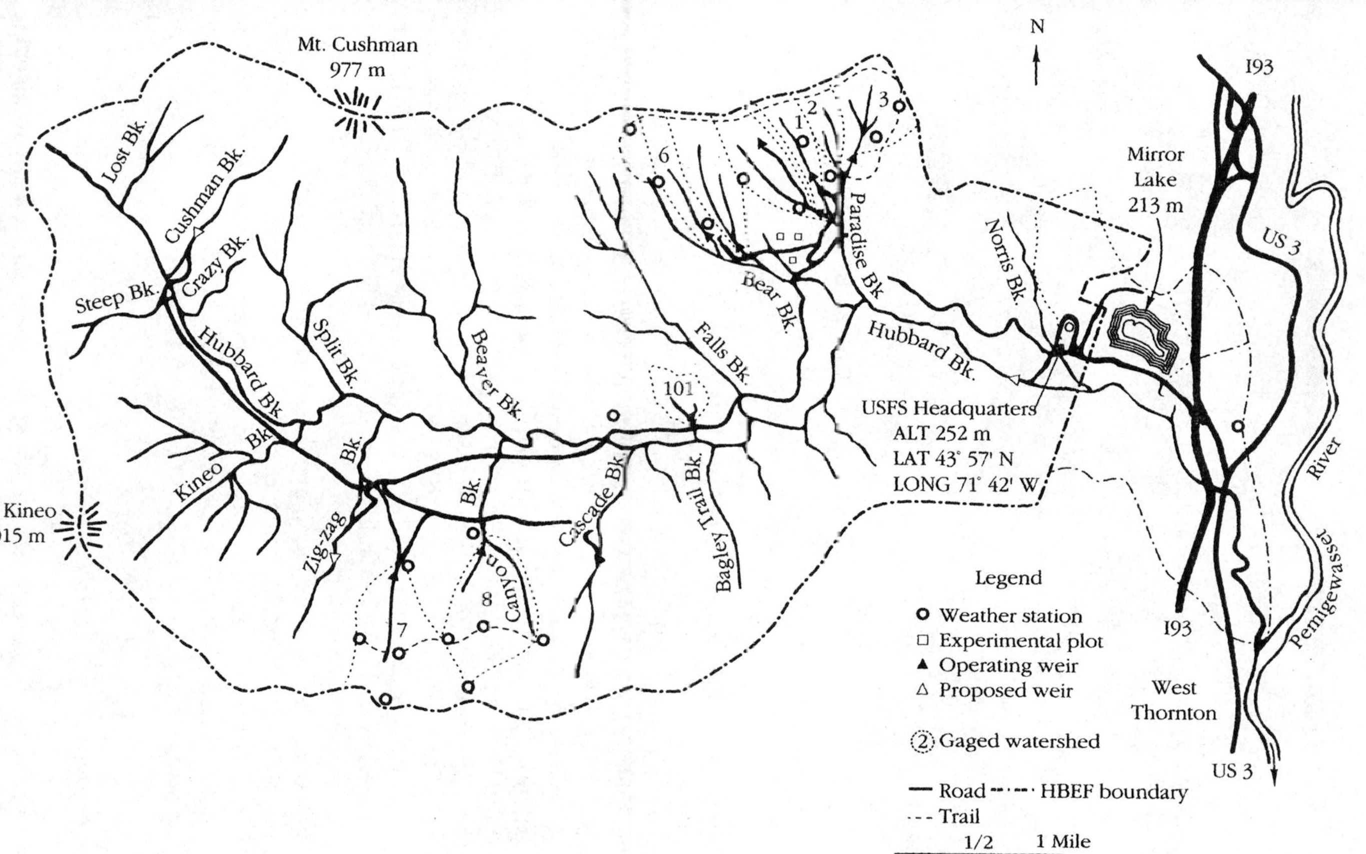

Figure 9-7 Map of Hubbard Brook Valley, New Hampshire. (Redrawn, with permission, from G. E. Likens (ed). 1985. *An Ecosystem Approach to Aquatic Ecology. Mirror Lake and Its Environment.* New York: Springer Verlag.)

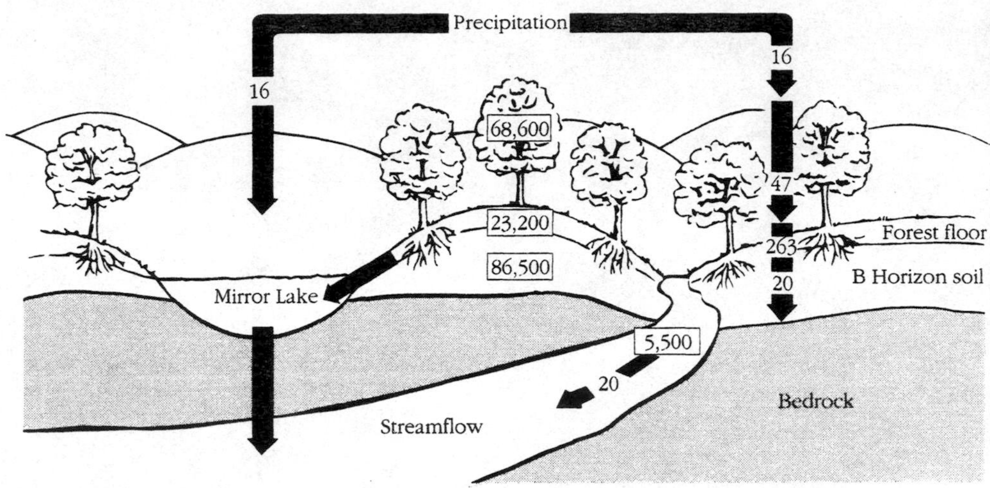

Figure 9-8 Standing stock of organic carbon (in blocks) and flux of dissolved organic carbon (circles), in kg/ha, in the Bear Brook watershed in Hubbard Brook Experimental Forest, New Hampshire. (Redrawn, with permission, from W. H. McDowell and G. E. Likens. 1988. *Ecological Monographs* 58:177–95.)

Figure 9-8 summarizes the standing crop and flux of DOC in Bear Brook watershed. Precipitation accounted for a small portion of the total flux (4 percent) and throughfall nearly three times more (11 percent). There was a large increase in the concentration and flux of DOC as throughfall percolated through the forest floor and into the A soil horizon, resulting in the annual flux that accounts for 62 percent of the total. As the DOC passes through the mineralized B horizon, the flux is reduced sharply by abiotic processes from 13 percent in the upper horizon to 5 percent at the 30 cm depth. the final component of the DOC flux is streamwater, which acounts for less than 5 percent of the total flux. It would appear that the flux of DOC in Bear Brook is regulated by abiotic-biotic linkages and by direct abiotic processes.

Based on this and other studies in the Hubbard Brook Valley, McDowell and Likens (1988) concluded that the mean age of organic matter in the soil is on the order of hundreds to thousands of years; of trees, on the order of tens to hundreds of years; of aquatic organisms, on the order of minutes to tens of years. In contrast, the turnover time of water in the stream is on the order of days, in the soil profile on the order of days to months, and in Mirror Lake (see Likens 1985) about 1 year.

Importance of Leaching in Ecosystems

For most ecosystems in a more or less steady state, nutrient losses due to leaching exceed inputs from the atmosphere (Likens et al. 1977, among others). This difference means that the excess must be compensated by the weathering of parent soil material. In successional situations, however, nutrient loss is usually less than input; in these situations, weathering of parent soil material may still be occurring, but the nutrients are taken up by the developing community. However, Jordan (1982, 1985) has shown that leaching from mature rain forests does not play an important role in the nutrient balance of such ecosystems. Lewis (1986), nonetheless, noted high losses of nitrogen and phosphorus in Venezuela's Caura River, which drains a large (47,000 km^2) undisturbed tropical moist forest.

In his study of an Amazonian rain forest in Venezuela between 1975 and 1980, Jordan (1982) found that precipitation input of nonvolatile minerals such as calcium, potassium, magnesium, and phosphorus exceeded or was equal to the input from the atmosphere (Figure 9-9). Leaching rates generally followed input rates during long-term trends (e.g., calcium in 1976) but not during short-term perturbations (e.g., magnesium in 1978). This rain forest thus maintains its nutrient balance on input from the atmosphere rather than depleting its parent soil materials through leaching.

Role of Fertilizers in Nutrient Budgets

Studies of watersheds in agroecosystems are important in relating land management to such external effects as the addition of agricultural nutrients in the form of fertilizers, as well as in enhancing understanding of nutrient cycles in such systems. In agroecosystems, nutrients enter via precipitation, fertilizers, nitrogen fixation, and irrigation and weathering of soil parent material; losses are derived through streamflow, subsurface flow, deep seepage, loss of volatile gases, and the harvest of plant and animal products.

In a study of several nutrient budgets in four watersheds of the Little River Watershed in Georgia, Lowrance et al. (1985) noted that 35 to 54 percent of the land was in agricultural uses (row crops and pastures). All the elements studied, except chlorine, had greater inputs than outputs each year, though by averaging the data from 3 years of the study, they find that all the elements, including chlorine, had greater inputs than outputs (Table 9-10). For all elements on all watersheds, fertilizer inputs exceeded those gained from precipitation, ranging from two times as much input for chlorine to twenty times as much for phosphorus and magnesium. Harvesting rather than streamflow accounted for the greater losses of nitrogen, phosphorus, and potassium, while the converse was the case for calcium, magnesium, and chlorine.

Not surprisingly, those watersheds with more agricultural land had consistently higher loads of all six nutrients in streamflow. Nutrient budgets for the upland portion of one of the watersheds indicated that large amounts of nitrogen, phosphorus, potassium, calcium, and magnesium were not accounted for and were being retained somewhere in the watershed or lost in some unquantified way (e.g., nitrogen via denitrification or ammonia volatization).

As the authors of this study note, the large quantities of nitrogen, calcium, magnesium, and potassium that enter these watersheds but are not removed in harvest represent an inefficient use of nutrient resources. Further, the economic effect of this inefficiency is especially significant for nitrogen due to the high cost of synthetic nitrogen fertilizers.

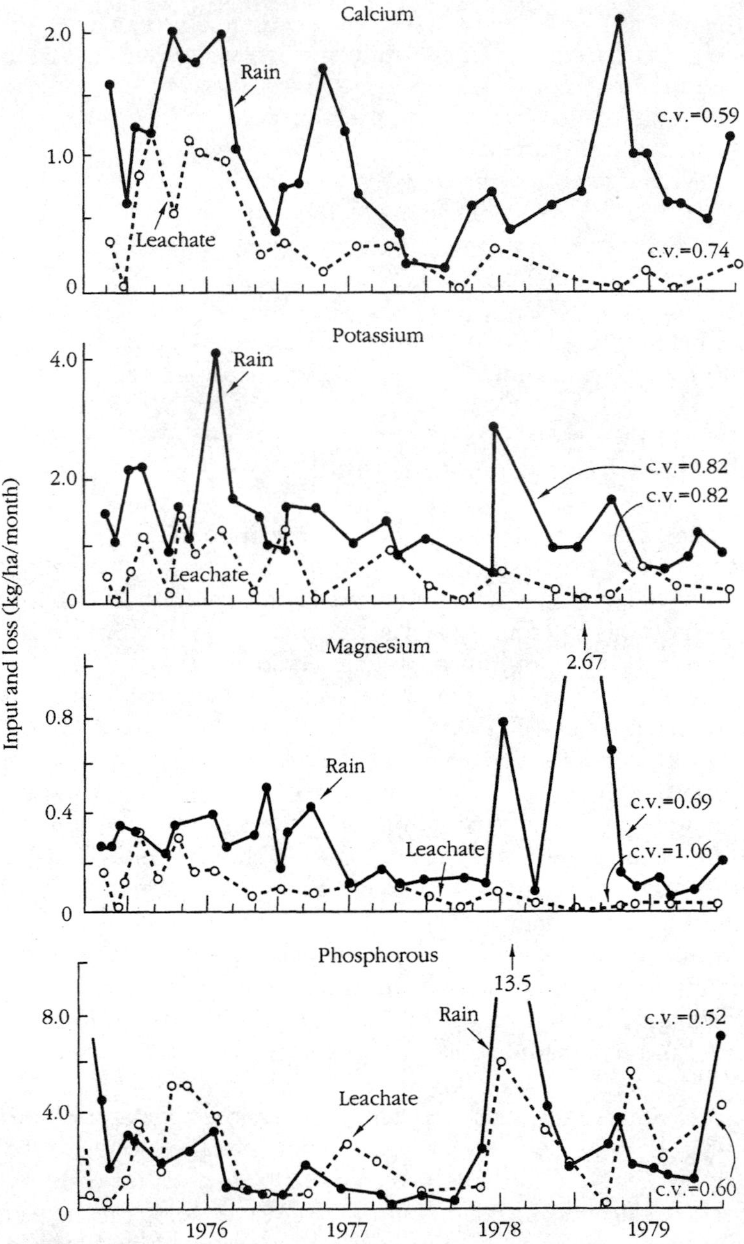

Figure 9-9 Precipitation input and leaching loss of calcium, potassium, magnesium, and phosphorus from a tropical rain forest in Venezuela. CV = coefficient of variation, the average CV of nutrient concentrations for all sampling dates. (Redrawn, with permission, from C. F. Jordan. 1982. *Ecology,* 63:647–54.)

TABLE 9-10 ANNUAL AVERAGE NUTRIENT BUDGETS FOR FOUR WATERSHEDS IN THE LITTLE RIVER WATERSHED IN GEORGIA (in kg/ha/yr)

	Inputs		Outputs		
	Precipitation	Fertilizer	Streamflow	Harvest	Balance
Nitrogen	11.9	55.2	2.5	23.6	41.0
Phosphorus	0.5	10.0	1.2	2.9	6.4
Potassium	4.3	25.3	5.2	6.1	18.4
Calcium	8.5	79.3	11.0	1.5	75.3
Magnesium	1.9	37.7	4.3	1.4	39.3
Chloride	10.7	19.0	20.9	0.8	8.0

Derived from data in R. R. Lowrance et al. 1985. *Ecology* 66:287—96.

West Africa: A Regional Nutrient Budget

An excellent synthesis of diverse data on the nitrogen budget for a developing region of the world, West Africa, is provided by Robertson and Rosswall (1986). Not only is West Africa, like other developing regions, an important part of the terrestrial biosphere, it is also one in which nitrogen fertilizers are not generally available, and, hence, the conservation of existing nitrogen resources is critical to improving crop and livestock production. This is particularly significant because population growth has been outstripping agricultural productivity.

West Africa is geographically discrete (Figure 9-10), with abrupt biogeochemical boundaries on three sides: to the north, the Sahara Desert, and to the west and south, the Atlantic Ocean (White 1983). Precipitation patterns result in well-defined vegetation zones similar to regions in many other tropical areas, ranging from desert vegetation to the north to lowland rain forest in the south.

Based on data from the late 1970s, Robertson and Rosswall constructed a nitrogen budget, the major features of which are shown in Table 9-11.

Nitrogen input. Biological nitrogen fixation is the dominant source of nitrogen in West Africa. Noncultivated systems account for 90 percent of the total annual nitrogen fixation, with more than half of it occurring in early successional rain forests. Of the nitrogen fixed by cultivated systems (millet, sorghum, maize, and rice, etc.), about 30 percent is contributed by leguminous crops such as soybeans, cowpeas, and groundnuts. Human (anthropic or anthropogenic) sources of nitrogen are relatively insignificant (about 2 percent): fertilizers 74×10^6 kg; imported agricultural products, 154×10^6 kg; and combustion of fossil fuel combustion, 41×10^6 kg. Precipitation also accounts for a relatively small source of nitrogen (less than 3 percent).

Nitrogen output. At least 73 percent of the nitrogen lost from West Africa is the result of volation by fire, with 66 percent of that occurring in cleared lowland

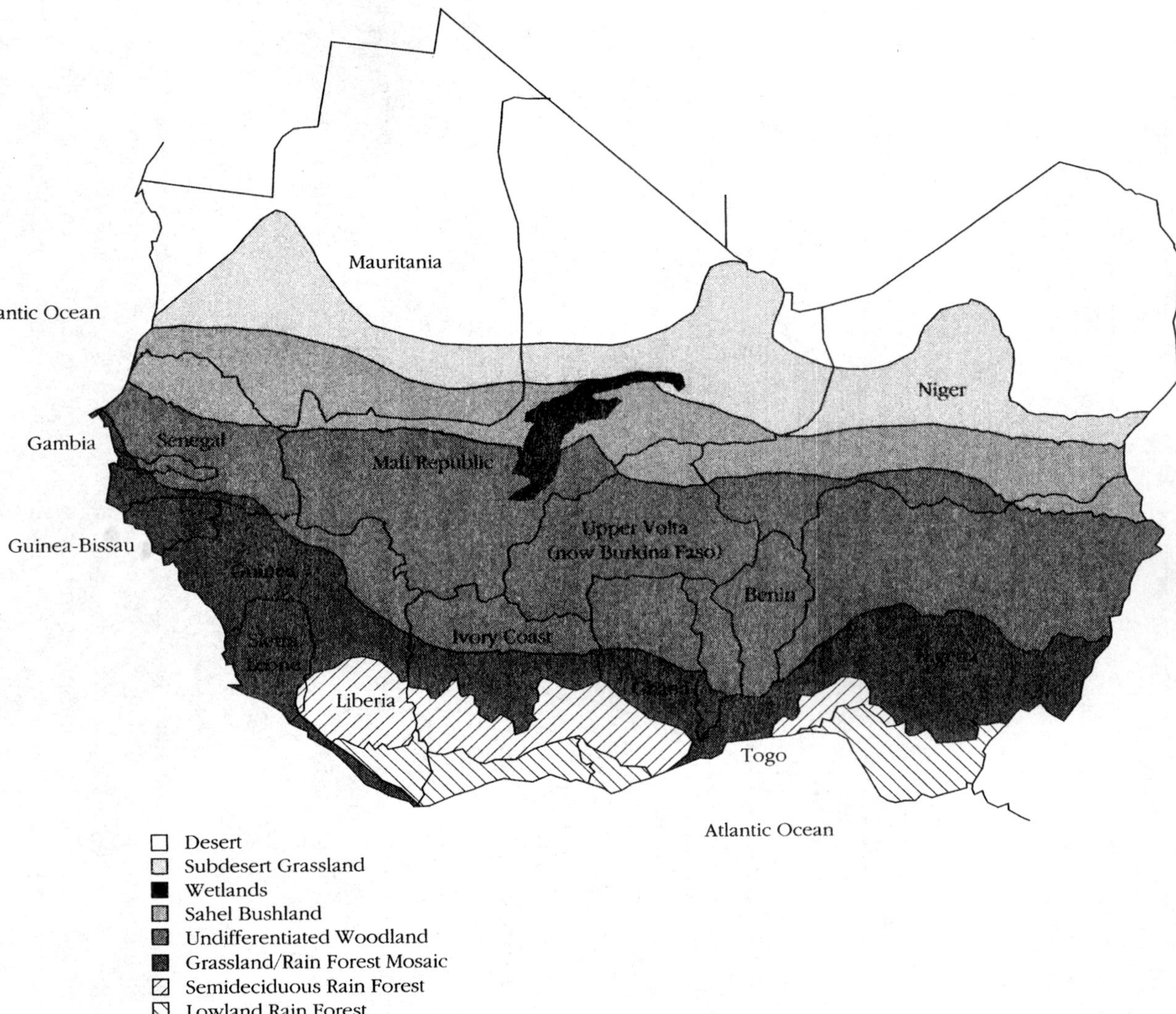

Figure 9-10 Geographic boundaries and major vegetation zones in West Africa. (After F. White 1983.)

TABLE 9-11 THE REGIONAL NITROGEN BUDGET (kg) IN WEST AFRICA

Annual nitrogen input	
Noncultivated systems	12.261×10^9
Cultivated systems	0.657×10^9
Human sources	0.269×10^9
Precipitation	3.900×10^9
Total	$\mathbf{17.1 \times 10^9}$
Annual nitrogen output	
Volatization by fire	8.340×10^9
Hydrologic export	1.480×10^9
Denitrification	1.140×10^9
Other losses	0.478×10^9
Total	$\mathbf{11.4 \times 10^9}$
Internal storage	
Noncultivated systems	5.60×10^9
Cultivated systems	$.03 \times 10^9$
Total	$\mathbf{5.6 \times 10^9}$
Total input	$\mathbf{17.1 \times 10^9}$
Total storage plus outputs	$\mathbf{17.0 \times 10^9}$

Data from G. P. Robertson and T. Rosswall. 1986. *Ecological Monographs* 56:43–72.

rain forests and moist semideciduous forests. Hydrological export to the Atlantic Ocean accounts for 13 percent of the loss, denitrification about 10 percent.

Nitrogen Budget. Because a considerable amount of nitrogen (5.6×10^9 kg) is immobilized in growing woody vegetation during the fallow phase of the crop rotation cycle common to West Africa, Robertson and Rosswall (1986) consider the overall nitrogen budget to be balanced within 1 percent, with the cycle dominated by pools and fluxes of nitrogen in noncropped systems. They acknowledge, however, that there are inherent inaccuracies, and hence uncertainties, associated with each value that limit precision; but presumably the uncertainties in one direction appear to have balanced out the uncertainties in the other direction (G. P. Robertson, personal communication).

Unlike nitrogen balances in developed regions, direct human fluxes in the form of fertilizers and fossil fuels play a very minor part in the West Africa nitrogen cycle.

The Hubbard Brook Studies

Among the most comprehensive studies of external nutrient budgets are those conducted by F. Herbert Bormann and Gene E. Likens on Hubbard Brook in the White Mountains of New Hampshire (Bormann and Likens 1967; Bormann, Likens, and Mellillo 1977; Likens et al. 1978; Bormann and Likens 1979; Likens 1981, 1985; McDowell and Likens 1988, among others).

Each of the six watersheds selected for study in the Hubbard Brook drainage (See Figure 9-7) is characterized by watertight bedrock and lateral boundaries that coincide with topographic divides. Thus each is discretely isolated from the waterborne output of adjacent watersheds and none is subject to any deep seepage or underground circulation. Furthermore, the isolation of the forest from sites of active agriculture minimizes the mineral contribution that windborne dust brings to many ecosystems. The homogeneous nature of the bedrock further reduces variability in the system. An additional and most propitious aspect of the choice of the watersheds is that they are within the Hubbard Brook Experimental Forest. This is an established hydrologic laboratory that continuously monitors precipitation and runoff by standard meteorological procedures. Weekly analysis of hydrologic input and output for particular nutrients by atomic absorption spectrophotometry and other techniques has permitted the establishment of annual budgets for the several significant cations and anions in the forest.

Nitrogen budget of one watershed. Before discussing the budgets of these various elements, it will be instructive to expand our consideration of the nitrogen cycle by examining the nitrogen budget in one of these forested watersheds (Figure 9-11). The particular hardwood forest studied was 55-years old and was dominated by sugar maple, American beech, and yellow birch. Annually most of the nitrogen is added via nitrogen fixation (14.2 kg/ha or 68 percent), the remainder coming from precipitation. Most of this added nitrogen is held by the ecosystem (16.7 kg/ha or 81 percent), the loss being by hydrologic outflow. Of the nitrogen held, 29 percent (4.8 kg/ha) is added to the aboveground living biomass, 25 percent (4.5 kg/ha) to the belowground living biomass, and the remainder is stored in the litter (7.7 kg/ha or 46 percent). Of the total nitrogen in this hardwood forest, nearly 90 percent is in soil organic matter (3700 kg/ha), 9.5 percent is in vegetation (532 kg/ha), with the remaining 0.5 percent being held by the soil (26 kg/ha).

Hydrologic cycle. With respect to the more comprehensive external nutrient budgets, it is necessary first to take into account the hydrologic cycle in these forested ecosystems. From 1955 to 1969 average annual precipitation in each of the watersheds was 123 cm, distributed quite evenly throughout the year (about 10 cm/ month). About one-fourth to one-third of the precipitation was in the form of snow. Of the total precipitation, 59 percent was lost through runoff and 41 percent by evapotranspiration (i.e., evaporation plus transpiration). Sixty-eight percent of the runoff occurred from March to May with April being the highest month (sixty-eight percent of total runoff). During the period of the study (1963–1969) precipitation was below average (123 cm) in 1964–1965 (95 cm) and above average in 1967–1968 (142 cm). The year of decreased precipitation was reflected in decreased input of calcium and potassium but not of magnesium and sodium. The investigators suggest that an intrusion of maritime air during 1964–1965 may have been responsible for the increased levels of magnesium and sodium, for these elements are abundant in seawater. It is also noteworthy that during the drought year evapotranspiration loss was about the same

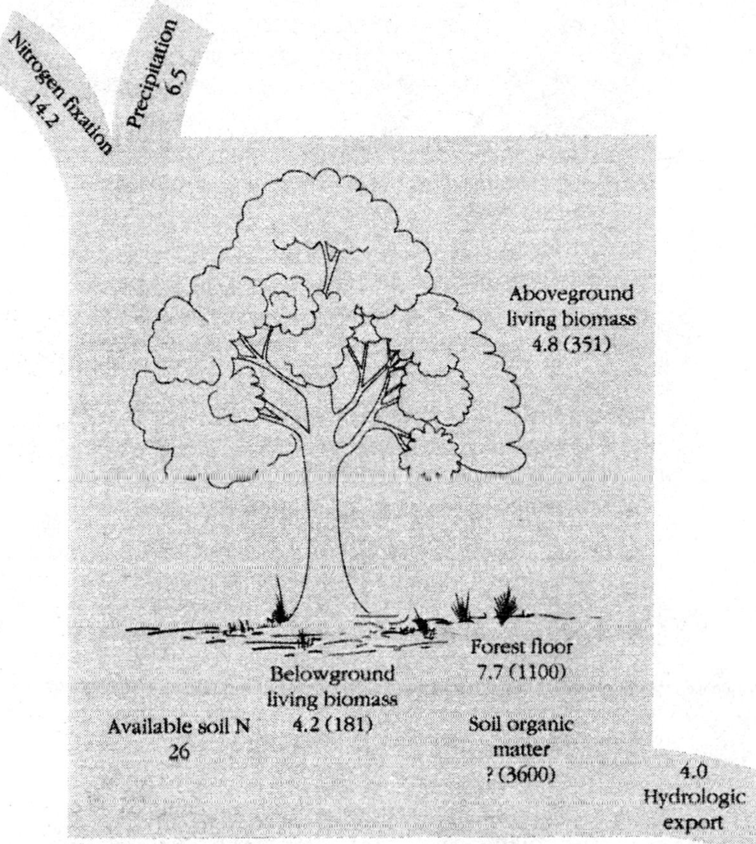

Figure 9-11 The annual nitrogen budget of a northern hardwood forest in kilograms of nitrogen per hectare. The numbers without parentheses represent the annual flow or transfer of nitrogen; the numbers within parentheses represent the total amount of nitrogen stored in those compartments. (Adapted by permission from F. H. Bormann et al. 1977. *Science* 196:981–3.)

as that in the above-average year (46 cm and 51 cm, respectively) but runoff was considerably less (49 cm and 91 cm, respectively). This observation suggests a greater constancy—and hence significance to the ecosystem—of the precipitation-evaporation ratio discussed earlier.

Nutrient losses. Although there are differences in the annual cation and anion budgets because of changes in the hydrologic cycle and the concentration in rainfall and runoff, there is a significant loss of nutrients from the ecosystem each year (Table 9-12). About 90 percent of this export is as dissolved substances and 10 percent as organic and inorganic particulate matter, the latter increasing with

TABLE 9-12 NUTRIENT BUDGETS (kg/ha) FOR SIX WATERSHEDS IN HUBBARD BROOK EXPERIMENTAL FOREST, NEW HAMPSHIRE, 1963–1969

	Input[a]		Output[b]		Mean net
	Mean	Range	Mean	Range	loss or gain
Cations					
Ca^{++}	2.6	(1.6–3.0)	11.7	(4.8–17.3)	−9.1
Na^{+}	1.5	(1.0–2.3)	6.8	(3.5–11.0)	−5.3
Mg^{++}	0.7	(0.3–1.2)	2.8	(1.6–3.7)	−2.1
Al^{+++}	x[c]		1.8	(0.9–3.2)	−1.8
K^{+}	1.1	(0.6–2.4)	1.7	(0.8–2.7)	−0.6
NH_4–N	2.1	(1.6–2.6)	0.3	(0.1–0.7)	+1.8
Anions					
SiO_2–Si	x		16.4	(9.7–22.3)	−16.4
SO_4–S	12.7	(9.7–16.0)	16.2	(9.8–19.6)	−3.5
HCO_3–C	x		2.9	(2.4–3.3)	−2.9
Cl^{-}	5.2	(2.6–6.9)	4.9	(4.2–5.3)	+0.3
NO_3–N	3.7	(1.5–5.2)	2.0	(1.1–2.9)	+1.7

[a]Input is the amount of dissolved cations entering the ecosystems in all forms of precipitation and equals concentration in milligrams per liter times volume of precipitation.

[b] Output is the amount of dissolved cations leaving the ecosystems in stream-water and equals concentration in milligrams per liter times volume of runoff.

[c]x = Not measured but small

Adapted by permission from G. E. Likens and F. H. Bormann. 1972. In *Ecosystem Structure and Function*, ed. J. A. Wiens. Corvallis, OR: Oregon State University Press.

increased stream discharge owing to increased erosion. Because some of the cation input becomes part of the resource pool available for biological synthesis, the total "output" is actually greater than that shown. What is being measured here is input and output from the whole system, not input and output within the system, which may or may not be completely balanced on any short-term basis, as in the nitrogen cycle just described.

There is, then, as shown for other ecosystems in the previous examples, a net annual loss of cations from Hubbard Brook. All these loss rates are consistent with the known activity for the geochemical cycle of these elements. In themselves, the losses from one ecosystem can constitute an external input to another ecosystem elsewhere and can be compensated for by weathering or chemical decomposition of the underlying substrate. As long as the substrate can supply the ecosystem with its requisite nutrients and also be somewhat profligate in the amount discharged in runoff, the ecosystem can be maintained. It is estimated that approximately 800 kg/ha of till and bedrock are weathered each year at Hubbard Brook, or some 11×10^6 kg/ha since the last glaciation some 14,000 years ago. If there is a net gain or depletion of one or more of these essential nutrients over a period of time, the system may well respond by a change in its composition or function.

Effects of clearcutting. After the pattern of nutrient cycles had been followed for several years, one of the six watersheds was experimentally subjected to clearcutting during the winter of 1965–1966, all the cuttings being left on the ground. In the following three summers a herbicide (Bromacil) was applied to inhibit regrowth of vegetation. Cation and anion gains and losses were monitored for both the cutover watershed and the five intact forested watershed ecosystems in the manner already described.

One expected result of clearcutting was an increase in runoff water owing to the loss of a tremendous amount of transpiration surface eliminated by the cutting. However, the order of magnitude of the increase for each of the three years exceeded by 40, 28, and 26 percent, respectively, that expected for the entire year (35 cm). The period of peak runoff (June through September) exceeded expectation by 418 percent in 1966 and 380 percent in 1967. Net losses of nutrients over the 3-year period from the cutover system increased more than eight times (Table 9-13). The loss was not the same for all nutrients, however. For example, the losses of sodium, magnesium, calcium, and potassium cations were, respectively, 2.5, 6, 8.5, and 20 times greater than in the undisturbed ecosystems, whereas NH_4^+ and $SO_4^=$ showed very little change. The increased losses result both from the greater quantities of drainage water moving through the system and from the absence of nutrient uptake by the vegetation.

With respect to the anions, the deforestation resulted in a major disturbance of the nitrogen cycle with spectacular increases in the loss of nitrate, 51 times the

TABLE 9-13 NET GAIN OR LOSS OF NUTRIENTS (kg/ha) FOR UNDISTURBED AND DEFORESTED WATERSHEDS, HUBBARD BROOK EXPERIMENTAL FOREST, 1966–1969

	Undisturbed	Deforested	Net gain or loss (deforested)
Cations			
Ca^{++}	−27.5	−233.6	−206.1
K^+	−4.4	−91.1	−86.7
Al^{+++}	−7.7	−61.9	−54.2
Mg^{++}	−7.8	−46.8	−39.0
Na^+	−18.3	−46.3	−28.0
NH4−N	+6.4	+4.9	+1.5
Anions			
SO_4−S	−12.3	−8.3	+4.0
HCO_3−C	−0.9	−0.2	+0.7
Cl^-	+3.5	−5.9	−9.4
SiO_2−Si	−48.0	−91.0	−43.0
NO_3−N	+6.7	−342.2	−348.9
Total	−110.3	−922.4	−812.1

Derived from data in G. E. Likens and F. H. Bormann. 1972. In *Ecosystem Structure and Function,* ed. J. A. Weins. Corvallis, OR: Oregon State University Press.

normal net loss. In the intact acid-soil forest the ammonia resulting from nitrification is used directly by the plants and little is converted into nitrate. In the absence of vegetation the ammonia is oxidized to readily dissolved nitrate and flushed from the system. A concomitant effect of nitrification is that the hydrogen ions produced replace metallic cations on soil exchange surfaces, thereby increasing their loss from the system. Thus the increased cation loss is due primarily to increased nitrification rather than to increased streamflow.

The widespread adoption of clearcutting as a harvest technique, but without the use of herbicides as in the foregoing experiment, has heightened concern about long-term productivity of forests and the untoward impact on water quality downstream. Although such harvests result in elevated nutrient losses, the extent of loss has been highly variable. Studies on nitrogen loss have received particular attention, partly because of the greater extent of its loss but also because nitrogen is frequently a limiting factor in plant growth. Furthermore, nitrogen has deleterious effects on downstream water quality by promoting increased plant growth, and the nitrate anion, as noted, promotes the loss of soil cations. Various studies show a wide range of nitrate losses after disturbance by clearcutting: from a low of 0.08 to 0.26 kg/ha/yr in an Oregon forest, a threefold increase; from 2.0 to 97 kg/ha/yr in Hubbard Brook, a 48-fold increase; from 0.05 to 7.3 kg/ha/yr in a North Carolina forest, a 146-fold increase.

Peter Vitousek and his associates (1979, 1982) attribute such differences to the resistance of the forested ecosystem to nitrogen loss. This resistance relates not only to a number of processes inhibiting various steps in the nitrogen cycle but also to the patterns of nitrogen circulation prior to the disturbance. Nitrogen-deficient forests are adapted to a tight circulation of nitrogen and experience little loss on being disturbed. These adaptations result from

1. Processes that prevent ammoniafication or ammonium accumulation (e.g., fixation of ammonium on clay lattices, volatization from near-neutral to basic soils, nitrogen uptake by plants).
2. Processes that prevent or delay nitrate accumulation (e.g., inhibition of nitrifying bacteria, denitrification to molecular nitrogen or nitrogen oxides).
3. Processes that prevent or delay nitrate mobility (e.g., lack of water to effect leaching, nitrate adsorption on iron or aluminum oxides).

Vitousek argues that resistance, if it occurs at more than one of these steps in the nitrogen cycle, will occur in the order just described. In contrast, a disturbance tends to result in significant loss in systems in which nitrogen is readily available and in which these resistance mechanisms are lacking. Increased soil temperature and moisture availability also tend to accelerate nitrogen loss by increasing mineralization, the decomposition of nitrogen-containing compounds with the release of nitrogen as ammonia or ammonium.

Effects of revegetation. A final and significant factor in nitrogen and other nutrient losses following clearcutting, fire, or other disturbances concerns the resilience of the forest ecosystem—that is, its rate of recovery through revegetation.

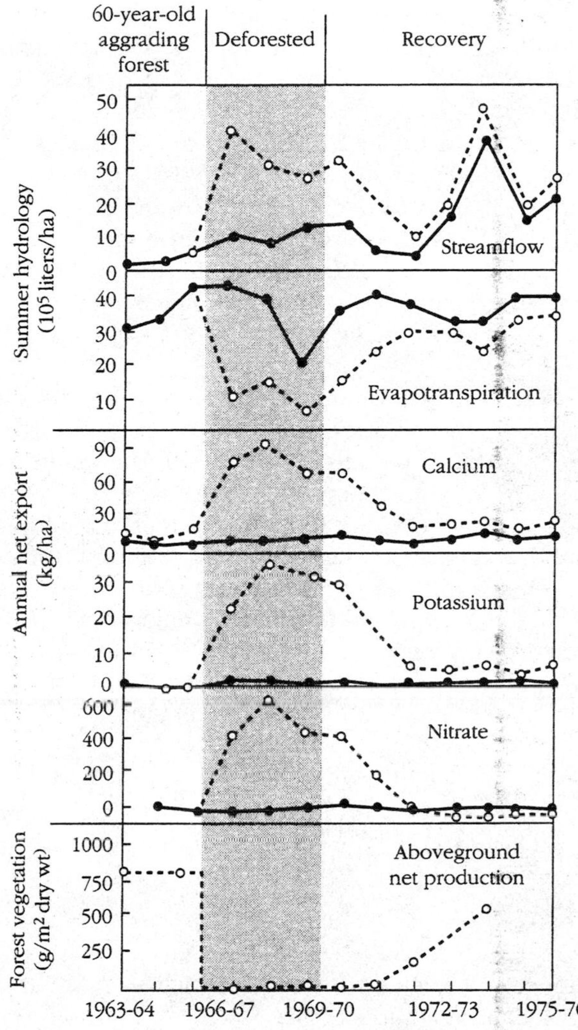

Figure 9-12 Effects of deforestation on hydrology, biogeochemistry, and above-ground net production in a northern hardwood forest ecosystem (○---○) compared with an undisturbed forest (●—●). (Modified by permission from G. E. Likens, et al. 1978. *Science* 199:492–6.)

Revegetation of the Hubbard Brook Experimental Forest plot provides a perspective on the shift in nutrient losses.

At the end of the deforestation experiment in the spring of 1969, the vegetation was allowed to regrow. Within 3 or 4 years, some of the biogeochemical and hydrological characteristics of the forest had returned to previous levels (Figure 9-12). As the graphs indicate, however, net losses of dissolved calcium, potassium, and nitrogen had begun to decline before regrowth, having peaked during the second year. Nevertheless, after 7 years, the levels of calcium and potassium had not returned to the pretreatment levels. Similarly, evapotranspiration was still below normal levels after 7 years. Although aboveground net production recovered to 5 percent of its previous level by the second summer, it had only approached about two-thirds of its previous level in five years.

The long-term losses of nutrients are also significant when compared to those of an adjacent undisturbed forest: 499 kg of nitrogen versus 40.3 kg; 450 kg calcium versus 131; and 166 kg potassium versus 21.7. These losses represent 28 percent of the total nitrogen stored within the forest before clearcutting, 88 percent of the calcium, and 58 percent of the potassium. Because the Hubbard Brook substrate is rich in calcium and potassium, weathering will eventually replace the losses of these nutrients. Because nitrogen is negligible in the substrate, however, it must be replaced by nitrogen fixation and precipitation. Because the net addition from precipitation is relatively small (Figure 9-12), it has been estimated that were this the only source of nitrogen, it would take 100 years to replace the amount lost. Because nitrogen fixation also occurs, the total recovery time would be somewhat less, depending on the conditions for nitrification.

It is estimated that 60 to 80 years will be required for full recovery of this disturbed hardwood forest. Such forests are prime sources for timber and pulp and so it is obvious that forest practices, in order to be ultimately cost effective, must be ecologically sound, so as to minimize long-term degradation for short-term gain. Such sound practices include using sites with the strongest regrowth and minimum erosion potentials, doing minimal damage to the forest floor and stream channels, ensuring the nearby availability of natural seed sources, and conserving those species, most of which are not of timber or pulp value, that play important roles in recovery. The integrity of a future forest is a mirror of its treatment during harvesting. Protection of our forest resources does not preclude their being used to meet society's needs. It does mean using the wisest techniques to ensure their being kept as continual resources.

SUMMARY

VOCABULARY

allochthonous
autochthonous
external nutrient budget
incipient precipitation
internal nutrient budget
stemflow
throughfall

KEY POINTS

- Internal and external nutrient budgets are interrelated, the former being dependent on inputs and the latter on outputs.
- Residence times, turnover, sites, and amounts of accumulations and other aspects of internal nutrient budgets vary with the kind and maturational stage of the biota, different parts of the ecosystem (e.g., aboveground, belowground, leaves, stems), latitude, season, soils, weathering, moisture levels, and the presence or absence of other nutrients.

- Cycling time tends to increase with increasing latitude, a correlate of primary productivity, nutrients being accumulated in boimass and detritus in the more northern environments.
- In contrast to most natural ecosystems, agroecosystems are well suited to nutrient budget studies because of their relative simplicity and defined boundaries.
- Precipitation accounts for the largest nutrient input to most ecosystems, being the major pathway in nutrient recycling in forests; by contrast, allochthonous nutrient inputs from adjacent terrestrial ecosystems are most significant in streams, rivers, estuaries, and coastal marine ecosystems.
- Residence time of organic matter in soils is hundreds to thousands of years, in trees is tens to hundreds of years, and in aquatic organisms is minutes to tens of years.
- Nutrient losses due to leaching exceed inputs from the atmosphere in most ecosystems.
- Six watersheds in the Hubbard Brook drainage in the White Mountains of New Hampshire have provided the most comprehensive study of external nutrient budgets, including the effects of weathering, clearcutting (which results in excess nutrient losses), and revegetation.

PART V

Population Ecology

CHAPTER 10

Population Growth and Structure

In a study of the common housefly some years ago, the American entomologist L. O. Howard determined that the female produces an average of 120 eggs at a time and that about one-half the eggs develop into females. He also observed that there were seven generations per year. Based on these data, of the 120 individuals produced by one female, 60 would be females, each capable of producing 120 eggs, and the number of offspring in the second generation would be 7,200. Assuming that all individuals of the reproducing generation die off after reproduction, we can see that the first fertile female would be responsible for the production of over 5.5 trillion flies in the seventh generation (Table 10-1). If we assume that all individuals survive a full year with the females producing only once, the population at the end of the year would be only about 1.7 percent greater. But, if we assume that all individuals survive and that all the females reproduce in each generation, the population would be increased by about 10 percent, exceeding 6 trillion flies by the seventh generation. In all three instances, the population would demonstrate a geometric increase, multiplying its size by a factor of 60 each generation.

At the other extreme of body size, according to calculations by Charles Darwin, a single pair of elephants would have over 19 million descendants alive after 750 years. This would be the case even with their long gestation period and small number of offspring produced in a lifetime.

In 1899, shortly after the English sparrow was introduced into the United States, it was estimated that in 10 years a single pair could give rise to 275,716,983,698 descendants and that by 1916 to 1920 there would be about 575 birds per 40 hectares. By 1916 to 1920, however, there were only 18 to 26 birds per 40 hectares, less than 5 percent of the expected number. Forces must have

TABLE 10-1 PRODUCTION OF HOUSEFILES (*Musca domestica*) IN ONE YEAR[a]

Generation	Total population: If all survive but 1 generation	If all survive 1 year but produce only once	If all survive 1 year and all females produce each generation
1	120	120	120
2	7,200	7,320	7,320
3	432,000	439,320	446,520
4	25,920,000	26,359,320	27,237,720
5	1,555,200,000	1,581,559,320	1,661,500,920
6	93,312,000,000	94,893,559,320	101,351,520,120
7	5,598,720,000,000	5,693,613,559,320	6,182,442,727,320

[a]Based on the assumptions that a female lays 120 eggs per generation, that half these eggs develop into females, and that there are seven generations per year.

acted against this geometric increase of sparrows. Similarly, forces must be operating against the increase potential in houseflies and elephants and other organisms as well. In the case of the sparrow, a number of factors seemed to have been involved, including the development of sparrow-eating habits by hawks and owls, changes in agricultural practices, and a decrease in the size of the horse population in cities (and thereby of the droppings on which the bird feeds). In addition, intraspecific competition for food and nesting sites may also have operated against the potential for geometric increase.

As these examples indicate, there are two opposing forces operating in the growth and development of every population. One is inherent in and characteristic of each species population—the ability to reproduce at a given rate. Opposing it is an inherent capacity for death or for physiological longevity. Opposition to growth also comes from all the forces of the physical and biological environment in which an organism exists. In the late 1920s, the American ecologist Royal Chapman (1928) referred to these forces respectively as *biotic potential* and *environmental resistance*, terms that still serve as useful handles for considering the complexities of population growth and regulation.

POPULATION GROWTH CURVES

The Sigmoid Growth Curve

Growth curve. To gain an understanding of the interaction of biotic potential and environmental resistance, we will turn from the theoretical growth of houseflies to empirical data on yeast. In the 1920s, Raymond Pearl, one of the primary contributors to our understanding of population ecology, used data obtained by a German investigator, T. Carlson, who had studied the growth of yeast by periodically centrifuging a culture and determining its volume and mass. Pearl converted these data to numbers of yeast cells, obtaining the fractional components that appear in Table 10-2. If the data of the first two columns (time and number of individuals present) are plotted and the points joined (Figure 10-1), the *growth curve* of the population can be seen to be *S-shaped*, or *sigmoidal.*

TABLE 10-2 GROWTH OF YEAST CELLS IN LABORATORY CULTURE

Time (t) (hours)	Number of individuals (N)	Number of individuals added per 2-hour period $\left(\frac{\Delta N}{\Delta t}\right)$
0	9.6	0
2	29.0	19.4
4	71.1	42.1
6	174.6	103.5
8	350.7	176.1
10	513.3	162.6
12	594.4	81.1
14	640.8	46.4
16	655.9	15.1
18	661.8	5.4

From *The Biology of Population Growth*, by Raymond Pearl. Copyright 1925 by Alfred A. Knopf Inc. and renewed 1953 by Maude de Witt Pearl. Reprinted by permission of the publisher.

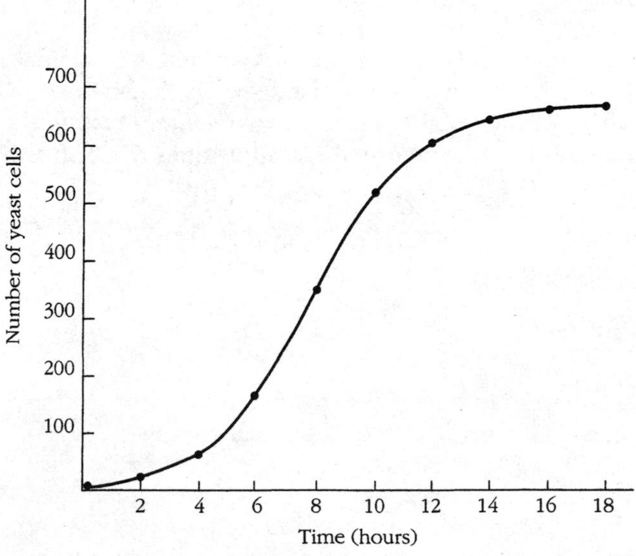

Figure 10-1 The growth curve of yeast cells in the laboratory. (Based on Table 10-2.)

Growth rate curve. In absolute numbers, the initial growth period is slow but is followed by a period of rapid increase and then by a slowing down at the upper level. It is difficult at first glance, however, to determine during what period the increase is greatest and, moreover, how the increase and decrease in rate compare. These shifts in rates can be more readily seen by constructing a *growth rate curve.* The change in the size of the yeast population (symbolized as ΔN, the Greek delta representing change) in a passage of time (Δt) can be derived directly from the tabulation in the second column of Table 10-2. During the passage of the first 2 hours the population tripled from 9.6 to 29.0 individuals, for an increase of 19.4 individuals. Similarly, in the next 2-hour period, the increase was from 29.0 to 71.1 individuals, a change of 42.1. The data so generated in the third column of Table 10-2 have been graphed in Figure 10-2, superimposed on the growth curve of the population (Figure 10-1).

Increase and equilibrium. Two major components of the growth curve—the period of increase and the period of equilibrium—can now be recognized, their interpretation being facilitated by the growth rate curve. In absolute numbers, during the period of increase (0 to 18 hours) there is a slow acceleration in the first 4 hours (the lag phase), followed by a period of rapid growth from 4 to 16 hours (the positive growth phase). The turning or inflection point of the positive growth phase is at 8 hours. Up to that point, population increase has been accelerating to its most rapid rate. This rapid growth portion (4 to 8 hours) of the positive growth phase is often referred to as the *logarithmic growth phase,* because a straight line would be produced if the data were plotted logarithmically. After this rapid log

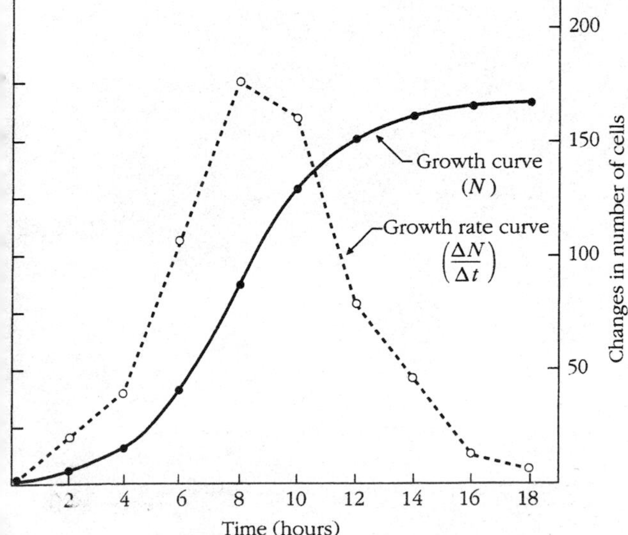

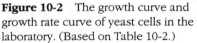

Figure 10-2 The growth curve and growth rate curve of yeast cells in the laboratory. (Based on Table 10-2.)

phase, the rate decelerates rapidly at first (10–12 hours) and then more gradually and rather uniformly to nearly zero (16–18 hours). There is no net change in the population at the zero point; it is at the stationary or equilibrium phase. That is, it has reached the maximum number that the environment can support, a limit sometimes referred to as the *carrying capacity* of the environment. If the environment is shifted, a different carrying capacity may result. Environmental shifts may be in the form of a chemical change, such as removing toxic wastes (Figure 10-3), or a physical change, such as an alteration of temperature (Figure 10-4), or still other changes to be discussed subsequently. The major point is that carrying capacity, like most other ecological attributes, is itself subject to change.

In the equilibrium or stationary phase of the yeast population there is no substantial net change in population. New individuals, however, are being continually introduced by budding and others are being continually lost through death. Still, these two forces—rate of birth or *natality* and rate of death or *mortality*—are in balance in the equilibrium period. The population increased during the positive growth phase when natality exceeded mortality; it would decrease if mortality exceeded natality. Fundamentally, then, natality tends to push the growth curve up and mortality to push it down.

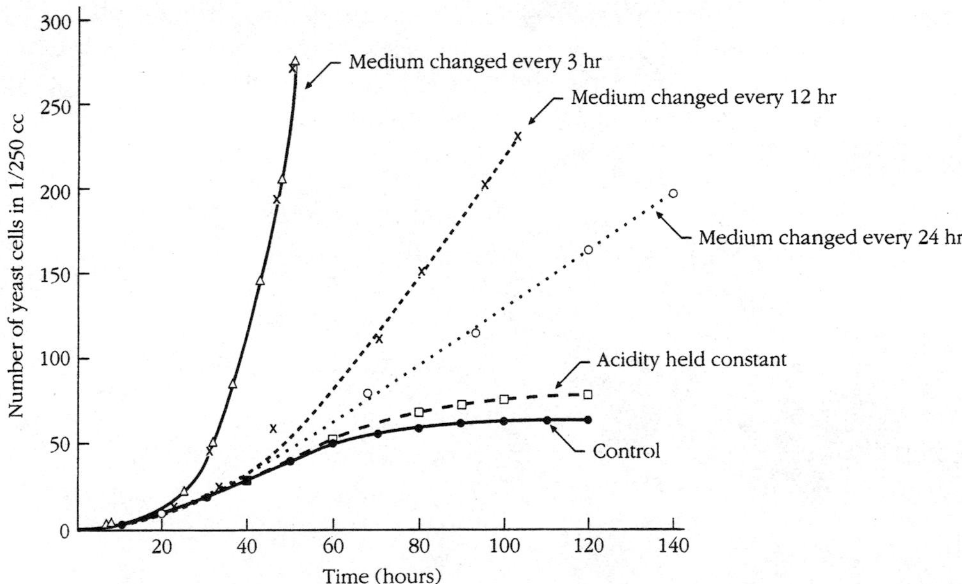

Figure 10-3 Growth curves of populations of the yeast, *Saccharomyces cerevisiae*, with modifications of the medium demonstrating changes in carrying capacity of the environment. The three curves on the left, if followed long enough, would level off. (Redrawn by permission from O. W. Richards. 1928. *Journal of General Physiology* 11: 525–38.)

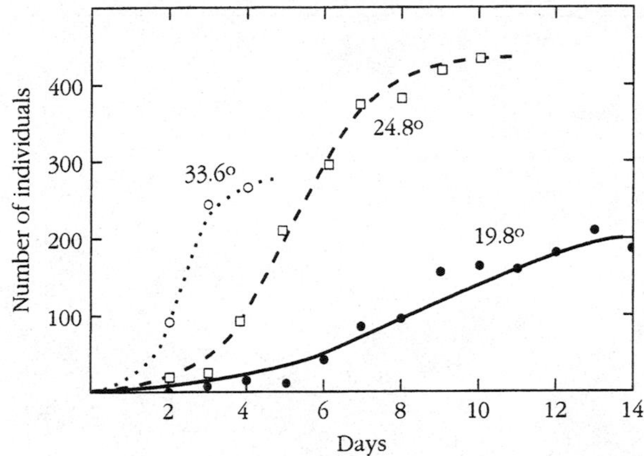

Figure 10-4 Growth of populations of the water flea, *Moina macrocopa*, at three different water temperatures (°C). (Redrawn by permission of A. Terao and T. Tanaka. 1928. *Proceedings of the Imperial Academy of Japan* 4:550–2.)

Generalizing the sigmoid growth curve. Before considering natality and mortality in more detail, it is important to discuss the sigmoid growth curve a bit further.

Of particular importance is the question of how the growth curve is exhibited in nature by populations of different species. After all, the curve presented here was generated by an asexually reproducing organism. Is it typical of populations in general? If it is a typical curve, then we need to see if it can be described in mathematical terms—that is, whether a model or paradigm can be constructed for it.

A good deal of information on population growth has been obtained. Yet only relatively few species have been studied and much of the data has been derived from laboratory study. However, there are enough studies of species populations under natural conditions and on a spectrum of different kinds of plants and animals to permit the statement that most species show a sigmoidal growth pattern when introduced to a fresh environment or following the relaxation of some factor that previously held the population in check. There is, in such cases, an initial slow rate of growth, followed by a logarithmic increase in rate to a maximum, at which point the growth rate curve begins to be deflected downward. The rate of change gradually lessens to zero as the population more or less initially stabilizes itself, that is, comes into equilibrium, with respect to its environment.

The J-Shape Growth Curve

In some cases, a population continues increasing at an accelerating rate and, instead of leveling off, precipitously decreases its rate to zero by a large dieoff of the population. In such instances, the initial portion of the growth curve has a *J-shaped growth curve* (Figure 10-5). J-shaped curves can be observed in the growth of insect populations that have but one generation a year or that depend on an annual crop (see Figure 10-17), and in a number of algae that undergo rapid and tremendous growth seasonally (a phenomenon referred to as a bloom), such as that shown

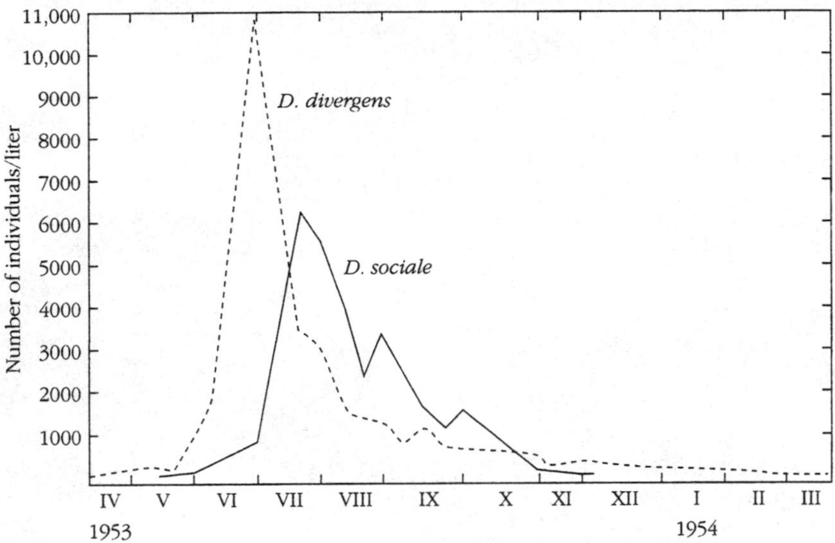

Figure 10-5 J-shaped growth curves of two species of golden-brown algae, *Dinobryon divergens* and *D. sociale.* (Modified by permission from S. Stankovic. 1960. *Monographiae Biologicae* 9:1–357.)

in Figure 10-5 (see also Figure 8-5). In each case, there is essentially unrestricted, rapid, exponential growth up to the limits of the environment. As these resources are exhausted (nutrients, space, etc.) or imposed (photoperiod shift, frost, drought, toxic wastes, etc.), the population dies off. In the case of two species of *Dinobryon*, golden-brown algae belonging to the Chrysophyta, *D. sociale* outcompetes *D. divergens* for environmental resources (Figure 10-5). In any event, as far as the latter species is concerned, a rapid dieoff occurs when its environmental resources are exhausted, effecting an initial J-shaped population curve, followed by a drop in the population to an equilibrium level about one-twentieth of its maximum.

THE "LAW" OF POPULATION GROWTH

In 1838 the French mathematician P. F. Verhulst wrote:

> I have long tried to determine by analysis what the probable law of population is; but I have abandoned this kind of research because the data provided by observation are too few to permit verification of the formulae in such a way as to leave no doubt as to their exactness. However, since the path I have followed seems to me to lead to the discovery of the true law, once there is sufficient data, and since the results I have obtained may be of interest, at least as a matter for speculation, I felt myself obliged ... to make them public.

What Verhulst then advanced was an equation that takes into account the fact that the instantaneous rate of growth in a limited environment is retarded by an increase in the number of inhabitants. A plot of the Verhulst equation yields a sigmoidal curve, which in mathematical lingo is termed a *logistic curve.* Verhulst then applied his mathematical formulation to available but very limited census data and observed a respectably good fit of this empirical data to the theoretical curve.

In 1918—80 years later and without knowledge of the Verhulst formulation—Raymond Pearl and Lowell J. Reed, both of The Johns Hopkins University, derived an equation yielding a logistic curve of population growth and applied it to census data of the United States (see Figure 17-8). Pearl and Reed saw the function of mathematical representation as "obviously the best general method of estimating population in intercensal years is that of fitting an appropriate curve to all the available data, and extrapolating for years beyond the last census...." Owing to shifts in factors that were not then foreseeable, the curve generated by the Pearl and Reed logistic equation has missed the mark: the population of the United States reached the 200 million level in November 1967, some 100 years prior to the date they projected.

The Mathematical Equations

Equation of geometrical increase. Using the data of the housefly, let us develop the mathematical description of the population curve. In our earlier discussion it was pointed out that each female housefly was capable of producing 120 eggs. Thus the potential total number of offspring that could be produced by a population of fertilized female houseflies would be 120 times the number of females present, N, or $120N$. The rate of change in the population in any one generation would then be given by the equation $\Delta N/\Delta t = 120N$. If $N = 60$ females, then $\Delta N/\Delta t = 7{,}200$ (see Table 10-1). Hence the rate of change is dependent both on the reproductive capacity of each individual, abbreviated r and in this case having the value 120, and the number of female individuals (N) present at the time. We thus have the general expression for rate of growth:

$$\frac{\Delta N}{\Delta t} = rN$$

This is an equation of geometric or exponential increase and would generate the growth pattern of the housefly (Table 10-1) as well as J-shaped population curves, such as that of the alga *Dinobyron* (Figure 10-5). Such populations continue to increase geometrically:

1. if r remains constant or changes only slightly,
2. if N continues to increase,
3. if the environment is unlimited or kept nearly so, as in changing the yeast medium every 3 hours (Figure 10-3), and
4. if there is neither immigration nor emigration.

In the foregoing example, N referred only to the females in the population. Typically N is used to refer to both males and females; in this case, if $N = 120$, r would equal 60.

The logistic equation. Environments are limited to varying degrees and thus there is a maximum population K that can exist in a given ecosystem. The difference between the maximum population and that already present would be expressed as $K - N$. As the population approaches the maximum, more "resistance" is encountered, for there are fewer unoccupied "places" $(K - N)$ in the system. This resistance effect can be expressed quantitatively as $(K - N)/K$. When N is small, the "resistance" value is nearly zero. That is, there are numerous "opportunities" for the population. But as N increases and approximates K, the value of the expression comes nearer to zero—there are fewer "opportunities," or "resistance" is higher. For example, if $K = 100$ and $N = 10$, the value of the expression is 0.9; however, if $K = 100$ and $N = 90$, the value of the expression is 0.1.

Because this resistance $(K - N)/K$ acts against the increase potential (rN) of the population, the product of the two would indicate what the rate of change in the population would be in a given time interval:

$$\frac{\Delta N}{\Delta t} = rN\left(\frac{K - N}{K}\right)$$

When N is small, this "resistance" value is, as shown, nearly zero and the biotic potential is essentially realized. But as N gets larger and approximates K, the value of the expression approaches zero. Because the value of r is assumed to be a constant, the increase in population size then becomes less and less as N gets larger and, finally, at carrying capacity there is no net change. In effect, then, this expression is a cybernetic or feedback factor in population growth.

Limitations of the logistic equation. The equation just presented is the logistic equation that was derived by Verhulst and subsequently and independently by Pearl and Reed. It would be too space consuming to show that the logistic curve does indeed fit quite well to data on population growth of many species. The data of the water fleas cultured at 24.8°C (Figure 10-4) and of yeast growth (Figure 10-1), for instance, fit a logistic curve almost perfectly. However, the ultimate value of the logistic expression is limited—it is not the complete or the only description of population growth. Furthermore, because of certain assumptions in the logistic equation, it is not a good fit in many populations. For example, it "works" for those populations in which there is no significant time lag between birth and reproductive maturity and where all individuals make identical contributions to population growth (i.e., where there are no differences in reproductive capacity). Thus the equation best describes population growth of those unicellular plants and animals that have short life cycles, reproduce asexually, and have high reproductive rates. Finally, the assumption that the population growth rate per individual decreases linearly as new individuals are added (the cybernetic factor) is not biologically sound, as Frederick Smith (1952) and others have noted.

Models other than the logistic can be applied to population growth data with as good or better fit. Many are based on the logistic equation with modifications to accommodate some of the limitations noted. Because these modifications are often specific to particular types of life cycles or organisms, the general utility of such models is much diminished. Although Pearl called his curve a "law of population," it does not serve as a generalization that is universally valid in explaining or predicting. Thus while it does have some general use as a mathematical description of population growth, it does not satisfy the criteria for being a scientific law.

BIOTIC POTENTIAL

Having described the two general patterns of initial population growth, the J- and S-shaped curves, we can turn to an analysis of the two major forces in the process—biotic potential, which pushes the curve upward, and environmental resistance, which pushes the curve downward.

We have already noted that the increase potential of a population at any moment is dependent on both r, the reproductive potential of each individual, and N, the number of individuals present at the time, and is expressed as $\Delta N/\Delta t = rN$ in unlimited situations. That is, the rate of change in the population is proportional to the number of individuals (e.g., females in sexually reproducing species). Under optimal conditions for the housefly, as we already have stated, r has the value 120 eggs per female generation. That is, each female can produce 120 eggs, and the actual change of the population is then determined by the number of females present (Table 10-1). The size of each succeeding generation in that hypothetical population is based on the assumption that r is a constant.

Under the variations of natural environment, however, the reproductive potential shows variation. Australian ecologist L. C. Birth (1957), for example, showed that a single female rice weevil (*Calandra oryzae*) produces 22.4 offspring/yr at 23°C, 30.6 offspring/yr at 29°C, and 6.2 offspring/yr at 33.5°C. A comparable situation is observed in the water flea (Figure 10-4), where the maximum reproductive potential is obtained at a temperature intermediate between 20 and 34° .

Because of differential tolerance to and optima for the wide variety of physical and chemical factors in the environment, it is reasonable to expect that there will be different reproductive capacities under different environmental conditions. In fact, the value of r for a species has different and unique values for each set of physical conditions. Moreover, it varies with age of the individuals in the population, the age of onset of reproduction, and the length of life. It also varies with density, generally declining with increased crowding. Finally, the value changes in direct relationship to the time interval: as the time interval shortens, the value of r lessens. For example, the value of r for the rice weevil at 29°C is 30.6 offspring per year but would be 0.084 per day. In the logistic equation, r values are small, reflecting instantaneous rates over very short time intervals.

Intrinsic Rate of Natural Increase

If the biotic potential of a population is measured under optimal growth conditions, it is referred to as the *intrinsic rate of natural increase*, the r of the logistic equation. These optimal conditions include not only a physiocochemical environment in which population growth is maximum but also one with a stable age distribution in which there is an absence of the negative feedback effects of increasing density [the $(K-N)/K$ of the logistic equation]. That is, the intrinsic rate is the maximum that is biologically possible. With these limitations, and others, it is little wonder that so little comparative information on this population parameter has been obtained. Francis Evans and Frederick Smith (1952), in a study reporting the value of r for the human louse (*Pediculus humanus*), also included the few available r values as well as the length of a generation for each of several species (Table 10-3). It can be seen that the intrinsic rate of natural increase, r for the population, is almost eight times greater for insects than for rodents but that among rodents and among insects there is not much difference. Of considerable interest, however, is the apparent inverse relationship between r and generation time. To remain in the game, it appears that for a decreased value of r compensations have been achieved by natural selection in increased generation time. In a subsequent study, Smith reaffirmed an inverse relationship not only between r and generation time but also between body size and r. That is, in general, the larger the animal, the smaller the value of r for the population.

As noted, there is no increase or decrease in a stabilized population—the rate of change is zero. Because the rate of change has been defined as $\Delta N/\Delta t = rN$

TABLE 10-3 INTRINSIC RATES OF NATURAL INCREASE FOR RODENTS AND INSECTS

	Intrinsic rate of natural increase (r/day)	Generation time (average in days)	Net reproduction rate (R_0)
Rodents			
English vole (*Microtus agrestis*)	0.0125	141.8	5.9
Norway rat (*Rattus norvegicus*)	0.0147	217.6	25.7
Insects			
Grain beetle (*Tribolium castaneum*)	0.101	55.6	275.0
Rice weevil (*Calandra oryzae*)	0.109	43.4	113.6
Human louse (*Pediculus humanis*)	0.111	30.9	30.9

Adapted by permission from F. C. Evans and F. E. Smith. 1952. *American Naturalist* 86:299–310.

and this has the value of zero in a stable population, the value of *r* under these conditions must be zero, for *N* is a positive integer. Because such is the case, Evans and Smith suggest that *r* measured under optimal conditions is a reflection of "the degree to which the natural environment departs from conditions most conducive to growth." That is, the value of *r* must be at a level high enough to keep the population in existence. Thus the larger the value of *r*, the more rigorous the environment can be assumed to be for the population. For this reason, Smith has suggested that the larger value of *r*, which is typical of small organisms, indicates that their environment is harsher than that of large organisms. A protozoan, for example, is more subject to diffusion as a lethal factor in its existence than is an elephant.

Net Reproductive Rate

There is yet one other aspect of reproductive potential that needs mention—*net reproductive rate* (R_0). It can be defined as the number of female offspring that replace each female of the previous generation. In our example of the housefly, the net reproductive rate would be 60; that is, each female of one generation is expected to produce 120 offspring, of which 60 are female. We are assuming here, of course, that all 60 females survive and are capable of reproducing. It follows, then, that in a stabilized population not only would r be zero, but also R_0 (net reproductive rate) would be 1—each female being replaced by only one female. This should be self-evident, for if R_0 were 2, each female would be replaced by two and so the population would increase. Net reproductive rate coupled with generation time, then, can also give an indication of environmental resistance. For example, according to the R_0 values in Table 10-3, rat replacement is five times that of mice. Because their *r* values are so close, it would seem that about five times as many rats fail to grow to maturity. When generation time is factored in, three times as many female rats as female mice are produced in a year. The life insurance of the Norway rat seems to be that of expending life through relatively high mortality. Similarly, the considerably lower R_0 of the human louse, compared to the grain and rice weevils, suggests, as Evans and Smith state, "that its natural environment is remarkable secure... [this] is indeed surprising, for a high rate is usually associated with a parasitic mode of life."

Shifting Values of r and R_0

That the values of *r* and R_0 for a given species are variable due to environmental circumstances is shown in a study of the aphid *Brevicoryne brassicae* by Richard Root and Annette Olson (1969). Aphids, or plant lice, derive their sustenance by inserting their syringelike mouth parts into the vascular system of a plant and sucking up the plant liquids. In this particular experiment, the aphids, which reproduce asexually, were confined in small wire cages on broccoli, collard, chinese cabbage, and the weed, yellow rocket. The results, as shown in Table 10-4, demonstrate that both *r* and R_0 were smallest on yellow rocket and highest on broccoli grown outdoors. Given the arguments advanced by Smith and Evans, the least and most adverse

TABLE 10-4 THE INTRINSIC RATE OF NATURAL INCREASE (r) AND NET REPRODUCTIVE RATE (R_0) OF THE APHID *Brevicornyne brassicae* ON DIFFERENT HOST PLANTS

Plant	r	R_0
Collard	0.179	31.64
Yellow rocket	0.094	5.58
Chinese cabbage	0.223	44.04
Broccoli	0.191	34.20
Broccoli (outside)	0.243	59.57

Data by permission from R. B. Root and A. M. Olson. 1969. *Canadian Entomologist* 101:768–773.

environments for these aphids are, respectively, the yellow rocket and broccoli grown outdoors. Put another way, the yellow rocket most resembles the "natural" state of affairs for the aphids inasmuch as the other three "habitats" are domesticated and cultivated varieties that are not native to the aphids. This interpretation is not so simple since about one-third of the aphids on yellow rocket developed into the winged condition, an adaptation that would enable the population to disperse to more favorable hosts under noncaged conditions. Development of the winged phase is thus a response to unfavorable conditions that hold down aphid reproduction (e.g., inhibitory chemicals). The low r value would then not reflect the least but rather the most adverse environment in this situation.

DEATH RATES IN POPULATIONS

Since the rate of increase in a population is effectively the difference between birth rate and death rate at any given time, now we need to turn our attention to mortality, or, if you prefer, survival.

Life Tables

The impetus to formulate "laws" of mortality commensurate with those of population growth also came from Raymond Pearl. Pearl and his students initiated the preparation of *life tables* patterned after the actuarial tables used to determine insurance rates for humans. Because of the interests of actuaries and other enterprises in human life expectancy and survivorship, the development and preparation of life tables had attained considerable sophistication by the 1920s. Pearl introduced the life table in 1921 as a way of assessing natural populations. It was not until much later that Edward S. Deevey, Jr. (1947) thoroughly reviewed the life table as a means of conveying survivorship information, thereby enabling the development of generalizations on mortality. Except for laboratory populations, or natural populations essentially under laboratory-type control, data on survival of natural populations are often indirectly assessed. Survivorship in birds and many other animals is usually estimated by banding-releasing-recapture methods. Needless to say, such a procedure gives less reliable data than a controlled population in a laboratory culture dish, but statistical reliability is possible by appropriate handling of the data.

A life table is basically a table of probabilities dealing with the rate of death and expectation of life at various time intervals over an organism's life span. To construct a life table or even to take the space here to introduce it in sufficient detail would be beyond the purpose of the chapter. What we can do, however, is present graphically that portion of the life table that gives the expected or probable number of survivors in a population at various age intervals.

Survivorship Curves

Survivorship curves for selected natural and laboratory animal populations are shown in Figure 10-6. To eliminate the confusion that would result from setting the horizontal coordinate to the actual life span of various organisms, which, as noted, extends from days for some to years for others, life span is expressed as a percentage of the maximum possible life span. The vertical coordinate is logarithmic. A straight line on such a graph indicates a constant rate with respect to the horizontal coordinate—in this case, age. The number of survivors, out of an initial population of 1,000, at any particular time (percentage) of the life span can be read directly from the graph. Thus in the case of the rotifer (Figure 10-6, Type I), about 920 individuals are still alive at the midpoint (50 percent) of the maximum possible lifespan, but only about 440 individuals are alive at the three-quarter mark.

With the wide variety of life phenomena, we would expect—and, indeed, we observe—a considerable variety of survivorship patterns. They extend from those with low mortality (high survivorship) throughout most of the life span to those with high initial mortality (low survivorship). Pearl (1928) identified three major types of survivorship curves, which are easily recognized in the actual curves shown in Figure 10-6 and which will be discussed in detail below:

- Type I — a convex curve in which mortality is low initially and throughout more than half the life span after which it increases markedly.
- Type II — a straight line in which mortality is more or less constant throughout the life span.
- Type III — a concave curve in which mortality is extremely high initially but decreases markedly for much of the remainder of the life span.

Type I survivorship curve: initial low mortality. Populations of rotifers (*Proales*), humans, and mountain sheep have a relatively low mortality until well past middle age, after which the death rate increases markedly and, in the case of the rotifer, almost precipitously. In passing, it should be noted that a curve almost identical to that of the rotifer can be generated by transferring pupae of the fruit fly

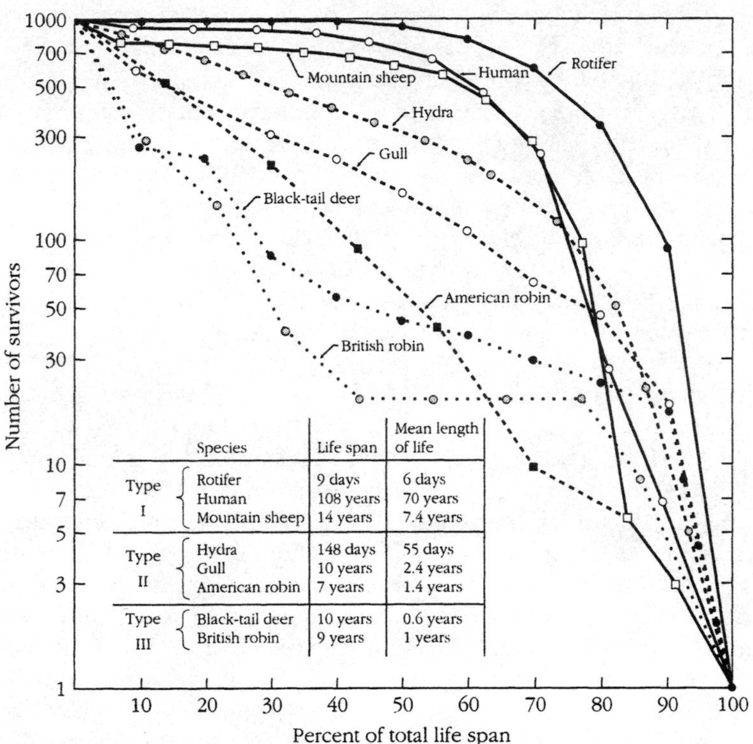

	Species	Life span	Mean length of life
Type I	Rotifer	9 days	6 days
	Human	108 years	70 years
	Mountain sheep	14 years	7.4 years
Type II	Hydra	148 days	55 days
	Gull	10 years	2.4 years
	American robin	7 years	1.4 years
Type III	Black-tail deer	10 years	0.6 years
	British robin	9 years	1 years

Figure 10-6 Survivorship curves for laboratory and natural populations; the vertical coordinate is logarithmic. Type I: Laboratory population of the rotifer, *Proales decipiens*, and natural populations of white American males, 1939–41, and of the Dall mountain sheep, *Ovis d. dalli.* Type II: Laboratory population of *Hydra fusca* and natural population of the gull, *Larus argentatus*, and the American robin, *Turdus m. migratorius.* Type III: Natural population of the black-tailed deer, *Odocoileus hemionus columbianus*, in managed open shrubland at high density of 223 deer/sq km, and natural population of the British robin, *Erithacus rubecula melophilus.* (Type I: Rotifer data of B. Noyes. 1922. *Journal of Experimental Zoology* 35: 225, reinterpreted by R. Pearl and C. R. Doering. 1923. *Science* 57: 209–12; human data from L. I. Dublin, A. J. Lotka, and M. Spiegelman. 1949. *Length of Life.* New York: Ronald Press; sheep data from E. S. Deevey. 1947. *Quarterly Review of Biology* 22: 282–314. Type II: Hydra data from R. Pearl and J. Miner. 1935. *Quarterly Review of Biology* 10: 60–79; gull and robin data from E. S. Deevey, op cit. Type III: Deer data from R. D. Taber and R. F. Dasmann. 1957. *Ecology* 38:233–46; robin data from E. S. Deevey, op cit.).

(*Drosophila*) to culture bottles lacking food. Under these conditions the longest-lived survivors last for only about 70 hours, but most of the population survives for some 70 percent of that time. In populations showing the rotifer-human-mountain sheep survivorship pattern, a given individual has a relatively long life expectancy at birth. Even for an individual at about one-third of the life span, the probability of living to a ripe old age is quite good.

The monarch butterfly also has a Type I survivorship (Wells and Wells 1992), as do a number of other animals. Among plants, where the procedures for obtaining data for life-table construction are much more laborious than for animals, range species in southeast California and several species of orchids of the genus *Dactylorchis* have been shown to have survivorship curves of the rotifer-human-mountain sheep type.

Another way of considering this type of survivorship is that the mean duration of life, or, as it is often called, *mean life expectancy*—the point at which one-half the population is still surviving—is much closer to the maximum duration of life (see tabulation in Figure 10-6, Type I). (To be correct statistically, *mean* life expectancy should read *median* life expectancy; however, very wide use of the former, as in other cases, prevails over technical accuracy.) It is of interest to note that what humans have been able to achieve through medical and environmental modification, particularly within the last century, is a shift of mean life expectancy closer to the maximum. At the time of the golden age of Greece, for example, life expectancy at birth was only about 30 years; by 1900 it had increased to 45 to 50 years, and at present it is more than 70 years. How close to the physiological limit of 100 to 110 years humans will be able to push mean life expectancy is intriguing both in the challenge of its possibility and in the concomitant problems of housing, feeding, and attending to the needs of a larger component of older persons.

Type II survivorship curve: constant mortality. The next cluster of survivorship curves is represented by the laboratory population of *Hydra* and the natural populations of the gull and American robin. In these populations there is a more or less constant rate of death throughout life (robin) or at all ages until near the end of the life span (*Hydra* and gull). In the case of *Hydra* and gulls, the rate is quite constant from birth through nearly three-fourths of the life span. For a given individual in populations of this sort, there is considerably less chance of survival to old age, at, say, the midpoint of the life span than for individuals in the rotifer-human-mountain sheep type of population. Also characteristic of this cluster of curves is the fact that the mean duration of life is about one-third of the total life span (*Hydra*, gull) or somewhat less (robin). If those starved fruit flies previously mentioned were permitted to feed, their survivorship curve would be intermediate, between that of *Hydra* and the mountain sheep, and with a lengthened life span.

Grizzly bears (Knight and Eberhardt 1985) and grey squirrels (Barkalow et al. 1970) also have a Type II survivorship as do many herbaceous annuals, such as the meadow grass *Poa annua* (Begon and Mortimer 1986) and *Phlox Drummondii* (Leverich and Levin 1979), as well as herbaceous perennials demonstrating Type II survivorship. Mediterranean fruit flies (*Ceratitis capitata*), while demonstrating Type II survivorship, actually have a reduced mortality at older ages (Carey et al. 1992).

Type III survivorship curve: initial high mortality. The third group of curves, represented by the black-tail deer and British robin, is one in which there is a high mortality at the beginning of the life span. Three-fourths of the initial population of the black-tail deer and of the British robin die off within the first 10 percent of the life span. In both populations the death rate does not change markedly during almost the first half of the life span. Populations of the Columbian ground squirrel (*Spermophilus columbianus*) in southern Alberta also show Type III survivorship (Zammuto 1987). Given individuals in populations of this type have a low life expectancy at birth, the mean duration of life being in the order of one-fifteenth of the life span.

Curves with an even more extreme deflection toward the lower left quadrant would be expected in profligate egg-laying groups like teleost fish, amphibians, and many mollusks, especially the open ocean species. Most annual as well as long-lived plants also demonstrate this more extreme initial mortality—consider trees with a prolific seed production coupled with a long life-span. One of the most complete analyses made of any plant population is that of two annuals, *Sedum smallii* and *Minuartia uniflora*, which grow on granite outcrops in Georgia and which exemplify this latter situation (Sharitz and McCormick 1973).

Bamboo, for example, has a natural life span of 60 to 80 years. During that interval, all the bamboo in an area will bloom, set seed, and then die; 10 years may elapse before the new seedlings grow to full size, during which time pandas, which are dependent on them for food, may starve. In May 1983, large areas of bamboo on mountains in the west of Sichuang Province and in the south of Shaanxi Province blossomed and withered (Zhou and Zhang 1991). Of the initial 22,450 hectares of bamboo, 77 percent died in 25 counties and up to 90 percent did so in the 11 worst hit counties. In less than a year 21 pandas were found starving and sick, only 9 of which survived after emergency rescue.

Species-Specific Mortality

These selected examples indicate that more or less distinct patterns of mortality not only exist in nature but also are species-specific—that is, they are as characteristic of a species as its birthrate, behavior, anatomy, and physiology.

Even within closely related species, however, differences are found in survivorship curves. An analysis of four different species of Pierid butterflies, the group that includes the cabbage whites, orange-tips, and sulfurs, showed that each demonstrated Type III survivorship but with species-specific differences in the mortality rate at each age interval (Courtney 1986).

Sex-Specific Mortality

Not only is mortality species-specific, it also appears to be sex-specific within a species (Figure 10-7). In general, females have the edge in both lessened mortality

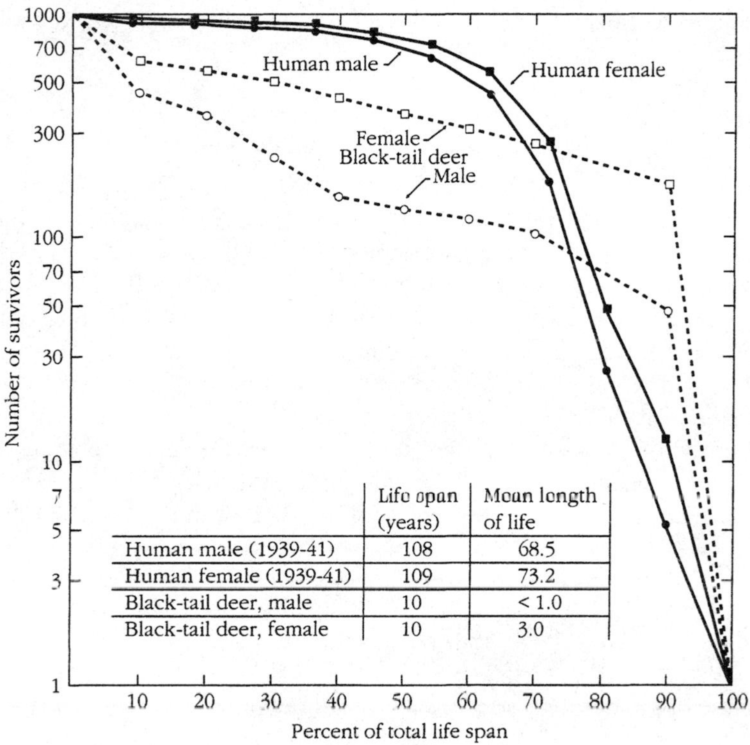

	Life span (years)	Mean length of life
Human male (1939-41)	108	68.5
Human female (1939-41)	109	73.2
Black-tail deer, male	10	< 1.0
Black-tail deer, female	10	3.0

Figure 10-7 Differences in survivorship based on sex. Universally the male of the species has a higher mortality throughout the life span. (Data on white males and females, 1939–41, from L. I. Dublin, A. J. Lotka, and M. Spiegelman. 1949. *Length of Life.* New York: Ronald Press; data on deer in unmanged chaparral at low density of 27 deer/sq km from R. D. Taber and R. F. Dasmann. 1957. *Ecology* 387: 233–46—cf Figure 10-6, Type III).

and a higher mean duration of life, both in humans and in a number of other species as well. In the case of the black-tailed deer of California, R. D. Taber and R. F. Dasmann (1957), whose study serves as basis for the graph, suggest that the higher mortality of males is related to their higher metabolic rate and nutrient requirement. Especially at 2 years of age, but also beyond, the male is more subject to predation (hunting) than the female. For the female, the constant rate of death over virtually the entire lifetime is doubtless related to the stress of reproduction.

In humans the evidence on mortality suggests that, in addition to the greater exposure of men to occupational and other environmental hazards, there is an inherent biological difference that favors a higher mortality in men. From birth on, and even during the intrauterine period, the ratio of male mortality to that of the female is always in excess of 1. According to studies by L. I. Dublin, A. J. Lotka, and M. Spiegelman (1949), the ratio peaks twice: at age 15 it is about 1.49:1; afterward it declines to allow of 1.27:1 at age 30, and then it rises again, reaching its maximum of about 1.54:1 at about age 55.

Intraspecific Mortality

Intraspecific differences other than gender also affect species-specific mortality. F. S. Bodenheimer (1958), of the Hebrew University of Jerusalem, who maintained a long, active research interest in problems of animal populations, prepared life table data on mutants of *Drosophila* from raw data collected by B. M. Gonzales. These data are presented as survivorship curves in Figure 10-8. The wild, or natural, population (nonstarved) has, as already indicated, a survivorship curve intermediate between the rotifer-human-mountain sheep type (Type I) and that of hydra-gull-American robin (Type II). Survivorship in the vestigial-winged form is of the hydra-gull-American robin type with a constant rate of death at all ages and a life expectancy about half that of the wild type. Although a natural population is generally assumed to be the best adapted because of the operation of natural selection

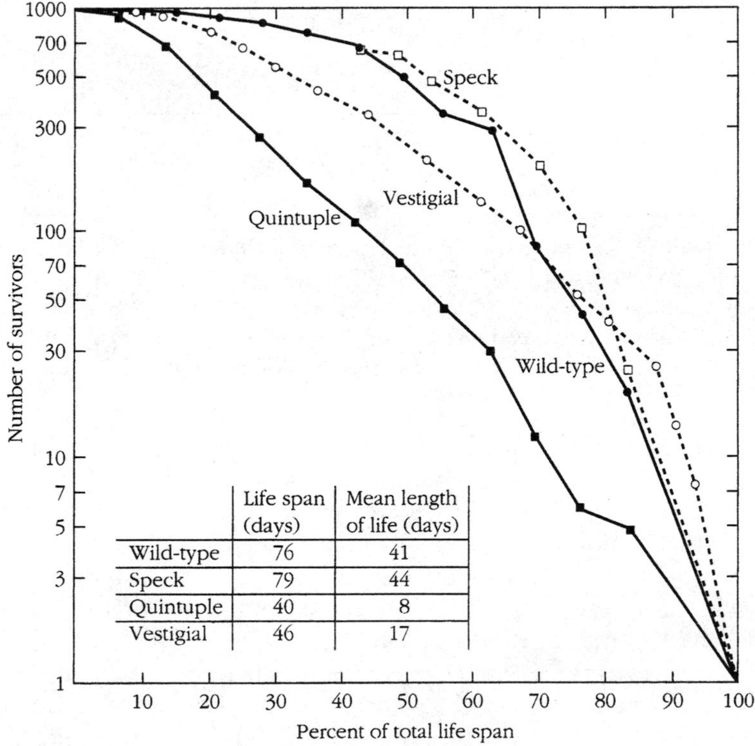

	Life span (days)	Mean length of life (days)
Wild-type	76	41
Speck	79	44
Quintuple	40	8
Vestigial	46	17

Figure 10-8 Differences in survivorship of genetic mutants of the fruit fly, *Drosophila melanogaster.* (Based on data of Gonzalez reinterpreted by F. S. Bodenheimer. 1958. *Monographiae Biologicae* 6:1–276.)

processes, it is of interest to note that the mutant known as speck has a lesser mortality, particularly in the latter half of life. Perhaps in nature such lessened mortality in the middle of old age in fruit flies is not so critical and hence not particularly advantageous or disadvantageous. Exemplary of the more unfavorable consequences generally attributed to mutation, the mutant quintuple not only has the highest mortality rate, but it has the shortest life span as well, with a life expectancy one-fifth that of the wild type.

Environmental Factors and Mortality

Death is, of course, a natural biological event, as natural as birth. As such, its occurrence in any given individual or population is subject to the vagaries of the milieu of the individual or population. It can be modified by environmental factors, as we have already observed for birthrate. The survivorship curves of black-tailed deer shown in Figures 10-6 and 10-7 are of the same species (*Odocoileus hemionus columbianus*), but the populations represented in each graph are from different habitats and have different densities. The deer population of Figure 10-6 occurred in an open shrubland maintained by fire and that of Figure 10-7 in a well-vegetated chaparral. The shrubland population was more than 2.5 times as dense as the chaparral population, 32 compared to 13 deer per square kilometer in July. The higher mortality of the shrubland population is partly due to greater success of predation, for there is less cover for deer. Interestingly, this is compensated by a reproductive rate that is more than twice that of the chaparral population.

The effect of density on mortality has also been shown for laboratory populations. Peter Frank and his colleagues (1957), for example, demonstrated that the optimum density for mortality rate, maximum length of life, and mean life expectancy in the water flea, *Daphnia pulex*, was at an initial density equivalent to eight individuals per cubic centimeter (Figure 10-9). Densities above and below this level were tolerable, but not as advantageous in regard to these parameters of population growth.

AGE STRUCTURE

The preceding analysis of death rate or survivorship should have suggested the substantial difference that may exist in the relative proportion of young, middle-aged, and older individuals in a population, depending on the type of survivorship it exhibits. A quick look at the various figures on mortality should indicate that if the various populations had continuous input at the left side of the graph, characteristic distributions of different age classes should be one of the results. For example, in those groups in which life expectancy approaches the maximum span of life (Type I, rotifer-human-mountain sheep), a greater proportion of the population would be in the older-age group than would be true of Type II (hydra-gull-American robin). If the population were modified by shifts in input (natality, immigration) and output

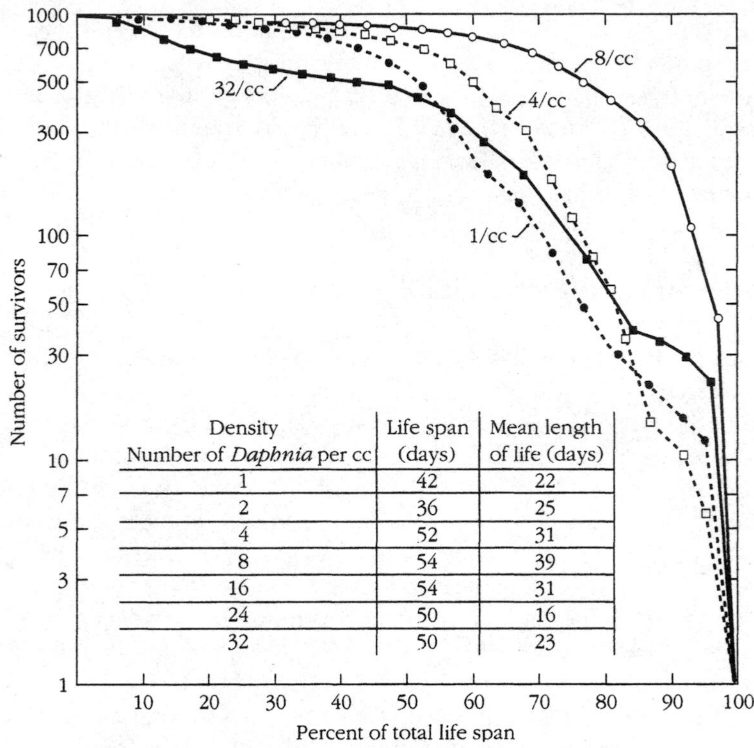

Density Number of *Daphnia* per cc	Life span (days)	Mean length of life (days)
1	42	22
2	36	25
4	52	31
8	54	39
16	54	31
24	50	16
32	50	23

Figure 10-9 Differences in survivorship as a function of initial density. (Redrawn by permission from P. W. Frank, C. D. Boll, and R. W. Kelley. 1957. *Physiological Zoology* 30:287–305.)

(mortality, emigration), however, the age structure would be expected to shift. These various inter-relationships of age structure in populations are worth considering briefly.

Major Age Groups

From an ecological point of view there are three major age groups in a population: *prereproductive*, *reproductive*, and *postreproductive*. These categories, proposed by Bodenheimer (1958), have considerable value as an aid in visualizing the status of a population (Figure 10-10).

Expanding population. In a rapidly growing population (Figure 10-10a), birthrate is high and population growth may be exponential, as in the case of the housefly, yeast, and alga considered previously. Under these conditions each successive generation will be more numerous than the preceding one; as a result, a pyramid-shaped age structure results.

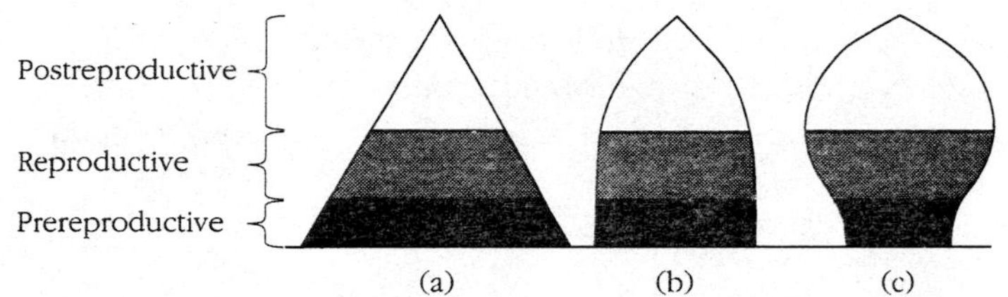

Figure 10-10 Age structure of different types of populations: (a) expanding population; (b) stable population; (c) diminishing population. (Adapted by permission from F. S. Bodenheimer. 1958. *Monographiae Biologicae* 6:1–276.)

A ready visualization of this effect can be seen by examination of Table 10-1 on the housefly. If the data of the third or fourth columns are plotted as successive layers in bar graph style, a pyramidal structure results. The oldest group in the population at the top of the pyramid would consist of 120 individuals; the base would consist of 5.7 to 6.7 trillion very young individuals.

Stable population. As the rate of growth slows and stabilizes (i.e., where *r* approaches zero and the net reproductive rate approaches one), the prereproductive and reproductive age groups become more or less equal in size; the postreproductive group remains as the smallest. The graphic representation of this stabilized population is that of a bell (Figure 10-10b).

Diminishing population. If the birthrate is drastically reduced, the prereproductive group dwindles in proportion to the reproductive and postreproductive groups, resulting in an urn-shaped age structure (Figure 10-10c). This is representative of a population that is dying off.

In a study of a hive of honey bees, where a seasonal trend in the population takes place, Bodenheimer observed these shifts in age structure (Figure 10-11).

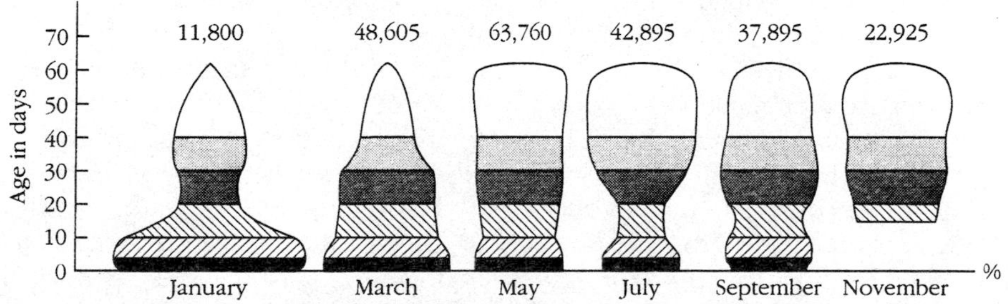

Figure 10-11 Changes in age structure in a hive of honey bees over one season. (Redrawn by permission from F. S. Bodenheimer. 1958. *Monographiae Biologicae* 6:1–276.)

There is a change from a rapidly growing population, with the triangular shape in January, to a stable population, represented by a bell shape in March, and a progressive dying off represented by a series of urn shapes from July through November.

Species Differences in Age Structure

The relative proportion of the life cycle occupied by each of the "ecological ages" differs from species to species. In humans, for example, the prereproductive period (birth to about age 15) occupies about 21 percent of the mean life expectancy, the reproductive, ages 15 to 45, about 42 percent, and the postreproductive, 37 percent. In the rat these proportions are about 25, 20, and 55 percent, respectively.

Animals. In contrast to these relatively short prereproductive stages, many, but by no means all, insects spend well more than half their life cycle in that stage. Some dragonflies spend 2 years in the egg and larval stages and survive barely 4 weeks as an adult, being capable of reproduction for as little as only 1 or 2 days of that brief span. Many birds and mammals have a very short, if any, postreproductive period. The chaparral black-tailed deer, for example, is capable of reproduction up until death at an age of 10 years. Because the 1-year-olds are not reproductive, the age structure in that population is a diminishing one with a peculiar pyramidal urn shape: about 42 percent of the population is in the prereproductive category (age 0–1); 58 percent is in the reproductive segment, equally divided into the age group 1–3 and age group 3–10. In their study of the human louse, Evans and Smith (1952) determined that the stable age distribution consisted of 5.69 percent adults, 26.43 percent larvae, and 67.88 percent eggs. Similar distribution exists in the grain beetle and rice weevil.

Plants. Among plants, a number of annuals start flowering and set their first seeds when very young. This situation is particularly true in extremely demanding and unpredictable environments, such as deserts. Other annuals developing in stable, predictable environments have a relatively long prereproductive (vegetative) stage whose change to the reproductive phase is often timed by *photoperiod*, that is, the relative proportion of daylight and darkness.

In biennials at least 1 year of prereproductive growth occurs before seed is produced and under stressful conditions flowering may be delayed several years.

Among herbaceous perennials, the prereproductive period appears to be quite variable: from 2 years in carpenter weeds (*Prunella vulgaris*) to 8 years in water avens (*Geum rivale*). Among bulbous species, the prereproductive period is from 1 (dahlias and gladiolus) to 4 to 7 years (daffodils, tulips). In woody perennials, conifers have earlier reproductive ages for a given length of life than angiosperms. Several conifers that survive over 200 years reproduce before age 10, whereas most angiosperms with a life span of over 200 years spend at least 20 years in the juvenile period. In general, the ratio of life span to prereproductive stage among angiosperms is 10:1. Plants with short prereproductive periods have short longevity and those with long prereproductive periods have long longevity and a long reproductive period.

Geographic Differences in Age Structure

The age structure of a population may differ geographically as a result of differential densities, as in the case of the two deer populations (Figures 10-6, Type III, and 10-7), or as an effect of a difference in an abiotic regulatory factor. The life history of the dragonfly *Epitheca cynosura*, for example, is so regulated by temperature that two years are required to complete development in southern Michigan, but only 1 year in Florida. Peculiarly, however, in North Carolina 90 to 95 percent of the population can complete development in 1 year; the remaining 5 to 10 percent of the population requires 2 years (Kormondy 1959). At any given time of the year the age structure in each of these populations is presumably stable but is very different. In late April, for example, the Michigan population would consist of a group of 2-year-old larvae about to transform to adults and a somewhat larger population of 1-year-old larvae. The Florida population would consist only of 1-year-old larvae about to transform to adults. The North Carolina population would consist of two chronological but three physiological age groups: about 90 percent of 1-year larvae about to transform to adults, about 10 percent of 1-year larvae that will carry over until the following year, and many fewer 2-year larvae, the 10 percent carried over from the previous year, now about to transform to adulthood.

Developmental Differences in Age Structure

Age structure also differs at different developmental stages of an ecosystem (Figure 10-12). In his study of spruce-fir forests of the coast of Maine, Ronald Davis (1966) recognized three major developmental stages: A (primary age class), in which

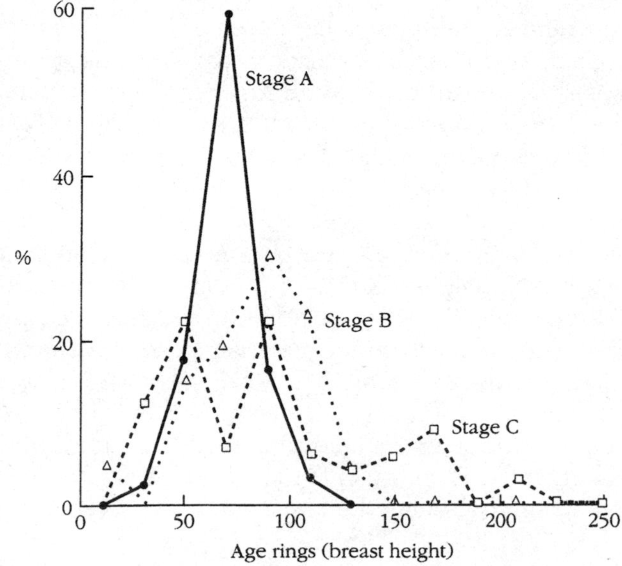

Figure 10-12 Age frequency distribution in percent for red spruce in different developmental stages of the spruce-fir ecosystem in Maine. (Redrawn by permission from R. B. Davis. 1966. *Ecological Monographs* 36:79–94.)

the trees had an average of 80 annual rings, an average height of 14.5 m, and a diameter (at breast height) of 10 to 20 cm; B (secondary age class), in which trees had an average of 112 annual rings, an average height of 19.5 m, and with a greater number of young trees with a diameter of 2.5 cm than of 10 cm—this stage arises when blowdown or death of older trees of stage A opens up the canopy, allowing germination and growth of new individuals; and C (virgin stand), in which trees had an average of 163 annual rings, an average height of 13.5 m, and a smaller density of trees of 10 cm diameter than the preceding stages.

Age Structure in Human Populations

Age structures of human populations in the major areas of the world are shown in Figure 10-13. By the bell shape it can easily be seen that the populations of Europe, North America, and the former USSR are relatively stabilized. Compare this situation with the wide-based triangles of the rapidly growing populations of South Asia, Africa, and Latin America. Among the significant revelations of these age structures is that the proportion of the population in the reproductive age is not strikingly different among the six populations (42 to 47 percent). The proportion of prereproductives to postreproductives is markedly different, however. Especially noteworthy in this connection is the extremely small proportion of older persons in Africa, South Asia, and Latin America. In these rapidly developing countries traditionally high birthrates are now combined with a quite drastically reduced and rapidly decelerating infant mortality. The result is extremely high proportions of younger persons. As medical advances reduce the high mortality in the older age groups and as contraception and family planning become more widely practiced, these pyramidal forms should tend toward a bell shape.

Comparison of the age structure in these six regions between 1965 and 1990 allows for several significant observations:

1. the age structure of Europe and North America has become discernably urn-shaped (i.e., a progressively more stabilized population edging toward diminishing size). These changes, more marked in North America's age structure, are the result of a considerable drop in birthrate, with the result that the prereproductive proportion of the populations has dropped from 25 to 20 percent in Europe and, more dramatically, from 32 to 21 percent in northern North America;
2. age structure in the USSR (data were projected just prior to the dissolution of that nation into its several nation states) has moved to a more definite bell-shape, symbolic of a stabilized or stabilizing population with a marked increase in postreproductives and decrease in prereproductives. The trend suggests the nations formerly composing the USSR are moving toward an urn-shaped age structure;
3. virtually no change has occurred in the populations of South Asia and Africa in spite of increased efforts at birth control, which, in some large measure, have been countered by increased survival of children because of improved nutrition and health care;

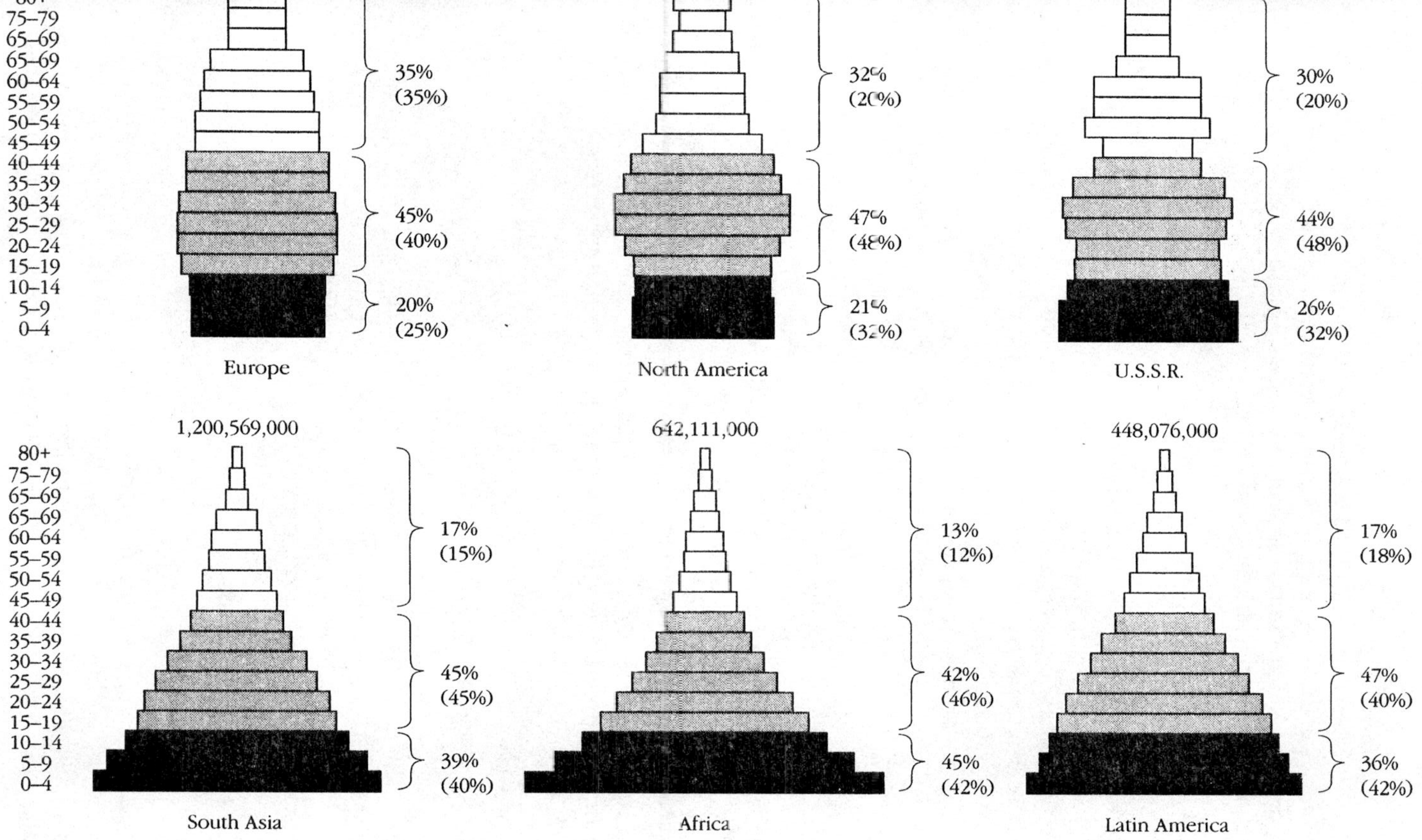

Figure 10-13 Age structure in human populations in the major areas of the world based on a medium projection by the United Nations from the estimated 1988 population. Each segment in the horizontal scale represents 1 percent of the total population; black bars are the prereproductive groups, stippled bars the reproductive groups, and clear bars the postreproductive groups. The numbers at the top of each age structure is the projected population for 1990, and the percentages within the parentheses are the proportions of the population as of 1965. (Adapted from United Nations Population Studies No. 122. 1991. *The Sex and Age Distributions of Population. The 1990 Revision of the United Nations Global Population Estimates and Projections.*)

4. although the reproductive and postreproductive groups are essentially the same in Latin America over the past 35 years, there has been a marked drop in prereproductives suggesting a progressive move toward a bell-shaped age structure indicative of a stabilizing population.

Stability of Age Distribution

Some years ago the biometrician A. J. Lotka (1922) advanced mathematical support for his contention that age distribution in any population is variable only within certain limits. Furthermore, he postulated that, subsequent to its being disrupted, the population will return to its previous and characteristic type of stable age distribution if both natality and mortality remain unchanged from the previous condition. The particular age distribution is, of course, not unrelated to the remaining components of the ecosystem. What Lotka indicated is that only if environmental conditions are restored will a disrupted and peculiar age distribution reinstate itself. Under a more or less permanently altered environment the stable age structure that develops would probably be different from the previous one.

Without exploring the mathematics from which this principle is deduced, it follows from the facts that the optimal value of r is dependent on stable age distribution (and the latter can be calculated if the value of r is known). Moreover, as we have pointed out, there is a unique value of r for each species for each set of physical and biological conditions. Thus altering conditions alters r, which, in turn, alters age structure. When conditions are returned to their previous state, age structure returns to its previous level and so does r. Age structure and r are mutually dependent.

FLUCTUATIONS AND EQUILIBRIUM

The S-shaped growth curve suggests a leveled-off stage in which the population and its environment are in some degree of equilibrium, that is at carrying capacity. However, as is true of virtually all biological processes, equilibrium is a dynamic state of fluctuation rather than an unwavering constant.

Minor and Major Fluctuations

In some populations under natural conditions the fluctuation around the equilibrium level is minimal, as in the classic case that occurred when sheep were introduced into Tasmania (Figure 10-14). In other populations, the amount of departure from the equilibrium level is not only greater but often irregular, as demonstrated in the case of laboratory-reared water fleas (Figure 10-15) and a desert rodent under natural conditions (Figure 10-16).

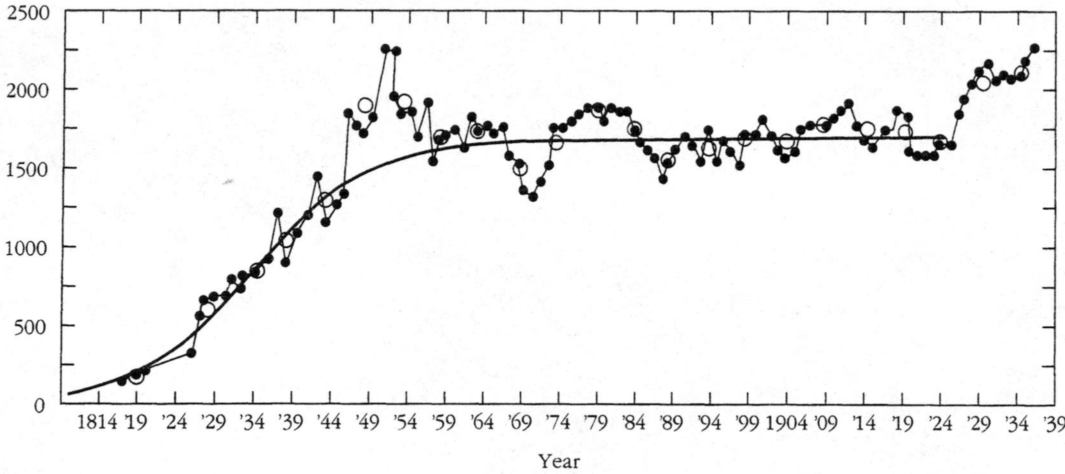

Figure 10-14 The growth curve of sheep subsequent to their introduction in Tasmania showing an initial sigmoidal pattern followed by semi-equilibrium. (Redrawn by permission from J. Davidson. 1938. *Transactions of the Royal Society of South Australia* 62: 342–6.)

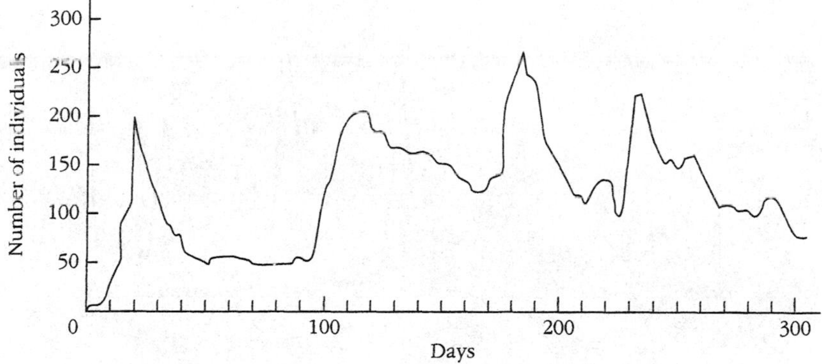

Figure 10-15 The growth curve of the water flea, *Daphnia obtusa*, in the laboratory, showing an initial J-shaped curve followed by irregular fluctuations. (Redrawn by permission from L. Slobodkin. 1954. *Ecological Monographs* 24:69–88.)

In yet other populations, such as the alga *Dinobryon* (Figure 10-5), the adult population nearly or actually dies out without showing any tendency to equilibrium; these are populations whose initial growth patterns are J-shaped. The pattern in blowflies (Figure 10-17) is a series of J-shaped curves. Although the large fluctuations cause the blowfly to come close to extinction, a sufficient number of young is introduced to maintain the population.

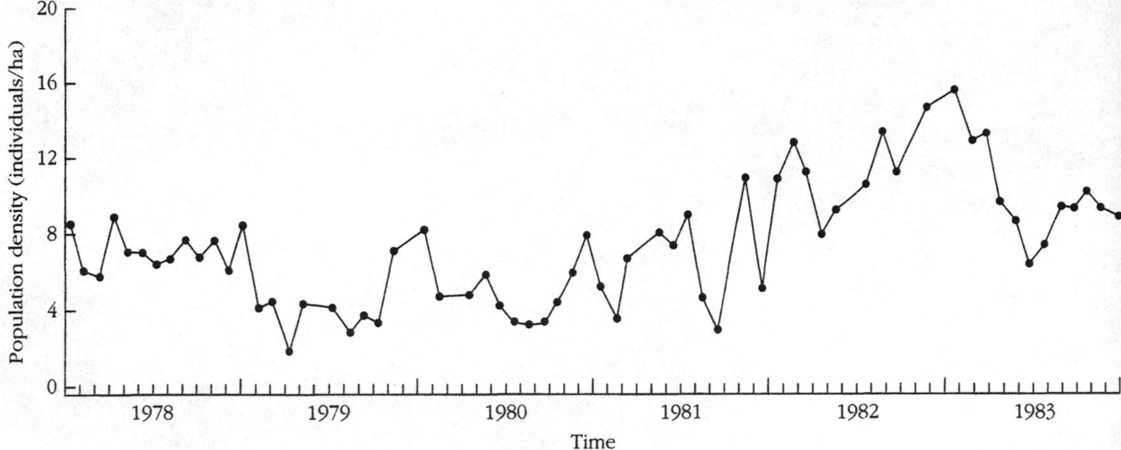

Figure 10-16 Variations in population density of the desert rodent, *Dipodomys merriami*, in the Chihuahuan Desert near Portal, Arizona. (Redrawn by permission from Z. Zeng and J. H. Brown. 1987. *Ecology* 68:1328–40.)

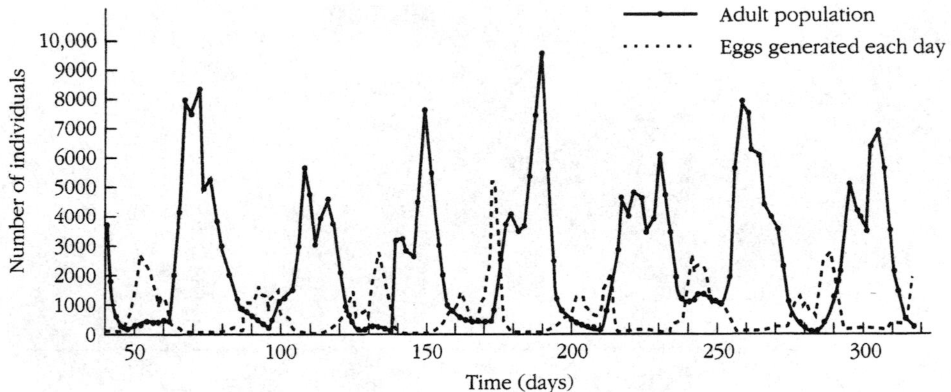

Figure 10-17 The growth curve of the blowfly, *Phaenicia cuprina*, in the laboratory, showing an initial J-shaped curve followed by rather regular fluctuations. (Redrawn by permission from A. J. Nicholson. 1955. *Australian Journal of Zoology* 2:9–65.)

Oscillations

In still other populations there is a regular oscillation with or without the near extinction exemplified by the blowflies. Certain predator-prey interactions, which will be discussed in a subsequent chapter, show such oscillations. Lemmings, meadow voles, and snowy owls show a 3- to 5-year oscillation, and ruffed grouse, ptarmigan, muskrat, snowshoe rabbit, and lynx show a 9 to 10-year oscillation. In the Province of New Brunswick, Canada, spruce budworm population cycles over the

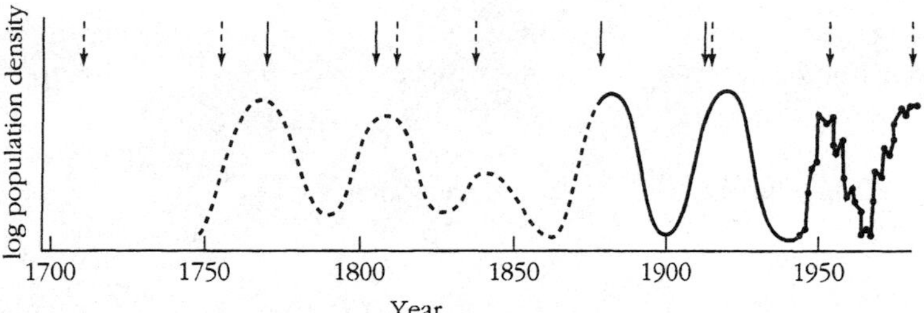

Figure 10-18 The 35-year oscillating population cycle of the spruce budworm, *Christoneura fumiferana*, over a 200-year period in the Province of New Brunswick, Canada (logarithmic scale). Graph based on sampling data since 1945, historical records since 1878, and radial growth-ring analysis of some surviving trees for the pre-1878 period. (Redrawn by permission from T. Royama. 1984. *Ecological Monographs* 54:429–62.)

past two centuries demonstrate a 35-year oscillation (Figure 10-18) with local populations in the province tending to oscillate in unison but not always with the same amplitudes and mean levels (Royama 1984). These oscillations appear to be independent of factors such as predation, food shortage, weather, and losses of young larvae during spring and fall dispersal, but rather are determined by parasitoids and infectious diseases, both of which are exacerbated by the number of larvae present; that is, they are dependent on the density of the population. Density-dependent regulation of populations will be discussed further in a subsequent chapter.

Population Regulation

In spite of the sharp differences in growth patterns typified by the postinitial growth phase or "equilibrium level" of sheep, water fleas, blowflies, and other species, populations seem to have an inherent ability to maintain themselves or to be so regulated as to be maintained. We have already shown that two major factors of a population's growth pattern, namely natality and mortality, are subject to both abiotic (Figure 10-4) and biotic (Figure 10-9) environmental influences. A more systematic consideration of the role played by these components in determining population size and distribution is the focus of the following chapter.

SUMMARY

VOCABULARY

biotic potential
carrying capacity
environmental resistance
growth curve
growth rate curve
intrinsic rate of natural increase
J-shaped growth curve

life tables
logarithmic growth phase
logistic curve
mean life expectancy
mortality
natality
net reproductive rate
postreproductive age group
prereproductive age group
reproductive age group
S-shaped or sigmoidal growth curve
survivorship curve

KEY POINTS

- The growth and development of a population is the result of two opposing forces, the capacity for reproduction, or biotic potential, and the capacity for death, or environmental resistance.
- Growth curves of populations demonstrate an S (sigmoidal) shape or a J shape, the former being more predominant; both can be described mathematically in the form of the logistic equation.
- The rapid phase of growth rate in both S- and J-shaped population curves is the logarithmic phase; the equilibrium phase of a S-shaped curve reflects the carrying capacity of the environment, a variable that results in oscillations and fluctuations of varying degrees.
- Under optimal conditions, a population's biotic potential is its intrinsic rate of natural increase, a characteristic that bears an inverse relationship to generation time.
- Life tables, constructed from probabilities of death rates and life expectancies, can be expressed graphically as survivorship curves, there being three major forms—convex, concave, and straight line.
- Mortality differs among and within species and by sex, resulting in the age structure of a population that can be expanding, stable, or diminishing, depending on the relative sizes of the ecological ages represented.
- Age structure is generally stable, being variable only within certain limits in most populations; age structure in human populations, however, is more variable and has been undergoing dramatic changes in recent times.

THE ROLE OF ABIOTIC FACTORS

Given the many interactions that occur between organisms and their environment, it is reasonable to expect that the latter can affect the size, distribution, and other characteristics of a population. As was noted in describing the nature of ecosystems, the environment consists of both abiotic and biotic components. In this chapter we will explore some of the ways in which abiotic factors such as nutrients, moisture, and food availability regulate populations and how regulation is effected by biotic factors such as competition, predation, and density.

Nutrients

In the earlier discussions of the role of nutrients (Chapter 5), the "laws" of minimum, limiting factors, and toleration (Chapter 3), as well as aspects of nutrient cycling (Chapters 8 and 9), it was noted that the supply and/or availability of particular nutrients can be a major limiting factor in an ecosystem. However, the role of nutrients in regulating a population's size, density, and distribution was only indirectly assessed. This role warrants further exploration, and thus we will consider the regulatory effect of sodium on meadow voles, phosphorus on woodland herbs, and nitrogen on insects.

Sodium and meadow voles. Published records of the meadow vole (*Microtus pennsylvanicus*) show a substantial correlation between the level of sodium in the soil and population density (Table 11-1). While correlations may give clues to actual causal relationships, they can also be misleading. A set of

TABLE 11-1 CORRELATIONS OF SODIUM LEVEL IN SOIL AND POPULATION DENSITY IN MEADOW VOLES (*Microtus pennsylvanicus*)

Sodium level	Highest density reported (number/acre)	Number and percentage of reports of more than 30 individuals per acre
Low	230	$\frac{12}{63}$ or 19%
Medium	400	$\frac{16}{22}$ or 72%
High	1,000+	$\frac{30}{33}$ or 91%

Data from G. D. Aumann and J. T. Emlen. 1965. *Nature* 208:198–99.

experiments was designed by G. D. Aumann and J. T. Emlen (1965) to determine the causal relationships between sodium and voles by offering restricted and unrestricted sodium chloride diets to a laboratory colony of voles established from a natural population. The experimental animals were offered distilled water and either unrestricted or restricted amounts of a 0.5-percent sodium chloride solution, both in standard water bottles; control animals had access only to distilled water. Some major findings of the study were as follows:

1. Experimental animals on an unrestricted sodium diet and in ratios of either 2♂:2♀ or 1♂:3♀ produced, respectively, 48 and 63 percent more young than the controls and maintained a significantly higher net population during the test period of 15 to 19 weeks.
2. High-density populations selected more salt solution than populations at one-fourth the density, the salt preference being higher in females than in males.
3. As density was increased sixfold and then decreased to the original density, the ratio of selection of salt solution to water correspondingly rose from 2:1 to 10:1 and dropped back to 4:1.

Aumann and Emlen suggest that the higher selection of sodium solution under crowding conditions was caused by a sodium deficiency in the animals. This condition perhaps resulted from inadequate adrenocortical regulation of sodium metabolism under the stress conditions of high density. (Later in this chapter we will consider some further evidence on adrenal interaction in population regulation.) From their experimental work and from the field data, Aumann and Emlen concluded that sodium is the critical factor limiting the population of meadow voles in many areas. Low soil-sodium levels would tend to restrict population growth by physiological (adrenocortical) responses associated with crowding; high soil-sodium levels would tend to unleash population growth to the limits imposed by some other factor.

Phosphorus and nettles. As in the correlation of sodium and vole populations, a number of general observations suggested that the distribution of the nettle

Urtica dioica was controlled by the supply of nitrogen. Using incubation techniques, for instance, it had been shown that nettles occurred in soils with a high intensity of nitrification along with high concentrations of other plant nutrients. Two British ecologists, C. D. Pigott and K. Taylor (1964), showed further that midsummer nitrogen and phosphorus levels in the aerial shoots of *Urtica* were about three times greater than in another broad-leaved, herbaceous perennial (*Mercurialis perennis*) that is commonly associated with it. Pigott and Taylor, however, found little difference in the amount of either total or inorganic nitrogen in soils directly associated with each of these two perennials. But as noted earlier, the amount available says little if anything about the rate of turnover of a nutrient. Under greenhouse conditions Pigott and Taylor found that additions of nitrogen as nitrate to various soils resulted in no greater growth of nettles and, in a few instances, produced a reduction. Addition of phosphate, with or without nitrogen, however, resulted in a 15- to 42-fold increase in growth on all soils. No significant growth response was obtained when other essential nutrients were used in the absence of phosphate.

Investigations of other herbs associated with nettles indicated that it was the most sensitive to phosphate deficiency, and, as might be anticipated, field experiments showed that there was a highly significant interaction between phosphate concentration and light intensity (Table 11-2). From these results Pigott and Taylor concluded that except in deep shade:

1. The failure of nettles to become established is attributable to a low supply of available phosphorus, hence phosphorus is a direct factor in nettle distribution.
2. The growth of the population is regulated by the amount of phosphorus available, not nitrogen, as original correlative findings suggested.

Nitrogen and insects. Although nitrogen does not play the limiting role in nettle populations, it is a regulating factor in other populations. Mark McClure (1980) has shown, for example, that the amount of nitrogen in leaves significantly affects the population success of the elongate hemlock scale insect *Fiorinia externa.* These plant pests, like aphids, feed on leaf sap and are capable of causing severe defoliation when present in large numbers. McClure found that the important factor in host suitability among some 14 coniferous species was the concentration of nitrogen in young foliage available to the nymphs during their development. Nymphs with a

TABLE 11-2 INTERACTION OF LIGHT INTENSITY AND PHOSPHATE SUPPLEMENT ON THE WEIGHT GAIN OF NETTLE *Urtica dioica*[a]

	No phosphate added	5 g/m^2 phosphate added
Low light intensity	2.0 ± 0.2	10.5 ± 1.1
High light intensity	15.2 ± 2.5	177.7 ± 24.0

[a]In the field between April 8 and September 1 (data in milligrams of dry weight, mean ± standard error). Modified by permission from C. D. Pigott and K. Taylor. 1964. *Journal of Ecology* 52 (Supplement):175–85.

TABLE 11-3 EFFECT OF FERTILIZATION ON NITROGEN CONCENTRATION OF YOUNG HEMLOCK NEEDLES AND SUBSEQUENT EFFECTS ON SURVIVAL AND FECUNDITY OF THE HEMLOCK SCALE INSECT, *Fiorinia externa*

Treatment	Total nitrogen (% dry mass)	Survival of nymphs (%)	Females with eggs (%)	Number of eggs per female
Unfertilized	4.28 ± 0.37	68.5 ± 7.9	13.5 ± 3.8	9.3 ± 1.9
Fertilized	5.64 ± 0.39	81.5 ± 4.6	42.2 ± 5.2	13.3 ± 2.1

Modified by permission from M. McClure. 1980. *Ecology* 61:72–9.

rich source of nitrogen experienced less mortality and developed at a faster rate; when transformed to adults, these nymphs produced more progeny. This situation would appear to be parallel to that of aphids on yellow rocket discussed earlier in which there is lowered reproduction on less favorable plants. In experiments involving the use of nitrogen fertilizer there was also a greater survival of nymphs and a higher percentage of females with eggs, the females also having larger numbers of eggs (Table 11-3).

David Lightfoot and Walter Whitford (1987) found that the number of arthropods on the foliage of creosotebush (*Larrea tridentata*) increased significantly in late spring on those plants that had been subject to experimental nitrogen fertilization. Water treatments alone had no effect on the number of foliar arthropods. Interestingly phytophagous (i.e., plant-eating), sap-sucking insects accounted for the majority of arthropods on the plants, and their densities varied most in relation to both foliage production and foliar nitrogen content. Lightfoot and Whitford suggest that much of the between-plant variation in densities of phytophagous insects that they observed within a stand of creosotebush may be due to sap-sucking insects tracking variable amounts of nitrogen in the foliage.

Moisture

Chemical interaction, as demonstrated in the case of meadow voles, nettles, and insects, is a major factor in regulating population growth and distribution. Physical factors that are characterized by gradations also play different roles in different species. For example, as would be expected, the incidence and quality of light have obvious effects, particularly on autotrophs; quantitative effects were shown on nettles (Table 11-2), and seasonal effects were discussed in Chapter 4. Since the majority of an organism's mass is water, it might well be anticipated that the degree of moisture might also be a population regulator. A few examples will explore that relationship.

Drought and birds. The long-tailed Hermit hummingbird (*Phaethornis superciliosus*) is characteristic of many birds of the humid tropics, in contrast to those of temperate-zones, in being at or near carrying capacity (hence at a more or less stable population size). In the Caribbean lowlands of Costa Rica, Stiles (1992) observed that the normal period and degree of mortality that occurs during the season of flower scarcity in late October–November was markedly increased by a period of unusual drought in 1973. The drought caused a severe flower shortage at the height of the breeding season. As a result, breeding behavior of males was reduced, successful breeding by females was drastically reduced, and the weights of both sexes dropped below the levels of the normal lean season. Survivorship of the breeding males dropped from the normal 90 percent to 60 percent, and the total breeding population of the following year was one-third lower than before the drought. It took 3 to 4 years for the population of breeding males to recover the numbers and age structure of the predrought period.

Not unexpectedly, George et al. (1992) found a similar drought effect on 15 species of grassland birds in western North Dakota. The total density of these grassland birds declined 61 percent between June 1987 and June 1988 as a result of one of the warmest and driest periods in the recorded history of the area. Densities of six of eight common species declined significantly during the spring-summer 1988 drought, but populations of all but two recovered in 1989, such that total bird density in June 1989 did not differ significantly from June 1987. George and his colleagues caution, however, that there could well be substantial changes in the species composition of the grasslands should drought conditions become more frequent, as some climate models predict.

Flooding and fish. At the opposite end of the moisture gradient is flooding, with flash floods of the mountainous, arid southwestern United States differing qualitatively and quantitatively from the spring floods of lowland and moister areas in central and eastern United States and along coastal areas such as the Gulf of Mexico. We might predict that the behavior of fish in each situation should have evolved over time to enable survival under each kind of flooding. Meffe (1984), in fact, found that to be the case. The live-bearing Sonoran topminnow (*Poeciliopsis occidentalis*), a native of the arid American Southwest, suffered lesser losses during flash flooding (as well as in laboratory simulations) than the morphologically similar mosquitofish (*Gambusia affinis*), a native of the more moist central and eastern United States, which was introduced to southwest streams in 1926.

This study suggests that the native topminnow apparently possesses such innate behaviors as quick response and proper orientation to the high discharges characteristic of flash floods, behaviors that the introduced mosquitofish lacks and that result in its displacement by the flooding. Meffe notes that, by predation, mosquitofish typically replace topminnows from native habitats within 1–3 years after introduction, the replacement being most rapid in localities that rarely or never flood. Long-term coexistence of the two species occurs in frequently flooded areas because of the differences in their evolved behaviors, the result of which is a periodic reduction in the population size of the predator before the native species is eliminated. Further exploration of the evolution of behavior and its impact on ecosystems will be found in Chapter 13.

Food

The degree of availability and accessibility of food would seem, at face value, to be an obvious population regulator. Its absence can lead to intensified competition, including aggressive behavior, or to starvation; its presence could lead to the other extreme of substantial population growth. In the instance of the long-tailed Hermit hummingbird described above, drought led to reduced floral production, the food source for the bird, and, in turn, to a reduction of the population; the population recovered the following year upon return of sufficient floral production. Also, increases in population numbers were observed in yeast when modifications of its food medium were manipulated (Figure 10-3).

Among many field and/or laboratory studies of the relationship of food to population growth and regulation, one by Stemberger and Gilbert (1985) can be cited. They demonstrated direct effects of five to seven algal food concentrations on population growth in eight species of planktonic rotifers with different body sizes. The threshold food level, that is, the food concentration required to maintain a population growth rate at zero, varied by a factor of 17 among the eight species, with a direct relationship between food concentration and body mass. The smallest species showed the lowest threshold food levels and the lowest food levels necessary to sustain growth at one-half the maximum growth rate. The smaller rotifers appear to be well adapted to living in food-poor environments, whereas larger species appear to be restricted to food-rich environments. These food-body size relationships could be significant factors in determining the species composition in planktonic communities.

Weather

While dramatic weather conditions such as hurricanes and tornados can decimate a population, such events are not regulatory in an equilibrium-oriented sense. However, periodic weather events associated with seasonal progression and involving changes in temperature and moisture may effectively regulate some populations. This would be particularly true for populations that depend on yet other populations attuned to those periodicities—for example, an insect population that feeds on a plant whose growth is subject to seasonal changes. A well-analyzed instance of this regulation involves thrips and roses in southern Australia.

Thrips and roses. In what is regarded as a classic study, J. Davidson and H. G. Andrewartha (1948) studied the relationship of food availability (roses) and population growth of Thysanopteran insects known as thrips (*Thrips imaginis*).

The number of adult thrips on roses is low at the beginning of spring (September in the Southern Hemisphere) (Figure 11-1). The period of increase to maximum density, about December 1, occurs over several months, but the decline is rather precipitous, producing essentially a J-shaped growth curve. There has been annual variation in the maximum size of the population, but the pattern was virtu-

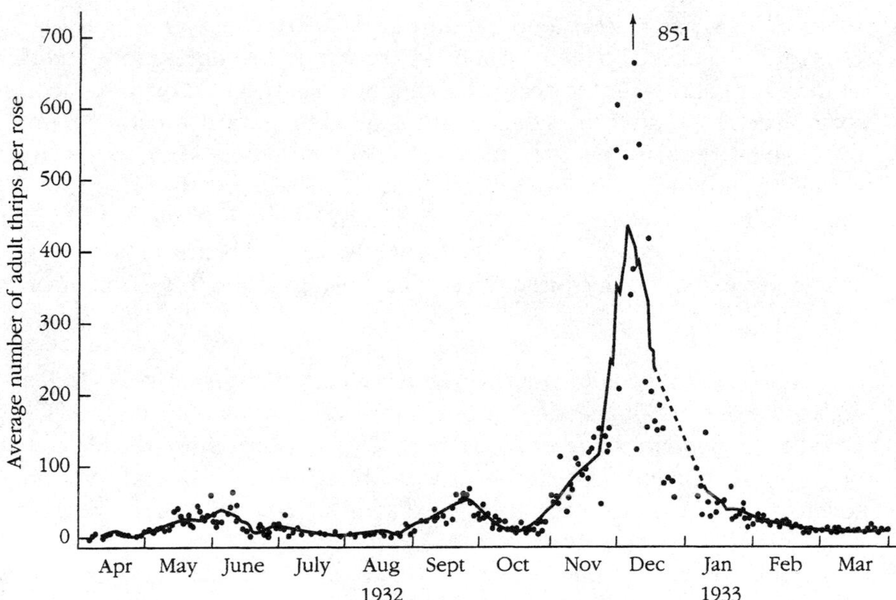

Figure 11-1 The population pattern of the thysanopteran insect, *Thrips imaginis*, on roses in South Australia. Each point represents a daily record; the curve is a 15-day moving average computed for a given date by averaging the data of that day with the 15 previous days. (Redrawn by permission from J. Davidson and H. G. Andrewartha. 1948. *Journal of Animal Ecology* 17:193–99.)

ally identical over a 7-year study period. This population pattern of the thrips parallels the pattern of flowering in the roses on which the thrips feed: during the dry summer few flowers are available, but in spring they are abundant. The few available but scattered rose flowers in the summer provide food and moisture for those thrips that then become the progenitors of the spring population. Breeding and development of this tiny insect are continuous throughout the year, but birthrate and survival rate are very low during both midsummer and midwinter. Eggs embedded in the tissues of the flower are relatively safe from desiccation during the summer, as are those nymphs and adults fortunate enough to be on living plants. The immobile pupa, however, is subject to desiccation and is the stage showing highest mortality during the dry summer. In contrast, the nymphal stages are more subject to mortality during the cold winter months, when flowers do not last long enough for them to complete their development.

Thus during the spring there is not only an abundance of food, it is also readily accessible—the population increases. During summer there are fewer flowers, which tend to be more widely scattered, and the thrips have more difficulty finding food—the population diminishes. But those thrips that do find flowers have a high survival potential; there is no absolute shortage of food for them. From these and other studies Davidson and Andrewartha and their colleague L. C. Birch (Andrewartha and Birch 1954; Birch 1957) contend that abundance and distribution of thrips are related to the seasonal change in availability as well as accessibility of food. Their food, in turn, is dependent on the influence of weather, particularly the effective degree-days between the first fall rain and the end of winter, and the total inches of rain during September and October. Their evidence that food is not limiting and that density levels of the thrips correlate with flower density led to their view that this is an animal population whose numbers are regulated by weather. They have extended this principle of weather regulation to a number of other natural populations as well.

Critical analysis of the thrips/rose/weather interaction. There is considerable debate about the degree to which weather directly regulates population growth independent of other biotic factors, particularly those related to the density of the population itself, a relationship to be discussed subsequently. Among the most severe critics of the Andrewartha-Davidson-Birch position that weather regulates populations was Frederick Smith (1952).

Smith subjected the rose/thrips data to extensive analysis and concluded that the analytical models used by the Australian ecologists were inappropriate to their interpretation. Moreover, Smith showed that direct analysis demonstrates an inverse correlation between population change and population size. This is the relationship expected when change depends on size—that is, when a density-dependence exists. There seems little question but that the mechanism responsible for actual regulation of the thrips populations is yet to be found and may well be in some interaction with other populations, such as those discussed subsequently.

Weather and muskrats. In spite of this reservation, it is certainly true that in those environments in which physical or even physicochemical factors show considerable seasonal fluctuation weather may play a critical role in determining the distribution and abundance of some organisms. Muskrats, for example, are continually subject to the vagaries of water-level fluctuations interacting with temperature. Low water levels in the fall, followed by extremely cold weather, result in unplugged burrow entrances, owing to a lack of available unfrozen mud. This condition, in turn, results in many animals freezing to death both in the burrows and as they search for unfrozen food. Low water levels also make their normally well-isolated houses more subject to predation by fox and mink. Conversely, high water levels in the spring flood the burrows, drown the young, and force the adults closer to shore, where they are again more subject to predation.

In less drastic environments, physicochemical factors certainly have some controlling, but not necessarily regulating, effect. By contrast, however, population patterns appear to be more subject to regulation through numerous biological intra- and interspecific interactions such as parasitism, predation, and competition, as well as by population density.

INTERSPECIFIC RELATIONSHIPS

Interactions between species involve a number of different relationships, all of which can be subsumed under the term *symbiosis*, which, from its Greek origins, means simply "living together." In some cases, the two symbiotic populations have no effect on each other (*neutralism*), whereas in most cases there is greater or lesser impact. *Competition* may involve a mutual inhibition or an indirect effect in situations in which a common resource is in short supply. In instances of *parasitism* and *predation* one population adversely affects the other by direct attack, parasites generally being smaller than their host and predators larger. Well-adapted parasites do not destroy their hosts whereas predators do; in this sense, as Charles Elton put it, predators destroy their capital while parasites live on the interest earned by the capital. In *commensalism*, which literally means "feeding at the table," one population is benefited while the other is not affected (a good example is the relationship between cockroaches and humans). *Protocooperation* is a situation in which both populations benefit but the relationship is not obligatory; by contrast, *mutualism* is a situation in which both benefit and neither can survive without the other (the relationship between nitrogen-fixing bacteria and leguminous plants is a case in point).

At face value, these different kinds of symbiotic relationships suggest that some may well play a regulatory role in population growth and development. In many cases, these relationships are direct, involving two interacting species; often, however, and thereby making the task of interpretation more difficult, the impact of one species on another requires the presence of a third species, a phenomenon recognized as an *indirect effect* (Wootton 1994). An exploration of that possible role is in order.

Parasitism

Parasites include a broad range of organisms including viruses, protozoans, fungi, helminths and numerous arthropods. Robert May (1983) has made a useful distinction between parasites based on their population biology rather than their taxonomy: *microparasites* (viruses, bacteria, fungi) have direct reproduction, usually at high rates, within the host; *macroparasites* (helminths, arthropods) have no direct reproduction within the host. Further, most viral, bacterial, and fungal microparasites tend to be small, have short generation times, and have a short duration of their infection relative to the average lifespan of the host; hosts that recover usually possess immunity against reinfection, and thus the infection is of a transient nature. By contrast, macroparasites are typically larger and have much longer generation times than their counterparts; their infections tend to be chronic or persistent with hosts being continually reinfected.

That parasites are a major contributor to many ecological interactions (Price et al. 1986) and especially to mortality in numerous species is well recognized; in humans, for example, it is generally agreed that parasitic infections are a major factor in infant mortality in developing countries, the survival curve being Type III (see

Chapter 10). However, the extent to which parasitic diseases have a regulatory effect on natural populations is more problematic to assess. Several case studies followed by a more theoretical consideration focus on this issue.

American chestnut and sac fungus. In 1904, the sac fungus *Cryphonectria (Endothia) parasitica* was accidentally introduced into the United States, the first infected trees being discovered in the Bronx Zoo. In its native China, *C. parasitica* is a parasite on the bark of the oriental chestnut and is kept in check by a variety of natural factors. In the United States, however, the majestic, towering American chestnut (*Castanea dentata*) proved no match for the miniscule fungus. The chestnut had been the most important, as well as most predominant, hardwood species in the forests of the Appalachian region of eastern North America. Its hardwood was used extensively for furniture, its tall straight timbers for telephone poles, its tannin for leather tanning, and its nuts used by both wildlife and people.

Cryphonectria parastica attacks the tree through such wounds as breaks in the bark or holes made by woodpeckers or bark-boring insects. The mycelium of the fungus grows into the bark and outer sapwood, forming a canker or spreading, oozing sore that eventually encircles or "girdles" the trunk or branch. This process precludes the normal flow of nutrients to and from the roots and foliage and eventually results in the death of the tree. By 1950 about 3.6 million hectares of American chestnut trees were dead or dying. Except for adventitious shoots from the roots, the American chestnut was virtually extinct (Anagnostakis 1982).

The spread of the fungus was not limited to the United States. In 1938 infected trees were discovered in northern Italy around Genoa, and by 1950 it had spread epidemically to the southern chestnut-growing regions as well. In 1951 an Italian plant pathologist, Antonio Biraghi, discovered a small grove of chestnuts that, although infected with the fungus, were unusually healthy. He isolated the fungus from the trees and inoculated other infected trees with it; the disease was not only arrested but the trees were cured as well. Subsequent investigations have shown that this particular strain of the parasite itself serves as a host for a parasitic virus. It is the presence of this virus, in the form of a double-stranded ribonucleic acid, that genetically alters the fungus's ability to kill susceptible chestnut tree hosts. This less virulent form of the virus has since been discovered in Michigan, Pennsylvania, Tennessee, and Virginia. Experimental work pursues ways of getting the less virulent form to disperse naturally and replace the more virulent form.

In addition, since the surviving root systems of blight-infested trees continue to produce sprouts throughout their natural range (Figure 11-2), it is possible that the release of genetically engineered, less virulent strains of the sac fungus could lead to restoration of this valuable forest species (Choi and Nuss 1992).

Adaptation of parasites and their hosts. Not atypically, the effect on population growth and development of parasitic relationships, which are numerous in recorded and even in fossil history, is comparable to that of the effect of a high flood—a population may be eliminated but not regulated by the parasite, as in the

Figure 11-2 American chestnut (*Castanea dentata*) sprouts in the Great Smoky Mountains, Tennessee. Root systems surviving the sac fungus (*Cryphonectria (Endothia) parasitica*) attack produce the sprouts, holding promise for eventual recovery of the species. (Photo by author.)

case of the American chestnut. As the presence of the hypovirulent form of the chestnut fungus suggests, however, adjustment of both parasite and host by way of adaptation and selection has enabled a tremendous number of parasite-host relationships to exist without host extinction. Impatient humans, however, have sidestepped the longer period of time generally required for such adjustments to occur and by chemotherapy have largely precluded the opportunity for adaptation in the case of their own pathogenic parasites.

Humans do, however, show a classic situation of adaptation to a parasite in the instance of the sickle-cell trait and malaria. Abnormal hemoglobin and the resultant sickle-shaped red blood cells, being less efficient and having less capacity for oxygen, are generally less advantageous to survival. Yet the presence of some of these abnormal, sickle-shaped, red blood cells in the blood of black Africans living in malarial areas confers a higher degree of resistance to the disease than the nonsickled condition. The history of malarial death and debilitation in nonadapted humans is a dramatic tale indeed. Outside malarial areas the sickle condition is less advantageous and tends to be selected against. Therefore it is not so frequent in later generations of African-American populations in the United States, for example. For nongenetically adapted humans, chemotherapy has conveyed a nonheritable or somatic adaptation to the protozoan parasites causing malaria.

Tsetse flies and game animals. Implicit in the case of the Chinese chestnut and its fungus and of humans (with sickle-cell) and malaria is the factor of regulation. Both host and parasite populations continue to exist, although the parasite has, or had at one time, the potential for completely exterminating its host. The parasite in a new association tends to devastate and debilitate rather than regulate in the equilibrium sense. Yet if the balance of the adapted host-parasite relationship is shifted, the consequences can be devastating.

In an experiment on game destruction in Africa, W. H. Potts and C. H. N. Jackson (1952), at the East African Tsetse and Trypanosomiasis Research and Reclamation Organization in Tanzania, followed the effects on three species of tsetse fly (*Glossina*), alternate hosts for African sleeping sickness, which is caused by the parasitic protozoans *Trypanosoma gambiense* and *T. rhodesiense.* In the area of the study, the tsetse fly feeds by sucking blood, mainly from hoofed game animals, including zebra, rhinoceros, antelope, giraffe, reedbuck, impala, and gazelle. As the number of animals was purposely reduced (some 8,500 animals killed in an area of 600 square miles over a period of 5 years), the population of the flies correspondingly dropped off in size. Comparable in a way to the thrips that were successful in finding a rose flower for food, those tsetse flies that happened to find one of the now-scarce game animals had an abundance of food. But most individuals were not so successful, with a consequent rise in population mortality. Furthermore, fewer individual flies meant fewer opportunities for encountering the opposite sex, leading to a lowering of birthrate. Continual analysis verified that the proportion of virgin females in the population increased as the population size decreased.

The combined effects of decreased natality occasioned by too few potential mates and increased mortality occasioned by relative scarcity of food resulted in complete extinction of the parasite in the region of the study. If it were possible to prevent their reintroduction from adjacent populations, which proved to be uncontrollable, the area would have remained free from the parasite. The investigators concluded that the game reduction approach to regulating the tsetse fly was both unnecessarily expensive and destructive to game animals and also ineffective because of continued immigration of game animals into the area.

Population regulation by the sterile male method. Population control of noxious parasites has been effectively accomplished by the sterile male method. This technique avoids the often unpredictable, let alone undesirable, side effects of using a wide-spectrum pesticide. In an experiment conducted in 1968 on Seashore Key off the west coast of Florida, several researchers from the U.S. Department of Agriculture

(Patterson et al. 1970) eliminated the indigenous population of the mosquito *Culex pipiens quinquefasciatus*, a vector of human filariasis, using the sterile male method developed in the 1950s by Edward Knipling and Raymond Bushling (see Knipling 1959). This was done by exposing the pupal stage to a sterilizing agent (thiotepa) and then releasing the adult males at rates of 8,400 to 18,000 per day for 10 weeks. These males proved to be as competitive in mating as their nonsterile counterparts so that there was a precipitous decline in the population. During the last two weeks of the study no larvae of the mosquito were found in any of the breeding sites. However, the method was not found economically feasible and alternative biological control methods have subsequently been found more effective and efficient.

The autocidal sterile male technique has also been used in the southern United States since the early 1950s to eradicate the screwworm, *Cochliomyia hominivorax* (Richardson et al. 1982). This blowfly lays its eggs on or near open wounds, including the fresh umbilical cord, of cattle, sheep, and other warm-blooded animals. The first instar larvae enter the wound and feed on fluids from the living tissues; this feeding enlarges the wounds, which thus attracts more ovipositing females. Mature third instar larvae drop from the wound and pupate in the ground. Millions of dollars of annual damage to livestock, game animals, and pets, have been caused by the parasitic screwworm. Damage was reduced dramatically by releasing males that had been subjected to radiation, the resulting dominant lethal mutations preventing viable embryos from forming.

Suppression of screwworm populations was highly effective except for major outbreaks in 1962, 1968, 1972–1976, and 1978. Technical problems, such as determining the number of sterile males needed in an area to be effective, partially account for the outbreaks. The real culprit, however, is genetic diversity. Investigators have determined that there are some nine genetically different types of screwworms, each adapted to somewhat different ecological environments. Thus although the release of sterile males of one type may eradicate that type in an area, immigration of other types from surrounding areas can sustain damaging levels of the parasite. Given the time frame involved, the evolution of the different types of screwworms is not a resistance response, as has been the case with human pathogens and antibiotics. Instead it is a case of adaptations over time by subpopulations to different environments. Thus the regulation of one subpopulation by humans is simultaneously countered by the immigration from other subpopulations of screwworms of other genetic types. This situation sustains an effective overall population level despite the attack on one subpopulation.

Regulation via differential susceptibility. An additional stabilizing mechanism in host-parasite interactions may occur through differential susceptibility of host individuals to a parasite. Such may be the case in the interaction of *Candida humicola*, a parasitic yeast, which occurs in the gut of the Green frog (*Rana clamitans*). Kurt Steinwascher (1979) has shown that low concentrations of the yeast increased the growth rate of Green frog tadpoles relative to tadpoles grown without the yeast. There is a differential response, varying inversely with the size of the tadpole, however. For small tadpoles, the yeast appears to act as a parasite, competing for food in the gut. For larger tadpoles, the situation appears to be mutualistic. Steinwascher reasons that because larger tadpoles are more likely to metamorphose and reproduce, and thereby contribute to continuation of the Green frog population, resistance to the parasite is not likely to develop. Thus the yeast may act to regulate the Green frog population size by altering the probability of metamorphosis of individual tadpoles.

Parasites and adaptation. Several instances of adaptation in host/parasite relationships have been described in this section. Generally speaking, parasitic infections appear to be more effective as regulatory agents among newly introduced species of plants and animals or when parasites are introduced into new regions (May 1983). Well-adapted parasites are relatively harmless to their hosts, an advantage to both symbionts. The development of an avirulent status can follow a number of coevolutionary pathways depending on the interplay of virulence and the ability of the parasite to be transmitted.

Adaptation of parasites is increasingly evident in attempts to chemically control pests that transmit the parasite. Pesticides that are initially effective in controlling pest species on plants and animals become less so as various resistance mechanisms come into play. The most important resistance mechanisms enhance the capacity to metabolically detoxify the pesticide and to produce alterations in target sites that prevent pesticides from binding to them (Brattsten et al. 1986). This phenomenon has resulted in a more comprehensive approach of integrated pest management that includes the use of chemical pesticides in combination with improved cultural (e.g., hygiene and sanitation) and biologically based techniques.

Predation

In many ways, host-parasite and predator-prey relationships are similar: the parasite and predator are benefited while the host and prey are adversely affected to greater or lesser degrees; also, as noted above, predators live on capital whereas parasites live on interest. A few examples of the numerous studies on predator-prey interactions should substantiate the accuracy of this capital-interest analogy.

***Didinium* and *Paramecium*.** Among a number of now classic studies by the Russian biologist G. F. Gause are those involving the interaction of two ciliated protozoans, *Paramecium caudatum*, which thrives well on yeast and bacteria, and *Didinium nasutum*, which thrives well on a diet of fresh paramecia (Gause 1934). In the first of these experiments (Figure 11-3) five paramecia were introduced into a small test tube containing an oat medium conducive to bacterial growth; 2 days later three didinia were introduced. Three and 4 days later, respectively, both the paramecium and didinium populations were extinct. Variations in the time of introducing the predator and even in the size of the container always produced the same result—the increase and decrease of prey was a phase ahead of the successive increase and decrease of the predator and both prey and predator were eliminated.

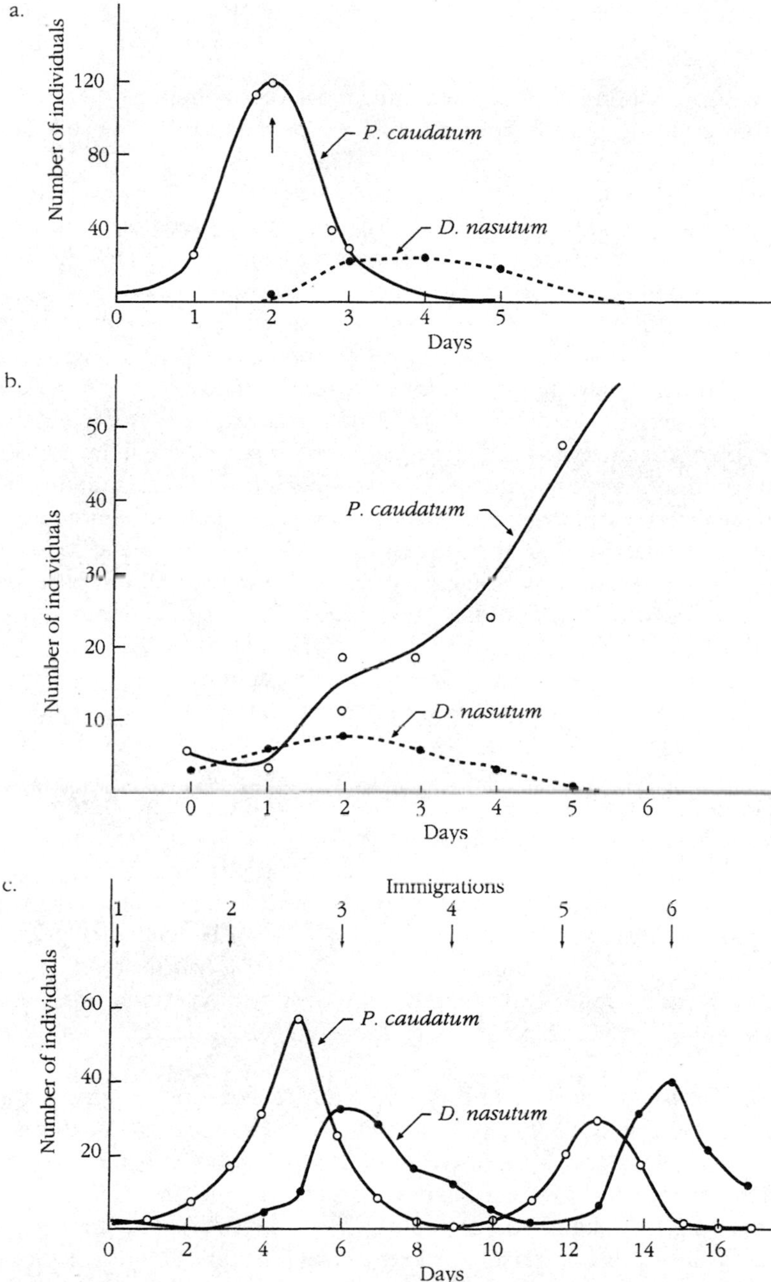

Figure 11-3 Prey-predator relationships in two ciliated protozoans, *Paramecium caudatum* and *Didinium nasutum*: a) in a homogeneous environment; b) in a heterogeneous environment; c) in a homogeneous environment with repeated immigrations of both predator and prey. (Redrawn by permission from G. F. Gause. 1934. *The Struggle for Existence.* Baltimore: Williams and Wilkins Co.)

Gause then modified the uniform or homogeneous oat medium environment by introducing some sediment that, in effect, created what he called "refuges" where paramecia could hide. The effect of this hiding is evident in the second graph (Figure 11-3b): the simultaneous introduction of predator and prey is followed by continued increase of prey and extinction of the predator. From replicates of this study Gause concluded that, as the microcosm approaches the natural heterogeneous conditions found in nature, population control is subject to a multiplicity of causes. In this particular instance, the simple predator-prey relationship (as in Figure 11-3a) becomes complicated by the accident of a heterogeneous environment in which paramecia are not so readily caught.

The extinction of the prey-predator relationship in a homogeneous environment can be mitigated to an oscillatory pattern by repeated immigration (Figure 11-3c). In this third series of experiments one paramecium and one didinium were introduced on the first day and every 3 days thereafter. At the outset the predator died out, probably from failure to find the too-few available prey. By the second immigration, the prey concentration was sufficiently high to permit successful predation, with a consequent growth of the predator population. By the fourth immigration, all the prey had been consumed and, in the absence of food shortly afterward, the predator population itself became extinguished. The fourth immigration of prey and fifth immigration of predator (like the first and second immigrations) initiated this cycle of oscillation again.

By experimentally manipulating the size of *Didinium nasutum* Hewett (1988) determined that larger species of *Paramecium* were encountered more frequently than smaller ones for both large- and small-sized *Didinium* but that larger *Didinium* captured more prey of either size per unit of time. Interestingly, the first capture of prey by either size *Didinium* was more difficult than subsequent captures, indicating behavioral adaptation. Among other findings, Hewett concluded that there were a number of advantages to large-sized *Didinium* feeding on larger-sized *Paramecium*, especially for survival under starvation conditions. We will have more to say about predator-prey size relationship subsequently.

Oscillations: hare and lynx. Gause's experiments give laboratory evidence to out-of-phase oscillations of interacting predator-prey populations. Such an oscillation pattern is obvious in the case of the snowshoe hare and the lynx (Figure 11-4), where the period of oscillation averages about 9.6 years (MacLulich 1937). Because the hare is the major food item of the lynx, the two cycles are certainly related. To what extent a direct and exclusively causal relationship exists is, however, more difficult to determine.

Environmental heterogeneity. Gause's experiments also point to the significance of heterogeneity in the enviornment as an important factor in population interaction. Welsh ecologist P. van den Ende (1973) has shown that even a seemingly homogeneous environment contains sufficient heterogeneity to allow populations to persist with out-of-phase oscillatory patterns. In experiments with the

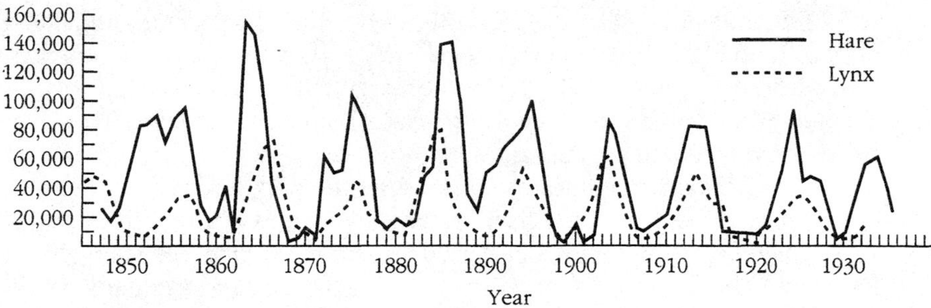

Figure 11-4 Oscillations in populations of snowshoe hare and lynx based on pelts received by the Hudson Bay Company. (Redrawn by permission from D. A. MacLulich. 1937. *University of Toronto Studies, Biological Series, No. 43.*)

bacterium *Klebsiella aerogenes* and its predator, the protozoan *Tetrahymena pyriformis*, mixed populations were maintained for more than 1,000 hours, covering 40 generations of the predator. During this period five oscillations occurred, each having a lower amplitude and frequency than the one before as the population appeared to approach an equilibrium. Both predator and prey showed adaptations to each other: the predator showed a marked reduction in size (fewer prey are required to sustain a smaller predator); the bacteria lost their polysaccharid, or mucoid, capsules (an economy permitting more rapid growth). More significantly, the seemingly homogeneous environment actually did provide a refuge in the form of the wall of the culture vessel to which some bacteria adhered. A tetrahymena feeds as it moves by creating feeding currents caused by beating its cilia. These ciliary currents, however, were not strong enough to dislodge the bacteria on the walls. The adhered bacteria thus acted as a source for immigration into the water medium and thereby contributed to the oscillatory, or possibly eventual equilibrium, pattern.

Effect of hunger. Other physiological and behavioral factors also contribute to stabilization of predator-prey populations. Hunger and food abundance constitute one such set of factors in the selection of prey, feeding selectivity being inversely proportional to the state of hunger. Under experimental conditions, for example, R. A. Pastorok (1981) has shown that dipteran *Chaoborus* larvae attack their prey, *Daphnia*, a cladoceran, and *Diaptomus*, a copepod, (both zooplankton) indiscriminately when the gut is empty. Because *Daphnia* swim twice as fast as *Diaptomus*, however, the Chaoborid larvae encounter and capture them more frequently. As the larval crop and midgut become filled with food, the predator avoids eating some *Daphnia* and specializes on *Diaptomus*. Thus the populations of prey would tend to oscillate while that of the predator would tend to remain at an equilibrium level.

Effect of size. The size relationship between cladocerans and their protozoan prey described above is paralleled by the relationship between fish and plankton. The introduction into a lake or pond of large fish that feed on zooplankton results in the elimination or reduction of large-sized planktonic species. These large species are selectively fed on because they are more easily located. Both the density of large species and the sizes of species constituting the zooplankton community fluctuate, however, as densities of their fish predators wax and wane (O'Brien 1979).

The relationship of prey and predator size involving, respectively, mussels and starfish was studied by Robert Paine (1976), who demonstrated that there is a maximum body-size limit of mussels (*Mytilus californianus*) that can be consumed by starfish (*Pisaster ochraceus*) of a given size (Figure 11-5). The figure also shows that there is a minimum size below which mussels are not attractive to a very large starfish. Under experimental conditions the starfish were allowed to reinvade an area from which they had been manually excluded for 5 years. During this interval the mussels had attained shell lengths of 8 to 10 cm and could not be consumed by starfish weighing 203 g or having a body diameter of 17.5 cm. Other experiments in which the mussels were protected and allowed to grow and were then subjected to predation confirmed that survival was greater in those expected to be too large to be eaten. Starfish weighing 500 g and with a diameter of 24.8 cm were unable to eat mussels larger than 12.5 cm.

These studies indicate that once a mussel has exceeded a certain size threshold, it cannot be effectively attacked by any individual predator. More broadly, size-limited predation suggests that predator and prey can coexist. A parallel is found in

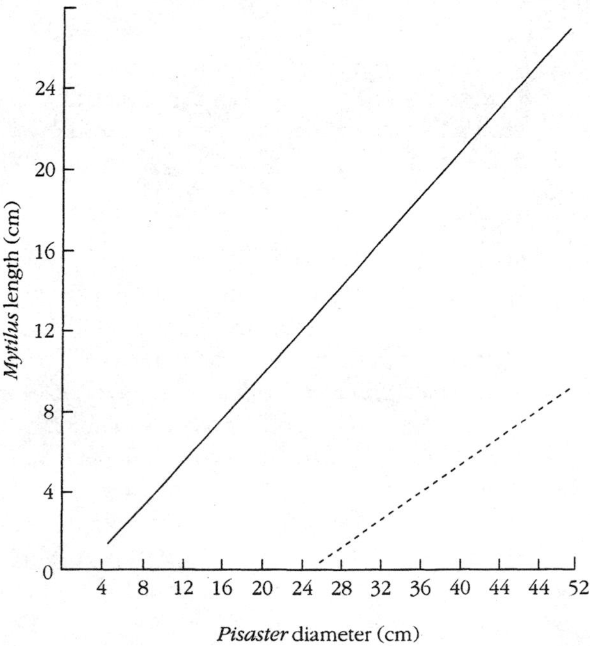

Figure 11-5 The shell length of mussels, *Mytilus californianus*, observed being eaten by starfish, *Pisaster ochracreus*, under field conditions. Solid line is the upper limit; dashed line is the lower limit; space between indicates prey sizes normally attacked by starfish of a given size. (Adapted by permission from R. T. Paine, 1976. *Ecology* 56:858–73.)

developing forests in which adult trees remain unharmed while seedlings are subject to grazing. Large surviving prey, such as trees and mussels, diversify the structure of the environment by providing a persistent habitat for associated organisms, creating spatial variation in the vertical profile of the habitat. The importance of this relationship in stabilizing an ecosystem is considered further in Chapter 13.

Effect of spatial distribution. The influence of spatial dimension on predator-prey interactions is well illustrated in experiments by C. B. Huffaker and his associates (1963) on an herbivorous mite (*Eotetranychus sexmaculatus*) that feeds on the skin of oranges and another mite (*Typhlodromus occidentalis*) that is its natural predator. It was already known from earlier studies that when the predator and prey are confined to a single orange, the predatory mites destroyed the herbivorous mites and hence themselves. This is like the interaction between *Paramecium* and *Didinium* shown in Figure 11-3.

Huffaker's first set of experiments was designed in two-dimensional space (a flat tray) such that as the number of oranges was increased, the amount of feeding space on each was decreased by coating the rest of the orange's surface with vaseline. In all these experiments the eventual outcome was that the predatory mite eliminated itself by destroying its prey. The introduction of the third dimension (Figure 11-6) was accomplished by arranging trays of oranges on grid-wire shelves supported by wooden dowels. In such a three-dimensional system involving 252 oranges, each with one-twentieth of its surface exposed (the equivalent of 12.6 whole oranges), there was considerable stability, with some oscillation, of both populations for 16 months. At that time a viral disease reduced the prey population below a maintenance level for the predator population.

The indirect effect of disease. During the late 1970s and 1980s on mainland Sweden, an outbreak of mange (*Sarcoptes scabiei*) severely affected the red fox (*Vulpes vulpes*), resulting in a dramatic reduction in their numbers (Lindstrom et al. 1994). The mange causes hair loss, skin deterioration, and ultimately death, the population of foxes declining by some 70 percent. The main prey of the red fox are voles, mountain hare, European hare, black grouse, hazel grouse, and roe deer. Through monitoring of the populations of both predator and prey, the investigators concluded that predation by red fox was a crucial factor in limiting the numbers of hares and grouse, as well as fawns per doe of roe deer in autumn, and in conveying the 3–4 year cyclic fluctuation pattern of voles to small game.

Effect of latitude. Intensity of predation is regarded by a number of ecologists to be greater on populations of both plants and animals in lower latitudes, consistent with the tenet that biotic factors play a greater role in evolution in the tropics than do physical factors. In a study of predation by ants on social wasp larvae, Jeanne (1979) established five study sites at latitudes ranging from 43°N to 2°S, extending from southern New Hampshire to western Para in Brazil. He found predation rates (measured as the time in which a food item was exploited) to be significantly greater in

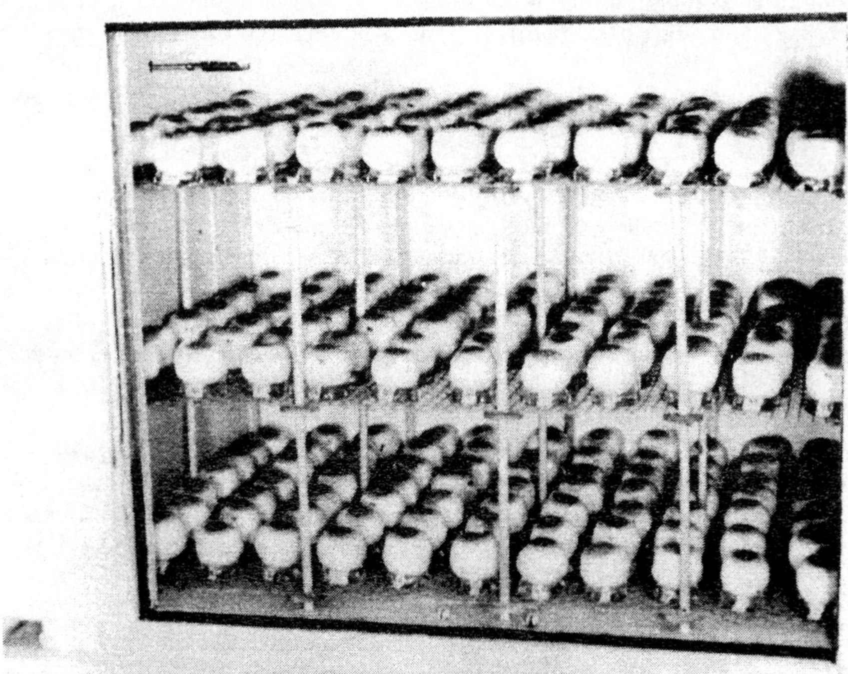

Figure 11-6 The laboratory cabinet (with doors open) showing the arrangment of oranges used in studies of the interaction between a predator, *Typhplodromus occidentalis*, and its prey, *Eotetranychus sexmaculatus*. Each orange rests on a glass furniture coaster. (Photograph courtesy of C. B. Huffaker.)

the tropics than in the temperate zone. Consistent with many other observations on biodiversity, he also found that the number of species of ants increased from 22 in the northernmost site to 74 at the southernmost, with the greatest numbers of species being found in forests rather than in fields, except in the northernmost location.

Although the limited number of studies that have been conducted on other predator-prey interactions tend to support this hypothesis of latitudinal impact on predation (e.g., octopuses and starfish preying on a marine snail (Fawcett 1984)), Wilson (1991) did not observe such variation in predation by numerous species of reptiles, birds, and mammals, among others, on the lizard *Uta stansburiana*. This is a topic in need of much more investigation.

Effects of human predation. From its earliest beginnings, the predatory impact of *Homo sapiens* has been singularly devastating. We need merely reflect on the extinction of the bison of western North America, among the many species humans have hunted to extinction. In Australia, for example, 17 mammalian species have become extinct since the arrival of the Europeans in the sixteenth century; these extinctions include the marsupial Tasmanian tiger, the last known of which died in 1936 in Hobart Zoo. Although the significance of hunting as against scavenging among our early ancestors has been called into question (Blumenschine and Cavallo 1992), there is no doubt about the degree of devastation of biodiversity human predation (let alone that by habitat destruction and degradation) has caused. The regulatory role of human predation under natural conditions is less well recognized, except for that which has occurred through the development of game preserves and game management.

Weddell seals (*Leptonychotes weddelli*), which occur mostly on shore-fast ice in Antarctica, have been subject to human predation since 1956 when both the United States and New Zealand established permanent research stations near McMurdo Sound (Testa and Siniff 1987). Both stations used Weddell seals to feed dog teams, the United States stopping the practice after the first few years, New Zealand continuing to do so but taking fewer animals. Although human predation severely depleted the seal population in the mid-1950s, it began to expand rapidly in the early 1960s, declined to low levels in the mid-1970s and has been fairly stable since 1979, although probably at a lower level than before human predation began.

Lesser snow geese (*Anser caerulescens caerulescens*) demonstrate differential age survival when hunting pressure is reduced. Based on hunter recoveries from over 80,000 of these geese banded between 1970 and 1988 near Churchill, Manitoba, Francis et al. (1992) found that adult survival increased from 78 percent in 1970 to nearly 88 percent in 1987, coinciding with a decline in the number of birds shot each year. By contrast, during the same time period, the survival rate of fledglings declined from 60 percent to 30 percent suggesting that mortality factors other than hunting pressure were involved, perhaps related to the density of the fledglings on the breeding grounds. In some other waterfowl (e.g., mallards), reduction of hunting pressure appears to be compensated by increased mortality from other causes. Ascertaining such species-specific differences in response to changes in hunting pressure, as well as to other mortality-inducing factors, makes the study of game management indeed challenging.

Classical predator-prey theory. Among ecological problems, none has occupied so dominant a position as the interactions between predators and prey, resulting in probably more publications on the subject than any other single topic, thus the development of many theories (Kuno 1987). For example, during the 1920s, the biometricians A. J. Lotka (1925) and V. Volterra (1928) showed that prey-predator interaction would result in oscillatory patterns. From the experiments of Gause it can be seen that such patterns are possible in a heterogeneous environment or one with repeated immigrations. We would also expect that natural selection would operate to increase the efficiency of the predator in capturing food, on the one hand, and simultaneously to increase the efficiency of the prey to avoid being captured. Thus, as shown by preceding examples, population patterns also depend on adaptive behavior. A "good" predator, in the sense of being well adapted, would be like a "good" parasite, living off the interest or at least the expendable capital of a population. The long-range trend of such prey-predator interaction would be a system of built-in checks and balances between the two populations, as well as between each of them and their myriad other relationships within the ecosytem they occupy.

David Pimental (1968) has shown that such a system can develop in a laboratory host-parasite experiment involving the housefly (*Musca domestica*) and the parasitic wasp (*Nasonia vitropennis*). The initial association of these two populations in a laboratory situation resulted in violent oscillations (Figure 11-7a) of both host and parasite. Populations derived from these initial associations achieved a more stable equilibrium in 2 years (Figure 11-7b). This factor is explained by genetic adaptation in which the increased resistance of the host is indicated by a reduction in parasite natality over the course of the experiment from 133 to 46 progeny per female (see discussion in Chapter 10 on the value of *r* in well-adapted parasites). This experiment demonstrates that the cybernetic feedback loop in ecological homeostasis can be genetic.

Yet another intriguing aspect of the theoretical-mathematical principles of prey-predator relations deduced by Volterra is that any factor that is moderately destructive to both predator and prey will increase the average prey population and decrease that of the predator. Thus, as Robert A. MacArthur and Joseph H. Connell (1966) have pointed out, the application of the general insecticide DDT to control scale insects (*Icerya purchasi*) that had already been brought under control by a natural predator, the ladybird beetle (*Novius cardinalis*), resulted in an increase of the scale insect, a parasite of citrus trees. Indeed, such complex interrelationships must be better understood before they are tinkered with.

Ratio-dependent predator-prey theory. The classic predator-prey model evolved from the Lotka-Volterra equations has two major shortcomings (Berryman 1992). First, the model predicts that an enrichment of the system will cause an increase in the equilibrium density of the predator but not in that of the prey and will destabilize community equilbrium. Second, the model predicts there cannot be a sit-

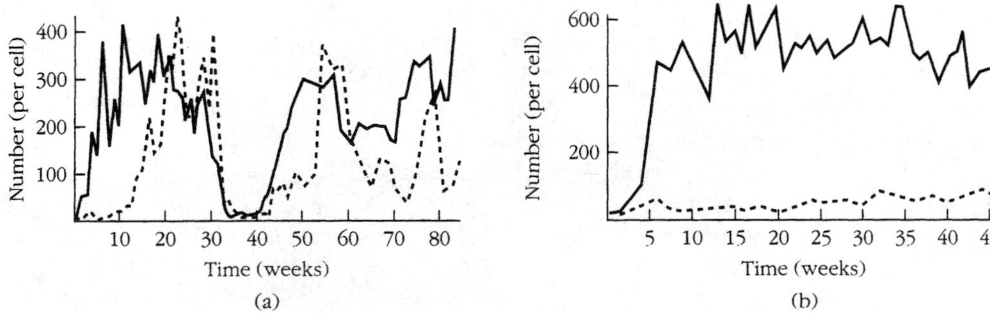

Figure 11-7 Population trends in the association of a new host-parasite relationship between the house fly, *Musca domestica*, and the parasitic wasp, *Nasonia vitropennis*; a) initial association; b) two years later. Solid line, the average number of flies; dashed line, the average number of parasites. (Reproduced by permission from D. Pimental, 1968. *Science* 159:1432–7. Copyright by the American Association for the Advancement of Science.)

uation of both a very low and a stable prey equilibrium density. Neither of these predictions is supported by field observations. As Berryman notes, these shortcomings of classical prey-predator theory result from the application of the principle of mass action to predator-prey interactions and can be overcome with models evolved from the Verhulst logistic theory discussed earlier or those that involve the ratio of prey to predators. The mathematics of these arguments are beyond the scope of this text but are well worth pursuing. For those inclined, there is an excellent discussion in a special feature section of the journal *Ecology* (Matson and Berryman 1992); also see critiques of ratio-dependence by Gleeson (1994), Sarnelle (1994), and Abrams (1994).

Concept of the Niche

Gause's principle, or the competitive exclusion principle, which will be discussed shortly, is sometimes stated in the form that no two species with identical niche requirements can continue to coexist. The key term in this statement is *niche*, an inclusive concept that means the sum total of all the ecological requisites and activities of a species, or what Eugene Odum has called its profession. It would follow that the likelihood of two species having identical niches is the likelihood of their being phylogenetically related—the closer the relationship, the more likely the identity of the niche. Thus Herbert Ross (1957) of the Illinois Natural History Survey noted that six species of the same genus of leafhoppers (*Erythroneura*) were all found feeding on the same part of sycamore trees at the same time and in the same way. Furthermore, they appeared to have identical species of predators. Doubtless the niche of each of these leafhoppers is extremely similar if not identical in most parameters. It is doubtful, however, that they are absolutely identical in all aspects; otherwise they would not be separable as species. Because of his observations, Ross concluded that no interspecific competition occurred during the period of study. Yet we cannot help speculating that in the period of study there was an abundance of and ready accessibility to food and a total environment conducive to or within the tolerance of each of the species. Interestingly, Ross noted that trees with small leafhopper populations harbored only two species, *E. lawsoni* and *E. acta*, whereas trees with large populations almost always supported all six species.

The term niche has been used in a variety of contexts, including spatial (distribution, habitat, physical location in an ecosystem), functional (trophic level, position in a food web), behavioral (its "way of life" as a predator, competitor, etc.) and abstractly. Because the latter definition is increasingly accepted, it would be well to consider briefly the niche as an abstract *n*-dimensional hypervolume, a concept developed by Yale ecologist G. Evelyn Hutchinson (1965). Assume that a particular population's tolerance to two important environmental variables, temperature and relative humidity, is known from laboratory studies. Assuming these variables to be independent, we could then represent the population's niche with respect to temperature and relative humidity by a two-dimensional box, as shown in Figure 11-8a. But because temperature and relative humidity are not, in fact, independent in their biological effects, we might find that tolerance to higher temperatures was tied to increases in relative humidity with the result that the population's niche with

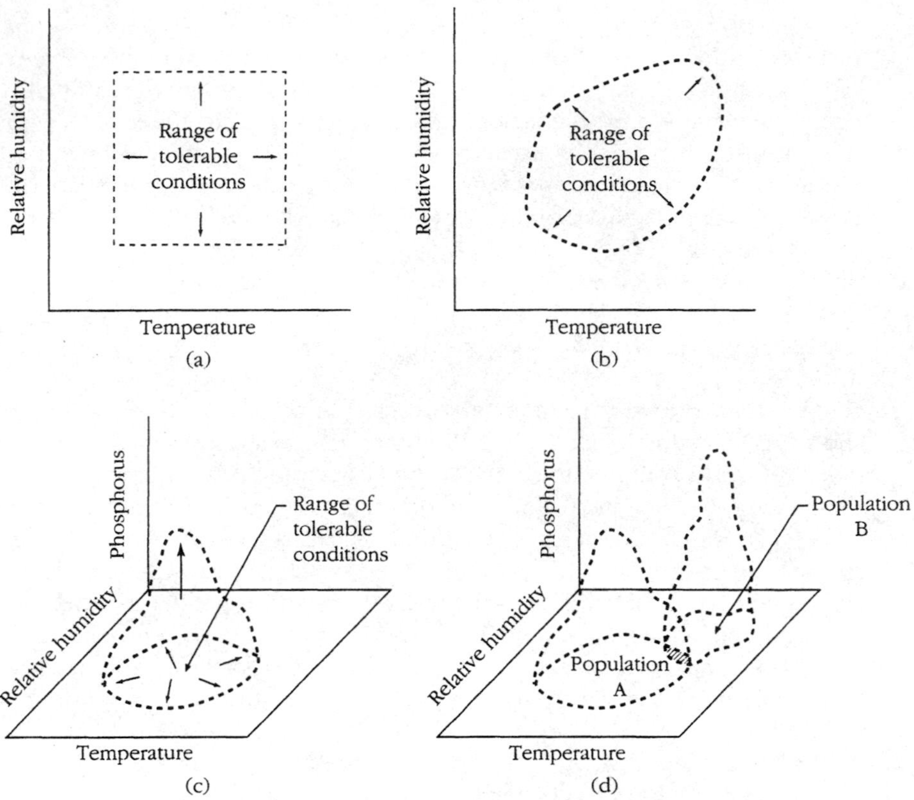

Figure 11-8 The ecological niche as an *n*-dimensional hypervolume; a) a two-dimensional niche with independent variables; b) with the variables interdependent; c) a three-dimensional niche; d) two populations with overlapping niches.

respect to these interacting variables would be better represented by an ellipse (Figure 11-8b). Suppose now that the tolerance to levels of available phosphorus was affected by interactions with both temperature and relative humidity. We now have a niche representing three variables or a three-dimensional, volumetric figure, such as shown in Figure 11-8c. Because a large number (n) of other environmental factors, both abiotic and biotic, affect the population, the niche is an *n*-dimensional hypervolume, an abstraction, for we are able to graph only with respect to three dimensions. Competitive exclusion mechanisms would be called into play only in the complete overlapping of *n*-dimensional niches (Figure 11-8d).

Competition

If two different species populations require a common resource, such as a nutrient, space, light, or moisture, that is potentially limited and actually becomes so, they are said to be in competition for it. Under such conditions can they exist together or will one displace the other? If they do exist together, how is this achieved, and can or do they maintain the same population size as when separate? Does competition play a comparable role to prey-predator relations in population regulation? These are some of the questions to be explored in this section.

Most ecologists hold that interspecific competition plays a significant role in population regulation and in setting distributional limits, thus directly affecting the structure of biological communities, and perhaps in adaptive radiation. As we shall see later, however, there is controversy over the relative importance of predation and competition in regulating populations.

Although competition is a widespread phenomenon exhibited by organisms from microbes to mammals, empirical evidence is elusive. Controlled laboratory experiments usually involve simplified competitive systems whose results cannot be applied to the complexity is obtained in natural ecosystems; and the complexity of variables operating in natural ecosystems confounds the interpretation of results, making alternate plausible explanations possible. An exploration among these kinds of studies is in order.

Resource Competition—the Competitive Exclusion Principle

The most studied of competitive interactions are those in which there is a common environmental requirement that is in short supply, resulting in either passive or active competition for that resource. Independent of each other, Lotka (1925) and Volterra (1928) developed mathematical formulations relating to competing populations that indicated that only one species will survive. Because of his empirical studies on competing populations that showed this effect, this principle is referred to as Gause's principle. It is also known as the *competitive exclusion principle*, so named by Garrett Hardin (1960).

Gause's studies on *Paramecium*. Using a culture medium conducive to the growth of two species of *Paramecium—P. aurelia* and *P. caudatum*—Gause determined the growth pattern of each when cultured separately and together (Figure 11-9). Each population has a typical sigmoidal growth pattern when grown separately as well as during the first 6 to 8 days when grown together. In the latter situation, however, there is a gradual diminishing of *caudatum* and a gradual increase in *aurelia* but not to the level attained when cultured separately. These results are consistent with the competitive exclusion principle predicted by mathematical theory. Comparable results have also been obtained in the laboratory with various competing populations, including other protozoa, yeasts, hydras, *Daphnia*, grain beetles, fruit flies, and, among plants, the small floating aquatic duckweed (*Lemna*).

Competitive exclusion in crayfish. Richard Bovbjerg (1970) studied competitive exclusion involving two species of crayfish in stream and pond habitats in

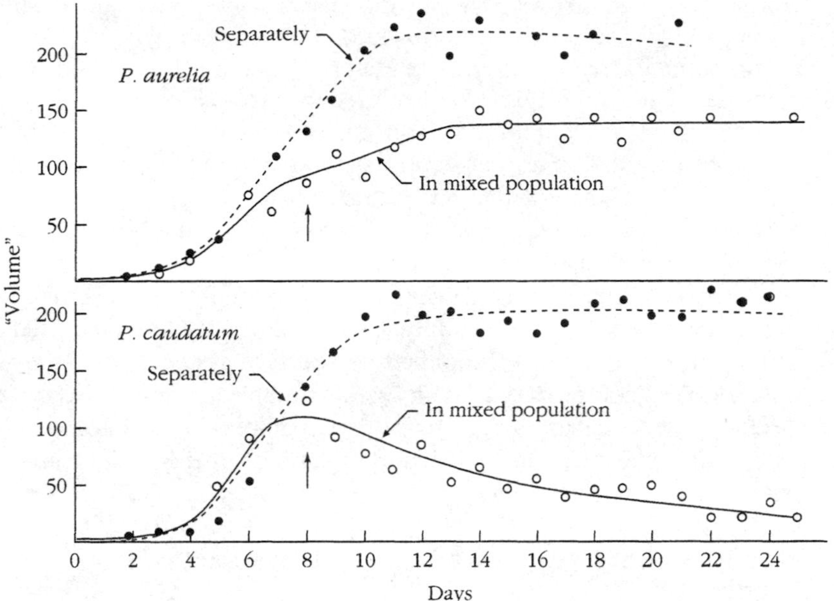

Figure 11-9 The growth of two closely related ciliated protozoans, *Paramecium aurelia* and *P. caudatum*, when grown separately and in mixed culture demonstrating competitive exclusion of *P. caudatum*. (Redrawn by permission from G. F. Gause. 1934. *The Struggle for Existence*. Baltimore: Williams and Wilkins Co.)

northern Iowa and southern Minnesota. One of the species, *Orconectes virilis*, is typically a stream or lake margin form and the other, *Orconectes immunis*, inhabits ponds. In the portion of the Little Sioux River in which the two species occur *virilis* is found on rocky riffles and *immunis* in muddy pools. After two summers of intense drought with much reduced riverflow, *immunis* occurred in large numbers in rocky sections of the river formerly occupied by *virilis*, which, however, by the end of the summer almost completely replaced *immunis*. Also, in the flood plain of the river that is normally occupied by *immunis*, spring floods introduced large numbers of *virilis*, which died as this pond dried up, while *immunis* survived by constructing burrows in the mud bottom.

These observations on displacement were studied experimentally in the laboratory by simulating pond-drying conditions. The experiments showed that *virilis* is excluded from the drying ponds behaviorally and physiologically. It showed poor if any burrowing capacity as well as a lack of tolerance to low oxygen levels that prevail in the drying condition—mean survival time was 4.2 hours and 75 percent of the population was dead within 5 hours. In contrast, *immunis* has a highly developed burrowing capacity and greater tolerance to oxygen depletion—mean survival time was 12.6 hours, and 20 hours elapsed before 75 percent of the population was dead.

In the simulated stream environment, however, both species showed nearly equal preferences for rock substrates (Figure 11-10) when tested alone; when tested together, preference for the rock substrate was enhanced in *virilis,* and *immunis* shifted to a strong preference for muck. Observations showed that this shift was the result of competitive exclusion of *immunis* by *virilis,* the latter being more aggressive in interspecific contacts. *O. virilis* evicted immunis from crevices in the substratum, places that normally provide protection during molting and egg carrying, and from predator attacks and flooding.

Bovbjerg's study indicates that as environmental conditions are modified, the competitive advantage (behaviorally and physiologically in the case of the drying pond and behaviorally in the case of the stream) may benefit the previously "disadvantaged" population. As another example, Gause showed that in a nonbacterial culture medium *P. caudatum* has a higher coefficient of increase than *P. aurelia* and would supplant it as the surviving population in joint culture. In the experiment described (Figure 11-9), however, bacteria were present, giving a competitive

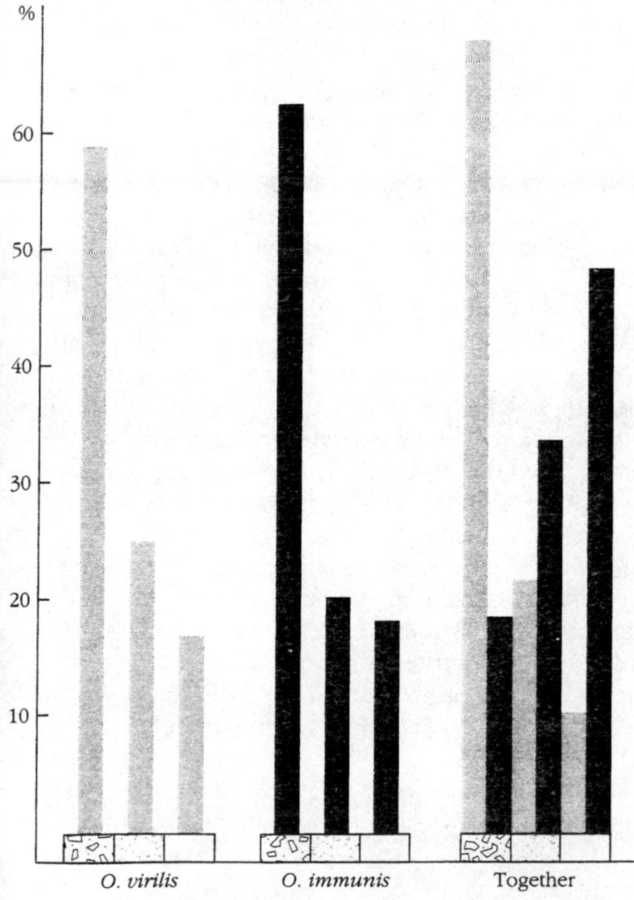

Figure 11-10 Preference of two species of the crayfish *Orconectes* to substrates consisting of rock, gravel, and muck (from left to right) when tested alone and together. (Reproduced by permisison from R. Bovbjerg. 1970. *Ecology* 51:225–36.)

advantage to *P. aurelia*, which is less sensitive to bacterial waste products. Thus as Gause himself noted in The Struggle for Existence

> ...there is in nature a great diversity of "niches" with different conditions, and in one niche the first competitor possessing advantages over the second will displace him, but in another niche with different conditions the advantages will belong to the second species which will completely displace the first. Therefore side by side in one community, but occupying somewhat different niches, two or more nearly related species will continue to live in a certain state of equilibrium (Gause 1934).

Darwin's finches. In his famous *On the Origin of Species,* Charles Darwin (1859) forcefully presented his case that the "struggle for existence" occurred where interspecific competition played a major role in determining the outcome of the struggle. Among other strong statements on this tenet, he noted, "We have reason to believe that species in a state of nature are limited in their ranges by the competition of other organic beings quite as much as, or more than, by adaptation to particular climates" (p. 140). Also, even more pointedly, "as species of the same genus have usually, though by no means invariably, some similarity in habits and constitution, and always in structure, the struggle will generally be more severe between species of the same genus, when they come into competition with each other, than between species of distinct genera" (p. 76). In contemporary terms, this is *niche overlap.*

Among the evidence advanced to support his thesis, Darwin offered the case of the 14 distinct species of finchs (Geospizidae) occurring on the Galapagos Islands. He contended that the existing species were all derived from a small number of generalized finchlike birds, evolving into different species occupying different niches. This cohort of species includes six ground feeders, several of which have heavy beaks used in crushing heavy-coated seeds, whereas the remaining eight species are adapted to perching in trees and feeding on insects, largely by removing the bark with their beaks (Figure 11-11). One species even drinks blood. Among the most specialized is the so-called cactus finch, which plucks a long spine from a cactus, holds it in its beak and uses it to probe crevices of the cactus stem for insects and to remove them. It is generally conceded that these 14 closely related species continue to live side by side, avoiding competition by *niche diversification* or *niche shift,* a change in some pattern or process in the niche.

In a series of studies, Peter Grant focused on the Galapagos finches to test the soundness of the competition hypothesis, concluding that interspecific competition was the major, but not the only, factor in establishing the finch communities observed today (Lewin 1983). According to Grant, differences in food supply serve as a complementary factor to interspecific competition in explaining the different finch communities found on the different islands. For example, of two particular species, *Geospiza difficilis* occurs on 20 of the 25 lowland islands in the archipelago, *G. fuliginosa* occurs on another 3, and the remaining 2 islands have neither species. Where they do occur together (on Isla Pinta), *G. difficilis* is principally in the highlands and *G. fuliginosa* in the lowlands with some overlap in the intermediate regions. It was found that although *G. difficilis* ate seeds in the highlands as it did on lowland

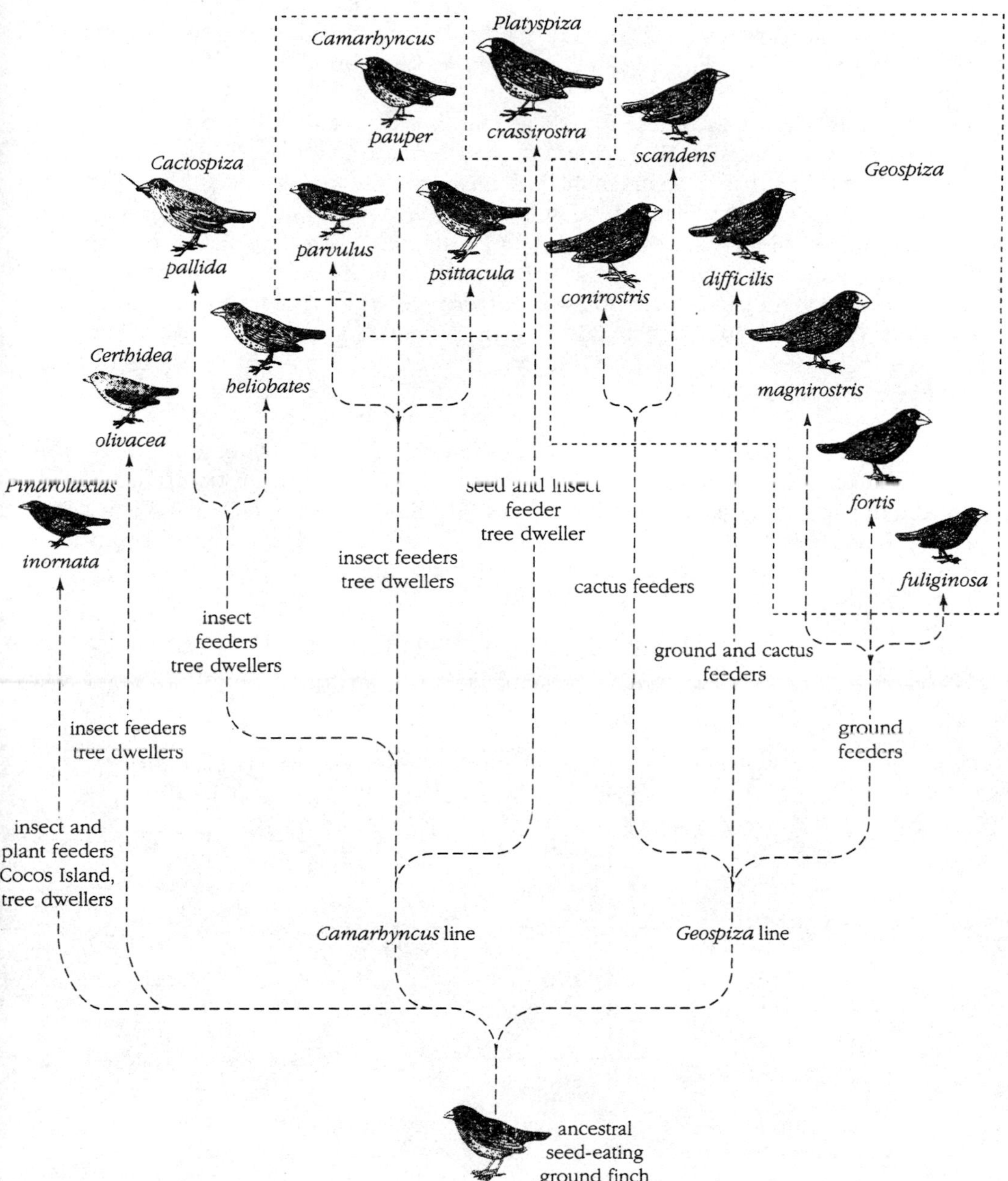

Figure 11-11 Adaptive radiation in Darwin's finches. (Redrawn by permission from E. J. Kormondy et al. 1977. *Biology: The Integrity of Organisms.* Copyright © Wadsworth Publishing Co., Belmont, CA.)

islands, much greater reliance was placed on insects thus, avoiding competition with *G. fuliginosa.* This shift in eating patterns can be referred to a niche shift.

Niche shift in warblers. Niche shift is seen dramatically in the case of two species of warblers on Mt. Karimuri in New Guinea (Diamond 1978). One warbler, *Crateroscelis murina,* occurs in increasing abundance from about 300 m to 1,642 m, where it is abruptly replaced by *C. robusta.* The latter species has a decreasing abundance from this level to the top of the mountain (2,432 m). Similarly, four species of fruit pigeons on the islands of the Bismarck Archipelago segregate in their niches according to the size of fruit they eat: larger pigeons eat larger fruit with the largest birds (average body weight 722 g) eating plum-sized fruits and the smallest birds (average weight 91 g) eating blueberry-sized fruits. Numerous examples of niche diversification abound.

Aggressive competition. As an example of a more aggressive type of competition, the kind most people seem to associate with the term (but that is relatively uncommon under natural conditions), a study that Joseph Connell (1961) conducted in Scotland on two species of barnacles is most instructive. Although larvae of *Chthamalus stellatus* can attach to rocks down to the mean tide level (Figure 11-12),

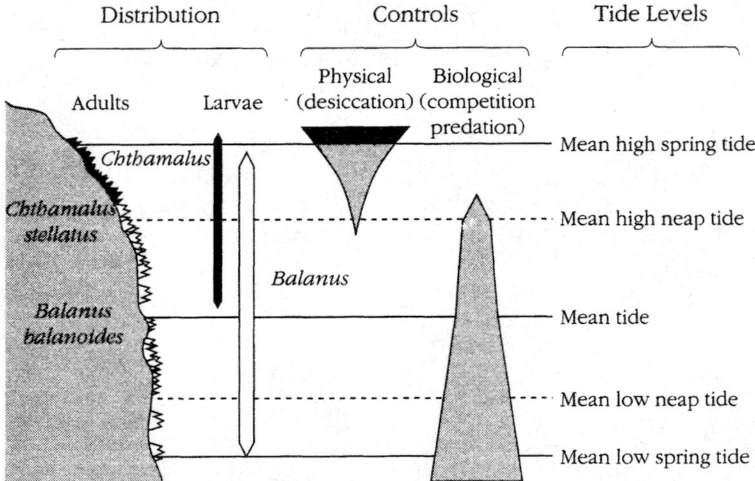

Figure 11-12 The potential and actual distribution of two species of barnacles, *Chthamalus stellatus* and *Balanus balanoides,* in the intertidal zone. Desiccation is the primary controlling factor in the upper part of the zone for *Balanus;* competition with *Balanus* is the primary controlling factor in the lower part of the upper half of the zone for *Chthamalus.* Predation by the whelk, *Thais lapillus,* and intraspecific competition are the primary controlling factors in the lower part of the zone for *Balanus.* (Redrawn by permission from E. P. Odum. 1963. *Ecology.* New York: Holt, Rinehart and Winston, and based on J. Connell. 1961. *Ecology* 42:710–23.)

in the presence of another barnacle, *Balanus balanoides*, they are successful in doing so only to about the level of the mean high neap tide. During neap tides the range between low and high water marks is least. Using a mapping technique, Connell showed that *Balanus*, which has a higher growth rate, actually pried *Chthamalus* larvae off the rocks or simply grew over them—competitive exclusion by direct, aggressive behavior for available space. When *Balanus* was removed from the area, *Chthamalus* populated the intertidal zone to the mean tide level. In the absence of a *Chthamalus* population, *Balanus* was found to be unsuccessful in maintaining populations above the mean high neap tide. Here control is effected by adverse weather, especially during the first year of life. It should be noted that, for this barnacle, adverse weather is warm and calm weather.

Below the mean high neap tide level, the size of the *Balanus* population is regulated by itself and by predation. As young *Balanus* grow, they compete for space—undercutting, displacing, growing over, and thereby smothering their own kind. This *intraspecific* competition, about which more will be said later, is one of the most significant factors in survival of the species during the first year. Individuals older than 6 months are preyed on by the whelk, *Thais lapillus*, accounting for virtually all the summer mortality. So as Figure 11-12 indicates, physical factors (weather) regulate zonation of barnacles at the upper part of the intertidal zone and have a greater effect on *Balanus* than on *Chthamalus*. Biological factors are significant in the lower part of the intertidal zone. Interspecific competition results in the exclusion of *Chthamalus* by *Balanus*. Intraspecific competition in *Balanus* results in elimination of less successful individuals and predation on *Balanus* by the whelk imposes a regulation on those survivors of the intraspecific competition.

While for barnacles biological factors control presence and distribution in the lower zones and physical factors in the upper, the converse is largely the case for salt marsh plants. For the latter, physical factors such as flooding and salinity, respectively, control presence and distribution in the lower and upper zones and biological factors such as competition do so in the upper, and or middle, zones (Bertness and Ellison 1987; Bertness, Gough, and Shumay 1992; Pennings and Callaway 1992).

Size advantage in competition. Although three species of Caribbean damselfish avoid direct intraspecific competition for the same benthic microalgae through territorial behavior, that is aggressive defense of feeding areas, there is direct competition between the two smaller species (*Stegastes dorsopunicans* and *S. planifrons*) and the larger *Microspathodon chrysurus* (Robertson 1984). The feeding areas of the two smaller species are distinct from one another, but those of the larger species overlap both of them. The three species have the same daily cycle of feeding activity and use the same feeding microhabitats, but because of the nonoverlap of their territories, the two smaller species avoid both inter- and intraspecific competition. However, the larger species aggressively dominates in the overlapping feeding areas, taking advantage of its size and thereby outcompeting the smaller species.

An intriguing case of size advantage in competitive exclusion operating in a parasitic relationship involves three species of the hymenopteran genus *Aphytis*, which are parasitic (actually hyperparasitic, that is a parasite on a parasite) on the homopteran California red scale (*Aonidiella aurantii*); the California red scale, in turn, is parasitic on California citrus plants (Luck and Podoler 1985). *Aphytis chrysomphali* was a natural biological control of the California red scale in coastal citrus groves where pesticide spraying or another natural biological control agent, the Argentine ant, were infrequent or absent. Introduced from China in 1947, *Aphytis lingnanensis* replaced *A. chrysomphali* in most of its range, providing economic control of the scale insect in citrus groves along the coast but not in the more interior groves. The third species, *A. melinus* was introduced from India and Pakistan in 1956–57 and became established in the interior citrus area where it competitively excluded *A. lingnanensis.* Although a number of factors are involved in the competitive advantage of *A. melinus* (e.g., differences in temperature tolerance and wandering behavior, among others) a key difference is obtained by the minimum size of the scale insect in which the two wasp species can deposit their eggs. *A. melinus* gets a head start in building its population because it can use smaller sizes of the scale insect than *A. lingnanensis.* This competitive advantage enabled *A. melinus* not only to outcompete *A. lingnanensis* in the interior areas but also to extend its range coastally.

Although it is generally held that competition is weak or absent in marsupial communities, Dickman (1986) has demonstrated competitive exclusion based on size relationship involving the dasyurid marsupials *Antechinus stuartii* and *A. swainsonii.* Under field conditions in which the marsupials were live-trapped, marked, and released in five enclosures, the larger and more territorial *A. swainsonii* outcompeted its smaller counterpart for food. The intensity of the competition was greater in the winter when surface litter invertebrates on which both species feed were in short supply. Evidence of the competitive advantage of *A. swainsonii* included increased numbers, enhanced survival of newly weaned young, increased extent of movement within and expansion of home range (territorial) areas, and decreased arborial activity.

Competition in plants. The foregoing examples of competition may well lead to the implication that competition occurs only in animals. This is not the case, however; for example, Fredrickson and Stephanopoulos (1981) describe a number of microbial competitive situations. From among many competitive interactions in the plant world, three will be discussed.

In the Piceance Basin of northwestern Colorado, Welden et al. (1988) studied competition among the shrubs *Amelanchier utahensis* (mountain serviceberry), *Artemesia tridentata* (big sagebrush), and *Symphoricarpos oreophilus* (Utah snowberry), and between the trees *Pinus edulis* (piñon pine) and *Juniperus osteosperma* (Utah juniper). Measurements by regression of the distance separating neighboring plants against the sum of their canopy areas demonstrated the presence of competition within the shrubs and the trees in all but one combination of species and in most of the 10 stands selected for study, 6 for shrubs and 4 for trees. No significant differences were found in the intensity of competition, that is in the absolute severity of physiological strain induced; however, significant differences were found in some of the shrub and tree stands in the importance of competition, that is the relative degree to which competition determines the physiological status or fitness of individuals, or growth rate of populations.

Testing the competitive effect of perennial grass on shrub seedling survival was carried out by Harrington (1991) on sandy loam soils in the semiarid grazing lands of eastern Australia. The interactants were a perennial grass, woollybutt (*Eragrostis eriopoda*), and a shrub, narrow-leaf hopbush (*Dodonaea attenuata*), the latter being an invasive shrub of perennial grasslands on sandy soils. Years of other observations had demonstrated that hopbush establishes infrequently in vigorous grasslands and more frequently where the perennial grass has been damaged by overgrazing. Harrington controlled moisture input in experimental plots and also controlled presence and abundance of the grass by clipping or killing. Differential response to moisture stress resulted in the grasses surviving without summer irrigation but all the shrub seedlings dying. An inverse relationship between shrub seedling survival and the amount of herbaceous growth demonstrated direct competition. Interestingly, irrigation during the summer allowed shrub seedlings to survive but also produced a large biomass of perennial grass, the presence of the latter being susceptible to fire to which the shrub seedlings are not tolerant. As Harrington notes, modern stock management has increased shrub establishment by overgrazing the grasslands thereby increasing the availability of summer soil moisture, as well as by suppressing grass fires. This practice has changed open grassland to dense shrubland, precisely the opposite of what stock management is supposed to achieve.

The perennial bluebunch wheatgrass (*Agropyron spicatum*) was the dominant plant species on the semiarid, intermontane region of the northwestern United States prior to the introduction of European agricultural practices (plowing, use of domestic livestock, fire control, etc.). Subsequent to these changes, annual grasses from Europe and Asia, and particularly cheatgrass (*Bromus tectorum*), invaded and dominated land previously held by wheatgrass. Efforts to restore the land to the more valuable perennial grasses have been singularly unsuccessful. Thus an understanding of the competitive relationships between the major annual cheatgrass and perennial wheatgrass becomes crucial.

Harris (1967) observed direct competitive effects in experimental seeding of wheatgrass in the presence of different densities of cheatgrass. For example, average survival of wheatgrass was 39, 69, and 86 percent, respectively, in dense, moderate, and sparse populations of cheatgrass. Moreover, the wheatgrass seedlings weighed nine times more when grown under a sparse, compared to a dense, competing population of cheatgrass. Among the competitive advantages of cheatgrass that Harris found are a greater seed production (65 to 200 times) and a more rapid seed germination in moist fall weather. Cheatgrass also germinates in the fall but to a lesser extent. But the most significant advantage of cheatgrass is its 50-percent faster rate of root elongation. Under field conditions cheatgrass roots grew throughout the winter, reaching average depths of 87 cm by March 9. In contrast, wheatgrass roots grew only slightly, reaching only 14 cm by the same date. Thus in spring upper soil moisture is robbed by cheatgrass, which capitalizes on its winter root growth before the

TABLE 11-4 AVERAGE ROOT LENGTHS OF WHEATGRASS (*Agropyron spicatus*) AND CHEATGRASS (*Bromus tectorum*) GROWN TOGETHER UNDER DIFFERENT DENSITIES OF EACH PLANT

	Wheatgrass (cm)	Cheatgrass (cm)
Grown separately	64.4	82.8
4 wheatgrass : 1 cheatgrass	56.9	85.4
1 wheatgrass : 1 cheatgrass	47.2	91.1
1 wheatgrass : 4 cheatgrass	41.0	82.1

Reproduced by permission from G. A. Harris. 1967. *Ecological Monographs* 37:89–111.

principal season of wheatgrass growth begins. In a parallel study of root growth under different competition densities, Harris found that increasing densities of cheatgrass decreased the average root length of wheatgrass (Table 11-4). Conversely cheatgrass roots grew deeper with increasing densities of wheatgrass. This situation suggests that there may be *intraspecific* competition in cheatgrass when it is grown alone.

Interference Competition

Interference with light. In his studies on duckweed (see Figure 11-13) British ecologist John Harper (1961) noted that in an uncrowded condition the sequence of intrinsic growth rates, in descending order, of the four species of *Lemna* cultured separately is *minor* > *natans* > *gibba* > *polyrrhiza*. Under crowded conditions, however, *minor* is least productive, the sequence of diminishing intrinsic growth rates being *natans* > *polyrrhiza* > *gibba* > *minor*. Because an addition of nutrients to the various cultures had no effect, Harper concluded that the shifts in growth rates were not caused by nutrient exhaustion but by competition for light. In *Lemna minor* this interspecific competition has a greater toll on growth rate than on the other species. Stunting of growth with increasing population density has also been recognized in other species. It is of interest to note that in his study Harper stated that "no single parameter of growth of two species in pure cultures was a reliable indicator of their fate in mixtures." This statement is also derivable from other findings in various population studies.

The reduced ability of *Lemna minor* to use light in the presence of other species of the same genus is an example of interference competition. In this case, there is physical intervention by one (or more) species on another, reducing the ability to use one or more common resources.

Chemical interference. Indirect competition can be caused by chemical as well as physical interference. The filtering rate of the copepod *Diaptomus tyrelli*, for example, is reduced by as much as 60 percent in the presence of another copepod, *Epischura nevadensis*, its potential competitor (and predator) (Folt and Goldman

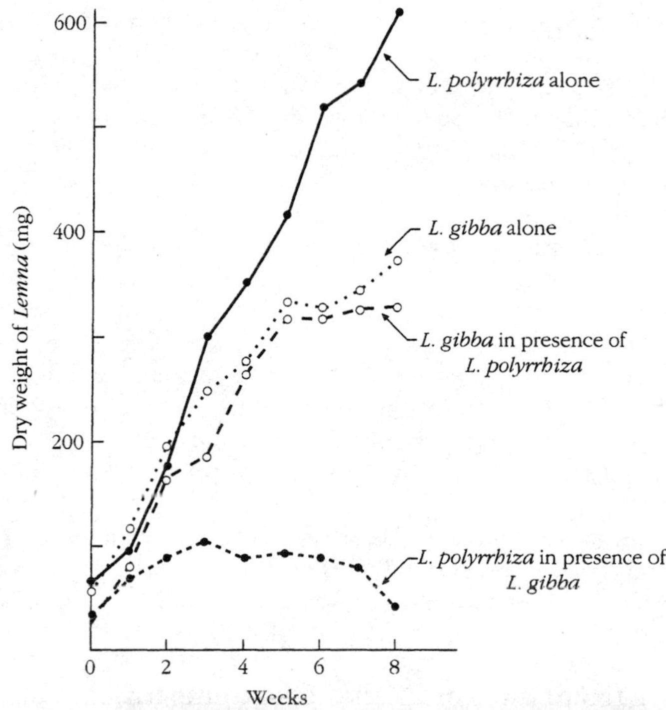

Figure 11-13 The growth of two species of duckweed, *Lemna polyrrhiza* and *L. gibba*, separately and in mixed culture, demonstrating competitive exclusion of *L. polyrrhiza*. (Redrawn by permission from J. L. Harper. 1961. *Symposium of the Society for Experimental Biology* 15:1–39.)

1981). In laboratory studies *Epischura* has been found to release a chemical of large molecular weight, whether *Diaptomus* is present or not. This chemical results in the weaker filter feeder (*Epischura*) having an inhibiting, or *allelopathic*, impact on the stronger filter feeder (*Diaptomus*). In aquatic algae the ability of blue-green algae (species of *Anabaena*) or their associated bacteria to produce chemical chelators that bind iron enable the blue-greens to grow while inhibiting the growth of other algae (Murphy et al. 1976). This chemical interference, which seems to be tied to nitrogen fixation, may be the causal factor in the sudden dominance or blooms of blue-green algae in enriched ponds and lakes.

In desert and chaparral communities, allelopathy has been invoked to explain the spaced distribution of many shrubs. For example, several species of sage (*Salvia* spp.) are found in clumps that are usually separated from grassy areas by bare areas, the presumed result of chemicals released by the sage (C. H. Muller 1969, among others). However, Bartholomew (1970) demonstrated that birds and rodents may be responsible for the bare spots since they removed seeds from the bare areas at a far faster rate than from surrounding grassland. In a study of chamise (*Adenostoma fasciculatum*), McPherson and Muller (1969) concluded that its seeds were prevented from germinating by a toxin (an allelopath) released from the shrub overstory, which is most abundant during the normal germination period. This toxin and its producing shrub overstory are destroyed by fire, a frequent occurrence in chaparral, thus releasing seeds from this allelopathic inhibition. However, Keeley and Keeley (1988) have critically called this concept of allelopathic control of germination into question.

Encounter interference. Another type of interference competition occurs in instances where a third factor unrelated to direct resource exploitation comes into play. For example, although rotifers and cladocerans compete for the same algal food source, the latter generally out-competes the former because of a size advantage (Gilbert 1985). However, the larger cladocerans can also mechanically interfere with the smaller rotifers as the latter are passed through and ejected from a cladoceran branchial chamber (Gilbert and Stemberger 1985). This encounter weakens the rotifer, putting it at a further disadvantage in exploitive competition for food.

Encounter interference also occurs among dragonflies by way of predation of larger upon smaller larval stages of the same or other species, the predation having greater impact on interspecific (as well as intraspecific) competition than direct exploitation of food by these cannibalistic insects (Johnson et al. 1985).

Relative Importance of Interspecific Competition

Although interspecific competition, like predation, has received considerable attention by ecologists both under field and laboratory conditions and in theoretical (mathematical) treatment, there is substantial disagreement as to its importance in nature (Schoener 1983, among others). The two opposing positions may be summarized as: an advocacy of strong biological interactions but an assertion that predation is more prevalent; a de-emphasis of strong biological interactions of any kind, and that, at best, competition is intermittent, perhaps even rare, and of minimal importance as an agent of natural selection.

Based on review of considerable studies, Schoener offers the following resolution: (1) during lean times, strong directional selection results from interspecific competition and produces adaptations most suited for resources used relatively exclusively by the species; and (2) during times of abundance, different types of resources increase differentially in abundance, and it then becomes advantageous for the species to use other strategies for survival.

It is certainly the case that ecological systems are complex and seldom explicable on the basis of a single interaction among its component parts. In the instance of the relative importance of predation versus competition, research most often centers on organisms that are either competitors or predators, seldom on species that are both. Wissinger (1989) studied two species of dragonfly larvae (*Libellula lydia* and *L. luctuosa*) that are both competitors for the same food sources (mostly various other invertebrates) and predators on each other's (as well as their own) larval forms. He concludes that competition and predation simultaneously affect coexistence between these two species, offering an explanation of how such ecologically similar species are able to coexist at relatively high densities.

Obviously interspecific competition, like predation, is deserving of considerably more attention in field and laboratory studies to bring resolution to the relative importance of each in population regulation.

INTRASPECIFIC RELATIONSHIPS

From the concept of the niche and Gause's principle of competitive exclusion it follows that competition would be expected to be most keen among individuals of the same species. Here, although they may vary with age, sex, and similar items, the niche requirements are indeed identical. We already noted that individual *Balanus* undercut or overgrow their own kind. The effect of increased crowding and/or the resulting increased competition on mortality and life expectancy was also previously shown for the black-tailed deer (Figures 11-6 and 11-7) and water fleas (Figure 11-10). Water fleas also show the effects of crowding on birthrate (Figure 11-14). In contrast to better survivorship at moderate densities, birthrates tend to show a consistent decline

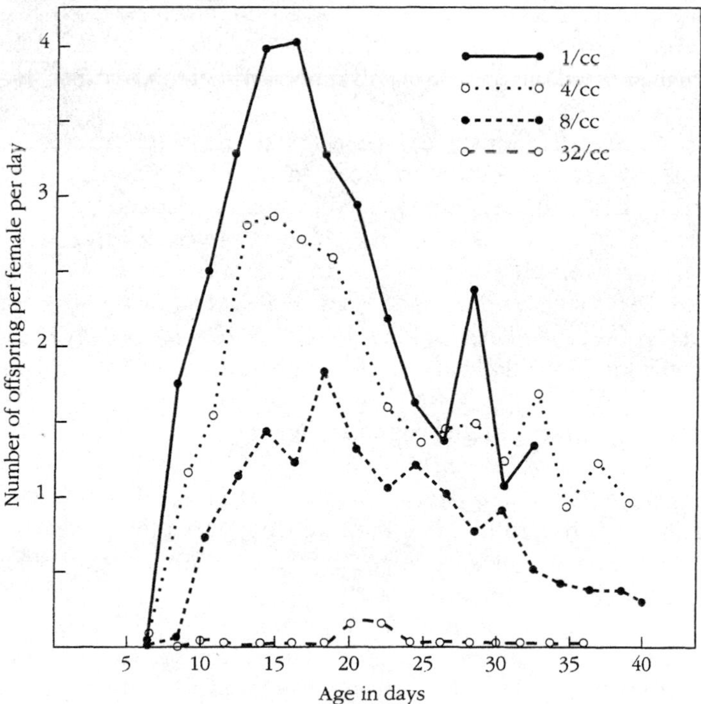

Figure 11-14 Average number of births per female per day in the water flea, *Daphnia pulex*, at different densities. (Redrawn by permission from P. Frank, C. D. Boll, and R. W. Kelly. 1957. *Physiological Zoology* 30:287–305.)

with increased crowding. Such effects have been observed with other well-studied laboratory populations. In the flour beetle, for example, Thomas Park of the University of Chicago (1933) showed that as the density of pairs increased along the gradient 1, 8, 40, and 80, there was a corresponding decrease in number of eggs per female per day of 10, 6, 2, and 1. This reduction in fecundity is apparently related to several factors: inimical excretory products in the culture, interference with copulation as well as feeding, and even an actual reduction in egg laying due to a reduced opportunity for doing so.

Density-dependent Population Regulation

Numerous studies on plants and animals have shown that the adverse impact of increased crowding and resulting interference competition for food and/or space or other vital factors appear to contribute to the regulation of populations. This phenomenon has led to a school of thought, dating to Thomas Malthus (1798), that maintains that populations of plants and animals are regulated by density-dependent factors, that is, processes that either increase mortality or decrease feduncity as the density of the population increases. A density-dependent feedback would, under such a thesis, hold a population within certain limits. The alternate school of population regulation holds that density-independent factors—that is, abiotic interactions and a number of interspecific situations previously described—are more important in population regulation. An assessment of this density-dependent viewpoint is warranted.

Some investigations like those on water fleas and flour beetles are conducted in laboratory settings; others are under natural conditions, and sometimes supplemented by laboratory experiments; yet others are based on modeling. A brief review of several of these studies, the subjects of which extend from arthropods and echinoderms to amphibians, reptiles, and mammals, as well as plants, generally support the view that crowding has negative effects on fecundity, mortality, and size in all species observed. These studies do not necessarily, however, prove a causal basis for population regulation nor do they preclude alternative explanations.

Density effects in arthropods. When cultured in separate cages, the terrestrial isopods *Armadillidium nasatum* and *A. vulgare* each showed reduction of growth rate, survivorship, and the size at first reproduction under conditions of increased density that resulted in interference competition (Ganter 1984). When cultured together, interspecific competition favored the growth rate, survivorship, and size at first reproduction of *A. nasatum* but the growth rate of only *A. vulgare.*

Although the initial population levels of the backswimming bug *Notonecta hoffmanni* were set above a putative equilibrium density in stock tanks, the population achieved equilibrium with the rate of food supply through a combination of reduced fecundity induced by interference competition for food and cannibalism, the latter not a surprising behavior among predators (Orr, Murdoch, and Bence 1990).

Using a derived mathematical model of a damselfly (*Ischnura elegans*) population, the parameters of which were in agreement with those in the literature, Crowley et al. (1987) examined four possible mechanisms of population regulation: food availability; feeding-related intraspecific competition; mortality-related intraspecific competition; and density-dependent predation. By varying the abundance of prey (i.e., adjusting the carrying capacity) the model demonstrated an increase in population density with increased availabilty of the food prey, with densities continuing to rise at high prey levels. However, the damselflies were unable to compete with each other for food since they were generally unable to deplete their prey food. Thus, feeding-related interference had essentially no effect on the population. Fecundity, however, was adversely affected by population density.

Intraspecific competition in the fruitfly *Rhagoletis pomonella* takes at least two forms (Averill and Prokopy 1987). If more than one larva develops in fruits of hawthorn (*Crataegus mollis*), survivorship drops sharply (Figure 11-15), the result of interference competition for food. In addition, multiple infestations cause a decreased pupae size which, in turn, results in prolonged maturation of females and lower fecundity. The second form of competition is, in effect, territorial: the female marks its oviposition site with a pheromone that is sufficient to deter most other females from laying more eggs in the fruit. This behavior may be viewed as adaptive, given that multiple egg-laying in one fruit results in the aforementioned decline in survivorship.

Population cycles in forest lepidoptera bear a relationship to density, but teasing apart direct cause and effect is problematic because of the many biological interactions that change with density (Myers 1988). The impact of predators, parasites, food quality, and weather may interact in different ways and at different times to cause these cycles. Myers noted that the deterioration in insect quality mediated by disease

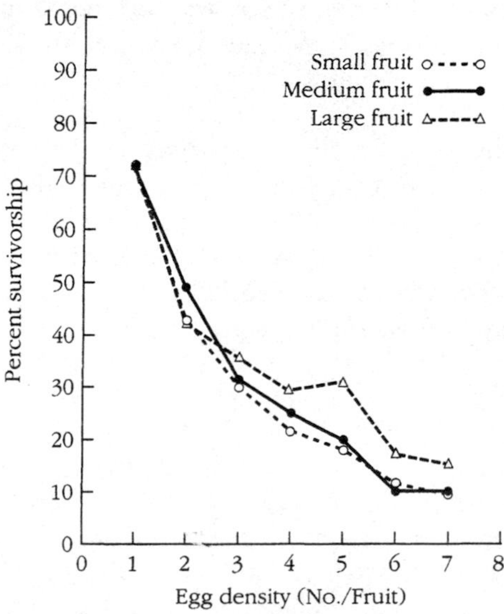

Figure 11-15 The effect of initial egg density of the fruitfly *Rhagoletis pomonella* on larval survivorship to puparium formation in three fruit sizes: small fruit (○---○), medium fruit (●—●), and large fruit (△---△). (Redrawn by permission from A. L. Averill and R. J. Prokopy. 1987. *Ecology* 68:878–87.)

as a causal factor in cyclic oscillations is often overlooked in field studies. Spruce budworm (*Choristoneura fumiferana*) populations have been oscillating for the last two centuries at an average interval of 35 years in the Province of New Brunswick (Royama 1984). Interestingly, localized populations throughout the Province tend to oscillate in unison. Royama concluded that the basic 35-year oscillation is not caused by predation, food shortage, weather, or dispersal of young larvae but by a combination of parasitoids and microsporidian infection (and possibly other factors), both of which lead to the deterioration in insect quality as suggested by Myers.

Density effects in echinoderms. Although competition for food can result in reduced fecundity and increased mortality, Levitan (1989) demonstrated that the sea urchin *Diadema antillarum* minimizes these effects by a reduction in body size. In field experiments, large (37.5 mm diameter) sea urchins at high densities reduced body size by about 7.5 mm, whereas small (27.5 mm diameter) urchins at low densities increased body size by 9.5 mm. Corroborating laboratory studies showed that urchin size was determined by food availability and not crowding. Levitan concluded that size regulation increases survivorship and reproduction under rapidly changing conditions of density and availability of resources.

Density effects in amphibians. By experimentally manipulating population densities of the salamander *Ambystoma laterale* in pools on the rocky shoreline of Isle Royale, Michigan, van Buskirk and Smith (1991) showed that increasing density signficantly reduced survival and growth (Figure 11-16a and b). Simultaneous observations of natural, undisturbed populations in other pools were consistent with the experimental results. Because there was no relationship between the level of food supply (plankton) and salamander density, the investigators believe reduced survivorship and growth are the result of interference competition by way of predation by the salamanders themselves, that is, cannibalism. The evidence for this conclusion was the widespread tail damage from attacks that were positively correlated with density.

Density effects in reptiles. Based on observations of the lizard *Anolis limifrons* over a 19-year period on Barro Colorado Island and a 4-year period at additional sites in central Panama, Andrews (1991) concluded that intraspecific competition for food increased with population density. Recruitment of new individuals, i.e., fecundity, was depressed when density was high and enhanced when food was relatively abundant. This density-dependent regulation of the population was, however, compounded by perturbations in the environment that resulted in relatively unstable population sizes.

Density effects in mammals. Among many studies on mammals, one by Southwick (1958) involving the house mouse (*Mus musculus*) living in corn stocks in England will suffice for the moment. As can be seen in Table 11-5, even in the absence of predators and any deprivation or limitation in the environment, there is reduced fecundity with increasing density. The effect is evident both on the percentage of females pregnant and on the average litter size.

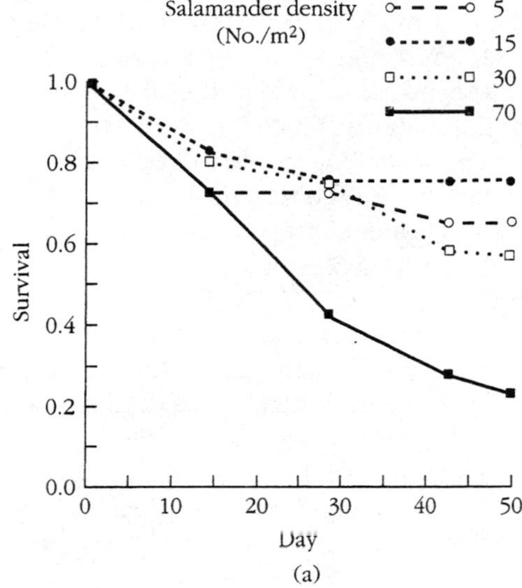

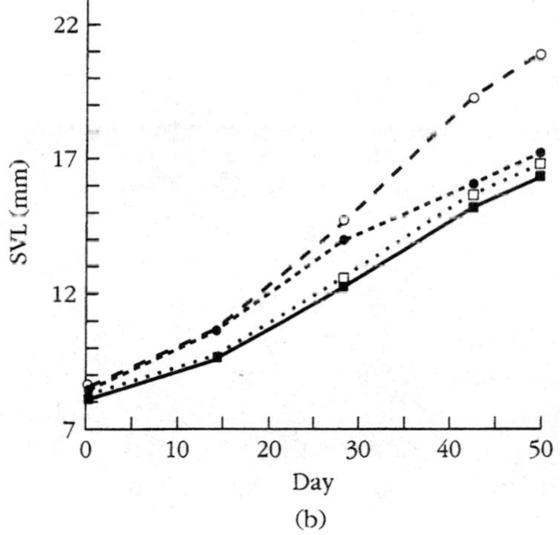

Figure 11-16 The effect of density of the salamander *Ambystoma laterale* larvae on (a) survival and (b) size as measured by mean snout-vent length (SVL). Each point represents the mean of three replicate populations. (Redrawn by permission from J. van Buskirk and D. C. Smith. 1991. *Ecology* 72:1747–56.)

TABLE 11-5 THE INFLUENCE OF DENSITY ON FECUNDITY IN THE HOUSE MOUSE, *Mus musculus*

	Sparse	Medium	Dense	Very dense
Average number/m^3	34	118	350	1,600
Average percent pregnant	58.3	49.4	51.0	43.4
Average number young per litter	6.2	5.7	5.6	5.1

Adapted by permission from C. Southwick. 1958. *Proceedings of the Zoological Society of London* 131:163–75.

Density effects in plants. Antonovics and Levin (1980) noted that various lines of evidence, though sparse, often circumstantial, and primarily from populations of dominant or abundant species, indicate that density-dependent processes have an impact on natural plant populations. Since the time of that observation, additional studies, still few in number, substantiate density-dependent impacts. For example, in monoculture, the annual plants *Trifolium incarnatum* and *Lolium multiflorum* both showed an increase in size inequality with increasing density suggesting an interference by the large plants that usurp resources and suppress the growth of smaller individuals (Weiner 1985).

Fowler (1986) investigated a natural population of the perennial bunchgrass *Bouteloua rigidiseta* in Texas, subjecting it to changes of density by the addition of seeds and removal of a third or so of the adult plants. He determined that survival, growth, and reproduction were all highly size-dependent, adult removal tending to increase the sizes of both new recruits and adults and seed additions decreasing the size of new recruits. Competition for resources among adults and among recruits, as well as by adults against recruits, affected the population in density-dependent fashion.

Density-dependent Population Self-regulation

It is suggestive, if not evident, from these few examples that increasing density tends to increase mortality and/or to decrease natality and thereby reduce population growth. This being so, the "pressure" on a population will vary with population size—as the population increases, the factors promoting a decrease tend to become increasingly effective and the result is some kind of fluctuation, great or small. In the case of the blowfly (Figure 10-17), the fluctuations are considerable in number, magnitude, and regularity. Interestingly, however, the population maintains itself without immigration, the mechanism that permits survival in thrips (Figure 11-1). Instead it does so by what Australian ecologist A. J. Nicholson referred to as self-adjustment to change—in this case, a change in the availability of food. Nicholson, along with others, maintained that the ability to adjust to changes in the environment is inherent in all populations. He put the position as follows (Nicholson 1957):

> Populations are self-governing systems. They regulate their densities in relation to their own properties and those of their environments. This they do by depleting or impairing essential things to the threshold of favourability, or by maintaining reactive inimical factors, such as the attack of natural enemies, at the limit of tolerance.
>
> The mechanism of density governance is almost always intraspecific competition, either amongst the animals for a specific requisite, or amongst natural enemies for which the animals concerned are requisites.

Self-regulation in blowflies. In the particular experiment of Nicholson's illustrated in Figure 10-17, the adjustment of population size is tied directly to egg production. At high densities competition among adults for food is so severe that no

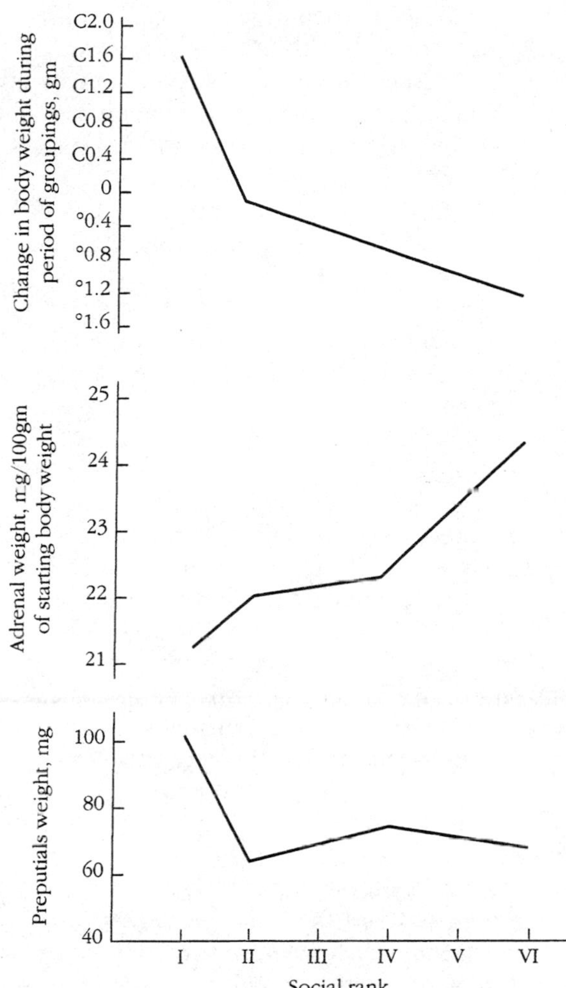

Figure 11-17 Changes in adrenal, body, and preputial weights with social rank (I being the most dominant, VI the least) for male and wild-stock house mice based on 14 populations of six mice each. (Modified by permission from J. J. Christian. 1961. *Proceedings of the National Academy of Sciences* 47:428–49.)

individuals are secure enough to enable the development of eggs. Mortality reduces the population of adults to the point where some individuals are able to obtain enough food to permit the development and laying of eggs. From that point until new adults hatch requires 16 days—2 for development of the egg in the female and 14 for development from the egg-larva-pupa stage to the adult. During this time the initial adult population continues to decline and is replaced by a new generation that increases in size, with the built-in time lag involved in growing up, to a point where competition again precludes egg production. Thus by compensatory reactions, the cycle continues.

Population self-adjustment in mammals due to stress. Self-adjustment of mammal populations to environmental changes may be regulated to a considerable extent by the endocrine system responding to sociopsychological factors in intraspecific competition. In laboratory experiments with house mice and voles John Christian (1961) used the weights of the adrenal, thymus, testes, and other sex accessory glands as measures of their function. Increases in population size in the same space resulted in increased adrenal weight and decreased thymus weight, an indication of increased adrenocortical activity. Concurrently, reproductive glands decreased in weight, an indication of inhibition of reproductive function. In a parallel set of experiments, other populations of males, of the same density as in the foregoing experiments, were placed in cages providing 42 times more area. Interestingly, the adrenocortical responses were comparable to the first findings. This factor strongly suggested to Christian that the response was not to density itself but rather to the presence of other mice—that is, a sociopsychological response.

To provide further verification of this relationship, advantage was taken of the dominant-subordinate relationship in which male mice rank themselves, the so-called peck-order phenomenon. As shown in Figure 11-17, adrenal weights were least in the dominant animals, greatest in the most subordinate, and more or less spread out in linear fashion in the intermediate social ranks. Importantly, Christian showed that these adrenal changes are not associated with fighting or with competition for food but with the social hierarchy. Comparable findings were obtained in studies on female mice. With increasing population size, reproduction declines, lactation is deficient, and maturation and growth are inhibited, many of the females never bearing young. Because the output of adrenal glucocorticoids is decreased—and they are a major source of resistance to disease by enhancing antibody formation and phagocytosis among other things—increased population density tends to reduce the population's major line of defense against infectious disease. This situation, together with inhibition of growth, maturation, and reproductive function, would collectively act as a damper on further population increases.

Whether the sociopsychological stress manifested in the laboratory house mice studied by Christian operated in the field study of the same species by Southwick (Table 11-5) is not known, for adrenal weights and other aspects of the stress syndrome were not measured. Field studies on snowshoe hares, voles in Wales and Germany, and Sika deer in Maryland, as well as a number of other laboratory and field studies, however, lend support to the sociopsychological stress hypothesis. On the other hand, studies on population irruptions of the brown lemming in northern Canada and Alaska and of the vole in Wisconsin do not show associated changes in adrenal weight as suggestive of deficiency or stress disease. Proponents of the social interaction-endocrine response mechanism espoused by Christian argue, however, that detection of the tissue changes involved requires the consideration of factors (microscopic examination, age, and maturation of tissues) that were not examined in these nonconfirming studies.

Population self-adjustment in mammals: an alternative to stress. Some investigators, notably Frank Pitelka (1964), tie these oscillations of herbivore populations not to an endocrine response but to periodic changes in the quantity and quality of forage and the rate of its decomposition in simple ecosystems, such as tundra. As a lemming population increases, grazing increases and more of the nutrients that are also required for plant growth are incorporated in the herbivores. Although excrement is then added to the soil at increasing rates with the increasing population size, overgrazing decreases the insulating quality of the vegetation cover and thereby adversely alters decomposition-mineralization rates. With dieoff of the lemmings, the vegetation cover increases, restoring, and improving soil insulation, and decomposition can then proceed at an accelerated pace, restoring nutrients to the soil. The cycle begins again with the next season's plant growth.

Population self-adjustment: natural selection. Based on his studies on voles in England, Dennis Chitty (1960, 1967) found little evidence to support the endocrine-stress or herbivore-vegetation hypotheses of population fluctuation and instead proposed that stress associated with conditions in peak populations might be genetically selective. He hypothesized that in high density situations social interactions might favor differential reproduction such that the genetic composition of the population would be altered. If the differential reproduction favored selection of certain behavioral traits, social interaction in the surviving community would be changed. In situations of increased density, for example, the more aggressive individuals are more likely to mate with a consequent increase in aggressive individuals in subsequent generations. Yet aggressiveness might be accompanied by selection against traits promoting viability. The result would lead to a decreasing population, a condition in which aggressiveness is no longer advantageous, thereby shifting the selective advantage to traits favoring reproductive capacity and viability with a consequent initiation of a new cycle of growth.

Chitty's hypothesis was initially questioned because it was believed that natural selection operates much too slowly to influence cycles of only a few years in duration, but confirming evidence has come from the work of Charles Krebs, among others. Krebs et al. (1969) has shown that the 3- to 4-year cycle of two species of voles, *Microtus pennsylvanicus* and *M. ochrogaster*, is accompanied by changes in social interactions, with voles being more aggressive in peak than in either increasing or declining populations. Moreover, he demonstrated that these behavioral changes are associated with genetic variations in the population brought about by changes in both birth and death rates throughout the cycle. Krebs postulates that rodent populations consist of two broad genotypes, one reproductively superior but intolerant of high density, the other tolerant of high density but having a lower reproductive rate. As the population increases, the individuals intolerant of high density leave (emigrate) whereas those tolerant of high density remain; of the remaining population, those that are most aggressive are then successful in reproduction and begin the cycle anew. Abramsky and Tracy (1979) have argued that emigration in voles regulates population size and immigration is responsible for the existence of the cycle.

Population self-regulation: social behavior. Although mechanisms might not be certain, there is little question that behavior has a population regulatory function. The effect of social hierarchy on mice has already been noted. In many social insects the removal of the queen liberates her previously subordinate associates from inhibition of egg laying and one or more begin to do so within a matter of hours. Not atypically, one of these subordinate egg-layers, by various means, gains queenly status and establishes a new regulatory-inhibitory role on the other potential queens in the population.

Among the more obvious kinds of social behavior that have regulatory effects on a population are the family and other social groupings, territoriality, cannibalism, and migration. The seasonal movement to and from given areas has a very significant regulatory role, both in the decimating effects of the migration movement itself and in avoiding an otherwise limiting environment during particular seasons. Migratory behavior and the intriguing navigational processes involved are factors that would take us beyond the scope of this chapter. Cannibalism, or intraspecific predation, is found in a wide variety of animals, as we have already seen, and functions as a means of interference competition, limiting population size before the resource itself becomes limiting (Fox 1975). As an example of but one type of social behavior in regulating population, let us consider territoriality.

Population self-adjustment: territorial behavior. Territoriality is space-oriented behavior and is evident in many invertebrates and vertebrates, with its greatest expression in birds and mammals. An individual, or a group, stakes out a geographic area, which it then defends in some manner. The extent of the territory is itself determined by a number of factors—vigor of the holder, amount of competition by would-be intruders, and so on. If the size of the territory is consistent with a sufficient food supply, the holder of the territory has greater assurance of survival and reproduction than do those excluded from the territory. Holding of the territory may, and often does, involve intense agonistic behavior. Sometimes, however, it is achieved by a conspicuous visual or acoustical display. As one example of the many that might be cited of the regulation of population size by territorial behavior, the study of the Australian black-backed magpie, *Gymnorhina tibicen*, by Robert Carrick (1963) is particularly telling.

Observations over 3 years on 1,345 banded birds in a study area of 13 square kilometers of savanna woodland in Canberra revealed two discrete but interacting subpopulations:

1. Territorial groups of two to eight birds, consisting of a pair or up to three adults, along with immature birds of each sex—this group occurs in the woodland areas, fights as a team, and never leaves the home territory.
2. Loose flocks of up to several hundred birds—the flock occurs in the open, treeless areas and is characterized by both sedentary birds and others showing limited movement of a few miles.

Breeding occurs only in the territorial groups. Immature birds remain as part of the group for 1 or 2 years before moving out into the flock. New groups, formed in the flock, continually attempt to take over an occupied territory but are seldom successful in doing so. The nonbreeding flocks contain a large proportion of immature individuals, plus females known to have bred when they were members of a territorial group. During the season when breeding occurred in the territorial groups the ovaries of these former territorial birds were found to be less well developed than their territorial counterparts. The end effect of this restriction of breeding to the territorial birds is a more or less steady equilibrium level in that subpopulation, a level that is unaffected by the density of the flock. This latter sub-population may and indeed does fluctuate markedly, especially due to its greater susceptibility to the spread of disease. Carrick reported two instances of heavy mortality in flock birds, one involving a contact-spread disease, the other a fungus associated with their food source. Strikingly, in neither instance were the territorial birds affected. Thus the territorial birds not only have the capacity for breeding but also a buffering against disease—two major selective advantages over the flock birds.

Population self-regulation: more on natural selection. According to these several instances, social organization does appear to be a significant population regulator in some species. V. C. Wynne-Edwards (1963, 1986) contends that the real test of sociality in a species is whether it has the potential of regulating its density by automatic and self-contained processes inherent in the population. According to this view, social populations are those that provide both for the exclusion of individuals from the group or habitat and for the recruiting or producing of new individuals into the population. Territorial species, typified here by the magpie, certainly demonstrate these criteria. Other social regulators, such as antibiosis, migration, and peck order, may be effective for other territorial forms. Among other mechanisms, Wynne-Edwards suggests that the intensity and volume of the dawn chrous of a territorial species might serve as an index of population density and appropriately stimulate or inhibit breeding behavior, thereby effectively regulating population size. Whether such behavioral communication is indeed operational and whether it involves interaction at the hormonal level remain to be established.

Wynne-Edwards's position that self-regulation is intrinsic within a population is in general agreement with those of Christian and Chitty, discussed above, as well as number of others. In his landmark book, *Animal Dispersion in Relation to Social Behavior*, Wynne-Edwards (1962) maintains that the response populations make to such factors as density or adverse environmental conditions are adaptations that have occurred through natural selection. An in-depth discussion of this position and its rationale as well as a thorough critique are beyond the scope of the text; but it is pertinent to note that his work has constituted a singular stimulus to the study of population biology and its relation to evolutionary theory. As might be expected in as dynamic a field as population ecology, Wynne-Edwards's position has adherents as well as vigorous opponents.

Concluding Comments

The foregoing discussion of the nature of population growth and regulation has barely scratched the surface of an exceedingly exciting and active area of ecology. In spite of these limitations, much has been suggested about the operation of different sorts of factors that may function in regulating populations. At a critical time in the life history of a given population, a physical factor, such as light or a nutrient, may be significant as a regulatory agent; at another time parasitism, predation, or competition, or even some other physical factor may become the operative factor. As complex and as variable as the niche of any species is, it is unlikely that this regulation occurs through any single agency (Murdoch 1994). Hestbeck (1986), for example, has shown in the California vole (*Microtus californicus*) that two separate mechanisms operate to regulate naturally occurring populations, behavioral and resource limitation. When occupation of the habitat is patchy, spacing behavior regulates density resulting in low survival due to high dispersal losses and a moderate mortality rate; individuals tend to be more reproductive. Under more uniform habitat occupancy, increased density reduces the dispersal rate thereby further increasing density against a constant resource base, the latter then becoming the limiting mechanism; survival is high due to low disperal losses and a moderate mortality rate, and individuals tend to be less reproductive. Matson and Hunter (1992) have referred to such interactions on population dynamics as "top-down" (e.g., predators, competitors, behavior) and "bottom-up" (e.g., resources and other abiotic factors).

There are numerous additional aspects of population ecology to which attention might still be given—the role of nonsocial and social behavior, including such fascinating topics as navigation and communication, among others. These topics, however, deserve a greater treatment than this volume permits; moreover, they are topics that have had recent elaboration in "small" books like this one. What we need to do now is integrate the preceding treatments of energetics, mineral cycling, and populations back into an ecosystem. As these last pages have suggested, populations are not isolated. They interact with their physicochemical environment and with other populations. They do so as associations of different populations, a community, interacting with each other and with their abiotic environment. Let us now turn to a consideration of the community and some aspects of its structure and function.

SUMMARY

VOCABULARY

allelopathic
commensalism
competition
competitive exclusion principle
indirect effect
macroparasites
microparasites
mutualism

neutralism	niche	predation
niche diversification/shift	niche overlap	protocooperation
	parasitism	symbiosis

KEY POINTS

- Abiotic factors such as nutrients (e.g., sodium, phosphorus, nitrogen), moisture (e.g., drought and flooding), food availability and accessibility, and seasonal weather patterns play a role in regulating some species populations.
- Among symbiotic relationships, parasitism, predation, and competition regulate some populations; some relationships are direct, others are indirect through a third species.
- In general, parasitism is more effective as a regulatory phenomenon in newly introduced relationships; well-adapted parasites are relatively harmless to their hosts, living off interest rather than capital; chemical control of pest parasites tends to lead to adaptation on the part of the latter to the former.
- Unlike well-adapted parasites, predators tend to live on their prey's capital rather than interest; this can result in oscillations of both predator and prey. Predator-prey interaction is affected by environmental heterogeneity, hunger and food abundance, size, spatial distribution, and disease; increase in predation with decreasing latitude has been proposed but not universally verified. The negative effects of human predation are significant.
- Classical theory predicts oscillatory predator-prey patterns; any moderately destructive factor affecting both predator and prey will increase the average prey population and decrease that of the predator. Shortcomings of the classical theory are addressed by involving the ratio of prey to predator.
- Historically, a population's niche has been defined spatially, functionally and behaviorally; it is best considered as an abstract *n*-dimensional hypervolume inclusive of all of a species' environmental factors and the degree of tolerance thereto.
- Competitive exclusion occurs where there is more or less complete overlapping of different species' niches and results in the elimination or niche diversification/shift of one of the competing species or of subgroups within a species.
- Competition can be overtly aggressive, as when there is a size advantage, or subtle, as in the case of allelopathy; it can also occur by interference with light and chemicals.
- Intraspecific relationships, because of the identical nature of the niche, are most keen, resulting in population regulation due to density, and the effects thereof (e.g., stress), in some populations; various adaptations to density, including but not limited to social and territorial behavior, result in the self-adjustment of population size in some species.

PART VI

Community Ecology

CHAPTER 12

The Structure and Function of Communities

INTRODUCTION

Energy flow, nutrient cycling, population growth and development—these are the significant properties of ecosystems discussed in the preceding chapters. The sharp focus on each of these broad concepts permitted some initial understanding of their operation but isolated phenomena or processes without relating them to the totality of structural and functional relationships. None of these ecological processes occurs in isolation, each is manifested by particular assemblages of different species populations in particular physicochemical environments. Thus the flow of energy, the cycling of nutrients, and the regulation of populations occur in an assemblage that may consist of broomsedge, field mouse, and weasel populations while a few feet distant the assemblage may consist of herbs, rabbits, and foxes. We now turn to consideration of the characteristics and properties of these assemblages of species populations—ecological communities.

An *ecological community* can be regarded as an assemblage of species populations that has the potential of interaction. More precisely, a community may be defined as an interactive assemblage of species occurring together within a particular geographical area, a set of species whose ecological function and dynamics are in some way interdependent (Putman 1994). These interactions include overt and critically determining competitive interaction and feeding relationships as well as more subtle manifestations, such as reliance of plants on animals for pollination and seed dispersal or of animals on plants for meeting habitat requirements. The range and nature of these interactions are myriad and can be but briefly addressed in this and the following chapter.

Ecological communities come in all sizes, shapes, and degrees of interaction of the constituent populations. The goal of this section is to examine the nature of some of the population interactions in communities and ascertain their effect on community structure and function. It is possible to conceptualize or even identify an assemblage of species populations in which there are so few interactions that any apparent organization arises primarily from independent autecological processes rather than from synecological ones. Such assemblages would just be collections of more or less autonomous populations in the same place at the same time (Strong et al. 1984); although these kinds of assemblages are worthy of study, they would not be germane to our purposes.

According to some ecologists, interactions among species populations that make up an ecological community give rise to properties of the community that are not inherent or readily recognized nor easily predicted from those properties of the species populations themselves. In this context, an ecological community is holistic, the totality of its properties being greater than the sum of its individual parts. Begon et al. (1986) have referred to these as *emergent properties*, noting in analogy that a cake has emergent properties of texture and flavor not apparent by consideration of its ingredients. For example, an aggregation of 1,000 trees clustered on a plot of land has a different set of properties than 1,000 trees in dispersed arrangement; the former results in reduced insolation to the forest floor and the accumulation of a layer of rotting leaves, neither of which characterizes widely dispersed trees.

That properties of a community (or, for that matter an ecosystem) are emergent, a derivative of being holistic, is not universally accepted. Part of the debate centers around the very nature of a community, which we will discuss later in the chapter (see section entitled "The Community: Individualistic or Organismic"), but part is also determined by the assignment of the term "emergent" to properties that are regarded by others as merely collective (Salt 1979, among others). An example of *collective properties* is the age structure of a populaton, which is the result of the combined ages of all the individuals making up the population. Likewise, community respiration, also to be discussed later in this chapter, can be regarded as a collective community characteristic since it is an aggregation of the totality of respiration carried on by the individuals in the community.

As we proceed through the following sections on the structure and function of communities we should keep in mind the distinction between emergent and collective properties, and thereby come to appreciate the oftentimes difficult discriminations ecologists make.

Robert May (1984) has wisely noted that ecology is a difficult science, partly because evolution has only given us one world, and it is not easy to perform controlled experiments. Yet, as Strong et al. (1984) note, the key to resolving the many intriguing conceptual issues facing community (and other) ecologists is the evidence derived from experimentation to corroborate or falsify a theory. While much progress has been made in experimentation on ecological communities, generally a formidable task, there is much yet to be done in this exciting field. We will explore some of the findings, both observational and experimental, in this chapter.

THE FORM AND STRUCTURE OF COMMUNITIES

Physiognomy

The form and structure, or *physiognomy*, of a community can be described in a number of ways, each assessing somewhat different aspects, each having certain advantages and limitations, and, as might be expected, each having its proponents in ecological circles.

Growth forms. Perhaps the least technical characterization of a community's structure, particularly as applied to plants, employs commonly recognized *growth forms* such as evergreens and deciduous trees, woody and herbaceous shrubs, herbs, and so on. Within each of these general categories we can distinguish further between needle-leaved, broad-leaved, and sclerophyll evergreens, or succulents, thorn, and rosette among shrubs or ferns, and grasses and forbs among herbs. Some of the numerous systems of plant-growth forms are straightforward and simple, some take into account a number of characteristics, providing an in-depth description. Dansereau (1951) proposed a system exemplary of the latter recognizing six growth forms (trees, shrubs, herbs, bryoids, epiphytes, and lianas); three sizes (tall, medium, low); four functions or leaf habits (deciduous, semideciduous, evergreen, evergreen-succulent or evergreen-leafless); six leaf shapes (needle or spine, graminoid, medium or small, broad, compound, thalloid); four leaf textures (filmy, membranous, sclerophyll, succulent or fungoid); and four aspects of coverage (barren or very sparse, discontinuous, in tufts or groups, continuous).

Life forms. A variation of plant-growth form, referred to as *life form*, was developed at the turn of the twentieth century by Danish botanist Raunkiaer (1934) based on the relation of the ground surface to the plant's embryonic or regenerating (meristematic) tissue. The latter remains inactive during unfavorable climatic regimes and resumes activity when favorable conditions return. The categories Raunkiaer recognized, based on the "amount and kind of protection afforded to the buds and shoot apices," are illustrated in Figure 12-1 and described as follows:

Phanerophytes: surviving buds or shoot-apices borne on negatively geotropic shoots that project into the air; shrubs and trees.

Chamaephytes: surviving buds or shoot-apices borne on shoots very close to but above the ground; creeping woody plants and herbs.

Hemicryptophytes: surviving buds or shoot-apices are situated at the soil surface protected by leaf and stem bases; herbs growing in rosettes and tussocks.

Cryptophytes: surviving buds or shoot-apices are buried in the ground at a distance from the surface that varies in different species; tuberous and bulbous herbs.

Therophytes: plants of the summer or of the favorable season; overwinter as seeds; annuals.

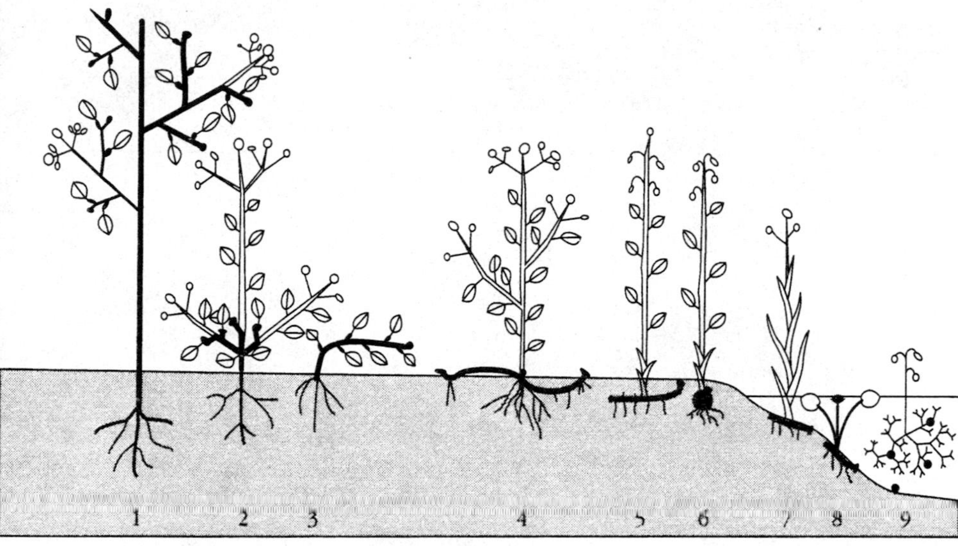

Figure 12-1 Four of Raunkiaer's five life forms: phanerophytes (1), chamaephytes (2–3), hemicryptophytes (4), and cryptophytes (5–9). The parts of the plant that die in the unfavorable seasons are unshaded; the persisting parts with surviving buds are shaded black. (Redrawn by permission from C. Raunkiaer. 1934. *The Life-Form of Plants and Statistical Plant Geography*. Oxford: Clarendon Press.)

Because the ability of a plant to survive in a given environment depends on its being adapted to that environment, and because the life form is a morphological adaptation, it should follow that given life forms would be more prevalent in some environments than in others. Certain grasses, for instance, predominate over herbs and shrubs in heavily grazed areas subject to fire through having their growing tips below the ground level (cryptophytes). In contrast, herbs and shrubs have theirs above the ground level (phanerophytes) where they are susceptible to both fire and grazing.

In analyzing the life forms of various regions and comparing them with a "normal" spectrum based on 1,000 species selected at random (Table 12-1, part A), Raunkiaer found a predominance of phanerophytes in tropical moist regions. This result is perhaps not unexpected, for the exposed buds are not subjected in this kind of environment to such adversities as low temperature or aridity. Desert areas would require, among other drought-resisting adaptations, maximum protection of buds and regenerating parts from high temperatures and aridity. Not surprisingly, a high preponderance of therophytes and not insignificant proportions of chamaephytes and hemicryptophytes are found in such environments. Intermediate between these climate extremes, and their corresponding sharp differences in amount of protection afforded the propagative organ, is the moist-temperate region in which hemicryptophytes predominate, thereby offering an intermediate amount of protection.

TABLE 12-1 THE PROPORTION OF LIFE FORMS, BASED ON THE RAUNKIAER SCHEME, IN VARIOUS ENVIRONMENTS

Location	Phanerophytes	Chamaephytes	Hemicryptophytes	Cryptophytes	Therophytes
A. Normal spectrum of 1,000 sp.	46	9	26	6	13
Tropical climate	61–74	6–16	4–12	1–5	5–16
Tundra climate	0–1	22–26	57–68	4–15	2–4
Temperate—moist climate	8–15	2–7	49–52	15–25	9–20
Desert climate	9–26	7–21	18–20	5–8	42–50
B. Cushetunk Mountain, New Jersey					
North slope	41.8	1.2	41.8	15. 2	0
South slope	32.1	1.8	46.4	12.5	7.1

Part A adapted by permission from P. Dansereau. *Biogeography—An Ecological Perspective.* 1957. New York: The Ronald Press Company. Part B from J. Cantlon. 1953. *Ecological Monographs* 23:241–70.

Relative proportions of different life forms also occur on a smaller scale, as in the instance of a north- and south-facing slope of a mountain (Table 12-1, part B). The north slope is more cool and moist than the south slope, and there is a vegetational response, in the way of life form, to this difference in microclimate. Finally, in yet another application, H. M. Hansen (1958) has applied the life form as an age indicator in considering fossil flora. He concluded that in the evolution of flowering plants the progression has been from taller to shorter phanerophytes, with a subsequent origin of hemicryptophytes and, most recently, the development of chamaephytes, therophytes, and cryptophytes.

Stratification. Both the life and growth form of plants emphasize height. Even a casual glance at a plant community conveys differences in the heights, or *stratification,* of the components. Trees are generally taller than shrubs, which are usually taller than herbs, and the latter taller than mosses and lichens. Tropical forests are characterized typically by a marked vertical stratification, especially in areas where the canopy is broken (see Figure 14-13). Stratification results in the upper strata, or canopy, receiving a larger proportion of solar energy, and in instances of particularly dense foliage, little sunlight reaches the ground, reducing photosynthesis and plant growth at that level.

Although we are more readily conscious of stratification in terrestrial plants, it also occurs in aquatic situations. However, it is to be noted that in terrestrial situations, variations in plant structure and stratification within a community lead to stratification of the environment (e.g., light penetration, humidity) and animal components (e.g., ground dwellers, arboreal dwellers), whereas in aquatic systems stratification of the environment (temperature differences, light penetration, etc.) lead to stratification of both the plant and animal assemblages. Among the algae, green algae tend to be nearer the surface, brown algae in deeper water, and some red algae at even deeper levels. Also, in aquatic environments, floating and motile plankton generally occur in the upper portions whereas free-swimming fish occur in the next lower segment followed by sedentary (e.g., corals) or bottom-dwelling (e.g., crabs, clams) forms, the latter in turn followed by those that burrow below the bottom surface. As will be described later, plankton may, under certain circumstances, become so dense as to result in oxygen depletion in bottom waters.

Again, less evident to the noncritical observer is the stratification that occurs among animals. Forest birds tend to be vertically stratified with some species feeding and nesting only in the canopy, others near or on the ground, and yet others at various levels in between. In that same forest, squirrels nest high above the ground, mice at or near the ground surface, and moles beneath the surface. Likewise, a bee hive is at a different vertical stratum than the mites at the surface level or the earthworms and nematodes in the soil.

Zonation. Horizontal changes in the physical environment are reflected in zonational changes in plant and animal components of ecological communities. As will be discussed in more detail in Chapter 15, as we proceed from a lake's edge, the first zone is that of vegetation rooted in the bottom but extended above the water level; this is followed by a zone of floating vegetation, subsequently by one of submerged vegetation, and finally one that is characterized by phytoplankton, which may be regarded as unrooted vegetation. Similar zonation is to be found in terrestrial situations where edaphic, or soil, characteristics change from, for example, more moist to drier conditions.

Horizontal dispersion. The horizontal spacing, or *dispersion*, of plants and/or animals can also be used to describe the structure of an ecological community. There are three basic patterns of dispersal (Figure 12-2): random, uniform or regular, and clumped or contagious; two additional patterns exist in combinations of random and clumped, and uniform and clumped.

Although a community may show truly random distribution (verifiable by its fit to a Poisson distribution), more often a number of factors results in various degrees of clumping (or contagion), a regularity in spacing, or a combination of both. Most often, a clumping of resources (e.g., water, nutrients) leads to clumping of organisms; reproductive patterns may also account for clumping. The availability of a resource, such as groundwater, or the release of biological inhibitors by sedentary

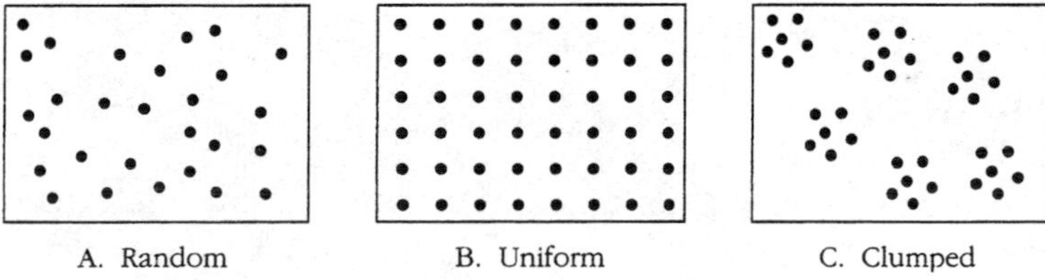

Figure 12-2 Basic patterns of dispersal.

plants (a phenomenon known as allelopathy (see Rice 1974) may inhibit other individuals of the species from developing within the radius of the inhibitor's effect; both situations can result in a more or less regular dispersal of that species. And, of course, in agricultural situations, regular spacing of orchard trees such as oranges, peaches, and macadamia nuts is quite common.

Guilds

Cornell ecologist Richard Root (1967) introduced the term *guild* with the following definition:

> A guild is defined as a group of species that exploit the same class of environmental resources in a similar way. This term groups together species without regard to taxonomic position that overlap significantly in their niche requirements. The guild has a position comparable in the classification of exploitation patterns to the genus in phylogenetic schemes.

In precise application of the term, Root noted that the major portion of the diet of a bird that is to be considered a member of the foliage-gleaning guild in the oak woodland must consist of arthropods obtained from the foliage zone of oaks. By contrast, birds that catch insects in the air would constitute another guild and those eating seeds yet another.

Although the intent of the term guild was to apply to a group of species using the same class of resources in a similar way, there is ambiguity as to what constitutes the "same class of resources" and the degree of similarity needed to constitute "a similar way." The result has been to use the term in less precise and more idiosyncratic ways than Root intended (Simberloff and Dayan 1991).

Guild denotes a functional attribute and thus may be, as it has been, easily confused with the term niche. It had been Root's hope that niche would continue to apply to the habitat requirements, due to function or role of an individual species, whereas guild would apply to the functional role of a group of species. To some extent that has taken place. Like the term niche, "guild" was a term needed in community ecology to apply to the group of species that comprise the potential competitors or partitioners of some resource; and had this term not been coined, an analogous word would have been used (Simberloff and Dayan 1991). To have fruitful application in research, Simberloff and Dayan identify two conditions that must be met: (1) the role of foraging method should be emphasized, and (2) a list of sympatric related biota should be included, especially in cases where the reasons for their exclusion would not be self-evident.

Trophic Structure

Yet another productive categorization of community structure is the use of trophic (or feeding) levels as discussed in Part III. Under this schema, the plant or animal community can be described by assignment to the category of producers, consumers (primary, secondary or tertiary) or decomposers. These trophic levels and their disposition into food webs, also discussed in detail in Chapter 7, carry influences and constraints on the organization and dynamics of communities (Pimm et al. 1991; Putman 1994; among others). For example, according to current food web theory, highly complex communities should be most sensitive to the loss of species from the top of the web because such extinctions propagate more widely than in simple communities; in contrast, simple communities should be more sensitive to the loss of plant species than complex ones because in the former the consumers are dependent on only a few species and cannot survive their loss.

Phylogenetic Composition

Perhaps one of the more obvious and easily comprehended means of describing a community is by identifying the constituent species assemblages or those of some other phylogenetic group. Thus a forest community might be described as spruce-fir, beech maple, or oak hickory, and an animal community as microarthropods of the soil, benthic fish, air-borne insects, and so on. Again, this structural characterization should be so well understood as to warrant no further explanation here.

Classification of Communities

The tendency to bring a semblance of organization to complex entities is commonplace: not content to call a rock a rock, mineralogists have developed hundreds of categories based, in large measure, on chemical composition; not content to call a chemical a chemical, chemists have identified more than 100 elements based on such factors as atomic structure and weight and hundreds of thousands of such element compounds; not content to call an animal an animal or a plant a plant, taxonomists have clustered them into kinds of similar nature called species, species into genera, genera into families and so on. Ecologists are not to be outdone, and thus a number of schema have been employed to bring a degree of order out of not so much disorder as the complexity of ecological communities. As in the case of taxonomists, there are "lumpers" and "splitters" among ecologists, referring respectively to those who tend to aggregate and those who tend to make finer discriminations among the entities being described.

Not surprisingly, classifications of communities, most notably of plants, are, for the most part, based on the structural characteristics described above. Physiognomy of the dominant growth-form is commonly employed as a classification criterion at both local levels and over broad geographic areas (e.g., evergreen or deciduous forests). Dominant species serve as another criterion of classification (e.g., spruce-fir forest, beech-maple forest). Further refinement can be obtained by classifying according to the dominant species at each vertical stratification (e.g., squirrel, mouse, mole). And even further discrimination can be obtained by considering the whole floristic composition of the community as developed by French ecologist Josias Braun-Blanquet (1932) and used extensively in Europe; a treatment of this approach is well beyond the scope of this text.

Classification into discrete categories implies that more or less sharp discontinuities are present among the entities being segregated. While such discrimination is possible in the classification of chemical elements, it becomes less feasible with biological entities and increasingly so as we move from distinguishing among species to differentiating among ecological communities. Except in some circumstances where a major physical discontinuity exists, as in the instance of a very high, precipitous cliff, community boundaries are more often blurred, one community's constituents extending into another and vice versa in a continuum (see Figure 14-15). This leads us to explore the historical development of the concept of a community and consider that concept and its concomitant of classification in contemporary terms.

THE COMMUNITY: INDIVIDUALISTIC OR ORGANISMIC

The Individualistic Nature of Communities

As we will learn later in greater detail (see Chapter 14), topographic differences in temperate deciduous forest are accompanied by environmental differences in insolation, precipitation, and soil characteristics. Further, a continuous gradient of microclimate and vegetation is observed instead of discontinuities in the plant communities. Gradient patterns imply a gradual and progressive change in the structure and composition of vegetation from one environmental extreme to another, a continuum in which discrete subdivisions do not, in reality, exist. Among many quantitative studies on temperate forests are those in Wisconsin by J. T. Curtis who concluded in one of his studies (Curtis and McIntosh 1951):

> ... tree species occur in a continuously shifting series of combinations with a definite sequence or pattern, the resultant of a limited floristic complement acting on, and acted upon, by a limited range of physical environmental potentialities. Such a gradient of communities is here called a *vegetational continuum* [italics added].

The nature of the community as implied by the continuum suggests that each species is distinct or individualistic and that each community is also distinct from all others—it is an *individualistic community*. An historically important statement of the individualistic nature of the plant community is that of its foremost proponent, the American ecologist Henry Gleason (1926):

> ... the vegetation of an area is merely the resultant of two factors, the fluctuating and fortuitous immigration of plants and an equally fluctuating and variable environment. As a result, there is no inherent reason why any two areas of the earth's surface should bear precisely the same vegetation, nor any reason for adhering to our old ideas of the definiteness and distinctness of plant associations. As a matter of fact, no two areas of the earth's surface do bear precisely the same vegetation, except as a matter of chance, and that chance may be broken in another year by a continuance of the same variable migration and fluctuating environment which produced it.

This statement has misled some observers into thinking that Gleason meant that communities were random mixtures of species undetermined by environment—a position that he did not hold. What is to be emphasized in the Gleason position is that species are individualistic and that no two are distributed alike. It follows, then, that because communities consist of individualistic and broadly overlapping species populations, they intergrade continuously, and each arbitrarily isolated community is, in fact, unique.

The Organismic Nature of Communities

As might be expected, unanimity among plant and animal ecologists about the nature of the community was not the case. There were those previous to and later than Gleason, including some ecologists today, who contend that communities are discrete, discontinuous, discernible and hence describable entities. In their view, analysis will reveal certain clusters of species reach their optimum in the same communities and, moreover, that the species of one such grouping would never occur as important members of another group. In the more extreme development of this view, plant associations are viewed as having objective reality like an organism or a species and as being capable of description in comparable ways. Major proponents of this *organismic community* view, which was pervasive in the nineteenth and early twentieth centuries and continues to date, were American ecologists such as Frederick Clements, S. A. Forbes, and, later, Warner Clyde Allee, and Eugene Odum. Among European ecologists holding the organismic view was Josias Braun-Blanquet, of the Zurich-Montpellier "school" of phytosociologists. For comparison with Gleason's statement above, that of Clements (1916), in which he regarded the mature form of a community to be an organic entity, poses a sharp contrast:

> As an organism the formation arises, grows, matures, and dies. Its response to the habitat is shown in processes or functions and in stuctures which are the record as well as the results of these functions. Furthermore, each climax formation is able to reproduce itself, repeating with essential fidelity the states of its development. The life history of a formation is a complex but definite process, comparable in its chief features with the life history of an individual plant...

R. F. Daubenmire (1968) of Washington State University, among others, was one of the most severe critics of the individualistic concept. While admitting that vegetation is a continuous variable, his studies in eastern Washington, the Bitterroot Mountains of Idaho, and elsewhere have identified points of rapid change in composition where some species drop out. Daubenmire believed that these points represent discontinuities and the identification of discrete rather than continuous communities. He argued that proponents of the individualistic school err in sampling and analytical techniques and in failing to consider the functional significance of the species in given systems. As is the nature of scientific controversy, proponents of the individualistic approach have, in turn, challenged Daubenmire's methodology.

In the intervening period of Clements and Gleason, numerous quantitative studies have stressed the importance of environmental gradients and their influence on inter- and intra-specific interactions and, in turn, on discontinuities/continuities of ecological communities. Those holding the individualistic view acknowledge that discontinuities between communities can be recognized; however, they note that such discontinuities are related to environmental discontinuities in abiotic factors such as moisture, acidity, nutrients, solar radiation, salinity, and so on, some of which can be the result of perturbations by humans or by natural occurrences (e.g., fire, volcanic eruptions, wave action). Organismic adherents regard such gradient analyses as representing disturbed stands or those not in equilibrium.

The organism/individualistic debate is far from settled, making the issue of "what is a community?" an exciting arena for investigation of both the factual and theoretical, like many other aspects of contemporary ecology. However, it is the case that the individualistic concept is more universally held today, largely as a result of the use of the increasingly sophisticated methods of quantitative techniques and analyses employed. To explore these in detail would go far beyond the scope of this introductory text, but interested students would be well advised to consult the extensive treatment of gradient analysis by Ter Braak and Prentice (1988), among others.

FUNCTIONAL ASPECTS OF COMMUNITIES

Because they consist of living organisms, communities are not static entities. Like the species that comprise them, ecological communities are dynamic and undergo more or less constant change in physiognomy and composition, owing to changes both in the environment and in the characteristics of the life cycles of the organisms themselves. For example, on the cooler and more moist north slope of Cushetunk Mountain (to be described in Chapter 14), many herbs have bloomed by late spring; however, on the drier and warmer south slope the maximum period of herbacous activity is during early summer. Likewise, insects overwintering as pupae emerge under certain conditions of temperature, moisture, and even the length of the photoperiod, one of many examples being the late summer cicadas so common in temperate climates (Williams et al. 1993).

Each species has its own characteristic pattern of sequential development that is largely attuned to and regulated by major environmental climatic gradients. It is not surprising, then, that whole communities demonstrate phenomena attuned to naturally recurring events. Thus in the deciduous forest biome spring brings release from dormancy and a resurgence of vegetative activity that culminates in leafing out in deciduous trees. Fall brings the onset of dormancy and a curtailment of activity manifested in leaf color change and fall. Because natural geophysical phenomena recur at different intervals, and because different species respond differentially to such stimuli, different rhythms would be expected, a priori, among different communities. Thus the sugar maple community that leafs out in spring and undergoes leaf fall in autumn is responding to an annual rhythm, that of shifting day-length. Similarly, this same group of plants responds during the growing season to the diurnal rhythm of alternating periods of light and darkness by undergoing photosynthesis during the day but not at night.

It is not intended here to explore the phenomenon of animal, plant, and community periodism in its myriad facets. The interest here is in considering the phenomenon with respect to the larger issue of the ecological community. Earlier, for example, we discussed the regulation of the distribution of two species of barnacles by tidal periodicity—that occurring daily and that occurring monthly. Our question now is the degree to which naturally recurring phenomena affect the structure and function of ecological communities.

As we proceed from this point, it will be instructive to keep in mind the individualistic/organism debate just discussed, as well as the distinction between emergent and collective properties described earlier in the chapter. Are the phenomena demonstrated analogous to that of an organism? To what extent do the processes described constitute emergent rather than collective properties?

Seasonal Changes in Various Ecological Parameters

Because of the change in inclination of the Earth's axis with respect to the Sun over the course of a year (see Figure 4-2), the total solar radiation received at a given geographical latitude varies at different times of the year (see Figure 4-3). As might be anticipated, such changes elicit a differential progression of responses by given species to both the quality and quantity of light (Figure 12-3), as well as to the temperature and other climatic differences that recur more or less periodically. In addition to changes in photoperiod, seasonal changes also occur often in moisture

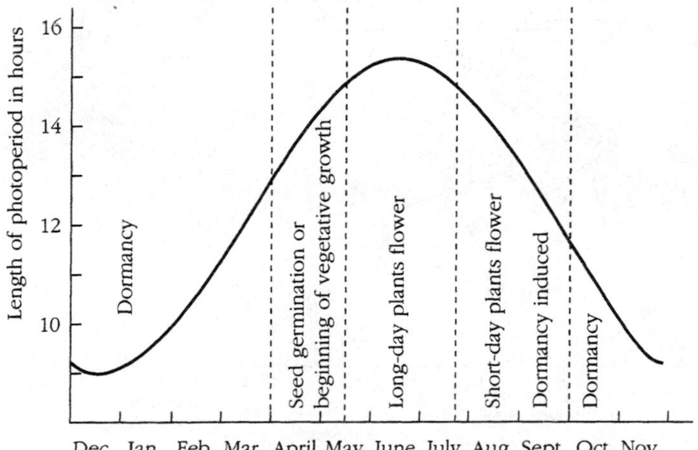

Figure 12-3 Yearly variation in length of day at 43° north latitude and the correspondence to photoperiodic responses of plants. Figure from *The Living Plant* by Peter M. Ray. Copyright © 1963 by Holt, Rinehart and Winston, Inc. and renewed 1991 by Peter M. Ray, Reprinted by permission of the publisher.

and temperature. In general, such periodic phenomena in organisms that are tied to periodic environmental change are referred to as *phenology*, a term that was once restricted to recurring events affecting only plants.

The phenology of sumacs. The sequence of phenological events in sumac (*Rhus glabra*) is shown in Figure 12-4. One striking characteristic of this onset stage of plant activity is that events that mark the initiation of development of an organ (e.g., bud elongation and opening, the appearance of the leaf, and inflorescence) are completed rather rapidly. In contrast, events dealing with completion of growth of an organ (cessation of growth of twig, leaf, and inflorescence) occur over a longer period of time. Comparisons of some nine clones of sumac showed great uniformity in the sequence of these phenological events, suggesting the most important control to be genetic. But the fact that slope exposure altered the time of onset implies an additional regulation by environmental gradients.

Synchrony and asynchrony in figs. Unlike temperate species like sumac that are attuned to periodic environmental changes, many tropical plants flower year round, a not surprising phenomenon when we consider the relative constancy of climatic conditions in the tropics. Monoecious figs (*Ficus*) provide perhaps the most extreme example of such flowering asynchrony (Bronstein et al. 1990). Phenological flowering is synchronous within a given fig tree but asynchronous within the population of fig trees, a seeming adaptation to enable pollinators (wasps) to persist locally over the course of the year. In a simulation model, Bronstein and her colleagues found that a median of 95 trees was required to produce an asynchronous sequence that could sustain the pollinator population for 4 years.

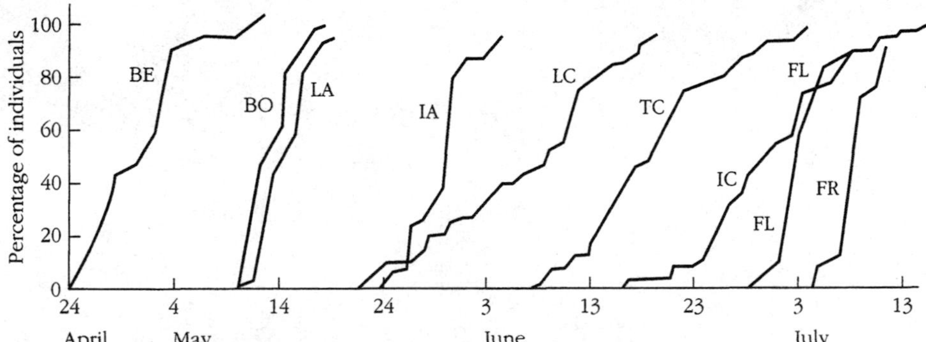

Figure 12-4 Progress of phenological events in sumac in southern Michigan. BE = bud elongaton; BO = bud opening; LA = leaf appearance; IA = inflorescence appearance; LC = cessation of leaf growth; TC = cessation of twig growth; FL = first flower; FR = first fruit; IC = cessation of inflorescence growth. (Redrawn by permission from E. F. Gilbert. 1961. *American Midland Naturalist* 66: 286–300.)

Wind and conifers. Subalpine fir (*Abies lasiocarpa*) and Englemann spruce (*Picea englemannii*) occurring at timberline (3,200 m) in southeastern Wyoming are subject to the drying effects of winter wind (Hadley and Smith 1986). Under such environmental conditions, there is a high occurrence of needle mortality as well as cuticle abrasion that removes the protective surface wax; needle death is frequent only in wind-exposed needles of trees showing the characteristic flag-form, known also as *krummholz*. (See Figure 3-4 for a similar effect from salt in the air.) If summer growth conditions are less than optimal, the drying effects are observed to be increased. Thus both summer and winter periods have impact on these conifers, contributing to their survival and dispersal.

Summer rainfall and a desert annual. The desert annual umbrella plant, *Eriogonum abertianum*, demonstrates a germination phenology tied exclusively to winter rains such that individuals may flower in spring, summer, spring or summer, or neither season (Fox 1989). Considerable mortality occurs between spring- and summer-flowering episodes, a phenomenon associated with the severe droughts that characterize the foresummer period. Of two desert sites in which the species was studied, higher mortality was observed in the Sonoran Desert, which has a lower summer rainfall and is hotter, than in the Chihuahuan Desert, which has greater summer rainfall and is cooler. Based on fossil and biogeographic evidence, Fox concludes that earlier flowering in the Sonoran Desert is an phenological adaptation to the lower summer rainfall and hotter temperatures there. Further, he suggests that this species' ancestral type was phenologically summer-flowering and invaded the Sonoran Desert when summer rainfall there was greater than it is now.

Seasonal drought and moist forest shrubs. Given the absence of significant photoperiod and temperature changes in tropical communities, it may not be surprising that tropical forest plants often show phenological responses such as vegetative and reproductive growth in short bursts associated with rainfall and moisture availability, both of which are more seasonal. By irrigating two 2.25 hectare plots of mature forest for three consecutive dry seasons on Barro Colorado Island, Panama, and comparing them with results in control plots, Wright (1991) sought to discriminate the role of moisture availability on seven species of understory shrubs (*Piper* and *Psychotria*). Although irrigation disrupted the phenological leafing response observed in the control plots, it had no effect on flowering in the four *Psychotria* species, suggesting that other phenological cues were operating.

Amazonian amphibia. Phenological responses among animals tend to be most easily observed in reproductive activities. Some of these periodic activities are triggered by hormonal rhythms, the rhythms themselves often being cued by changes in environmental factors such as light, temperature, and moisture. Many insect species respond to recurring temperature and photoperiod changes in their development rates and adult reproductive activity (Wolda 1988). Among the many examples of phenological responses in animals that could be cited, a study in Manaus,

Brazil, on amphibian breeding will suffice. In Manau, where rainfall is seasonal, being wetter in midwinter and late spring, Gascon (1991) observed two distinct phenological patterns in an amphibian community: a group of 5 frog species bred during the entire rainy season and another group of some 18 species bred only after heavy rains. Further, most of the latter group adjusted their breeding to the earlier rains that occurred in the second year of the study. Gascon noted that these two patterns of phenology occur in other tropical amphibian communities as well. Interestingly, two species not only bred independently of rainfall but at chronologically the same time each year suggesting the operation of some other environmental cue and/or an internal biological rhythm.

Tropical forests. Most tropical woody plants produce new leaves and flowers in bursts rather than continuously, and most forest communities display seasonal variation in the presence of new leaves, flowers, and fruits (van Schaik et al. 1993). Although this synchronizing may be, in part, an adaptation to reducing predation, it appears that climate is primarily responsible. Peaks in flowering accompany peaks in irradiance if water is sufficient. Thus, in seasonally dry forests, the start of the rainy season finds a concentration of leafing, flowering, and fruiting/seeding.

Periodism and Temperature

Aquatic communities. Because of the high specific heat of water (see Chapter 3), daily changes in ambient temperature have only modest effects on freshwater and marine communities, whereas the more prolonged temperature changes associated with the seasons, particularly in temperate zones, have a more marked effect (see Figures 4-8 and 4-9). In temperate freshwater ponds and even large temperate lakes the fall turnover (see Chapter 4) results in a significant diminution of biological activity which resumes after the spring turnover. As should be anticipated, the seasonal changes in boreal forest aquatic communities are much more marked.

Terrestrial communities. Unlike aquatic environments wherein temperature change is mitigated by the specific heat of water, ambient temperature changes can result in significant periodic activity within a community. A most dramatic example of such *periodism* is found in desert situations where the difference between the warmest and coolest parts of a day may be in the order of 60°C or more (see Figure 4-7). In addition to a number of diurnal periodic physiological changes in plants (e.g., closure of stomata), most animals tend to be inactive during the heat of the day, often by burrowing into the ground, and to be active nocturnally.

Photoperiodism

Long-day and short-day plants. The regulatory influence of photoperiod—that is, the differential sensitivity of species to the length of the day (*photoperi-*

odism)—was first demonstrated in 1918 by W. W. Garner and H. A. Allard (1923). Their initial studies on tobacco were augmented by work on other species and subsequently by many other investigators on yet many other species. From their studies Garner and Allard concluded that there are *long-day* and *short-day plants.* The former flower when the photoperiod exceeds a given minimum (usually 12 to 14 hours), the latter when the photoperiod is below a given maximum (usually 11 to 15 hours). Some plants show no photoperiodic response; they are usually referred to as neutral-day or day-neutral plants.

The effect of this differential response by given species to temperature and to the quality and quantity of light in a given ecosystem is reflected in the seasonal progression of major events (Figure 14-3). Thus we have come to another basis of ecological, or ecophysiological, categorizing of vegetation, not by their taxonomic status but by their functional status.

From this discussion it might be assumed that a given species has as specific a physiological response to photoperiod as it has a specific morphological adaptation to another environmental parameter. So we might anticipate that, over its extensive range (Figure 14-16), sugar maple would leaf-flower-fruit, etc. in any given area only when the critical temperature and photoperiod levels were reached. It would then follow that there should be a south-to-north progression in these phenological events, corresponding to the progression in these environmental gradients. In a general way, such patterns are recognized, for a 1° northward shift in latitude (about 110 km) amounts to a difference of about 4 days in the onset of a given phenological event. Spring is "earlier" to the south. This same effect is also generated by about a 125 m change in elevation.

Ecotypes: prairie grass. There is ample evidence that widespread species are not physiologically, or even morphologically, identical in all parts of their range. These ecological races are referred to as *ecotypes.* For example, Charles Olmsted (1944) demonstrated that different populations of a common prairie grass known as side-oats grama (*Bouteloua curtipendula*) did not have identical responses to given photoperiods. Although essentially identical morphologically, populations from such northern areas as Canada and the Dakotas were shown to be long-day plants, demonstrating normal vegetative and flowering behavior only with photoperiods of 14 hours or longer. In contrast, populations from southern areas such as Texas and Arizona were shown to be short-day plants, functioning normally under 13 hours of light but failing to flower on photoperiods longer than 14 hours.

Eoctypes: alpine sorrel. According to an extensive study of alpine sorrel (*Oxyria digyna*), H. A. Mooney and W. D. Billings (1961) showed that populations from California, Colorado, and Wyoming require about 15 hours of light for flowering. Photoperiods of over 20 hours are necessary for flowering north of the Arctic Circle. These investigators also reported other physiological differences in populations at different latitudes, the more northern populations having more chlorophyll, higher respiratory rates at the same temperature, and an attainment of peak photosynthesis at lower temperatures.

Friend and Woodward (1990) note further that mountain plants have an increased ability to fix CO_2 from lower than sea-level concentrations, use high irradiance levels more efficiently, photosynthesize and grow at lower temperatures, and have intrinsically lower growth rates than their lowland counterparts. These responses of mountain plants to their environment are due to a complex mixture of genetic and environmental influences, all of which are not easily distinguished one from the other. Nonetheless, based on model simulations, they conclude that there is some evidence that some mountain plants have evolved in response to their particular altitudinal environment.

Ecotypes: yarrow. An example of ecotypes that show little or no moprhological variation is observed in the variation of range and average height of different populations of a perennial herb, the yarrow (*Achillea*) (Figure 12-5). In extensive

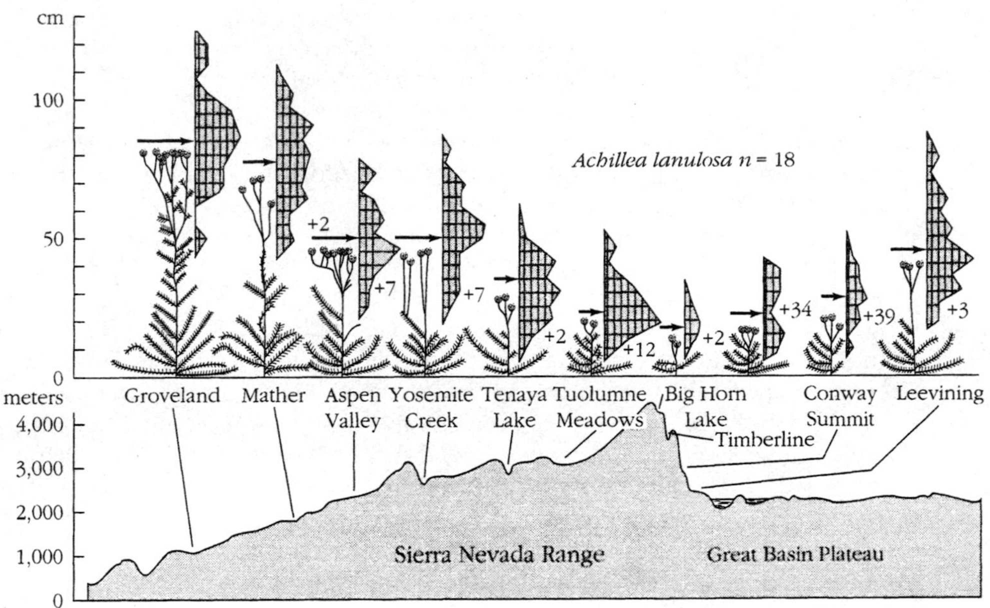

Figure 12-5 Ecotypic variation in the height of *Achillea lanulosa* grown in a uniform garden at Stanford, California. The plants originated in the localities shown in the profile, a 100-mile transect across central California at approximately 38° north latitude. Each population sample represents about 60 individuals. The frequency diagrams show variation in height within each population: the horizontal lines separate class intervals of 5 cm according to the marginal scale, and the distance between vertical lines represents 2 individuals. The number to the right of some frequency diagrams indicate the nonflowering plants. The specimens represent plants of average height, the arrows point to mean heights. (Redrawn by permission from J. Clausen, D. D. Keck, and W. M. Hiesey, 1948. *Carnegie Insitute of Washington Publication No. 581*:1–129.)

studies Jens Clausen, David Keck, and William Hiesey (1948) compared populations of *Achillea* from widely different habitats by transplanting and growing them under one uniform condition. In other studies they compared some 14 populations from different habitats under uniform growing conditions of three different climatic types. As a result, they were able to show the existence of considerable genetic variation and racial differences among plants of a single species.

Ecotypes: grasses. Using the transplant garden procedure of Clausen, Keck, and Hiesey, Calvin McMillan (1959) studied the role of ecotypic variation in distribution of the central grasslands of North America. As many as 12 species of grass, including side-oats grama, in which Olmsted had previously reported photoperiod ecotypes, were studied from more than 40 sample sites ranging over the central grasslands from Montana, North Dakota, and Minnesota in the north, to Oklahoma in the south. The transplant garden was at Lincoln, Nebraska, where responses under both natural and experimental light regimes were observed. McMillan observed not only geographic variation in photoperiod response in a number of species but also both latitudinal and elevational gradients (Figure 12-6).

In progressing southward some 1,300 km, but with a change of only 150 m in elevation, from Devil's Lake, North Dakota, to Ponca City, Oklahoma, the period during which initial flowering occurred in an experimental community consisting of six species increased from about 40 to 125 days. In contrast, the period of initial flowering over the same range of latitude is strongly influenced by elevation. This is seen in the progression from Miles City, Montana, at 715 m, to Wheatland, Wyoming, at 1,480 m, and Peyton, Colorado, at 2,200 m. All these communities show an initial flowering period of 40 days, like that of Devil's Lake to the north, but at different latitudes. At essentially the same latitude, however, altitude also affects the situation. There is a west-to-east elevational gradient at the same latitude comparable to a north-to-south latitudinal gradient at the same altitude. For example, the west-to-east transect involves a decrease in elevation from 2,200 m at Peyton, Colorado, to 1,500 m at Flagler, Colorado, to 825 m at Hoxie, Kansas, and to 310 m at Manhattan, Kansas, and a corresponding lengthening of the period of initial flowering. This has, then, the same effect as a latitudinal gradient at the same elevation, as demonstrated by the Devil's Lake, North Dakota, to Ponca City, Oklahoma, transect.

Stating these results in another way and using other data obtained, a given species of grass in the northern part of the grassland biome consists of long-day plants that flower during the long days of midsummer. In the southern part of their range the same species consists of populations of a given species occurring in the western part of the range at high altitude flower earlier than their eastern counterparts at lower altitudes. In each instance, McMillan sees these ecotypes as attunements to the peculiar climatic conditions of their natural regions—a more northern community, for example, being attuned to a shorter frost-free period by a shorter period of flowering. Thus clearly there has been natural selection for aggregations of individuals of different species with characteristics of early maturity that allow survival under such conditions—a selection for entire communities. It would thus be the case that the

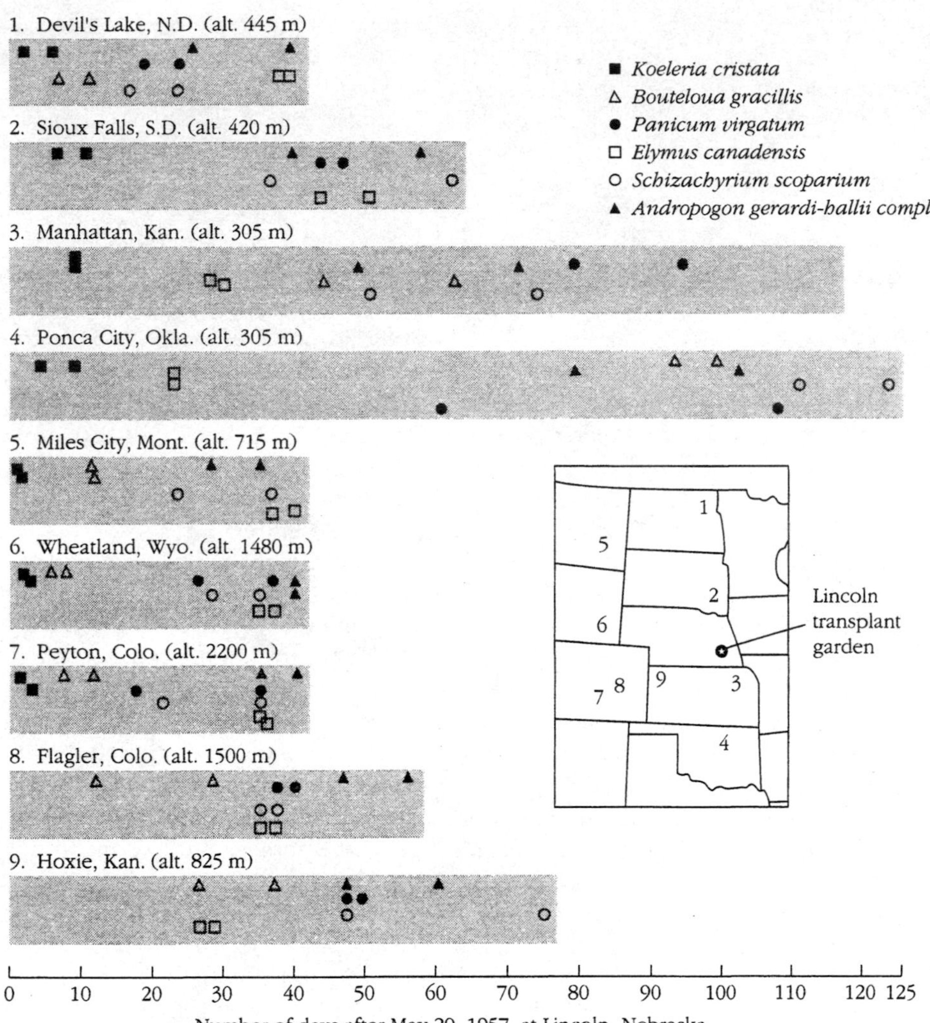

Figure 12-6 Comparison of community responses of various grassland species in a transplant garden at Lincoln, Nebraska, in 1957. The symbols indicate the extremes of initial flowering time of population samples transplanted from the locations shown on the inset map. (Redrawn by permission from C. McMillan. 1959. *Ecological Monographs* 29:285–308.)

continuity of the grassland over a broad geographic area of admittedly nonuniform habitats is related to the ecotypic variation inherent in grasses, a variation that, in turn, is regulated by genetic gradients. Because the likelihood of identical habitats is low, a repetition of combinations of certain species over broad geographic ranges is almost certain to involve ecotypes of the species involved. The great central grassland from Texas to Alberta thus appears to be a physiognomically uniform vegetational unit in spite of nonuniformity of habitat, primarily because grasses are grasses—because of their wide range of genetic and physiological adaptation.

Periodism and Metabolism

Turning to yet another aspect of community dynamics, we can consider two similar but opposite processes of considerable importance, respiration and photosynthesis. All producers in the community contribute to the total or gross photosynthesis and the consequent output of both stored chemical energy in the form of biomass and the significant byproduct, oxygen. All organisms in the community, producers included, use this stored energy and oxygen in metabolic activities associated with growth and reproduction, a complex of events measured as respiration. But whereas photosynthesis is dependent on light, respiration generally is not. In some plants increasing light intensity increases respiration, a phenomenon known as photorespiration. In such situations, glycolic acid, an intermediate product of photosynthesis, is oxidized more rapidly, thereby reducing the overall efficiency of photosynthesis. With or without plants that undergo photorespiration, for a community or an ecosystem to maintain itself—that is, to be balanced or show a net gain—the output of photosynthesis in a closed system must be at least equal to the respiratory demands of the system. Therefore it is important to consider the periodic characteristics of these two opposing metabolic processes.

Metabolism in microecosystems. Robert Beyers (1966) has investigated the metabolism of a number of highly diverse aquatic ecosystems under laboratory conditions, small counterparts of natural, and unnatural, systems that are conveniently referred to as microcosms. In his studies using a 12-hour photoperiod, he demonstrated that nighttime respiration is at a maximum within a few hours of darkness and drops progressively thereafter. Photosynthesis shows a similar pattern during the period of light, reaching its maximum within the first few hours and decreasing thereafter through the remainder of the day (Figure 12-7). A number of investigations under natural light conditions have shown that photosynthetic maxima occur in midmorning, in agreement with Beyers's laboratory results. It is important to note, however, that under natural conditions the photosynthetic maximum occurs several hours before the time of day (noon) when energy input is at maximum. The lack of conformity of maximum energy input and rate of photosynthesis appears related to an actual inhibition of photosynthesis at higher light intensities, a phenomenon that has been reported by a number of investigators.

In his study of microcosm metabolism Beyers found that the average ratio of gross photosynthesis to respiration was 1.05. That is, there was a slight excess of photosynthetic activity sufficient to meet the respiratory demands of the systems—the systems were in balance. Evidence suggests a rather close coupling of these processes in balanced or steady-state systems. On cloudy days, for example, not only does photosynthesis decrease, respiration does also.

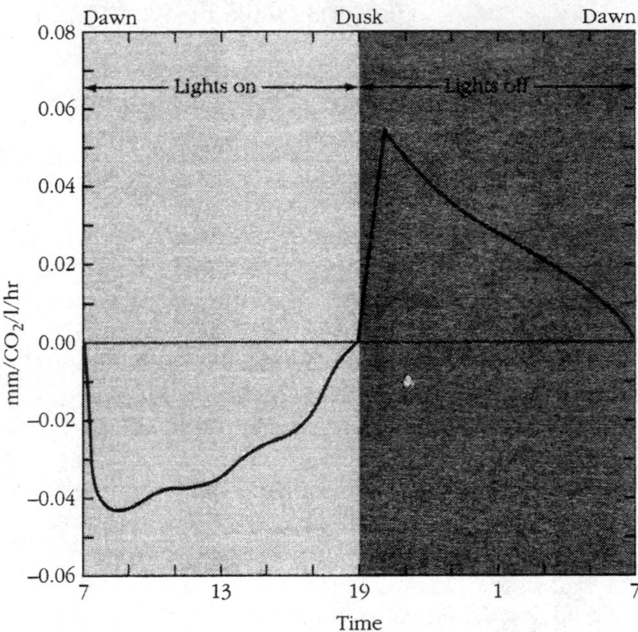

Figure 12-7 The diurnal pattern of photosynthesis and respiration in freshwater ecosystems as reflected by the rate of change of dissolved carbon dioxide. (Redrawn by permission from R. J. Beyers. 1966. *Primary Productivity in Aquatic Environments*, ed. C. S. Goldman. Berkeley, CA: University of California Press.)

Diurnal metabolism in seasonal pools. A physiological metabolic adaptation to diurnal variations in the availability of carbon may involve the use of an alternate mechanism of carbon assimilation known as CAM, or crassulacean acid metabolism (Monson 1989). Shallow seasonal pools are typically so densely vegetated that photosynthetic demand can greatly reduce the concentration of CO_2 early in the morning and thereby reduce the rate of photosynthesis for the entire day, a pattern different from that shown by Beyers. In such pools, Keeley and Sandquist (1991) have shown that the quillwort *Isoetes howellii*, a primitive annual vascular plant, is able to assimilate carbon by CAM at night when the CO_2 level is high as a result of respiration by community constituents and diffusion from the atmosphere. This overnight accumulation of the acid amounts to between one-third and one-half of the total daily carbon assimilation in *Isoetes*. They conclude that the diurnal changes in carbon availability selected for CAM photosynthesis in this case.

As an aside, it is pertinent to note that the CAM pathway is also of adaptive significance in desert ecosystems where it has been studied more extensively. The CAM pathway is the photosynthetic route characteristic of succulents that demonstrate high-temperature physiological performance and high water-use efficiency (MacMahon 1988). In transect studies, as aridity decreases there is a decrease in the number of CAM species and an increase in those using the regular, or C_3 pathway.

Community classification based on metabolism. Howard Odum (1956) has proposed a classification of ecosystems based on community metabolism, on the ratio of photosynthesis (*P*) to respiration (*R*). According to this scheme, a stabilized system is one in which the ratio of photosynthesis to respiration is unity—that is, where $P/R = 1$. A system in which photosynthesis exceeds respriation, $P/R > 1$, is considered autotrophic; and one in which respiratory demand exceeds photosynthesis—that is, $P/R < 1$—is said to be heterotrophic (Figure 12-8). In the autotrophic system biological fertility is based on current production; in the heterotrophic system it is based on past production, organic matter accumulated over a period of time and often imported from another ecosystem.

Note, however, that stability is not contingent on a given habitat (desert, grassland, forest, etc.) but rather on the dynamic interaction of the forces involved in promoting given photosynthetic and respiratory rates. Nonetheless, the tendency is for systems to proceed toward stability (P/R =1) and thus to maintain themselves—that is, the diagonal line of the graph—over both the short and long term. It was noted that even on a diurnal basis a reduction in photosynthesis is coupled with reduced community respiration. On an annual basis in temperate systems, the spring-summer autotrophism is offset to varying degrees by fall-winter heterotrophism.

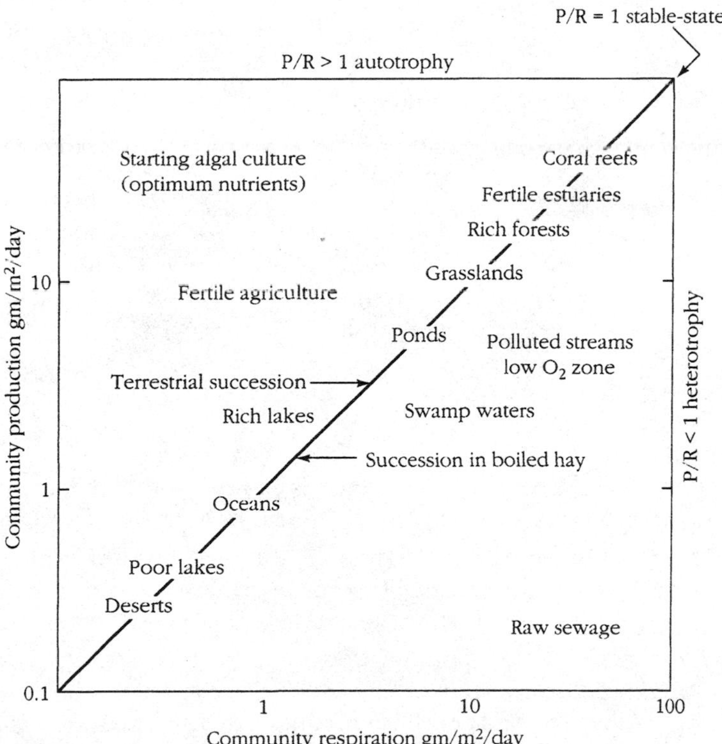

Figure 12-8 Various types of communities classified on a basis of community metabolism. (Redrawn by permission from E. P. Odum. 1959. *Fundamentals of Ecology.* Philadelphia: W. B. Saunders Co.)

The tendency to stability and the adjustments involved therein have been demonstrated in a rather simple but elegant experiment by B. J. Copeland (1965), who showed that an 85 percent reduction in light intensity was followed, as expected, by a reduction in metabolism. The drop was about 83 percent. Interestingly, even though the reduced light intensity persisted, the system recuperated, gradually increasing its metabolism until it reached the previous balanced level, all in a period of about 60 days. The recuperation, however, was accomplished by a change in the dominant producer, from turtle grass to blue-green algae. Copeland's ecosystem was disturbed but restabilized at about its previous level. Other systems also show this tendency to stability, although they may do so at different levels along the diagonal, depending on the relative amounts of photosynthesis and respiration. This tendency to stability and its regulation constitutes the focus of the next chapter.

SUMMARY

VOCABULARY

collective properties
dispersion
ecological community
ecotype
emergent properties
growth form
guild
individualistic community
krummholz
life form
long-day plant
organismic community
periodism
phenology
photoperiodism
physiognomy
short-day plant
stratification
vegetational continuum

KEY POINTS

- Whether an ecological community is holistic with emergent properties or an entity with collective properties is a matter of debate/interpretation.
- The physiognomy of a community can be described and communities classified on the basis of growth and life forms, stratification, zonation, dispersion, presence of guilds, trophic structure, phylogenetic composition, and metabolism.
- The nature of a community is regarded by most ecologists as individualistic, represented by a continuum in which discrete subdivisions do not occur; an alternative view is that each community is analogous to an organism, being discrete and discernible entities.
- Functionally, communities demonstrate phenological events in development and reproductive activity in response to seasonal climate changes, diurnal periodism phenomena in response to light (photoperiodism and metabolism), and the development of ecotypes.

CHAPTER

13

Stability and Change in Communities

The aphorism "nature abhors a vacuum," expressed by the seventeenth century Dutch exponent of pantheism, Baruch Spinoza, aptly describes a major ecological phenomenon, that of *succession.* Bare ground does not remain that way for long because "Nature" in the form of vegetation moves in with dispatch. The initial colonization is followed by a series of sequential vegetational (as well as animal) replacements, or successional changes, that are often predictable in a given area or site. For example, in a series of studies, among which is a classic one published in 1899, American ecologist Henry C. Cowles showed that the different communities initially present in Indiana in such diverse habitats as flood plains, sand ridges, shallow and deep ponds, and clay banks, all demonstrated a predictable sequential series of structural changes; further, each culminated in a more or less stable community of beech-maple forest. Although the degree of such regularity and predictability was disputed by a number of later ecologists (including Jackson, Futyma, and Wilcox 1988), it is the case that structural and compositional changes in a community over time, both in the present and in the distant past, are recognizable. The factors that regulate such changes, predictable or not, are, as we shall see, complex.

The phenomenon of succession is not accepted without argument; these arguments are not about whether succession occurs, but rather what is occurring. Are the sequential changes occurring within a community or between communities? Are whole clusters of species replaced by other whole clusters of species, or is it a case of a simple species-by-species replacement? Are we observing an organism that is developing from an incipient state through a number of determinate stages to maturity or a sequence of increasingly complex biological communities related to but more or less independent of those that preceded and those that follow? These are among the contending views on succession, not entirely unrelated to the contention about individualistic and organismic views of ecological communities. There are no easy answers, and that is one of the exciting and demanding challenges of contemporary community ecology.

Before exploring the regulation of communities, including their successional characteristics, it will be instructive to gain some understanding of the nature of succession, both structurally and functionally, in a variety of communities and the effect of environmental perturbations on that process.

SUCCESSION IN THE PRESENT

Succession in Beach Ponds

Along the south shore of Lake Erie, at Erie, Pennsylvania, is a peninsula some 4 miles long known as Presque Isle (see Figure 18-1). Owing to a combination of its sandy shore and its exposure to the violent storms and heavy wave action for which Lake Erie is notorious, the peninsula is subject to the frequent establishment of small beach ponds (Figure 13-1a). These ponds are created when an elevated bar of sand develops, thereby isolating a small portion of the lake. The ponds are seldom more than 100 to 200 m long, 10 to 20 m wide, and 1 m deep. Some ponds are wiped out days, months, or even a few years later by subsequent storms that either break down the sand bar or blow in enough sand to fill in the basin. A better-protected pond survives this geological fate only to be immediately subject to its biological fate, ecological succession (Figure 13-1b). Because of the geological history of the peninsula and the persistence of some beach ponds, it was possible to identify ponds of different chronological ages and therefore to study some structural and functional changes that occurred as the ponds progressed from an open water to a vegetated semiterrestrial state (Kormondy 1969).

Structural changes. From a structural standpoint two major changes were observed in successively older ponds: (1) a change in species composition and (2) a change in variety or diversity (Figure 13-2). Some plant species that were present during the initial stages are missing from the more advanced stages. Stonewort alga (*Chara*) and cattail (*Typha*) demonstrate this kind of change. Similarly, some species are not present until fairly late in succession, the yellow water lily (*Nuphar advena*), for example. Yet other species, of which the bulrushes (*Scirpus*) are prime examples, persist as major components of the system for much of its history. These differences are doubtless tied to the tolerance limits characteristic of each species as well as to the subtleties of interspecific competition. Comparable shifts of species composition in phytoplankton and, to a less marked degree, in the animals were also observed.

In addition to these changes in composition, there is a marked increase in the number and kind of both autotrophs and heterotrophs. Sparseness of distribution and limitation of kinds mark the early ponds; increased density and heterogeneity characterize the older ponds. The relative simplicity of structure of the young pond

A.

B.

Figure 13-1 Beach ponds on Presque Isle at Erie, Pennsylvania. (a) A newly formed beach pond less than 1 year old; (b) a beach pond about 50 years old. (Photos by author.)

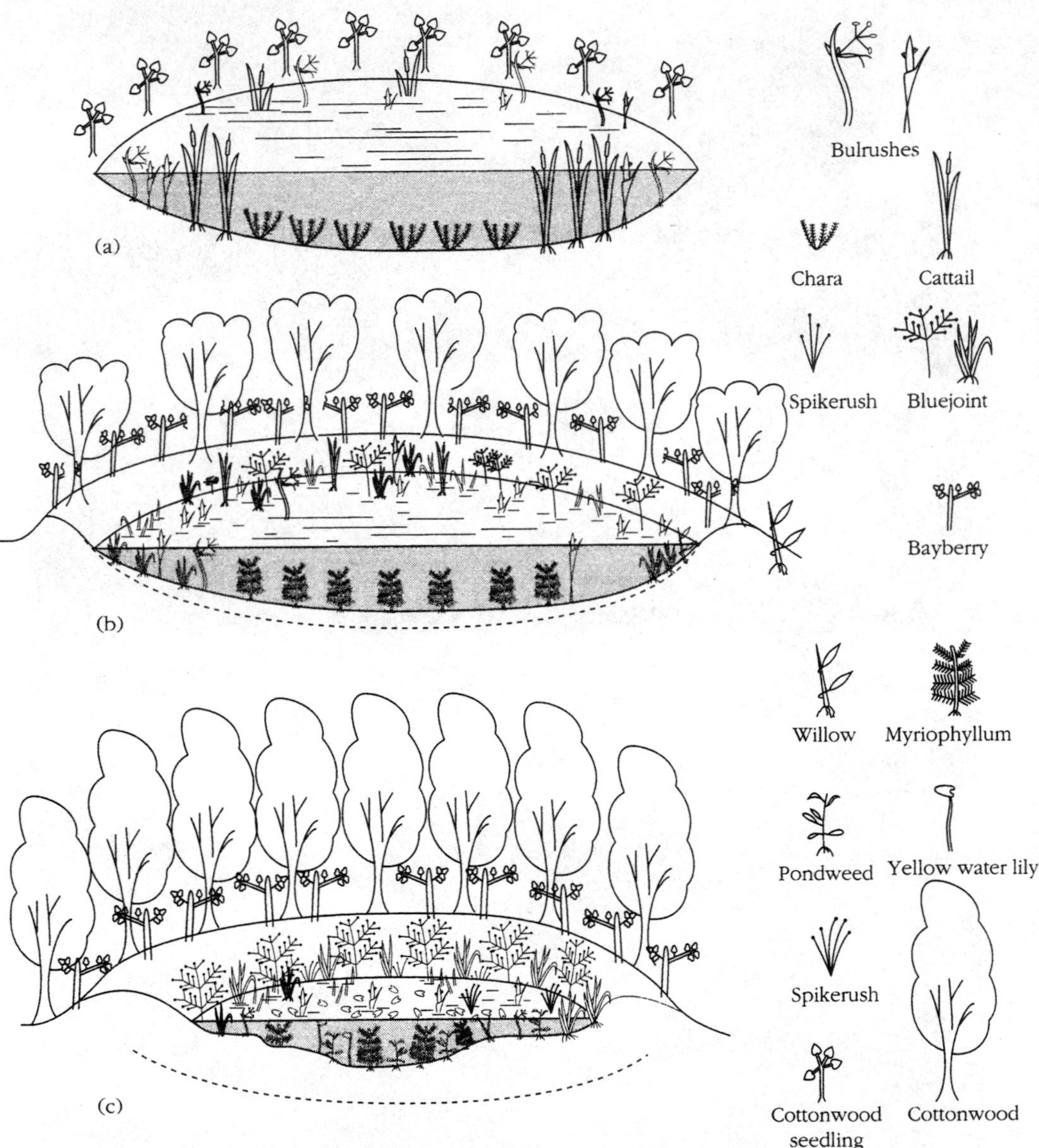

Figure 13-2 Generalized stereo-profiles of beach ponds on Presque Isle at Erie, Pennsylvania, at (a) 4 years, (b) 50 years, and (c) 100 years of age. Note the changes in kind, distribution, and abundance of different species and the accumulation of organic matter filling in the bottom.

is in striking contrast to the relative complexity of the more advanced stages in succession (Figure 13-2).

Functional changes. On the functional side, two major changes can be noted. One is a progressive increase in the amount of both living and dead organic matter. In part, this biomass increase is expected from the already noted increase in numbers; and because the amount of living organic matter shows an increase, there would be an eventual increase in dead organic matter. The accumulation of organic matter on the bottom reaches the point, it may be noted, that the pond is progressively filled in. Measurements of other organic components, such as plant pigments, notably chlorophyll, also show a progressive increase with more advanced stages of succession.

The second major functional change is a shift in community metabolism from young ponds, which are autotrophic ($P/R > 1$), to those that are stabilized ($P/R = 1$), and, finally, heterotrophic ($P/R < 1$). Although the number and kind of primary producers increase and result in increased gross production, at least initially, the increasing respiratory demand of the system results in a progressive decrease in the metabolic ratio. The increasing heterotrophy of these aquatic systems applies to that phase of their succession in which they achieve a terrestrial condition. Although no studies were conducted beyond this transitional stage, evidence elsewhere strongly indicates that such systems would continue to undergo terrestrial succession with an accompanying trend toward stability of their metabolic ratio.

Succession on New Islands

The opportunity to study succession in newly created environments is limited to situations such as beach ponds, recent lava flows (Figure 13-3) and newly emerged islands. An opportunity for the latter was provided between 1882 and 1985 when the water level of Lake Hjalmaren in Sweden was lowered about 130 cm, resulting in the emergence of a large number of new islands. Beginning in 1886 and extending through 1985 a series of plant surveys was conducted on the islands at several time intervals (1886, 1892, 1903–04, 1927–28, and 1984–85) by various investigators; these were analyzed by Rydin and Borgegard (1991), who conducted the last survey. Their report focuses on the 30 islands for which a complete 100-year set of data was obtained.

Successional stages. Of the 112 plant species that occurred on at least 10 islands or were abundant on at least 3 islands in any one of the surveys, six major categories or stages of species composition were recognized based on frequency or abundance:

1. Pioneers (13 species) reached their maximum at the first survey, decreasing drastically after the first or second survey;
2. Pioneer Stayers (6 species) reached many islands by the first survey and remained at similar levels until the fourth or fifth survey;
3. Early Successionals (16 species) showed a distinct increase between surveys one and two, a maximum in the second or third and declined thereafter;

Figure 13-3 Succession on a lava flow in Volcanoes National Park, Island of Hawaii. Pioneers are ferns (a) followed by low shrubs (b) and finally by *Ohia* trees (c). (Photos by author.)

4. Early Successional Stayers (10 species) had a distinct increase between surveys one and two, but no subsequent decline;
5. Mid Successionals (20 species) had a maximum at survey three and/or four and declined after the latter;
6. Mid Successional Stayers and Late Species (47 species) reached their maximum between survey three and/or four but maintained their importance in the fifth survey or increased between survey four and five, reaching a maximum in the latter.

Examples of each of these categories is shown in Figure 13-4.

Structural characteristics. The Pioneers formed a distinct group and had a high proportion (8 of 13 species) of therophytes (annuals) that were autogamous (self-pollinating), along with insect-pollinating species. The 6 species that lacked animal,

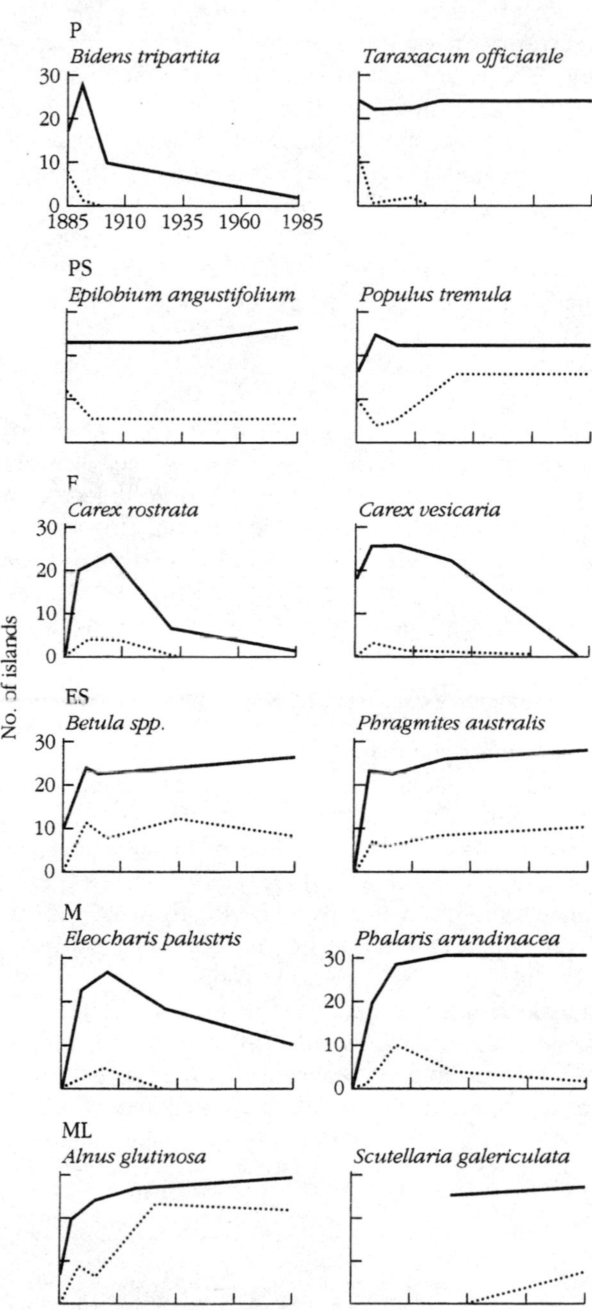

Figure 13-4 Occurrence of species during a century of succession on islands in Lake Hjalmaren, Sweden. Solid line shows the number of islands where the species was found (frequency); dotted line shows the number of islands with larger populations (abundance). The groups, based on abundance, are: P, Pioneers; PS, Pioneer Stayers; E, Early Successionals; ES, Early Successional Stayers; M, Mid Successionals; ML, Mid Successional Stayers and Late Successional Stayers. (Adapted with permission from H. Rydin and S.-O. Borgegard. 1991. *Ecology* 72:1089–1101.)

wind, or water dispersal ability all showed late arrival times on the islands. Pioneer Stayers and Early Successional Stayers were on the average taller than Pioneers, and both groups had higher seed numbers (fecundity) than the others. Pioneers and Early Successionals had higher light requirements than the other groups. Seeds in Pioneers were not smaller than in Early Succesionals or Mid Successionals and did not differ between annuals and perennials, but species with the heaviest seeds arrived late, these being dependent on dispersal by birds. The observations that light requirements, as well as a preference for open habitats, decreased over time suggests that physiological factors may play a more important role than inherent life history patterns in the replacement of species after the Pioneer stage.

Remote Sensing of Succession

A major problem in observing succession is that ecologists don't live long enough! As we have seen in the two foregoing examples, the time-span of succession was in the order of a century. The observations on Lake Hjalmaren were conducted by a number of different ecologists who, fortunately, carefully documented their work so that Rydin and Borgegard could provide a description of succession over 100 years. In the instance of the beach ponds, it was assumed that the differences in structure in ponds of different age were so slight that an inference of successional stages was permissable. Since succession is a long-term process, long-term studies are in order and, to be most accurate, should have continuity of personnel and equipment; this translates to studies over a 20 to 30 year period at best. The newly developing technique of remote sensing, however, offers the possibility not only of reliably determining the course of succession over a period of time but the encompassing of much larger areas than can be studied by an individual or even a team of ecologists. Other applications of remote sensing in ecological studies are discussed by Matson (1991).

Forrest Hall and a group of associates (Hall et al. 1991) studied landscape units of an area of 3600 m^2 using satellite remote sensing over a period from 1973 to 1983. Changes in these large areas were supplemented and corroborated by ground-observed and photo-interpreted plots in 1983. Over the 10-year period it was possible to observe sizeable values of transition rates and changes in structure using the remote-sensing data. Based on these observations, this technique would permit monitoring of forest succession over as much as a 20-year period, an important benchmark for identifying the effects of air pollution and other environmental perturbations. Importantly, it would not be necessary for the same individual to make those observations over those lengthy periods of time.

PERTURBATIONS AND SUCCESSION

The series of changes in communities observed in newly created environments such as beach ponds and new islands are referred to as *primary succession*. *Secondary succession* refers to those sequential changes that occur when an established ecological community is disturbed and subsequently undergoes changes in its structure and function. Whether the secondary sequence follows the pattern of the primary sequence depends on a host of abiotic and biotic factors, both as agents of the disturbance and as components of the redeveloping community. Among the former are fires, ice storms, floods, drought, high winds, landslides, and large waves. Among the latter are such direct interactions as grazing, predation, competition and infection (disease) and the indirect effects of habitat modification caused by activities such as burrow digging by earthworms and rodents or stream damming by beavers, let alone numerous human actions such as bulldozing, construction, and pollution.

The extent of the impact of a disturbance on an established community depends, among other factors, on the severity of the disturbance, the time of its occurrence in the successional process, the multiple interactions that occur among the members of the surviving community, and the proximity to other communities of like or dissimilar nature. The role of disturbance in ecological communities has been an arena of increased ecological interest and has led to the phrase *patch dynamics* (Thompson 1978) used to describe the common focus of such efforts on the conditions created by disturbance (frequency, severity, intensity, and predictabliity) and the responses of organisms to the disturbances. Excellent reviews of the role of disturbance on ecological communities are to be found in Sousa (1984) and Pickett and White (1985). A few examples of perturbations and their effects on succession are in order to give meaning to these comments.

Perturbations in Terrestrial Communities

Fire ants. During the 1940s, the fire ant *Solenopsis invicta* (its name derives from the severe burning sensation its bite creates) invaded the southwestern United States from South America, creating a wave of destruction of naturally occurring communities during this time. At a field station in central Texas, Porter and Savignano (1990) observed that an invasion of fire ants resulted in a 70 percent drop in the number of native ant species and a drop in the total population of native ants of 90 percent. Similarly, the number of species of non-ant arthropods (isopods, mites, and certain scarab beetles) dropped 30 percent and their population numbers 75 percent. The primary causal factor appears to have been competition, another result of which was that there was a 10- to 30-fold increase in the number of ants at the infested sites, more than 99 percent of which were fire ants. The likelihood of this arthropod community reestablishing itself is indeed remote.

Oak wilt. At the time of European settlement in southern Wisconsin, namely about 1840, an area known as Noe Woods was an oak savanna that soon developed into a fairly pure black oak–white oak (*Quercus velutina* and *Q. alba*, respectively) forest of a type common to the region (McCune and Cottam 1985). Between 1956 and 1983, oak wilt disease (caused by *Ceratocystis fagacearum*) caused a 15 to 20 percent rate of death for the black oaks during each of the periods between the 5-year censuses of the forest. White oak increased in basal area but declined in numbers, and the black cherry, boxelder, and American elm that replaced the oaks are, for various reasons, unlikely to fill the latter's self-replacing role in the forest. Here, again, is a community so disturbed that it is unlikely to again have its former characteristic structure.

Depending on their severity, both crown and ground fires can be significant perturbing agents in terrestrial communities (see Chapter 5), as can be parasites, a case in point being that of the sac fungus parasitic on the American chestnut (see Chapter 11).

Perturbations in Freshwater Environments

Flash flooding. We would ordinarily expect that a sudden flash flood, or spate, would likely wipe out an existing community in a stream, as has been observed in a number of cases. However, Boulton et al. (1992) have shown that in a Sonoran Desert stream in central Arizona, a consistent cycle of seasonal changes in the structure of the stream community occurred each year with little alteration by the flooding, the changes being those of relative presence or absence rather than of abundance. For example, seasonal abundance of oligochaetes and mayflies peaked in spring, and a gastropod and caddisflies were dominant in summer. By contrast, there were major changes in the fall community structure influenced, apparently, by drying rather than flooding. The rate of change in species composition declined during succession in all but the spring period, but two summer sequences showed the highest intitial rates of change. Thus unlike the foregoing examples, this community was resilient to the disruptions caused by flooding.

Volcanic intrusion. The once pristine Spirit Lake, lying at the base of Mount St. Helens in the State of Washington and surrounded by steep and densely forested mountain slopes, was drastically perturbed on May 18, 1980, when the volcanic mountain exploded with the force of a 10-megaton nuclear explosion. The resulting pyroclastic lava flow carried immense quantities of debris, old and/or burnt trees, and mineral-containing rock into the lake raising the temperature from about 10°C to 30°C and filling extensive portions of its basin. Extensive and intensive studies have been conducted on Spirit Lake leading to the conclusion that it will never return to its pre-eruption condition (Larson 1993). Significant changes have and are occurring in its flora and fauna and in its chemical composition; its surrounding environs harbor pathogenic bacteria and prolific noxious plants. This is an instance where secondary succession is leading to a markedly different commmunity than existed before an event, as well as being a commentary on nature's power to dramatically alter the landscape.

Given the extensive perturbation caused by the eruption, sites on Mount St. Helens are providing optimum opportunities to study primary succession (del Moral and Bliss 1993), while other, less violently altered sites, allow secondary succession, including recolonization by the fallout of arthropods (Edwards and Sugg 1993).

Perturbations in Marine Environments

Sea grass. Not unlike many introductions of non-native species in marine environments, the sea grass *Zostera japonica* apparently came accidentally to the west coast of North America from Asia in the 1950s by ship, in this case with shipments of oysters used to support the local shellfish industry (Posey 1988). It quickly expanded its range over the coastal area of northwestern United States and continues to spread southward. Using a combination of observations and transplant experiments as the grass expanded to new areas, Posey showed that this foreign or non-native species (technically an "exotic") resulted in changes in the community of bottom-dwelling organisms in midintertidal tideflats with the number of taxa present increasing and the densities of several common organisms changed, some becoming greater, others lesser. Most of the changes are associated not with direct exclusion of resident fauna but rather with modification of the habitat. For example, the blades of the seagrass provide hard surfaces for attachment; the blades also baffle currents, thus increasing filtering ability (and densities) of some suspension feeders.

El Niño. In Spanish, the Christ child is known as *El Niño*, a term that has been applied to a change in the Peruvian and Ecuadorian coastal areas that occurs at Christmastime. Annually during this period, the normally cold water of the north-flowing Peru Current is displaced by a warm, southward current that is considerably less rich in nutrients. On occasion, however, the temperature of the southward current increases, the nutrient content lessens further, and the duration of the displacement lengthens, sometimes as long as a year. Besides the perturbation El Niño creates in the world's weather pattern (Kerr 1988a), a subject of considerable interest but beyond our scope in this text, it can and has had significant detrimental effects on the Peruvian anchovy population and those species that, in turn, are dependent on the anchovies—higher-order fish, guano birds, marine mammals, and even human fishermen (Tegner and Dayton 1987; Glynn 1988). Two of the most severe El Niños of the century (each technically referred to as the El Niño–Southern Oscillation, or ENSO) occurred in 1982-83 and in 1986–87. The impact of the 1982–83 El Niño on a kelp forest community near San Diego, California, was studied by Dayton et al. (1992) who reported significant changes in mortality patterns of the kelp (*Macrocystis pyrifera*) and the first large-scale understory mortality in several decades, obliterating much of the structure of the kelp forest. It also led to outbreaks of understory algae, intraspecific competition, and changes in grazing patterns that resulted in variation in recovery rates of the kelp forest in different areas. All this notwithstanding, there is sufficient survival to allow for prompt recovery to preexisting patterns, except in one southern site where sea urchin grazing had an additional impact and slowed the recovery rate.

SUCCESSION IN THE PAST

Succession in the present represents what occurs over a short period of time and can be studied by direct observation of existing communities, making certain assumptions and inferences in doing so. Succession in the past has occurred over long periods of time, namely thousands of years, and must be inferred from direct observations of relics of the past, such as pollen, seeds, and tree rings. Whereas succession in the present focuses on changes in structure, as well as function in some cases, in more or less circumscribed locations, succession in the past tends to focus on large geographical regions and seeks an identification of particular communities with particular climates. Thus we describe the predominant composition and distribution of the major contemporary biomes (see Chapter 14), tying their characteristics to the major climatic patterns of the region in which they occur. Since there have been major climatic changes during Earth's history, major changes in the boundaries of biomes have also changed. This exciting dimension of ecology is generally referred to as *paleoecology*, which deals with determining major ecological communities from their fossil record and thereby dating the climate in those particular regions. The assumption is that, with minor adjustments, particular kinds of communities observed today in particular climates existed in the past in similar climates.

Postglacial Succession in Eastern North America

As was noted in the discussion of beach pond succession, there is a progressive filling in by organic matter. Among the constituents of this organic matter are pollen grains released both from the community in the pond and the surrounding area. By removing a core of the sediment and tabulating the percentages of each species' pollen grains in successive levels of the deposit, the community composition that prevailed at the time each level was deposited can be deduced. The climate corresponding to each level is then inferred from the particular community composition, on the assumption that the climatic requirements of the species have not significantly changed in the interim.

The principle of determination of the chronology of forest composition and climate by pollen analysis, or *palynology*, had its initial development in Europe about 1915. One of its first applications in the United States was made in the 1920s and 1930s by Paul Sears (1930) (Figure 13-5). The reconstruction of this postglacial forest in northern Ohio enabled Sears to infer the following climatic sequence:

Depth in meters	Climate
4.2-3.6	Cold, wet of northern Labrador
3.3-2.1	Gradual shift from oceanic to continental climate
1.8-1.2	Cool, dry climate of southern Manitoba
0.9-0.6	Period of maximum desiccation
0.6	Abrupt increase in humidity. Cool, moist climate of northern Great Lakes
0.3	Moderation of temperature and continued increase in humidity; present climate of north-central Ohio

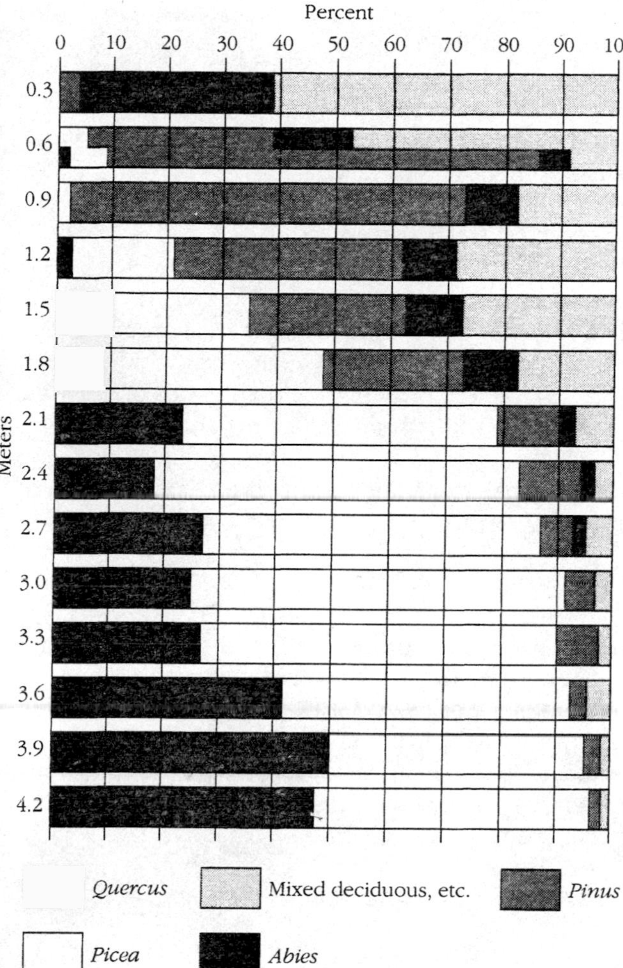

Figure 13-5 Pollen diagram of Bucyrus Bog in northern Ohio showing the proportions of the major genera at each level. According to the investigator, Dr. Paul Sears, "What is labeled *Abies* was probably *Picea canadensis*, illustrating one of the difficulties encountered in early work, namely, the lack of reliable literature for pollen identification." (Redrawn by permission from P. B. Sears. 1930. *Ohio Journal of Science* 30:205–17.)

Additional studies by Sears and by the many students and colleagues he stimulated to engage in this type of investigation have sharpened the picture of postglacial vegetation and climate in North America, in both glaciated and nonglaciated regions. The subsequent application, beginning in the 1950s, of radiocarbon dating of pollen and associated sediments has permitted the assignment of a more precise time scale to the chronology. Based on numerous studies, Sears (1932) advanced the following postglacial sequence for eastern North America:

I. A moist, cool period; maximum of spruce (*Picea*) and fir (*Abies*). Ended about 11,000 years ago.

II. A dry, warmer period; maximum of pine (*Pinus*) with oaks (*Quercus*) almost always an accessory. About 9,000–11,000 years ago.

III. A more humid, warm period; maximum of beech (*Fagus*), with hemlock (*Tsuga*) in some places. About 6,000–9,000 years ago.
IV. A dry, warm period: maximum of oaks and hickory (*Carya*). About 3,500–6,000 years ago.
V. A more moist and cool period; increase of more mesic genera [maples (*Acer*), hemlocks, chestnut (*Castanea*), pine]. The present.

Postglacial Succession in Western North America

Sediment cores from three ponds on the south side of Mount Rainier, Washington, at an elevation of 1,300 to 1,500 m conform to findings east of the Cascade Range in showing little vegetational change during the last 5,000 years (Dunwiddie 1986). Earlier, however, 6,000 to 3,400 B.P. (Before the Present), the dominance of fir (*Abies lasiocarpa* and *A. procera*), Douglas fir (*Pseudotsuga menziesii*) and pines (*Pinus monticola* and *P. contorta*) suggest a warmer and drier climate than we have today, as well as one that favored frequent fires, a condition verified by the presence of charcoal in the sediments.

Postglacial Succession in the Tropics

A number of studies have shown that even the tropics were not immune to the impact of the last glacial advance and subsequent retreat that affected North America's vegetational distributions. For example, analysis of Late Pleistocene (14,350 to 11,050 B.P.) sediments of a lowland tropical lake in Panama indicated a relatively high abundance of montane forest elements such as oak (*Quercus*) and magnolia (*Magnolia*) existing some 900 m below their present range. This suggests a cooling on the order of about 5°C since that time (Bush et al. 1992).

Preglacial Succession in Midwest North America

The ecological communities existing in western Illinois preceding the Pleistocene (about 14,500 B.P.), a period known as the Late Wisconsin and extending back to 28,000 B.P., were determined by an analysis of pollen, vascular plant macrofossils, bryophytes, insects and stratigraphy (Baker et al. 1989). The earlier period, from 27,900 to 26,000 B.P. was dominated by spruce (*Picea*) and jack pine (*Pinus banksiana*); but, during the next period (26,600 to 22,700 B.P.), jack pine had all but disappeared and was replaced by larch (*Larix laricina*), suggesting a climatic cooling as the glaciers began to advance into the Midwest. Further cooling in the next period is suggested by the subsequent decrease in spruce and an increase in sedges (*Cyperaceae*) and spikemoss (*Selaginella*).

Reverse Succession in the Central Yukon

Based on pollen and plant macrofossil evidence, it appears that shrub tundra with groves of balsam poplar (*Populus balsamifera*) occupied the central Yukon between

10,000 and 8,000 B.P. and was gradually replaced beginning about 9,400 B.P. with white spruce (*Picea glabra*) (Cwynar and Spear 1991). About 6,500 B.P., black spruce (*Picea mariana*) and green alder (*Alnus viridis*) increased, characterizing the landscape in open spruce woodlands not unlike today's northern boreal forest. About 5,000 B.P., the forest began to revert to shrub tundra with relict populations of spruce surviving in favorable situations. These changes suggest that Alaska and northwest Canada experienced warmer summers than today from about 10,000 to 6,000 B.P.

40,000 Years of Successional History in Florida

Spanning the geological period of the Middle Wisconsin (40,000 to 29,000 B.P.), the Wisconsin (29,000 to 14,000 B.P.) and the postglacial period (14,000 B.P. to today) is a study on Camel Lake in northwest Florida (Watts et al. 1992). The earliest of these periods was dominated by forests with abundance of pine, oak, and other deciduous tree species including chestnut (*Castanea*), but showed great variation in species composition probably because of climatic variation. The Late Wisconsin period had a species-poor pine forest, and the period from 14,000 to 12,000 B.P., a period that is believed to be the coldest phase at Camel Lake with a suggested climate of −5°C mean January temperature, was marked by spruce and hickory (*Carya*). After 12,000 B.P., oaks and various other deciduous hardwoods were dominant. The present-day turkey oak and long-leaf pine forests began replacing the spruce hickory after about 7,000 to 8,000 B.P.

Interpreting the Fossil Pollen Record

Margaret Davis (1969) has shown that the nature of past (and present) vegetation cannot always be deduced intuitively on a one-to-one correlation from the frequencies of pollen in an assemblage. This is so because of the differences in amount of pollen produced by different species but also because of differences in their dispersal by wind or resistance to decay. Thus it is necessary to determine pollen accumulation rates by relating the pollen concentration in sediments to time intervals and also to take into account the relative pollen production of different species. As a result of such efforts, what was originally thought to have been a forest exclusively made up of pine some 8,000 to 9,500 B.P., because of a pine pollen maximum at that time (see Stage II in Sears's sequence above), was a mixed-deciduous coniferous forest, including pine, such as modern forests of east-central Ohio (see Figure 14-16).

As would be expected from even limited acquaintance with the distribution of contemporary ecosystems, there are differences in the particular sequence of ecosystems in different locations and some disagreement among investigators about given events and interpretation. Anlayses of pollen and macrofossils, as seen in Camel Lake, have shown that the southern limit of spruce in the eastern United States reached what is the present border of Florida and Georgia and that the spruce was replaced by pine about 22,000 B.P. Margaret Davis (1969) argues further that the modern boreal forest biome of Canada does not have an antecedent in the plant communities that existed south of the great ice sheets or that developed as the ice retreated, as would be interpreted from Sears's postglacial sequence (Stage I). Instead this boreal forest appears to be an association established quite recently, an adjustment in a particular environment rather than a product representing continuity in time.

The use of pollen-based methods to reconstruct climates is based on the *dynamic equilibrium hypothesis* (Solomon et al. 1981), which proposes that vegetation changes in response to continuous climatic pressures. This is opposed by the *disequilibrium hypothesis* (Iversen 1960 and Davis 1976, among others) that attributes the different rates and directions of movement of different taxa to the existence of different distribution areas at the time of glacial maximum as well as to different intrinsic rates of spread. In support of the former hypothesis, Prentice et al. (1991) concluded from a study of pollen in eastern North America since 18,000 B.P. measured at intervals of 3,000 years, that the changes in continental-scale vegetation patterns have responded to continuous changes in climate, with lags of about 1,500 years. Their findings do not put other hypotheses to rest but rather add fuel to the debate for current and future ecologists.

Paleoecology Beyond Palynology

Investigations in paleoecology are much broader in scope than might be implied by this discussion of pollen analysis. The principle of the relationship of climatic adaptation and community composition cannot be approached by a study of pollen in a marine stiuation, for example. Here the relative distribution and abundance of key invertebrates, quite often foraminifera and mollusks, serve the same function as pollen in a bog sediment. Dendrochronology, the analysis of growth rings in trees, particularly long-lived species, also provides a key to past climate (Fritts and Swetnam 1989). However, it is not the purpose here to explore these exciting fields in detail but rather to indicate two general points: (1) that the ecology of the past can be studied and offers as many if not more challenges to the investigator than a study of the ecology of the present—and certainly each will make and has already made substantial contributions to the other; and (2) communities are indeed dynamic, ever changing in composition and function in response to both immediate and long-term environmental changes. The seeming stability of present-day communities is indeed deceptive and is likely to give a false sense of security. As a biological unit, they remain stable by not being static.

MATURATION OF COMMUNITIES

The maturation and development of communities have been characterized by Eugene Odum (1969) as a "strategy" of increased control of the physical environment that provides maximum protection from environmental perturbations. The culmination of this dynamic and largely predictable process of community-controlled modification of the environment is a stabilized ecosystem. Odum has identified a number of trends to be expected in the development of ecosystems (Table 13-1), only a few of which have been singled out in the preceding discussion. Of these, the four major structural and functional attributes of ecological succession (increase in species diversity, increase in structural complexity, increase in organic matter, and tendency toward metabolic stability) are, in a sense, both causes and effects of the very processes of change and the eventual stability that generally characterize ecosystems. A newly available environment, like a beach pond, a plowed but unplanted field, a talus after an avalance, or a new island, is fair game for spores and seeds. But not all these spores and seeds can "play the game" in such a rigorous environment, where temperature extremes are intensified, solar exposure is considerable, and moisture is restricted. Those that succeed in becoming established also immediately chemically and physically begin to modify their environment by their metabolic and behavioral activity. They add their waste products and thereby chemically change the nature of the substrate. They may bore into the soil and thereby physically affect drainage. Spinoza, to whom we referred at the start of this section, also noted a basic ecological principle (out of his context, to be sure) in stating that "nothing exists from whose nature some effect does not follow." Thus it is that each species alters its own environment and that of its associates such that it eventually precludes its own and sometimes others' existence; but in so doing, it provides a new set of conditions within the tolerance range of yet other species. Thus organic matter and various metabolites that increase are both results of previous ecological activity and causative agents of subsequent changes.

The Climax Community

If all this is so, then why does an ecosystem eventually achieve a kind of steady state? This is a condition referred to as a *climax community,* one that is self-perpetuating for relatively long periods and one in which the dynamic changes not only occur but are necessary for the maintenance of the community. On Presque Isle the beach ponds ultimately fill in, undergo terrestrial succession, and terminate in a climax community, a beech-maple forest. The climax community results when no other combination of species is successful in replacing it. In part, this situation can be explained by the tolerance limits and optimum requirements inherent in each species.

But stability of the climax is not so simply explained. Nor is there agreement among ecologists that an organismic entity, the climax, exists. As representative of the organismic school, of which he was an early exponent, Clements (1916) contended that there was only one true climax in a given climatic region. This concept is usually referred to as the *monoclimax theory* in which the end stage of succession is reached after an orderly progression to a predictable end point, the *climatic climax.* According to this theory, biotic effects are believed to be the major force in structuring the community, the pattern of which is the result of interspecific interactions.

TABLE 13-1 TRENDS TO BE EXPECTED IN THE DEVELOPMENT OF COMMUNITIES

Ecosystem attributes	Developmental stages	Mature stages
	Community energetics	
1. Gross production/community respiration (*P/R* ratio)	Greater or less than 1	Approaches 1
2. Gross production/standing crop biomass (*P/B* ratio)	High	Low
3. Biomass supported/unit energy flow (*B/E* ratio)	Low	High
4. Net community production (yield)	High	Low
5. Food chains	Linear, predominantly grazing	Weblike, predominantly detritus
	Community structure	
6. Total organic matter	Small	Large
7. Inorganic nutrients	Extrabiotic	Intrabiotic
8. Species diversity—variety component	Low	High
9. Species diversity—equitability component	Low	High
10. Biochemical diversity	Low	High
11. Stratification and spatial heterogeneity (pattern diversity)	Poorly organized	Well-organized
	Life history	
12. Niche specialization	Broad	Narrow
13. Size of organism	Small	Large
14. Life cycles	Short, simple	Long, complex
	Nutrient cycling	
15. Mineral cycles	Open	Closed
16. Nutrient exchange rate, between organisms and environment	Rapid	Slow
17. Role of detritus in nutrient regeneration	Unimportant	Important
	Selection pressure	
18. Growth form	For rapid growth	For feedback control
19. Production	Quantity	Quality
	Overall homeostasis	
20. Internal symbiosis	Undeveloped	Developed
21. Nutrient conservation	Poor	Good
22. Stability (resistance to external perturbations)	Poor	Good
23. Entropy	High	Low
24. Information	Low	High

As noted earlier, the organismic view has been challenged by numerous ecologists of the individualistic school of thought, most prominently Britain's Arthur Tansley (1939), and America's Henry Gleason (1926) and Robert Whittaker (1953), among others. In this view, communities are the result of species-specific responses to particular environmental factors, the resulting species assemblages being largely the happenstance of innate life histories that happened to be in the same place at the same time. As a result, a given climatic area can contain a number of climax types forming a mosaic governed by the interaction of climate, soil, and topography, as well as by biotic factors (See Figure 14-18). This concept is usually referred to as the *polyclimax theory,* which conveys a continuity of climax communities varying with the prevailing environmental gradients. Although there are contemporary adherents to the monoclimax view, most ecologists today hold to the polyclimax theory or some modification of it.

REGULATION OF COMMUNITIES

Whether the monoclimax or polyclimax theory is espoused, it is acknowledged that communities do undergo succession, and mature communities do exhibit a degree of stability for varying periods of time. Whether this stability is a property of the individual species composing the community or of the community itself is, as noted at the beginning of the chapter, debatable.

Connell and Slayter (1977) outlined three different models regulating succession: the facilitation model, inhibition model, and tolerance model (Figure 13-6). The *facilitation model* characterizes the monoclimax or climatic climax view, each species making the environment more suitable for the next species or group of species. The *inhibition model* applies to situations in which initial colonists prevent subsequent species from colonizing, a model in which succession depends on who arrives first and in which the successional pattern is not orderly. In the *tolerance model,* which is intermediate between the other two models, any species can start the process of succession but a somewhat orderly process ensues, leading to the eventual climax community.

An examination of some of the factors regulating community development and structure will indicate the complexity of this exciting arena of contemporary ecology. In some large measure the various factors to be considered can be regarded as the relative contributions of "bottom-up" forces (e.g., resources, climate, and other abiotic factors) and "top-down" forces (e.g., biotic interactions) (Matson 1992).

Role of Species Diversity

One mechanism of community regulation that has held considerable persuasion—and generated considerable debate—is species diversity. G. Evelyn Hutchinson (1959), in an essay with the intriguing title, "Homage to Santa Rosalia, or Why Are There So Many Kinds of Animals?", posed the question well. He also suggested an answer, as did Robert MacArthur (1965), who formalized the hypothesis

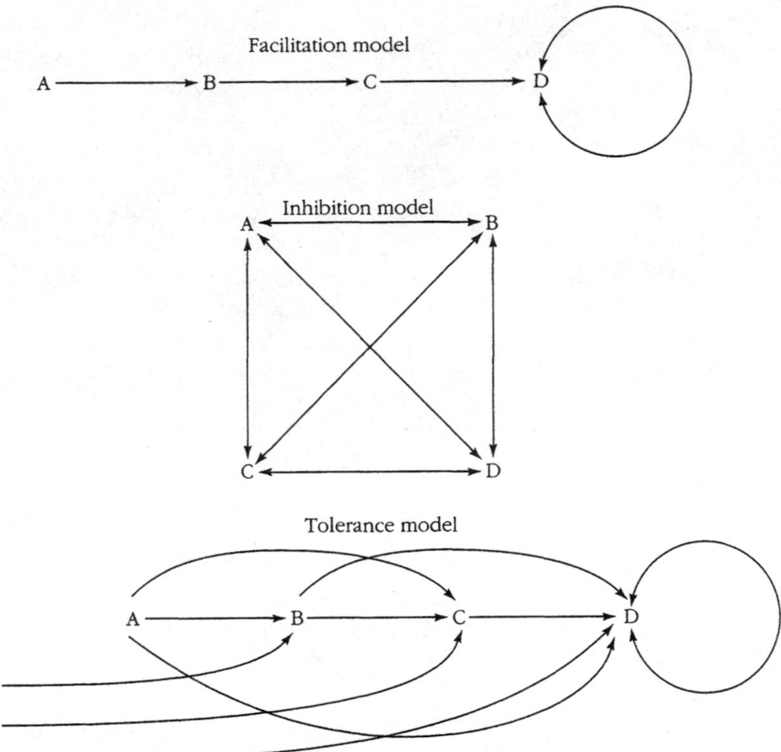

Figure 13-6 Three models of succession proposed by Connell and Slayter. Four species are represented by A, B, C, and D. An arrow indicates "is replaced by." The facilitation model is the classic model of succession. In the inhibition model, all replacements are possible, and much depends on who gets there first. The tolerance model is represented by a competitive hierarchy in which later species can outcompete earlier species but can also invade in their absence. (Reproduced with permission from P. Stilling. © 1992. p. 344. *Introductory Ecology*. Prentice Hall Inc. Upper Saddle River, New Jersey.)

mathematically with particular reference to the number of trophic pathways in an ecosystem.

The species diversity argument. The argument for species diversity as the regulator of community development and stability runs like this: the more advanced the system, the more complex its function by virtue of the increased number and availability of different ecological niches; the more niches, the more diversified the flora and fauna; the more diversified the biota, the less likelihood that a shift in one component would adversely affect the system as a whole. Thus if there are many kinds of producers, each overlapping to various degrees in their respective ability to use various wavelengths and intensities of light, a shift in spectral quality or quantity would tend to affect some but not all producers. The ecosystem would therefore be able to continue with little if any detectable disruption. The persistence of a major change would probably result, however, in new equilibrium levels and new combinations of associations of the "survivors."

Copeland's (1965) study of the shift of the dominant producers from turtle grass to blue-green algae with a shift in light intensity is a case in point in the opposite direction (see Chapter 12). The absence of immediately available alternate routing of production resulted in a dramatic drop in output; recovery eventually came but at the expense of some period of time. Highly integrated and complexly interrelated communities, like those of the climax, possess considerable stability and are therefore capable of responding quite readily to various kinds of short-term environmental insults that are not too severe.

Accordingly, the stability of the climax community is related to its species diversity and, moreover, the aging phenomenon or succession is best described as development toward high diversity—a large number of ecological niches and its counterpart of a large number of species. The latter follows from the Gausian competitive-exclusion principle (see Chapter 11) that only one distinct species population can occupy one distinct niche at any one time. The number of ecological niches in a given ecosystem is a function of the history and development of the system and of the multiplicity of energy flow routes through the system. Evidence presented earlier (see Figures 6-4, 6-5, and 14-19) shows the tendency of productivity to increase from the poles to the tropics. This factor is partly related to differences in abiotic characteristics, such as incident radiation (Figure 4-3), but it may eventually also prove to be related to the higher diversity of species in the tropics (Table 13-2). It should follow, then, that a climax tropical forest should, by virtue of its higher species diversity, be relatively more stable than a climax deciduous forest and the latter more stable than a tundra climax.

Critique of the species diversity argument. Although the hypothesis of a positive relationship between the stability of an ecosystem and its species diversity has a ring of intuitive plausibility to it, empirical verification is largely wanting (May 1984, Goodman 1975). In addition to raising a legitimate criticism that discussions of stability-diversity carry an imprecision of definition of key terms, Nelson Hairston and associates (1968) noted from their laboratory study involving bacteria and grazing and predatory protozoa that stability on a higher trophic level was increased by increased diversity on a lower trophic level (a finding supportive of the Hutchinson-MacArthur hypothesis) but also that stability is not predictable from consideration of only the number of energy pathways in the system.

Luckinbill (1979) has shown that in the interaction of the predator *Didinium nasutum* and its prey *Colpidium campylum* an increase in species diversity caused by adding alternative prey destabilized the association, thus causing extinction of the community. By experimental manipulation using a predatory fish, Zaret (1982) demonstrated that the predator destabilized the community in a lake in Panama by

TABLE 13-2 THE NUMBER OF SPECIES IN DIFFERENT GEOGRAPHIC REGIONS SHOWING A GENERAL INCREASE IN DIVERSITY WITH DECREASING LATITUDE

Insects	Beetles[b]		Ants[a]		Dragonflies[c]	
	Labrador	169	Alaska	7	Nearctic	59
	Massachusetts	2,000	Iowa	73	Neotropical	135
	Florida	4,000	Trinidad	134		
Mollusks	Nudibranchs[a]		Land snails[b]		Lamellibranchs[b]	
	Arctic latitudes	10	Labrador	25	Newfoundland	30
	Temperate latitudes	90	Massachusetts	100	Cape Hatteras	150
	Tropical latitudes	130	Florida	250	South Florida	200
Vertebrates	Coastal marine fish[b]		Snakes[a]		Nesting birds[a]	
	Labrador	75	Canada	22	Greenland	56
	Massachusetts	225	U.S.A.	126	New York	195
	Florida	650	Mexico	293	Colombia	1,395
Plants	Flowering plants[b]		Ferns and clubmosses[b]		Endemic genera of flowering plants[d]	
	Labrador	390	Baffin Land	11	North temperate zone	127
	Massachusetts	1,650	Labrador	31	Tropical zones	465
	Florida	2,500	Massachusetts	70	South temperate zone	55

Data adapted from:
[a] A. G. Fisher. 1960. *Evolution* 14:64–81.
[b] G. L. Clarke. 1954. *Elements of Ecology*. New York: John Wiley & Sons.
[c] C. B. Williams. 1964. *Patterns in the Balance of Nature*. New York: Academic Press.
[d] R. Good. 1953. *The Geography of Flowering Plants*, 2nd ed. London: Longmans, Green and Company, Ltd.

exterminating 13 of 17 native fish, whereas no extermination resulted from the introduction in the adjacent and relatively less environmentally stable river system.

Work by Robert May (1973), among others, on diversity-stability modeling have not supported the idea; and in his studies on forest insect pest species Kenneth Watt (1965) failed to find a positive relationship between stability and diversity. Watt concluded that the factors leading to stability of a pest species included the proportion of the total environment containing food for the pest, an increase in the presence of other species competing for the same host, and a smaller number of enemies.

If species diversity is not the factor or major factor in stabilizing community patterns and processes, then what does govern? Predators, as suggested by the experiments of Luckinbill and Zaret cited above? Competitors, as suggested by Watt? Food webs as suggested by Pimm et al. (1991)? (Also see the discussion of food webs in Chapters 7 and 12.) Abiotic factors such as fire or the availability of nutrients? A brief consideration of alternatives to species diversity will indicate the complexity of this issue.

Role of Predators

A number of field experiments have shown that changes in the densities of predatory (and other) species affect the structure of their ecological communities. Stream-inhabiting midge larvae that weave filamentous algae into tufts that serve as retreats are preyed upon by invertebrate predators such as dragonfly larvae. If the latter are preyed upon by fish, the pressure on the midges is lessened, and they graze down the algae (Power et al. 1992). By experimentally adding, separately and in combination, mantids (*Tenordera sinensis*) and a wolf spider (*Lycosa rabida*), both predators, into a terrestrial old-field community, Hurd and Eisenberg (1990) demonstrated differential effects on the native arthropod community. The mantids depressed not only the total numbers of resident arthropods by 10 to 15 percent but also reduced their biomass by 50 percent, an effect that was confined to the largest-size categories of arthropods. By contrast the wolf spiders had no measurable effect.

Aquatic communities. Diehl (1992) conducted an experimental study of the effect of an introduced omnivorous predator, perch (*Perca fluviatilis*), on a freshwater macroinvertebrate community in the littoral zone (that section containing rooted vegetation and in which light penetrates to the bottom). The results showed that perch consumed predatory and nonpredatory macroinvertebrates almost exclusively, but, in the absence of vegetation, the consumption of the nonpredatory microcrustaceans was increased. An increased density of perch caused a decrease in biomass of predatory invertebrates and the abundance of the dominant predatory invertebrate, a megalopteran, but had no effect on the biomass of nonmolluscan herbivores and detritivores. He showed further that relatively large predatory invertebrates can coexist with fish in the littoral zone and play an important role in the structure of that community. Comparably, Hixon and Beets (1993) found that predation, along with competition, the rate of colonization, availability of naturally occurring refuges (e.g., holes), and other factors, plays a significant role in the structure of coral-reef fish communities.

The Role of Competition

Barnacles. Connell's study of aggressive competition between two species of barnacles (see Chapter 11 and Figure 11-12) was pursued from a successional perspective by Farrell (1991). This experimental study showed that the first colonizer on a cleared plot, *Chthamalus dalli*, was replaced by *Balanus glandula* according to the tolerance model of succession described earlier, in this instance by competition for space. By contrast, the several species of macroalgae that came next in the successional sequence did so only after *Balanus* was established, an instance of the facilitative model of succession. However, the study also demonstrated that a significant role in the rate of succession was played by consumers, namely limpets, which delayed the establishment of *Balanus* and of the macroalgae.

Seedling growth. In a study of primary succession on an Alaskan floodplain, Walker and Chapin (1986) showed that competitive interactions played a significant role in controlling seedling growth. In a greenhouse, alder (*Alnus tenuifolia*), through its associated nitrogen-fixing bacteria, facilitated the growth of seedlings of willow (*Salix alaxensis*) and poplar (*Populus balsamifera*). However, under natural conditions of alder stands in which nitrogen levels were high, seedlings of naturally occurring spruce (*Picea glauca*) and transplanted seedlings of alder and poplar did not grow as well. Experiments employing clearcutting and trenching showed that the alder stands inhibit the growth of seedlings of other alders, as well as spruce and poplar, by root competition and shading.

The Role of Nutrients

The availability of nitrogen cited in the case of alder inhibition of seedling growth suggests, as does the discussion of nutrients in Chapters 5 and 8, that nutrients play a role in succession and in community regulation. A few examples will confirm that position.

McLendon and Redente (1991) studied the effects of five annual applications of nitrogen and phosphorus to a sagebrush community in northwestern Colorado that had been disturbed, for experimental purposes, in 1984. Nitrogen, but not phosphorus, significantly affected the secondary succession for all years except the first by allowing annuals to remain as dominants through the fifth year of the study.

In an old-field community in Ohio, Carson and Barrett (1988) showed that fertilizer enrichment (nitrogen, phosphorus, and potassium) led to dominance by summer annuals during a 3-year study of succession in a 1-year-old experimental plot. Species richness was greater in the enriched plots for only the first year and was significantly lower than the control plots. In a 4-year-old fertilized plot, two summer annuals and a biennial replaced the dominant perennial grasses.

Additions of nitrogen and phosphorus in two freshwater lakes in Michigan resulted in increases in bacterial abundance and chlorophyll levels in phytoplankton (Pace and Funke 1991), altering the structure of the community. But foraging by introduced *Daphnia* on protozoa, which were not affected by the nutrient enrichment, also played an important role in altering community dynamics.

An enriched atmosphere of CO_2 interacting with nutrients and low light resulted in the greatest stimulation of seedling growth of six species of deciduous trees in a Massachusetts forest (Bazzaz and Miao 1993). The early successional species (gray birch, ash, and red maple) showed increased seedling growth with elevated CO_2 only in the presence of high nutrients; by contrast, the later successional species (red oak, yellow birch, and striped maple) showed the greatest increase in seedling growth with low light. Nutrients play a more critical role in the earlier stages of succession, whereas light intensity becomes more significant subsequently.

In these experimental studies, it is clear that nutrients play an important role in community dynamics (see also Vitousek and Reiners 1975); thus, the processes of decomposition and mineralization under natural conditions warrant due attention in understanding community regulation.

Other Factors

Fire. Among other factors involved in regulating communities is fire, a significant nutrient resource-releasing agent and, in some instances, a requisite to forest regeneration by releasing seeds from fire-dependent cones. Because the role of fire as an abiotic factor was described in some detail in Chapter 5, no further discussion is given here.

Alien species. Introduced species of plants and animals, generally referred to as *alien species*, generally alter ecological communities (Vitousek 1992). At the highly disruptive end of the spectrum of mild to strong alteration, communities may become so transformed by alien species as to be scarcely recognizable (Loope 1992). By the same token, Merlin and Juvik (1992) have shown that some island plant species competitively resist alien invaders in the absence of ungulates like wild pigs. In a study of benthic community changes resulting from the invasion of the seagrass *Zostera japonica*, Posey (1988) showed that the alien species changed the physical habitat as well as the richness and densities of the resident fauna. Thus, alien introductions can lead to increased stabilization of an existing community or a change in community composition that, in turn, leads to a new level of stabilization.

Spatial factors. The role of spatial location and spatial differences in community dynamics is another factor receiving increased attention (Karelra 1994). For example, Kratz and DeWitt (1986) demonstrated the importance of spatial dynamics in the development of peatlands, in particular in a peatland lake system in northern Wisconsin. Near the lake edge organic matter accumulates vertically, thickening the floating mat. In the next zone organic matter also accumulates vertically compacting the underlying peat. In the zone farthest from the shore the peat reaches its maximum density and none accumulates. In a study in subalpine habitats on Mount St. Helens in the sixth year following the 1980 eruption, Wood and del Moral (1987) concluded that the characteristics of early succession depended on the spatial position and dispersal abilities of species in the seed pool and not on environmental gradients such as moisture, temperature, and nutrients. And, Lively et al. (1993) found that most of the variation in the percentage of barnacles and brown encrusting algae covering a rocky intertidal community in the northern Gulf of California was due to microspatial effects caused by the foraging patterns of a common predatory snail.

Historical antecedents. Yet another factor to be considered in understanding the diversity of species in given communities and the regulatory factors that govern community structure and function is geological and evolutionary history (Signor 1990). As Brown and Maurer (1987) note, variation on large spatial and long temporal scales determines the biogeographic and evolutionary processes that shape the composition of the pool of species that historically have had access to a geographical region. But also, variation of small spatial scales and short time scales influences the dynamics of local populations of interacting species and determines the combination of species that constitute local communities. These two processes are not independent inasmuch as species cannot occur in local communities unless historical events have given them access to the region, and species cannot remain in the pool unless they are able to maintain populations within local communities.

Concluding Comment

At the outset of this discussion of community regulation the role of species diversity, so intuitively attractive, was found to be wanting. Further exploration indicated that a variety of factors, biotic and abiotic, may be involved at different stages of the successional process including that of the so-called climax. Wilson and Agnew (1992) review the myriad ways in which vegetation in a community provides positive-feedback switches, processes that modify the environment, making it more suitable for that community. Simultaneously or sequentially, predation, competition, herbivory, food webs, pathogens, and other biotic interactions have their regulatory impact (Putman 1994). Succession itself is a complex process driven by many processes acting simultaneously in any given situation (Walker and Chapin 1986). To seek for a single, simple answer to the governance of communities is akin to searching for the Holy Grail. As philosopher Alfred North Whitehead is reported to have noted, "Seek simplicity, and distrust it."

AN EPILOGUE ON BIODIVERSITY AND EVOLUTION

Woven throughout the text have been two topics that warrant further attention: *biodiversity*, the number and abundance of species, and the interaction of ecology and evolution. Each is sufficient for extended treatment, but given the purpose of this book, only a brief introductory discussion can be undertaken.

Biodiversity

It is a paradox of contemporary science that we know how many stars there are in the Milky Way, namely some 10^{11}, the mass of an electron, namely 9.1×10^{28}, and the diameter of the Earth, namely 12,752 km or 7,926 miles, but do not know how many species, let alone organisms, there are. Different projections have been made by a number of ecologists, taxonomists, and evolutionists based on different assumptions and yielding disparate numbers. Among those who have given the matter critical attention are Princeton ecologist Robert May (1988, 1992, among others) and Harvard ecologist E. O. Wilson (1988, 1992, among others). But once having answered that conundrum, why is such information of value? A brief exploration of both questions is in order.

The number of species. Estimates of the total number of species range from 3 to 70 million, meaning that the presently known 1.4 million species constitute only about 2 percent of the total (Raven and Wilson 1992, Groombridge 1992, among others)! Among the many problems encountered in ascertaining the number of species is that different groups of animals have held preeminence among naturalists. Birds and mammals are so well studied that the likelihood of discovering new species is remote. However, 11 of 80 known living species of whales and porpoises were discovered in this century, the most recent in 1991 (Raven and Wilson 1992). In 1992 a new genus of a large land mammal (now named Vu Quang ox) was found in the rugged region dividing Vietnam and Laos, the fourth new genus found in this century; this is a region that subsequently yielded two new species of deerlike animals and a new species of fish resembling carp (Linden 1994).

As well studied as butterflies and beetles have been, only a portion of their actual numbers are probably known; terrestrial vascular plants are pretty well catalogued but not so for their marine counterparts—however, botanists recently discovered three new families of flowering plants in Central America and southern Mexico (Raven and Wilson 1992). When we turn to bacteria, fungi, and even protozoa the unknown far outnumbers the recorded. Likewise, some environments are well studied (e.g., temperate forests and grasslands) and the component species are well cataloged; however, organisms of the polar regions, the ocean benthos, coral reefs, and soils are not.

Without delving into the various methodologies that have been employed to arrive at estimates, an exploration that would take us too far afield for the immediate purposes (but which are described in the articles by May and Wilson cited above), a conservative comparison of the estimated number of species will be instructive.

Among the many phenomena in biological diversity are two worthy of note: the relative contributions different groups of organisms contribute to diversity (Table 13-3) and the relative abundance of species in different geographical regions (Table 13-2). Of the some 1.4 million recorded species, groups like mammals, amphibians, bacteria, and echinoderms represent only a small proportion of the whole. The 4,000 species of mammals, for example, constitute only 0.25 percent of the total. Algae, protozoa, fungi, and mollusks have considerably more species described than the preceding group, but less than noninsect arthropods and trees; insect species are the most numerous, constituting more than half of all known species. But, again, these numbers are deceptive in that a number of species at least equal to or even twice or many times more than those identified have not yet been discovered, described, or recorded; this is especially so among those groups that have been given the least attention.

Patterns in diversity. Beyond the comprehensive viewpoint of global diversity, that is the total number of species on Earth, ecologists are interested in the configurations or patterns in which diverse groups of organisms exist. The most common perspective is that of *point* or *α-diversity,* that is the number of species in a given area and the geographical distribution patterns that result. As we shall see momentarily, point diversity is relatively well documented for a wide variety of organisms and has led to such generalizations as an increase in diversity with decreasing latitude. Much less is known about the other kind of diversity, known as *β-diversity* (Pimm and Gittleman 1992), which is the turnover of species across space, including the dimensions of a species' range. As Pimm and Gittleman note,

TABLE 13-3 NUMBERS OF DESCRIBED SPECIES OF LIVING ORGANISMS

Kingdom and major subdivision	Common name	No. of described species	Totals
Virus			
	Viruses	1,000 (order of magnitude only)	1,000
Monera			
Bacteria	Bacteria	3,000	
Myxoplasma	Bacteria	60	
Cyanophycota	Blue-green algae	1,700	4,760
Fungi			
Zygomycota	Zygomycete fungi	665	
Ascomycota (including 18,000 lichen fungi)	Cup fungi	28,650	
Basidomycota	Basidomycete fungi	16,000	
Oomycota	Water molds	580	
Chytridomycota	Chytrids	575	
Acrasiomycota	Cellular slime molds	13	
Myxomycota	Plasmodial slime molds	500	46,983
Algae			
Chlorophyta	Green algae	7,000	
Phaeophyta	Brown algae	1,500	
Rhodophyta	Red algae	4,000	
Chrysophyta	Chrysophyte algae	12,500	
Pyrrophtya	Dinoflagellates	1,100	
Euglenophyta	Euglenoids	800	26,900
Plantae			
Bryophyta	Mosses, liverworts, hornworts	16,600	
Psilophyta	Psilopsids	9	
Lycopodiophyta	Lycophytes	1,275	
Equisetophyta	Horsetails	15	
Filicophyta	Ferns	10,000	
Gymnosperma	Gymnosperms	529	
Dicotolydonae	Dicots	170,000	
Monocotolydonae	Monocots	50,000	248,428
Protozoa			
	Protozoans: Sarcomastigophorans, ciliates, and smaller groups	30,800	30,800
Animalia			
Porifera	Sponges	5,000	
Cnidaria, Ctenophora	Jellyfish, corals, comb jellies	9,000	

TABLE 13-3 (CONTINUED).

Kingdom and major subdivision	Common name	No. of described species	Totals
Animalia			
Platyhelminthes	Flatworms	12,200	
Nematoda	Nematodes (roundworms)	12,000	
Annelida	Annelids (earthworms and relatives)	12,000	
Mollusca	Mollusks	50,000	
Echinodermata	Echinoderms (starfish and relatives)	6,100	
Arthropoda	Arthropods		
Insecta	Insects	751,000	
Other arthropods		123,161	
Minor invertebrate phyla		9,300	989,761
Chordata			
Tunicata	Tunicates	1,250	
Cephalochordata	Acorn worms	23	
Vertebrata	Vertebrates		
Agnatha	Lampreys and other jawless fishes	63	
Chrondrichthyes	Sharks and other cartilaginous fishes	843	
Osteichthyes	Bony fishes	18,150	
Amphibia	Amphibians	4,184	
Reptilia	Reptiles	6,300	
Aves	Birds	9,040	
Mammalia	Mammals	4,000	43,853
TOTAL, all organisms			1,392,485

Reproduced with permission from E. O. Wilson (ed). 1988. *Biodiversity*. Washington, DC: National Academy Press.

if the range of a species is large, then α-diversity is almost independent of the area sampled; contrarily, total diversity may be high while α-diversity is low, with species' ranges being small and adjoining rather than overlapping.

Geographical distribution of biodiversity. As just noted, the geographic distribution of diversity is of the α-diversity type, the number of species in a given area. In part this has already been pointed out earlier in this chapter (see Table 13-2), with the discussion of the trend to increasing diversity with decreasing latitude, that is from the Arctic/Antarctic through the temperate to the tropical. Among the few statements to which all ecologists appear to agree is that **although tropical rainforests occupy only about 7 percent of the land surface, they contain more than half the species of the entire biota of the world.**

Wilson (1988), for example, notes that he recovered 43 species of ants belonging to 26 genera from one leguminous tree in a tropical rain forest in Peru, a number equal to the entire ant fauna of the British Isles, and that Peter Ashton of Harvard's Arnold Arboretum found 700 species of trees in ten 1-hectare plots in Borneo forests, the same number as in all of North America! As Wilson further notes, it is not unusual for a square kilometer of forest in Central or South America to contain several hundred species of birds and many thousands of species of butterflies, beetles, and other insects.

By contrast, drylands have been found to be the richest of South America's six major macrohabitats, including rainforests, in the number of mammals and to be more biologically diverse (Mares 1992). Conservation efforts that address only tropical rainforests, the major focus today, will result in short-shrift to the diversity of fur-bearers.

Using 94 data sets from around the globe, Scheiner and Rey-Benayas (1994) found that: (1) more plant species are found in well-lit, well-watered places, that is, sites of high levels of photosynthesis (perhaps not a surprising finding); (2) more species are found in places where the temperature between summer and winter varies widely than in those with relatively even seasons, a finding somewhat contrary to what some ecologists have held, namely that equitable climates make it easier for new species to evolve and persist; and (3) not only do warm locations with large seasonal temperature fluctuations have more species, they also are more likely to have different varieties at different sampling sites.

Supporting part of the foregoing findings Reice (1994) has suggested that perturbations and heterogeneity, rather than equilibrium, generate biodiversity. He notes that ecological communities are always recovering from the last disturbance, the biota adapting to predictable disturbances, such as annual snowfalls and snowmelts, but losing out when the disturbance is erratic.

The fragility of biodiversity. Although, as we have seen, naturally occurring phenomena such as fire, volcanic action, and floods can adversely affect community structure and thereby diversity, it is without question that human activity, especially on fragile environments, has far greater impact on the maintenance of biodiversity, primarily through habitat destruction.

Tropical rain forests, presumably the richest source of biodiversity, are also among the most fragile and hence vulnerable habitats for several reasons. First, the predominant soils are oxisols (see Table 5-1), typically acidic, nutrient poor, and rich in iron oxides. These soils tend to harden, forming lateritic boundaries, from which vital elements such as calcium and potassium are leached and in which phosphorus forms insoluble compounds with iron and alumninum, thus reducing its availability. Secondly, the high rates of decomposition and primary production mean most of the carbon and nutrients are bound up in the biomass, thus precluding rapid regeneration in the case of clearcutting. Finally, because seeds of most rain forest species germinate within a few days or weeks, most sprout and die in the hot, nutrient poor oxisols (Gomez-Pompa et al. 1972); thus, along with the foregoing factors, the regeneration of a mature rain forest takes centuries, and natural restoration may never occur.

Given this fragility, it is significant to note that tropical deforestation is reducing species by one-half of a percent per year; likewise, coral reefs, the marine counterpart to tropical rain forests, are also being subjected to degradation, largely through human activity. Although the topic of deforestation is explored more fully later, here it is pertinent to note that about 1 percent of the total tropical forests of the world are being permanently cleared or converted to slash-burn agriculture each year. If continued at this rate all tropical rainforests will be gone by the year 2135. And, of course, with that devastation would come the total destruction of the largest pool of biodiversity in the world.

The values of biodiversity. As described in more detail later, tropical rain forests are the reservoirs of many resources of great utilitarian value. These include oils, gums, rubber, fibers, dyes, tannin, resins, and turpentine, along with a wide variety of fruits, root crops (most of which have not been exploited to meet human food needs), and ornamental plants, as well as a number of modern medicines (Cox and Balick 1994). Geographic varieties have provided the materials for agricultural manipulation of more productive and disease-resistant strains. Deforestation reduces the uptake of carbon dioxide in photosynthesis, thereby increasing the presence of this global-warming, Greenhouse gas (see Chapter 20) but also decreasing the amount of oxygen contributed to the atmosphere.

As Ehrenfeld (1988) points out, these utilitarian values placed on biodiversity justify it in economic terms, and a cost/benefit analysis might well support full exploitation of tropical rain forests for maximum economic gain. Further, he notes that although a great many chemical compounds of potential benefit to human health may probably be obtained from tropical rain forests, pharmaceutical researchers believe they can get new drugs faster and cheaper by computer modeling of the molecular structures found promising on theoretical grounds. If these utilitarian and economic values are not sufficient, what remains then is the inherent and intrinsic value of diversity as its own warrant for survival—that is, biodiversity has value in and of itself, and it is inherently wrong to destroy it. Precisely what is to be preserved and how is the subject of considerable debate (Ehrlich and Wilson 1991, Levin 1993, Solbrig 1992, among many others). We shall explore further the valuing of the environment toward the end of the book.

Ecology and Evolution

Throughout the foregoing chapters there has been an underlying and sometimes explicit theme of adaptation—predators to prey, parasites to hosts, physiology to climatic and edaphic factors, behavior to both abiotic and biotic factors, communities to habitats, and so on. Although the study of the interrelationships of ecology and evolution is a highly active one resulting in an abundance of literature, a full exploration would require considerably more time and space than is intended for this introductory text. However, some brief comments are in order.

Darwinian evolution. As we all know, the founder of modern evolutionary theory was Charles Darwin, whose famous voyage on the ship *The Beagle*, which began in 1831 and lasted for 5 years (see Moorehead 1969 for an excellent account of this journey), made him aware of the tremendous variety of both living and fossil organisms. This coupled with his subsequent reading of Malthus's treatise on human population growth led to the basic tenets of the theory of evolution by natural selection, or "survival of the fittest." He finally published the *Origin of Species* in 1859, some 20 years after its formulation, prompted, in part, by the development of virtually identical ideas by Alfred Russel Wallace. In brief, the formulation of natural selection can be reduced to three basic steps: (1) more organisms are born than can survive given the limitations or carrying capacity of the environment, (2) organisms vary in a variety of structural and functional ways, and (3) organisms with variations that are favorable in given environment survive and reproduce; others die off.

The major development in evolutionary theory subsequent to Darwin's formulation was its marriage to genetics (a field that was initiated some 40 years after Darwin's major publication), and particularly with population genetics. Genetics provided the explanation of the origin of variation and population genetics the basis of isolating mechanisms that lead to speciation. The most widely accepted of such mechanisms today is referred to as *allopatric speciation* (literally "other country"), which involves the separation of populations by a geographic barrier. The converse, or *sympatric speciation* (literally "same country"), applies to populations that are not geographically isolated, a mechanism whose models include self-fertilization and polyploidy and which is debatable.

Given the tremendous literature on natural selection, speciation, and the interrelationships of ecology and evolution, selecting a topic to discuss here is difficult. However, the islands of Hawaii, a part of the world largely unknown to most of the world, serve as a unique natural laboratory for evolutionary studies and provide some equally unique insights into this fascinating, fundamental field of contemporary biology.

Evolution in Hawaii

Gulick and *Achatinella*. It has often been said that had Darwin visited the Hawaiian Islands, then known as the Sandwich Islands after the Earl of Sandwich, he might well have observed the effects of geographical isolation on speciation and considerably advanced his hypothesis of natural selection. Actually it was the second edition of Darwin's published observations during the 1845 voyage of *The Beagle* that stimulated the first true student of evolution in Hawaii, John T. Gulick (1832–1923). Gulick, born into a missionary family in Hawaii, came to advocate the importance of isolation in the evolution of new species, a position that was in the minority in the 1880s (Amundson 1994). Darwin himself, by the time of the publication of the *Origin of Species*, had come to doubt the importance of isolation as a condition for natural selection to operate and held instead to the view of environmentally controlled speciation (Amundson 1994). The basis of Gulick's tenet was his observations and collections of arboreal snails, mostly of the genus *Achatinella* (Figure 13-7). Based on his mammoth collection of these snails, whose finely graded variations were found

Figure 13-7 *Achatinella* and *Partulina* tree snails in Hawaii showing some of the variation between species: (a) *Achatinella pulcherrima*; (b) *A. mustelina*; (c) *A. leucorraphe*; (d) *Partulina mighelsiana bella*; (e) *P. semicarinata*; (f) *P. proxima*. (Photos by William P. Mull.)

across the ridges and valleys of Oahu, Gulick is reported to have noted, "... all these *Achatinellae* never came from Noah's Ark." Gulick maintained that natural selection was inadequate to explain the geographical observations he had observed in tree snails, or, in other words, that biological variations overshot environmental differences.

Hawaii's unique flora and fauna. In the more than the century since Gulick's work, substantial efforts have yielded a considerable body of information about Hawaiian evolutionary biology, excellent summaries of which appeared in volumes of *Trends in Ecology & Evolution* in July 1987 and *Natural History* in December 1982. In an introductory essay, Simon (1987) encapsulates some of the salient aspects of the uniqueness Hawaii offers for the study of the interrelationships of ecology and evolution as well as the major contributions to understanding that have resulted from such studies.

Even though it is the most isolated oceanic island group in the world, the Hawaiian Archipelago, volcanic in origin and dated with considerable accuracy (hence contributing to determining rates of speciation), has a spectacular array of species and unique ecosystems that have evolved with a level of *endemism* (i.e., found nowhere else) higher than any other region in the world: some 1,000 species of flowering plants, 2,000 lower plants, 7,000 to 8,000 insects, 1,000 land snails, 1,500 marine mollusks, more than 100 birds, 680 fish, 3 sea turtles, 1 bat, and 1 seal. There are no native amphibians nor terrestrial reptiles in Hawaii.

These endemic species are largely related to Indo-Pacific fauna and flora and arrived over hundreds of thousands of years by various means of overseas dispersal. The large numbers of potential habitats created by varied topography and by steep gradients in rainfall, climate, and altitude, provided unfilled ecological niches that are among the major contributing factors to speciation, making Hawaii the world's premier showcase for examples of adaptive radiation (Tangley 1988; Kaneshiro 1988). The best known of these radiations include *Drosphila,* tree snails, honeycreepers, and several groups of plants.

There are more than 700 species of endemic Drosophilidae in Hawaii, constituting about one-fourth of the world's known Drosophilidae described. The differences among species involve internal and external morphology and some astonishing patterns of mating behavior (Kaneshiro and Boake 1987). Among land snails, the arboreal *Partulina* has 42 species and *Achatinella* 41 species, all of which show variation with respect to shape, sculpture, and size (Figure 13-7). Darwin's finch advocates to the contrary, Hawaiian honeycreepers, with 43 species, are a better example of radiation than the 14 Galapagos finches due to numbers of species, bill morphology (Figure 13-8), plumage coloration, and behavior, as well as interspecific flocking, male coloration relative to females, and nest morphology and position (Freed, Conant, and Fleischer 1987).

Fragility of Hawaii's biota. As Simon (1987) and others have underscored, the extraordinary characteristics of island biotas such as those of Hawaii, which are of great interest, are also the characteristics that create their fragility: specialization, limited ranges, loss of dispersal ability, flightlessness, and lack of defenses (e.g., tox-

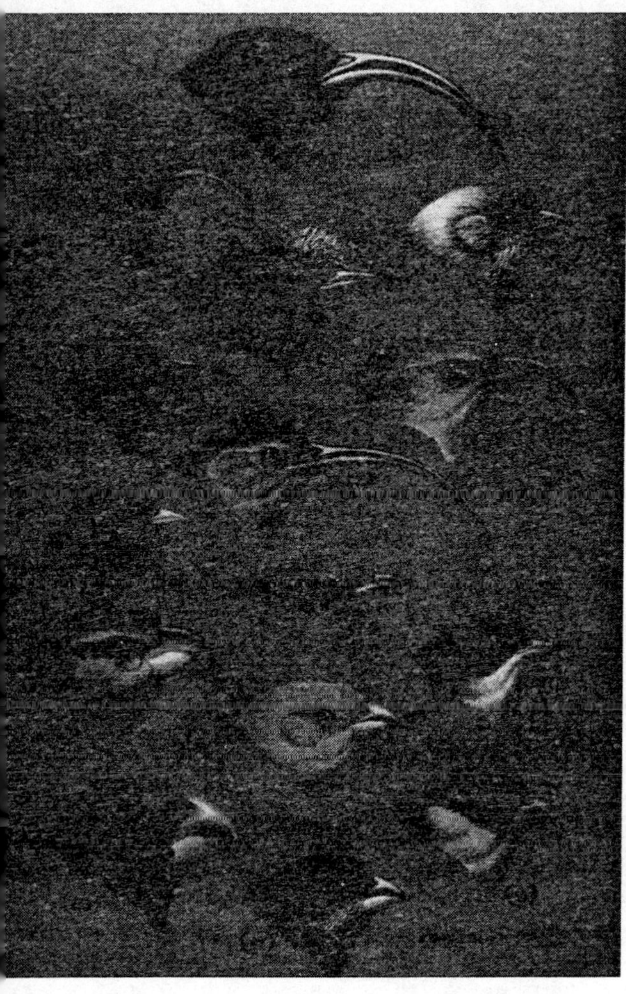

Figure 13-8 Variation in bill morphology and color patterns of selected Hawaiian honeycreepers: (a) mamo (*Drepanis pacifica*); (b) iiwi (*Vestiaria coccinea*); (c) crested honeycreeper (*Palmeria dolei*); (d) ula-ai-hawane (*Ciridops anna*); (e) apapane (*Himatione s. sanguinea*); (f) akiapolaau (*Hemignathus monroi*); (g) Kauai akialoa (*Hemignathus procerus*); (h) Hawaii akepa (male, *Loxops c. coccinea*); (i) Hawaii amakihi (*Hemignathus v. virens*); (j) Kauai creeper (*Oreomystis bairdi*); (k) Maui parrotbill (*Pseudonestor xanthophrys*); (l) ou (male, *Psittirostra psittacea*); (m) grosbeak finch (*Chloridops kona*); (n) Nihoa finch (female, *Telespyza ultima*); (o) poouli (*Melamprosops phaeosoma*). Painting by and courtesy of H. Douglas Pratt, Jr.

ins and thorns are largely absent!). Native snails, for example, are being decimated by introduced predators, and it is estimated that nearly 70 percent of the original avian land fauna is extinct; nearly one-third of the native species of flowering plants are endangered. The causes are basically twofold: conversion of habitats (e.g., shoreline modifications for beaches and harbors, housing developments, agriculture, forestry) and the introduction of alien predators, competitors, and disease. We can also consider such environmental perturbations as volcanic eruptions on the largest island of the archipelago, Hawaii, and the devastating hurricanes such as those that beset the islands in 1982 (Iwa) and 1992 (Iniki).

From the first Polynesian settlements in the third or fourth century A.D. to Captain Cook's arrival in 1778, dogs, pigs, rats, chickens, and about 30 kinds of plants for cultivation (e.g., taro, breadfruit) were introduced. Since 1778, Stone and Scott (1985) estimate that more than 2,000 arthropods, 50 land birds, 18 mammals, and

600 plants have been introduced to Hawaii. Those doing the most damage include feral pigs, feral goats, cattle, and mongooses among mammals (in addition to humans themselves, of course), along with the western yellowjacket, strawberry guava, and certain grasses (Vitousek, Loope, and Stone 1987). The mongoose introduction is one of the most inept: mongoose were introduced to control rats, but rats are nocturnal, and mongooses are diurnal. Instead of preying on rats, which continue to thrive, mongooses have devastated ground-nesting birds.

These largely human interferences with natural ecological systems are explored in greater detail in the next and final section of this book.

SUMMARY

VOCABULARY

alien species
allopatric speciation
β-diversity
biodiversity
climatic climax
climax community
disequilibrium hypothesis
dynamic equilibrium hypothesis
el Niño
endemism
facilitation model
inhibition model
monoclimax theory
paleoecology
palynology
patch dynamics
point/α-diversity
polyclimax theory
primary succession
secondary succession
succession
sympatric speciation
tolerance model

KEY POINTS

- The series of structural and functional changes in succession that ensue in an ecological community are either primary, that is initiating on or in a newly created environment, or secondary, reinitiating in a previously established ecosystem that has been disturbed, usually by abiotic forces such as fire, flood, and volcanic eruptions.
- Major ecological communities existing in the past can be determined from their fossil record; pollen analysis has been one of the major vehicles for such studies, all of which are based on the assumption that the climatic requirements of species populations have not changed significantly in the intervening period.
- Four major structural and functional attributes of succession (increase in species diversity, structural complexity, and organic matter, and tendency toward metabolic stability) can be regarded as both causes and effects in leading to a more or less steady community state known as the climax community.

- "Bottom-up" (e.g., resources, climate) and "top-down" (biotic interactions) factors are involved in the regulation of communities. Species diversity, a seemingly attractive agent in stabilization, is nonetheless suspect because of the lack of empirical verification.
- Predators, competitors, fire, nutrients, spatial factors, and food webs all appear to play some role in the regulation of some communities or the members of same.
- The number and abundance of species in the world, biodiversity, are only best estimates. Diversity tends to increase with decreasing latitude. Tropical rainforests, the most fragile of ecosystems, are the most diverse, containing half the known species of the world's entire biota and serving as reservoirs of resources of great utilitarian value.
- Hawaii's relative isolation and its topography have enabled the evolution of a unique flora and fauna that is fragile in the face of both natural processes and human intervention

PART VII

Major Ecosystems of the World

Biomes: The Major Terrestrial Ecosystems

The major ecological processes of energy flow and nutrient cycling occur within the context of biological assemblages of producers, consumers, and decomposers in a physical and chemical environment, an ecosystem. The abiotic forces that interact with and affect organisms, species, populations, and communities, and hence ecosystems, were the subject of previous chapters. We now turn to an emphasis on the biotic component as we consider the major ecosystems of the world.

BIOMES

In the broadest sense, there are two major types of ecosystems—aquatic and terrestrial. In turn, subdivisions can be recognized in each type, thus we can distinguish freshwater, estuarine, and marine aquatic ecosystems and several major types of terrestrial ecosystems, such as grassland, forest, and tundra. The former are distinguished on the basis of a major chemical difference (i.e., salt content), the latter generally on the basis of the predominant type of vegetation (grass, trees, etc.). The major terrestrial ecosystems occur on a major regional or subcontinental level and are most often referred to as *biomes*, although terms like provinces, biochores, regions, and formations are used by some ecologists. The major biomes of North America and the world distribution of similar biomes are shown in Figure 14-1.

Biome Distribution

One of the striking aspects of biome distribution is that some of the division lines tend to parallel the lines of latitude. This is especially evident in the Old World. In the Americas the latitudinal divisions are evident in eastern North and South

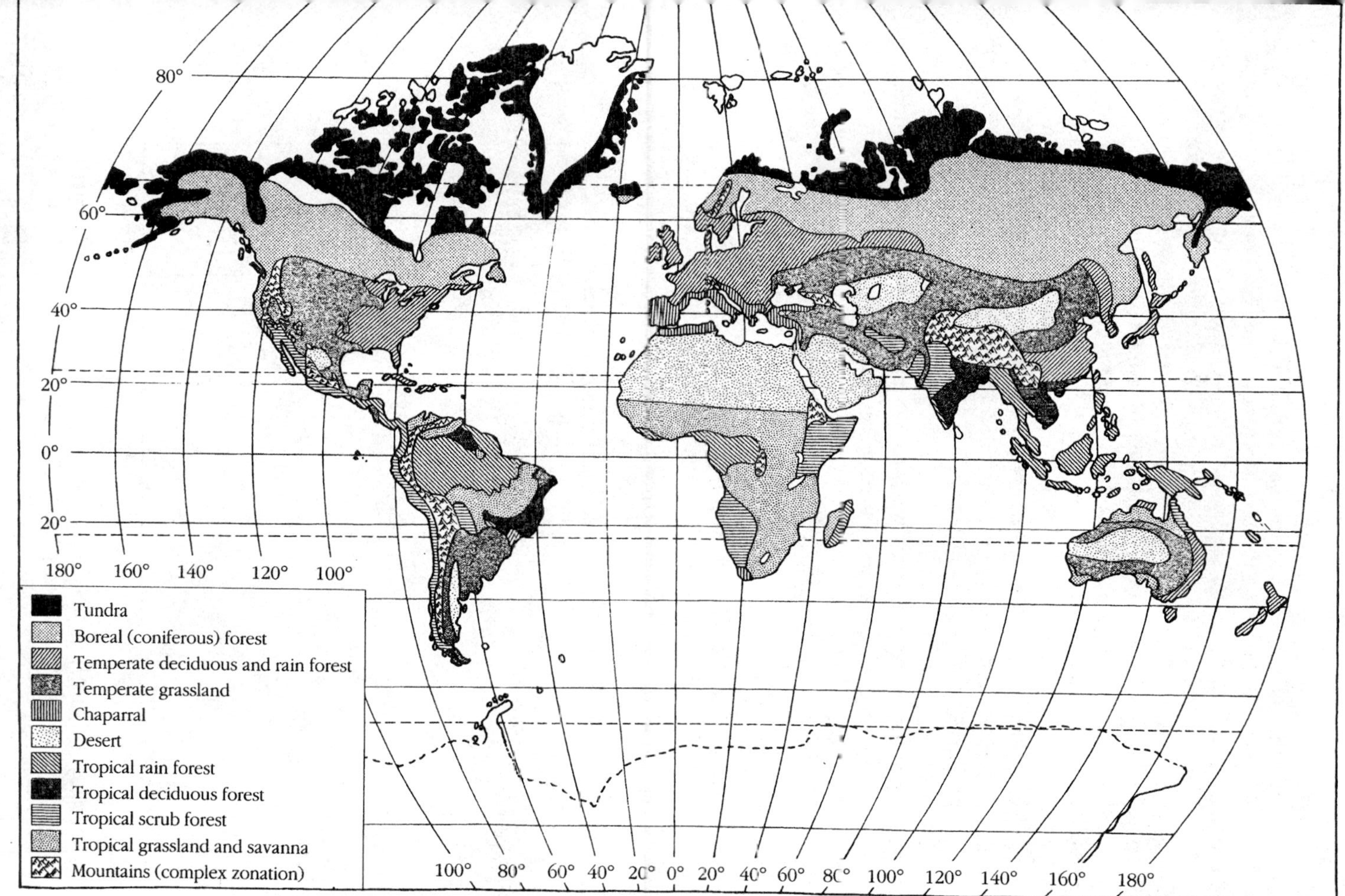

Figure 14-1 The major biome types of the world. (Redrawn by permission from E. P. Odum. 1959. *Fundamentals of Ecology.* Philadelphia W. B. Sunders Co.)

America but not in the high mountain regions that form the western "backbone" of the two continents. Not only are demarcations of biome types more or less latitudinal, but, more strikingly, the same type of biome is found within the same general latitudes. This is particularly evident in the case of tundra and boreal forest in the Old and New Worlds. Finally, in tall mountains such as the Rockies, Andes, and Himalayas, the division lines between biomes are elevational rather than latitudinal (Figure 14-2). The particular biomes to be found at a given altitude also vary with latitudes, however, a given zone occurs at progressively lower altitudes at progressively more northern latitudes (Table 14-1). The zone of Douglas fir–ponderosa pine, for example, lies between 675 to 2,000 m in the Cascade Mountains of the State of Washington, 1,350 to 2,300 m in the central Sierra Nevadas, and 1,650 to 2,700 m in the southern Sierra Nevadas in California.

Soil, climate, and biome distribution. Temperature is largely dependent on the incidence of solar radiation, which is, in turn, directly associated with latitude (Figures 4-2 and 4-3). Because precipitation patterns (Figure 4-13) are strongly influenced by major wind patterns, which are also associated with latitude (Figure 4-11), it can be inferred that climate is important in the distribution of biomes (Woodward 1987). Inferences of this kind appear in the earliest natural history writings. It is important to note that it is the interaction of temperature and rainfall that is of climatic significance in biome distribution: a low rainfall coupled with a low temper-

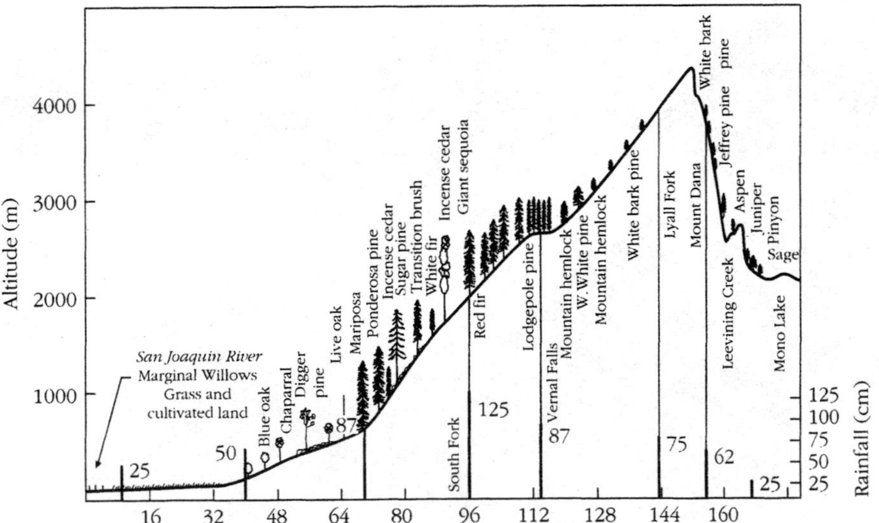

Figure 14-2 Profile of central Sierra Nevadas showing altitudinal distribution of principal forest types. (Redrawn from B. O. Hughes and D. Dunning. 1949. *Pine Forests of California.* U.S. Department of Agriculture Yearbook.)

TABLE 14-1 THE CHANGE IN VEGETATION WITH ALTITUDE AT DIFFERENT LATITUDES

Latitudinal zonation	Altitudinal zonation			
	0–1000 m	1000–2000 m	2000–4000 m	4000–6000 m
Tropical (0° – 20°)	tropical	subtropical	temperate	arctic-alpine
Subtropical (20° – 40°)	subtropical	temperate	arctic-alpine	—
Temperate (40° – 60°)	temperate	arctic-alpine	—	—
Arctic and Antarctic (60° – 80°)	arctic-alpine	—	—	—

Adapted by permission from R. Good. 1953. *The Geography of Flowering Plants*, 2d ed. London: Longmans, Green and Company, Inc.

ature or higher rainfall coupled with higher temperature might sustain a grassland (Figure 14-3).

Because a given biome is recognized by the type of vegetation and all but a few exceptional species root in the soil, there would seem to be substantial reason to infer that soil type might also be a major regulatory factor in biome distribution. A comparison of the soil map of the United States (Figure 5-3) and of the world (Figure 5-4) with the world distribution of biomes (Figure 14-1) lends further circumstantial evidence to such a conclusion.

At this juncture, it would be advantageous to put these questions of causality to laboratory investigations, but obviously doing so is out of the question. Can you imagine the size of a growth chamber or modern phytotron that would be required to handle such an investigation? Thus the ecologist is beset again with a natural phenomenon about which the best that can be done is to make inferences from abundant but noncontrolled (as well as noncontrollable) observations.

Cause and effect in biome distribution are confounded because of the complexity of interaction among the components—vegetation, climate, and soil. Soil and vegetation are parts of the same ecosystem. They develop in parallel, influencing each other and influencing and being influenced by climate. In the earlier discussion on soils and fire (Chapter 5), the influence on weathering, erosion, and mineral loss caused by the absence of vegetation and, conversely, the influence of soil on vegetation were noted. Additional examples were also described in Part IV.

Influence in another direction occurs between soil and climate, and then on vegetation, in that high evaporation rates, with or without high precipitation, tend to keep nutrients near the surface. This situation is evident in the rich soils of the prairies, which result in the high rates of production characteristic of the central grasslands of North America, and the rather rapid decomposition made possible by suitable climate facilitates release of nutrients to sustain these high yields. In contrast, under conditions of high rainfall with reduced evaporation, particularly where vegetation is either sparse or shallowly rooted, nutrients may be leached from the soil and make their way out of the ecosystem through circulating groundwater. This is the situation in the highly mineral deficient podzol soils of the northern part of the boreal forest.

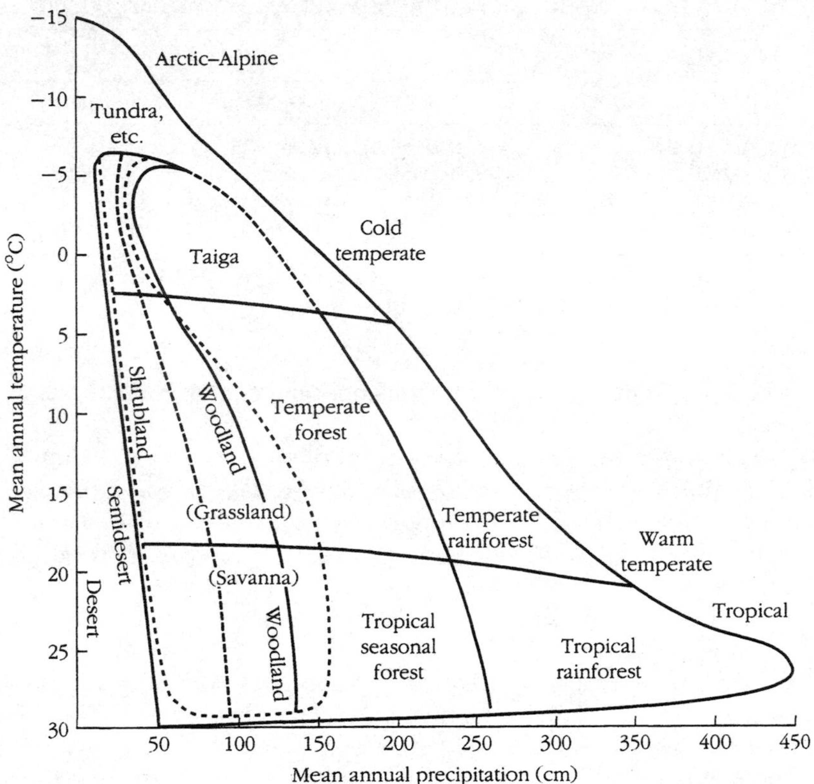

Figure 14-3 Distribution of biome types in relation to precipitation and temperature. The dot-and-dash line encloses a wide range of environments in which grassland or woody biome types may prevail. (Redrawn by permission from R. H. Whittaker. © 1975. *Communities and Ecosystems*, 2d ed. Macmillan Publishing Co., Inc. p. 167.)

Although these well-recognized soil-climate-vegetation interactions do obviate precise causal analysis, the prevailing opinion is that the distribution of biomes is primarily controlled by climate. Excellent correlations can be derived from our present knowledge of distribution of biome types and patterns of temperature, precipitation, and evaporation.

Fossils and biome distribution. There is also evidence in support of climatic influence on biome distribution recorded in the growth of tree rings and fossilized plant macrofossils and pollen. For general accounts of determining biome distribution by their fossil record, see Davis (1969), Fritts (1972), and LaMarche (1974). For vegetation histories of specific areas, the following are among many that might be cited: Cumberland Plateau of Tennessee (Delcourt 1979); Great Basin of the western

United States (Wells 1983); Labrador (Lamb 1985); northern Florida (Watts and Stuiver 1980); White Mountains of New Hampshire (Spear 1989). For a review of long-term changes in land-based ecosystems, see Behrensmeyer et al. (1992). One specific example is in order.

According to a major study by Watts (1979) on the vegetation of central Appalachia and the New Jersey coastal plain at the time of the maximum extent of the Wisconsin glacier (circa 15,000 years B.P.), grass-dominated tundra occurred in central Pennsylvania, sedge-dominated tundra covered the higher mountains of central Appalachia, while southern Pennsylvania, New Jersey, and northern Virginia were dominated by forest tundra, consisting of spruce (*Picea*), dwarf birch (*Betula glandulosa*), and tall herbs. Forests of spruce and jack pine (*Pinus banksiana*) covered North and South Carolina and northwest Georgia. Broad-leaved trees were not the predominant vegetation anywhere on the coastal plain or in the Florida peninsula. These Pleistocene biome distributions are in striking contrast with the situation today (Figure 14-1), wherein tundra is confined to arctic (and alpine) regions, and broad-leaved deciduous forest is the prevailing biome in the eastern half of the United States.

The Biomes of North America

To characterize adequately each of the major biomes would be beyond the scope and purpose of this book. Not to treat them at all would leave a major ecological concept like a skeleton, having no real character without its flesh. As a compromise between these extremes, the vegetation, fauna, soil, and climate of each of the major biomes in North America will be briefly characterized and illustrated. In addition, two significant biomes that do not occur in North America will be briefly characterized—tropical rainforest and tropical savanna.

Although the biomes of North America are singled out, comparable situations occur in other parts of the world that happen to have the same climate (Figure 14-1). Interestingly, however, the vegetation is usually only distantly if at all related taxonomically but shows similar physiognomy or life form. The reader is referred to the extensive treatment of this topic in advanced texts, such as those by Barbour and Billings (1988), Eyre (1963), Oosting (1950), Shelford (1963), and Walter (1973).

Tundra. *Tundra,* which means "marshy plain," lies largely north of latitude 60° North and constitutes about 20 percent of North America, including about 2.5 million km^2 in Canada, 2.0 million km^2 in Greenland, and 0.3 million km^2 in Alaska (Bliss 1988). Although there are considerable variations in climate, ice cover, soils, size of the flora, and composition of plant communities, arctic tundra can be characterized by the absence of trees, the predominance of dwarfed plants (5–20 cm high), and an upper ground surface that is spongy and uneven, or hummocky, as a result of freezing and thawing of this poorly drained land (Figure 14-4). Spodosols, inceptisols, and histosols are the characteristic soils of arctic tundra.

Also characteristic of the tundra is the presence of a permanently frozen soil (permafrost) at a depth of a few centimeters to several meters. The permafrost may

Figure 14-4 Tundra biome: patterned ground near Point Barrow, Alaska. The polygons, 5 to 8 m across, result from winter freezing; the cracks are filled with ice, which thaws at the surface in summer. Note where a tracked vehicle, a "weasel," has passed, lower left; the tracks of a "weasel" are nearly 2 m at the outer edges and hence make an excellent scale in the photograph. (Photo courtesy of W. C. Steere.)

be relatively dry in the case of coarse gravels or largely filled with ice in less porous soils. According to Bliss (1988), thickness of the permafrost is about 400 m at Barrow, Alaska, 500–600 m in the Mackenzie River Delta region, and 400–650 m+ throughout the High Arctic. The permafrost line is the ultimate limit of plant root growth, but the immediate control is the depth to which soil is thawed in summer.

Although there is variation from place to place within the biome, temperature, precipitation, and evaporation are characteristically low, the warmest months averaging below 10°C and the wettest with about 25 mm of precipitation (Figure 4-15e).

The vegetation consists of relatively few species: a few grasses but mostly sedges are characteristic of the numerous marshes and poorly drained areas; large areas consist of ericaceous heath plants (bilberries and dwarf huckleberries), low flowering herbs, and lichens. Perhaps the most characteristic arctic tundra plant is the lichen known as "reindeer moss" (*Cladonia*), a symbiotic mutualism of a fungus and an alga. The principal herbivorous mammals are caribou (reindeer in Eurasia), musk ox, arctic hare, voles, and lemmings. Carnivores include the arctic fox and wolves. Characteristic birds are longspurs, plovers, snow bunting, snowy owls, and horned larks. Reptiles and amphibians are few or absent, and insects are rare with the exception of blackflies and mosquitoes during the summer when they are extremely abundant.

Figure 14-5 Alpine tundra in Yellowstone National Park. (Photo by author.)

Alpine tundra is quite similar to some arctic tundra but differs in the absence of permafrost and in the presence of better drainage and generally a longer growing season. Mosses and lichens are less prominent, flowering plants more so (Figure 14-5). Also, whereas the transition between alpine tundra and boreal forest is charactreistically abrupt, in the Arctic it is gradual and of wide dimension, up to 160 km or so. Lichens tend to be less abundant in alpine vegetation; sedges, dwarf willows, and grasses are more common. Elk, deer, and bighorn sheep use alpine tundra for grazing in the summer. Mountain goats, pikas, and marmots are also common herbivores. Birds and insects tend to be more common and diverse than in arctic tundra.

For a succinct review of North American arctic environments, see Bliss (1988); for a discussion of alpine vegetation, see Billings (1988); for broad comparisons of arctic and alpine environments, see Ives and Barry (1974); and for discussion of tropical alpine plants, see Smith and Young (1987).

Boreal coniferous forest. This moist-cool, transcontinental *boreal coniferous forest*, or *taiga*, biome (Figure 14-6), sometimes referred to in literary parlance as "the great north woods," lies largely between 45 and 57 degrees north latitude. It also extends south at higher elevations in the eastern United States. The climate is cool to cold with more precipitation than in the tundra and occurring mostly in the summer (Figure 4-15a).

Figure 14-6 Boreal forest biome: western red cedar and Alaska cedar on Skowl Arm, near Old Kasaan, Prince of Wales Island, South Tongass National Forest, Alaska. (Courtesy of U.S. Forest Service.)

The predominant vegetation is of the needle-leaf, evergreen variety, notably: white spruce (*Picea glauca*) and balsam fir (*Abies balsamea*) east of the Rocky Mountains; red pine (*Pinus resinosa*), white pine (*Pinus strobus*), and hemlock (*Tsuga canadensis*) in the Great Lakes region; black spruce (*Picea mariana*) west of Hudson Bay; and white spruce–black spruce in Yukon and Alaska. The coniferous montane forests of the Rocky Mountains, the Cascades, and Sierra Nevadas are not considered part of the boreal coniferous forest. Red spruce (*Picea rubens*) and balsam fir extend south to northern Georgia, balsam fir being replaced by Fraser fir (*Abies fraseri*) in the Appalachians. In the east, tamarack (*Larix laricina*) a deciduous conifer, and other conifers, such as black spruce, are common in moister situations and jack pine (*Pinus banksiana*), which also extends to Alaska, in drier and fire-burned areas. Dominance of jack pine in the latter situation results from the nature of the cone, which remains unopened until subjected to burning. Thus after a fire that may wipe out mature trees, as well as seedlings and seeds of other conifers, the seeds newly released from fire-scorched jack pine cones can establish a new community. Quaking aspen (*Populus tremuloides*), balsam poplar (*Populus balsamifera*), and paper birch (*Betula papyrifera*)—all nonconiferous—are also characteristic of burned areas in the eastern portion of this biome. Generally they occur on more moist soils, such as along streams and in wet valleys, although aspen also occurs on dry sands. These nonconiferous trees return after fire by sprouting from their stumps and roots.

The understory is relatively limited as a result of the continual low light penetration. However, among common understory associates are orchids and ericaceous shrubs such as blueberry and cranberry, which also occur in bogs.

Mammals include moose, bear, deer, wolverine, marten, lynx, sable, wolf, snowshoe hare, voles, chipmunks, shrews, and bats. A considerable variety of birds, such as nuthatches, juncos, and warblers, inhabit coniferous forests. Amphibians are relatively common in the southern parts of the taiga; reptiles are relatively rare. A wide variety of insects is found, among them tree pests, such as bark beetles, and defoliating insects, such as sawflies and budworms.

Boreal forest soils are spodosols, characterized by a deep litter layer and slow decomposition. They are generally acidic and mineral deficient, the result of the movement of a large amount of water through the soil. In the absence of a significant counter-upward movement of evaporation, soluble essential nutrients such as calcium, nitrogen, and potassium are sometimes leached beyond the reach of roots. As a result, no alkaline cations are left to counter the organic acids of the accumulating litter.

For succinct, recent discussions of this biome, see Elliott-Fisk (1988) and Bonan and Shugart (1989).

Temperate deciduous forest biome. As its name indicates, the *temperate deciduous forest* region is characterized by a moderate climate (Figure 4-15b) and deciduous trees (Figure 14-7). It occupies most of the eastern half of the United States and has been extensively affected by human activity.

Among the predominant genera in upland areas are maples (*Acer*) and beech (*Fagus*) on moist soils, and oak (*Quercus*) and hickory (*Carya*) on drier soils. Basswood (*Tilia*) is also common, as was chestnut (*Castanea*) prior to the 1920s. Along stream bottoms, cottonwood (*Populus*), sycamore (*Platanus*), elm (*Ulmus*), and willow (*Salix*) are common. In some locations coniferous vegetation may be quite predominant, and among such elements we may find eastern white pine (*Pinus strobus*), eastern hemlock (*Tsuga canadensis*), and eastern red cedar (*Juniperus virginianus*).

The understory of shrubs and herbs in a mature deciduous forest is typically well developed and richly diversified. Here a considerable portion of the photosynthesis and flowering is attuned to the short days of the spring season, prior to the leafing out of and consequent shading by the tree canopy.

The largest herbivore and carnivore are, respectively, white-tailed deer and black bear. The black bear is actually an omnivore. Formerly the mountain lion and wolf were the largest carnivores, but their populations were decimated by human activity. Other mammals include red and gray fox, bobcat, weasel, opossum, raccoon, and small herbivores, such as voles, mice, squirrels, and chipmunks. A

Figure 14-7 Deciduous forest biome: 120-year-old stand of sugar maple, beech, and hemlock in Pennsylvania. (Courtesy of U.S. Forest Service.)

wide variety of birds inhabits deciduous forests, including red-eyed vireo, wood thrush, oven bird, ruffed grouse, tufted titmouse, wild turkey, and various woodpeckers. Amphibians, reptiles, and insects are both abundant and diverse.

Inceptisols (brown forest soils) characterize deciduous forests. The surface litter layer is thin due to rapid decomposition, earthworms being active in this process as well as in mixing the soil. The upper soil layer is mildly acidic because of the rich organic litter added yearly and because it tends to be leached of calcium. Acidity decreases with depth, however. Brown forest soils are more fertile than those of the taiga because of their higher levels of nitrates and other soil nutrients that are held by clays and organic particles, making these nutrients accessible to plant roots. Climate in this region, although varying considerably from north to south and east to west, is moderate with a definite winter period, characterized by snow and frozen soil and lakes in the northern portions, rain and coolness in the southern portions. Precipitation is evenly distributed throughout the year.

For further discussion of the deciduous forest biome, see Ovington (1983) and Greller (1988).

Grassland. In central North America are the *grasslands*, the tallgrass prairie toward the east and the shortgrass prairie, or plains, westward. In Europe and Asia some grasslands are called steppes; in South America, grasslands are called pampas, or llanos. Prior to its conversion to agriculture and urban development, the tallgrass prairie was dominated by species of bluestem (*Andropogon*) forming dense covers 1 to 2 m tall. Westward, buffalo grass (*Buchloe dactyloides*) and other grasses a few inches high dominated the landscape. Flowering herbs, including many kinds of composites and legumes, are common but much less important than the grasses. Trees are largely limited to stream valleys, where cottonwood is dominant (Figure 14-8), and to low mountains, such as the Black Hills of South Dakota.

The transition westward from tall-to shortgrass is correlated with increasing aridity, a result of reduced rainfall (Figures 4-13) coupled with increased evaporation. The tallgrass prairie is characterized by irregular rainfall, which is reduced toward the end of summer. In the eastern portion the pattern of temperature and rainfall (Figure 4-15d) is comparable to that of the deciduous forest (Figure 4-15b), but precipitation drops progressively westward.

Because the precipitation-evaporation ratio is below 1 in the grassland, leaching is considerably less than in the eastern soils. The mollisols (chernozem coils or black earths) of the tallgrass prairie are among the richest in nutrients and consequently the most fertile in the world. Organic matter accumulates in the upper portion of the soil, rendering it dark. This upper portion remains neutral to slightly alkaline because of the continued replenishment of cations like calcium and potassium through the upward movement associated with evaporation. Westward, the conditions in the direction of shorter grass are not so much due to a limitation of nutrients, whose rather complete cycling parallels that of the chernozem soils, as they are due to increasing aridity.

Figure 14-8 Grassland biome: shortgrass on the sand hills of north central Nebraska, near Valentine; note the presence of trees only in the valley. (Photo by author.)

Grassland mammals are dominated by smaller burrowing herbivores (prairie dogs, jack rabbits, ground squirrels, and gophers) and larger running herbivores, such as bison, pronghorn antelope, and elk, along with such carnivores as badger, coyote, ferret, red and gray wolves, and cougar. Before the westward movement in the nineteenth century, it is estimated that 75 million bison inhabited North America's grasslands; by 1888 only 88 bison survived within the western United States. Grassland birds, provided with only one vegetation layer, are more limited in kind but include meadowlark, prairie chicken, longspur, horned lark, dickcissel, rodent-eating hawks, and grasshopper sparrow, named for its song, not for feeding on grasshoppers, the most abundant arthropod.

For further discussion of the grassland biome, see Coupland (1979) and Sims (1988). For a detailed analysis of grasslands on a more global basis, see Yang (1990).

Desert. In its most typical form, the *desert* consists of bush-covered land in which plants are dispersed with much bare ground between them. For example, creosote bush (*Larrea divaricata*), the dominant plant of the warm deserts (Mojave, Sonoran, Chihuahuan) of the southwest United States (southeast California, southern Nevada, southern and western Arizona, and southern New Mexico and Texas), can be regularly spaced at intervals of 5 to 10 m (Figure 14-9). In the cold Great Basin Desert of the Northwest between the Cascade-Sierra chain and the Rocky Mountains (see Figure 4-13), and including Nevada, western Utah, and the bordering parts of California, Oregon, Idaho, and Wyoming, the predominant sagebrush (*Artemisia tridentata*) forms somewhat denser semidesert scrub. The vegetation of the Great Basin Desert and the Mojave is less complex than that of the Sonoran, which has well developed communities. The greater number of species and the warmer winter temperatures account for much of this difference.

The cold Great Basin Desert constitutes about 32 percent of North America's total desert area of 1,277,000 km^2; the warm deserts constitute the remaining 68 percent (Sonoran 21.5, Mojave 11.0, and Chihuahuan 35.5) (MacMahon 1988).

Low erratic precipitation (Figure 4-15f) coupled with soil and air temperatures that are extremely high by day and drop abruptly by night (Figure 4-7), low humidity, and high insolation are the major desiccating environmental factors to which desert vegetation has adapted. Desert plants, generally *xerophytes* (from the Greek *xeros*, dry, and *phyteis*, to grow), are adapted to drought through reduced leaf size and the dropping of leaves in dry conditions, both processes reducing water loss via evapotranspiration. To some extent, reduced leaf size and leaf dropping may increase water loss from the soil due to a loss of shade. Other adaptations occur in the root systems. Nearly 90 percent of the root system of the Saguaro cactus, for example, is in the top meter of the soil, an optimum situation in which to take advantage of any rainfall. Moreover, the root hairs on many desert plants are ephemeral, drying back under drought conditions and thereby reducing potential water loss by osmosis. Yet other species are short-lived annuals that complete their life cycle during the short moist period. Finally, particularly in the warm southern

Figure 14-9 Desert biome: Saguaro cactus and creosote bush in the Sonoran Desert in Saguaro National Monument, near Tucson, Arizona. (Courtesy of U.S. Forest Service.)

deserts, water-storing succulents, such as the giant cacti of the Sonoran desert, are adapted by their protoplasmic colloids, which permit the accumulation of substantial water reserves, as well as by a reduced leaf surface, which obviates water loss via evapotranspiration.

Animals similarly face and have adapted to low water levels through drought avoidance (e.g., impervious integuments), water conservation (e.g., excreting uric acid instead of urea), and reduction in direct water intake (e.g., using and conserving water produced in metabolism) along with behavioral adaptations (Figure 4-8). Most mammals tend to be nocturnal, many burrowing during the heat of the day, or are crepuscular—that is, active only at sunrise and sunset. Mammals include collared peccary, gray fox, kangaroo rat, pocket mouse, and, in cooler areas, jack rabbit and pronghorn antelope. Birds are less common and diverse but include roadrunner, cactus wren, thrashers, and doves; in cooler areas there are sage grouse, sage sparrow, and sage thrush. There is a rich lizard and snake fauna. Among insects the desert locust is the most notorious, being given to large outbreaks or "plagues" such as those recounted in biblical accounts of Egypt during the period of Jewish captivity and that occurred in the Great Salt Lake basin as recorded during the early Mormon settlements.

Camels of the great hot deserts of the world, such as the Negev in Israel, are uniquely adapted physiologically and warrant brief mention. Contrary to popular belief, camels have no water storage places in their bodies (that includes their well known hump(s)) and can lose up to 27 percent of their body water without damage (Evenari et al. 1971). Thanks to their enormous drinking capacity, they can restore this water in a few minutes. They can also tolerate as much as a 6°C daily fluctuation in their body temperature without increasing their rate of evaporation; they also produce very concentrated urine and dry feces, thus conserving considerable amounts of water.

Because productivity is low, the litter layer is comparatively limited; as a result, the organic content of these aridisols is very low. Evaporation tends to concentrate soluble materials near the soil surface, which is thus highly alkaline. Soils are generally relatively rich in nutrients, except for nitrogen. In certain situations, the soils may be strongly saline, producing salt deserts with a whitish or gray crust on the soil.

For discussion of the cold, intermountain deserts, including the Great Basin Desert, see West (1988); for further discussion of warm deserts, see MacMahon (1988) and Evenari et al. (1985); for discussion of the Sahara Desert, see Cloudley-Thompson (1984); and for discussion of deserts in general, see Allan and Warren (1993).

Lesser Biomes

Eastern pine–oak biome. In addition to the major biomes, there are several others whose land area in the United States is considerably less but that deserve brief mention. One of them, the eastern pine-oak, which is often considered as part of the deciduous forest biome, occupies the coastal plain from Long Island and New Jersey to the border of Texas. It is dominated in upland, well-drained sites by extensive pine forests—pitch pine (*Pinus rigida*) to the north and loblolly (*P. taeda*), slash (*P. caribaea*), and longleaf (*P. australis*) farther south (Figure 14-10). The soil is sandy and thus low both in nutrients and in water-retention properties. Considerable evidence indicates that these coastal pines depend on recurrent fires for their maintenance; under conditions in which fire has been curtailed, hardwood forest invariably develops. As in the case of jack pine noted earlier, fire hastens the opening of cones of some species, releasing seeds that are able to germinate on soil laid bare by the burning off of organic matter.

For a succinct discussion of this biome and the remainder of the southeastern coastal plain, see Christensen (1988).

Figure 14-10 Eastern pine-oak biome: the pine barrens with a young stand of pitch pine with an understory of oak and sassafras, Ocean County, New Jersey. (Courtesy of U.S. Forest Service.)

Subtropical biome. This biome is at the very southern tip of Florida and westward on the Florida Keys; it is subject to a uniformly warm annual temperature and abundant precipitation (Figure 4-15c). Driving through the Everglades, for example, we first observe the freshwater marshes dominated by tall sawgrass (*Cladium jamaicensis*), but as salinity increases coastward in the tidal and subtidal areas, there is a gradual transition to virtually pure dense swamps of red mangrove (*Rhizopora mangle*) (Figure 14-11). Scattered islands of trees, or hammocks, occur in the freshwater marshes, and they constitute, in miniature, tropical forests such as those encountered farther south. These hammocks are characterized by both evergreen and deciduous trees, notably mahogany, gumbo limbo, bays, and palms laden with epiphytes (largely members of the pineapple and orchid families), ferns, and vines, or lianas, including the well-known strangler fig (*Ficus aureus*).

Broad-sclerophyll biome. To the west, and best developed on the coastal ranges of southern California, is a region characterized by species with thick, hard evergreen leaves, the *broad-sclerophyll biome* (Figure 14-12). *Sclerophyll* tissue is a drought adaptation, strengthening tissues otherwise held turgid by water. North-facing and moister slopes are characteristically dominated by small trees, including several species of evergreen oaks. South-facing and drier slopes are dominated by low

Figure 14-11 Mangrove, Everglades National Park. The prop roots provide attachment surfaces for sessile organisms. In coastal areas the prop roots reduce tidal currents and promote sedimentation. (Photo by author.)

dense thickets of evergreen shrubs, such as chamise (*Adenostoma*) and manzanita (*Arctostaphylos*). Such formations are referred to as *chaparral*. The climate is temperate to subtropical, rainfall occurring largely during the winter months when it is too cold for plants to grow. The growing season is very short because as the temperature rises to a favorable level for plant growth, the rainfall decreases, causing moisture stress in the plants. The long dry summer, together with the nature of the vegetation, makes the region subject to frequent fire, an event that generally favors the extension of chaparral at the expense of the sclerophyll forest.

For further discussion of chaparral, see Keeley and Keeley (1988).

Figure 14-12 Broad sclerophyll biome: Chaparral dominated by red shank, ceanothus, and sage, with Mt. Palomar in background. Cleveland National Forest, California. (Courtesy of U.S. Forest Service.)

Tropical Biomes

Inasmuch as most readers of this book will be students in North American colleges and universities, this text provides a succinct introduction to the biomes in this part of the world. Although a discussion of all biomes in the world is beyond the scope of this book, much of what has been described for North American biomes will apply to comparable biomes elsewhere, differences being largely in the makeup of the plant and animal communities. However, because of their extensive size and ecological significance, there are two other biomes that warrant brief discussion—the tropical rainforest and the tropical savanna.

Tropical rainforest. *Tropical rainforests* occur 10° or more north and south of the equator in Central and South America, central and western Africa, Southeast Asia, the East Indies, and northeast Australia, in addition to the oceanic islands generally within these same latitudes. About 40 percent of the Earth's tropical and subtropical landmass is dominated by open or closed forest: of this, 42 percent is dry forest, 33 percent is moist forest, and only 25 percent is wet and rain forest (Murphy and Lugo 1986).

Tropical rainforests are among the most ancient of ecosystems. Fossil evidence in Malaysia and elsewhere suggests that tropical rainforests have existed continuously for more than 60 million years (Richards 1973). They hold considerable ecological significance because of their influence on climate, the global balances of carbon and atmospheric pollutants, and their highly diverse pools of species with considerable potential for new sources of food, fiber, and medicinal and industrial products (Jordan 1985).

In tropical rainforests, the annual rainfall, which exceeds 200 to 225 cm, is generally evenly distributed throughout the year. Temperature and humidity are also relatively high throughout the year.

The flora is highly diversified: 2.5 square kilometers may contain 300 different species of trees, a diversity unparalleled in any other biome. Tropical rainforests are also typically well stratified into a continuous canopy of trees 25 to 35 m tall, with an interrupted emergent layer of very tall trees with buttressed bases (Figure 14-13). At the edges along streams, or where the canopy is broken, there is a well developed understory of trees and shrubs. Epiphytic orchids and bromedliads, as well as vines (lianas), are characteristic as are ferns and palms. Most plants are evergreen with large, dark green, leathery leaves.

Decomposition is rapid and the soils, oxisols throughout much of the biome, are subject to heavy leaching and so tend to be acidic and nutrient poor (Lucas et al. 1993). The potential nutrient loss is offset by extremely rapid uptake by the plants (i.e., high productivity), with most of the nutrients being tied up in the biomass. It is the nature of the soil, both its potential for high leaching as well as its chemical composition, that promotes a rocklike quality (laterites) when exposed to air that largely has prevented western-style agriculture from being applied to the tropical forests.

For further discussion of tropical rainforests, see Collins (1990), Golley (1983), Jacobs and Oldeham (1987), Leith and Werger (1989), Webb and Kikkawa (1990), and Whitmore (1990), among others.

Tropical savanna. *Tropical savannas* are grasslands with scattered, drought-resistant trees that generally do not exceed about 10 m in height (Figure 14-14). Savannas constitute extensive areas in eastern Africa, areas that support the richest diversity of grazing mammals in the world, and also occur in Australia, South America, and Asia.

The climate is generally characterized by a rainy season (May through October) and dry season (November through April). In the llanos of Venezuela, for example, nearly 90 percent of the annual rainfall of 130 cm falls during the rainy season.

The soils are nutrient poor due to heavy leaching. Quite widely distributed in savannas are laterite soils, which, when dried (purposefully or accidentally), harden to rocklike consistency, thereby precluding their use in western-style agriculture.

Although climate and soil are significant regulating factors in this biome, the controlling factor appears to be fire, to which the grasses and relatively few trees are well adapted. As a result of this latter factor, species diversity is quite low in comparison to adjacent tropical forests; in some situations, a single species of grass and tree may be dominant over large areas. There is considerable discussion, and even

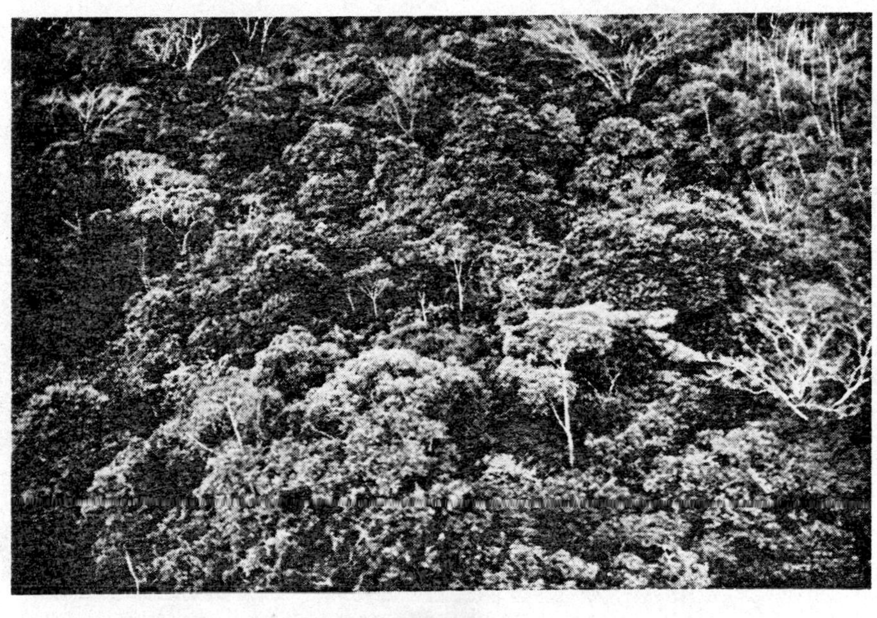

A.

B.

Figure 14-13 Tropical rain forest in Panama showing tiered structure of vegetation. (a) Aerial view near Canal Zone. (b) Clearing for pipeline, Canal Zone. (Photos courtesy of Gerrit Davidse, Missouri Botanical Garden.)

Figure 14-14 Tropical savanna in eastern Venezuela between Anaco and Maturin; the dominant grass is *Trachypogon.* (Photo by author.)

some experimentation, on managing the savannas for food production—the llanos of Venezuela for cattle and capybara, a large rodent, and those of East Africa for the natural grazers (antelope, wildebeest).

For further discussion of tropical savannas, see Boulier and Hadley (1970), Boulier (1983), and Tothill and Mott (1984).

GENERAL CONSIDERATIONS

In a more thorough discussion of each of the biomes of North America, or of the tropical forest and savanna, it is possible to recognize finer subdivisions of each—for example, the wetter hemlock-fir and drier redwood sequoia regions of the west. Such finer discrimination would be in order in a more extended monograph. As it is, our discussion has been limited to a brief description of vegetation, fauna, soil, and climate and has been subject to considerable generalization. It must be recognized that these "typical" situations are abstractions, yet not without basis in fact. What has been described is the predominant vegetational life form of a given area, and the biome-type map (Figure 14-1) suggests sharp, discrete separations of the various vegetation zones. Anyone who has traveled even moderately has recognized that this is not so and that the transition zones are sometimes of considerable magnitude. Occasionally the transitions may be quite abrupt, as between the deciduous forest and coastal pine; here the boundary is largely related to a sharp discontinuity in the type of soil.

It is true that within any given biome we may find areas that do not seem to belong. In certain parts of northern Indiana and Ohio, for example, we find prairie vegetation and mollisols (chernozem soil), although this area lies within deciduous forest biome and should be characterized by a different soil type. In this particular instance, the presence of this prairie peninsula, as it is called, is an historical accident, a remnant of a different climatic regime. In other cases the "misfit" is the result of local topographic and/or microclimatological influences (Figure 14-15). In spite of their internal variability and lack of discrete boundaries, together with the hazards of broad generalization, the concept of the biome as a major ecological system has its worth.

Variation in the Temperate Deciduous Forest Biome

Although the deciduous forest biome is dominated by deciduous trees—a tautology, of course—no given species is characteristic of the entire biome. Sugar maple, for example, is widely distributed in the eastern half of North America, but it is dominant only in the northern portion, sharing this status with basswood (*Tilia*) to the west and beech (*Fagus*) to the east. Although many of the deciduous tree species of the biome are widely distributed, their area of dominance is restricted and typically shared with one or more other species. This situation is well illustrated in the forest regions of eastern North America, as described by E. Lucy Braun (Figure 14-16). The map suggests sharp discontinuities between forest types, but how discrete are these boundaries and how uniform are the communities within each of these forest regions? What regulative factors are involved in these patterns? It is obvious that we need to turn from the more generalized vegetative and physiognomic description of structure to a consideration of species composition and the peculiar factors that dispose a particular organization of types to occur in particular regions.

Even though each major forest ecosystem is typified by its characteristic assemblage of species, variation within the region may be considerable. An excellent illustration of the local regulation of community composition was shown in Figure 14-15. Although the portion of southern Ontario bordering the Great Lakes is dominated by beech and maple, relatively slight variations in temperature and moisture and the interaction of these two gradients on the nature of the soil result in quite different species assemblages. Similarly, Grace Bush (1982) has shown that the spatial and temporal patterns of water availability, which are controlled principally by the substrate, are the major factors determining the composition of forest communities in the State of Maryland. Even casual observations, either by the inveterate woods tramper or by a sedentary auto passenger, reveal the variability in vegetation observed in passing over even moderately rolling country interspersed with an occasional stream valley. Local topographic and microclimatological factors do seem to be involved in determining localized assemblages.

Cushetunk Mountain. In an excellent study on this phenomenon John Cantlon (1953) showed a decided vegetational difference on the north and south

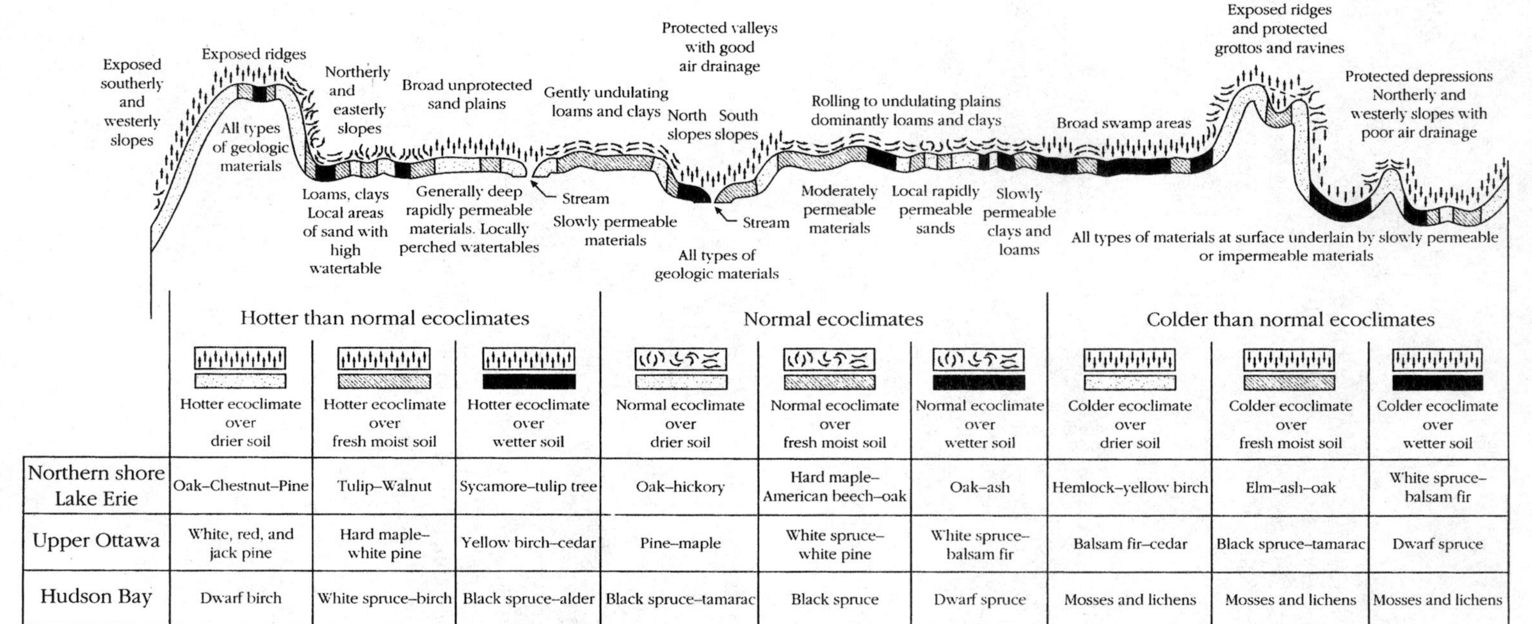

	Hotter than normal ecoclimates			Normal ecoclimates			Colder than normal ecoclimates		
	Hotter ecoclimate over drier soil	Hotter ecoclimate over fresh moist soil	Hotter ecoclimate over wetter soil	Normal ecoclimate over drier soil	Normal ecoclimate over fresh moist soil	Normal ecoclimate over wetter soil	Colder ecoclimate over drier soil	Colder ecoclimate over fresh moist soil	Colder ecoclimate over wetter soil
Northern shore Lake Erie	Oak–Chestnut–Pine	Tulip–Walnut	Sycamore–tulip tree	Oak–hickory	Hard maple–American beech–oak	Oak–ash	Hemlock–yellow birch	Elm–ash–oak	White spruce–balsam fir
Upper Ottawa	White, red, and jack pine	Hard maple–white pine	Yellow birch–cedar	Pine–maple	White spruce–white pine	White spruce–balsam fir	Balsam fir–cedar	Black spruce–tamarac	Dwarf spruce
Hudson Bay	Dwarf birch	White spruce–birch	Black spruce–alder	Black spruce–tamarac	Black spruce	Dwarf spruce	Mosses and lichens	Mosses and lichens	Mosses and lichens

Figure 14-15 The interrelation of microclimate, soil, and forest distributionin selected regions of Ontario. (Redrawn by permission from G. S. Hills. 1952. *Ontario: Department of Lands and Forest, Research Report No. 24*:1–41.)

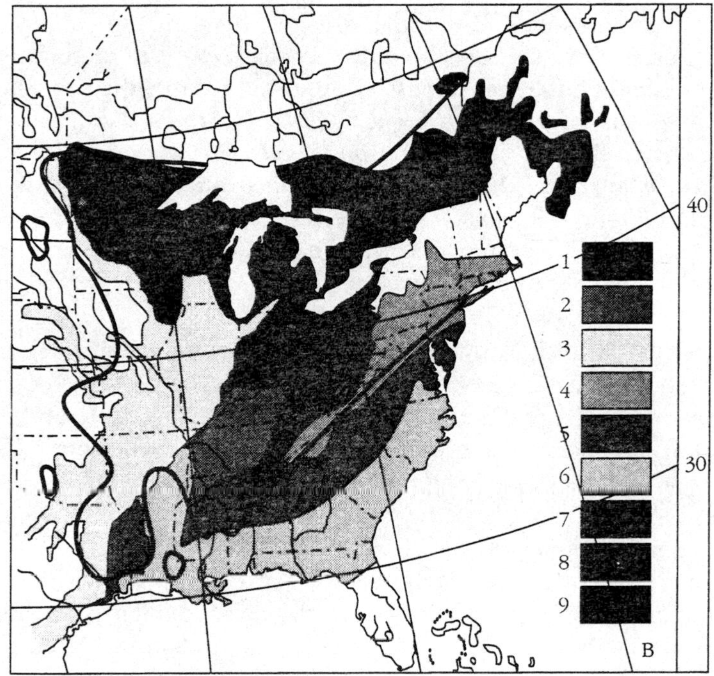

Figure 14-16 Bioclimatic limits of sugar maple (heavy line) showing major forest regions recognized by E. Lucy Braun: (1) mixed mesophytic; (2) western mesophytic; (3) oak-hickory; (4) oak-chestnut; (5) oak-pine; (6) southeastern evergreen; (7) beech-maple; (8) maple-basswood; (9) hemlock–white pine–northern hardwoods. (Redrawn by permission from P. Dansereau. *Biogeography–An Ecological Perspective.* 1957. New York: The Ronald Press Company.)

slopes of Cushetunk Mountain in New Jersey, a difference tied directly to micro-climate. Cantlon found that the south slope had higher air and soil temperatures and a larger vapor pressure deficit, the latter being a measure of water availability and evaporation. The average difference in the mean monthly temperature on the two slopes, for instance, was 2.5°C at 4 cm below the soil surface, 2.3°C at 5 cm above the soil surface, and 2.0°C at 20 cm above the soil surface. These seemingly minor differences contributed, however, to a striking difference in the vegetation of the two slopes. In general, vegetation differences were greatest near the ground with the bryophyte or ground cover showing the most marked difference. The more moist and cool north slope had the greater amount of ground cover and higher occurrence of individuals of given species.

For a somewhat detailed example of the differences in slopes, a consideration of only the tree layer is instructive. First, the south slope had a greater density with 21,987 trees/3500 m^2 versus 15,759 trees/3500 m^2 on the north slope. However, the total basal area of the south slope was less, 6.7 m^2/3500 m^2 versus 8.7 m^2/3500 m^2.

This situation resulted from the north slope's having a greater proportion of larger individuals. Next, the most abundant species, in order of decreasing density, were not the same on the two slopes. On the south slope, the sequence of the first four was dogwood (*Cornus florida*), ash (*Fraxinus americana*), sassafras (*Sassafras albidum*), and tuliptree (*Liriodendron tulipfera*). On the north slope, the sequence was ash, witch hazel (*Hamamelis virginiana*), birch (*Betula lenta*), and maple (*Acer rubrum*). This particular species of maple was fourteenth in abundance on the south slope, the birch was ninth, and witch hazel did not attain sufficient height (1 m) to be included in the tree canopy.

Great Smoky Mountains. This contrast between the vegetation of a north and a south slope implies a sharp discontinuity, based on exposure or orientation. The corresponding discontinuities observed in temperature and moisture may not, therefore, be so surprising. But if we were to conduct a study following a transect from one of these slopes to the other, a continuous gradient of microclimate and vegetation rather than a discontinuity would be observed. In the Great Smoky Mountains of Tennessee and North Carolina mean temperatures decrease with elevation at an average of 1.2°C for each 300 m. Thus high elevations in these mountains average 5.6 to 8.3°C cooler than the base. Similarly, annual precipitation increases from about 125 cm in the lower valleys to about 200 cm at high elevations. These major environmental gradients dictate, to a considerable extent, the distribution and relative abundance of individual species and, in consequence, affect the composition of the plant communities and their associated ecosystem components. In a major study of this problem Robert Whittaker (1956) plotted the distribution and abundance of the vegetation along a moisture gradient. Almost all species showed a bell-shaped curve of population distribution along the gradient, with broad overlap among the different populations (Figure 14-17). Nonetheless, the existence of these gradations does not preclude the recognition of major vegetation types over a broader area of the same region (Figure 14-18) or on the very broad scale of the biome (Figure 14-1).

Biomass and Biomes

The accumulation, productivity, and distribution of boimass in the major biomes are illustrated in Figure 14-19. It can be easily noted that tropical biomes are more productive than temperate ones, and the latter more so than arctic biomes. The greatest accumulation of organic matter is not only in forests, as contrasted with other ecosystems, but is progressively greater toward the equator. Within the boreal forest, for instance, total accumulation increases from 100,000 kg/ha in northern taiga spruce forests to 330,000 kg/ha in southern taiga spruce forests. The more northerly deciduous forests of beech have accumulations around 370,000 kg/ha in comparison with 400,000 kg/ha in more southerly deciduous forests of oak. Comparable trends can also be noted in the rate at which the organic matter accumulates.

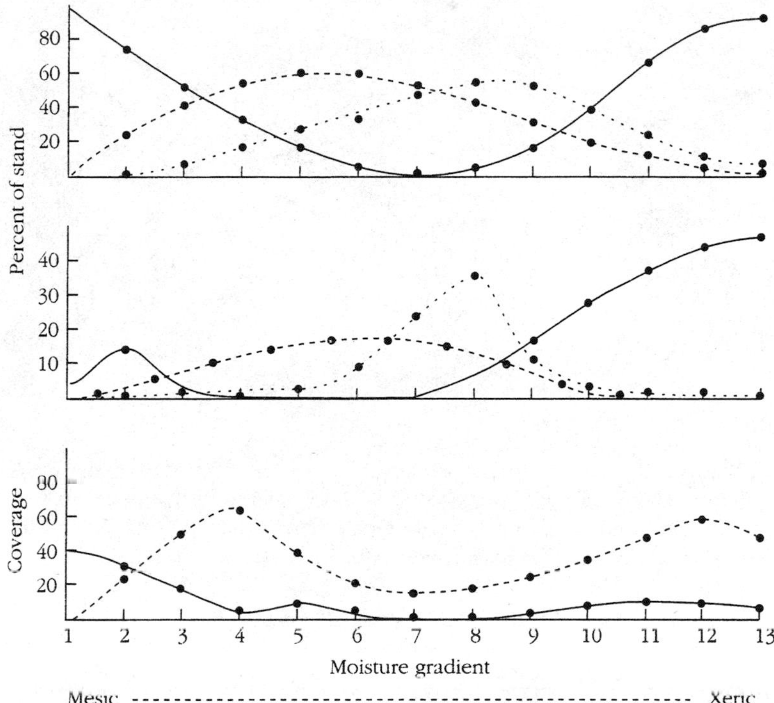

Figure 14-17 Transect of the moisture gradient in the Great Smoky Mountains at 500 and 830 meters. Top curves are for tree classes: (a) mesic; (b) submesic; (c) subxeric; (d) xeric. Middle curves are for tree species: (a) birch (*Betula allegheniensis*); (b) dogwood (*Cornus florida*); (c) oak (*Quercus prinus*); (d) pine (*Pinus virginiana*). Bottom curves are for the understory: (a) herbs; (b) shrubs. (Redrawn by permission from R. H. Whittaker. 1956. *Ecological Monographs* 26:1–80).

Biomes and Mineral Cycling

Inferences about mineral cycling in each of these biomes can be made from considerations of the annual increment in the litter, the amount of humus, or dead organic residues, and the ratio of the latter to fresh, or green, litter (Table 14-2). Although the trend is toward an increase in the annual net increment in litter toward the equator, this seems to be more a reflection of total productivity, inasmuch as the percentage of annual productivity that it represents is about the same for each major forest biomass. However, the retention of this organic matter and its bound minerals is considerably less toward the equator, as can be seen both by the diminution of humus and the ratio of humus to green litter. The accumulation of litter and retention of humus indicate low rates of decomposition—the tundra situation illustrates this condition. In contrast, the absence of humus (deserts) or its low accumulation (tropical forests) indicates high rates of decomposition and rapid mineral cycling (Table 14-3).

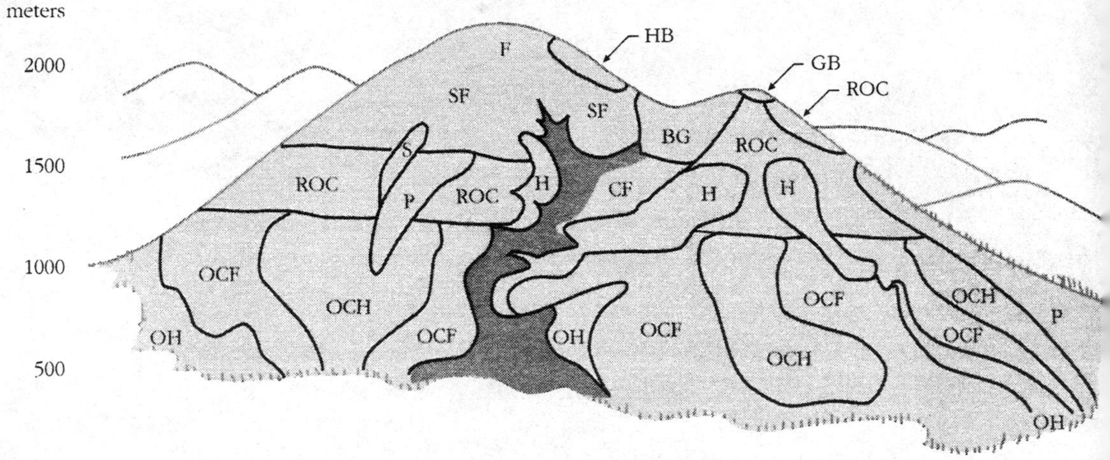

Figure 14-18 Topographic distribution of vegetation types on an idealized west-facing mountain and valley in the Great Smoky Mountains. Vegetation types: BG, beech gap; CF, cove forest; F, Fraser fir forest; GB, grassy bald; H, hemlock forest; HB, heath bald; OCF, chestnut oak–chestnut forest; OCH, chestnut oak–chestnut heath; OH, oak–hickory forest; P, pine forest and pine heath; ROC, red oak–chestnut forest; S, spruce forest; SF, spruce–fir forest; WOC, white oak–chestnut forest. (Redrawn by permission from R. H. Whittaker. 1956. *Ecological Monographs* 26:1–80.)

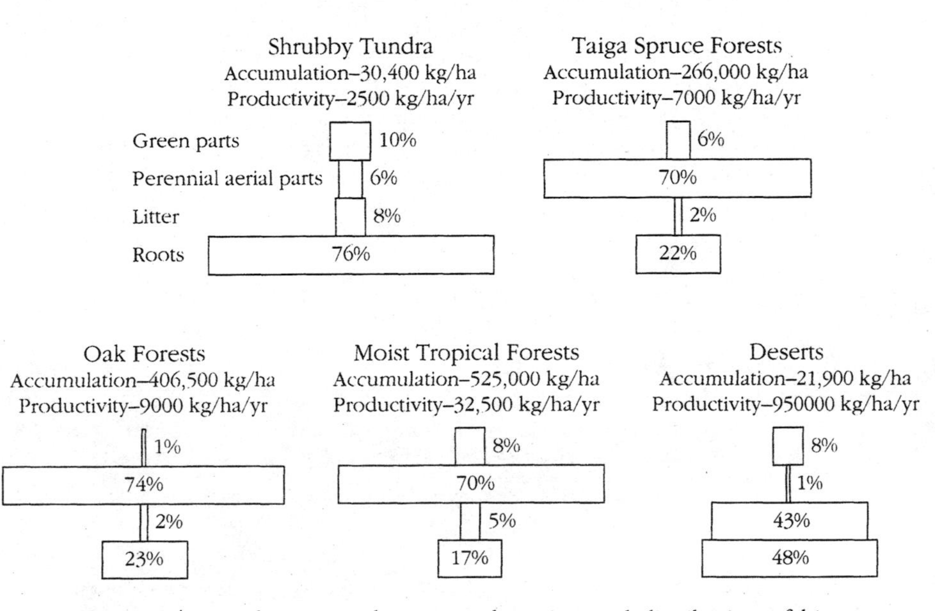

Figure 14-19 The accumulation, productivity, and distribution of biomass in selected major biomes. (Based on data from L. E. Rodin and N. I. Bazilevic. 1964. *Doklady Akademii Nauk SSSR* 157: 215–8.)

TABLE 14-2 THE ACCUMULATION AND ANNUAL CHANGES IN THE LITTER AND HUMUS OF THE MAJOR BIOMES (in kg/ha)

	Shrubby tundra	Boreal spruce forests	Oak forests	Moist tropical forests	Deserts
Litter	2,400	5,000	6,500	25,000	9,400
Annual net increment in litter in litter	100	2,000	2,500	7,500	100
Percent of productivity	4	29	28	23	1
Dead organic residues (humus)	83,500	45,000	15,000	2,000	—
Organic residues: green litter	92:1	15:1	4:1	0.1:1	—

Data from L. E. Rodin and N. I. Bazilevic. 1964. *Doklady Akademii Nauk SSSR* 157:215–8.

TABLE 14-3 TOTAL DETRITUS IN SELECTED BIOME SOILS (in kg/m^2)

Tundra and alpine	21.6	Temperate grassland	19.2
Boreal forest	14.9	Tropical savanna	3.7
Temperate forest	11.8	Desert scrub	5.6
Tropical forest	10.4		

Data from W. H. Schlesinger. 1977. *Annual Review of Ecology and Systematics* 8:51–81.

SUMMARY

VOCABULARY

biome
boreal coniferous forest
chaparral
desert
grassland
sclerophyll
taiga
temperate deciduous forest
tropical rainforest
tropical savanna
tundra

KEY POINTS

- The division line of some biomes tend to parallel lines of latitude; biomes at a given altitude also vary with latitude, a given zone occurring at progressively lower altitudes at progressively more northern latitudes.
- Soil and climate are the major factors in determining the distribution of biomes; climatic influences are supported by paleoecological studies.

- Each of the major biomes is characterized by a more or less unique combination of vegetation and fauna adapted to the predominant soil(s) and climate, ranging from cold and dry to the north (e.g., tundra) to hot and wet toward the equator (e.g., tropical rainforest).
- Within each biome considerable variation exists because of local conditions of soil and climate and the attendant biota; variations in temperature and moisture produce microclimates, resulting in different assemblages of species in given locations.
- The greatest accumulation of biomass is in forests, increasing in amount toward the equator. Retention of organic matter and its bound minerals lessens toward the equator due to high rates of decomposition and rapid mineral cycling.

Aquatic Ecosystems

As noted at the beginning of the section on biomes, the broad categories of aquatic ecosystems—freshwater, marine, estuarine—are distinguished on the basis of differences in their salt content. Composing more than 70 percent of the Earth's surface, aquatic ecosystems are not only the dominant feature of Earth but are also very diverse in species and complexity of interaction among their physical, chemical, and biological components.

As in the preceding treatment of terrestrial ecosystems (Chapter 14), an extended discussion of aquatic environments is beyond the scope of this book. In addition, the literature on these systems is tremendous, and thus only a limited number of seminal references is provided in the following sections.

RESHWATER ECOSYSTEMS

Freshwater ecosystems cover about 2 percent of the Earth's surface, an area of about 2.5 million km^2. Although referred to as freshwater, a number of lakes contain a variety of chemicals referred to collectively as salts. The volume of inland saline waters (containing more than 3 grams of salt per liter) is estimated to be 104×10^3 km^3 in contrast to a total volume of 125×10^3 km^3 for nonsaline waters (Burgis and Morris 1987).

With a volume of 23×10^3 km^3, a maximum depth of 1620 m (the deepest in the world) and a mean depth of 740 m, Lake Baikal in Siberia contains about one-fifth the Earth's freshwater. The surface area of the Caspian Sea of 371×10^3 km^2

makes it the largest single freshwater lake, though it does contain saline water. At 83×10^3 km^2 Lake Superior is the second largest lake on Earth, and the largest in North America. Collectively, the North American Great Lakes, with a surface area of 245×10^3 km^2 and volume of 24.6×10^3 km^3, constitute the largest continuous mass of freshwater in the world (Figure 15-1).

Although the Nile River at 6,698 m is the longest river on Earth, the Amazon, next longest at 6,440 m, carries the largest volume of water, delivering an average of 643×10^6 m^3 of water every hour to the Atlantic Ocean, 60 times the flow of the Nile. Others of the longest rivers in the world, in descending order, are the Chang in China (6,382 m), Ob-Irtysh in the Soviet Union (3,362 m), Huang (or Yellow) in China (4,674 m), and the Congo in Africa (4,669 m). In North America, the longest rivers, in descending order, are the Mackenzie (4,242 m), Missouri (4,078 m), Mississippi (3,780 m), Nelson (2,576 m), Arkansas (2,349 m), Colorado (2,335 m), Columbia (2,001 m), and Snake (1,671 m).

Freshwater ecosystems are characterized as having running water (*lotic ecosystems*) or still water (*lentic ecosystems*). Lentic ecosystems are hardly still, subject as they are to winds, tidal effects in large bodies of water, and currents created by en-

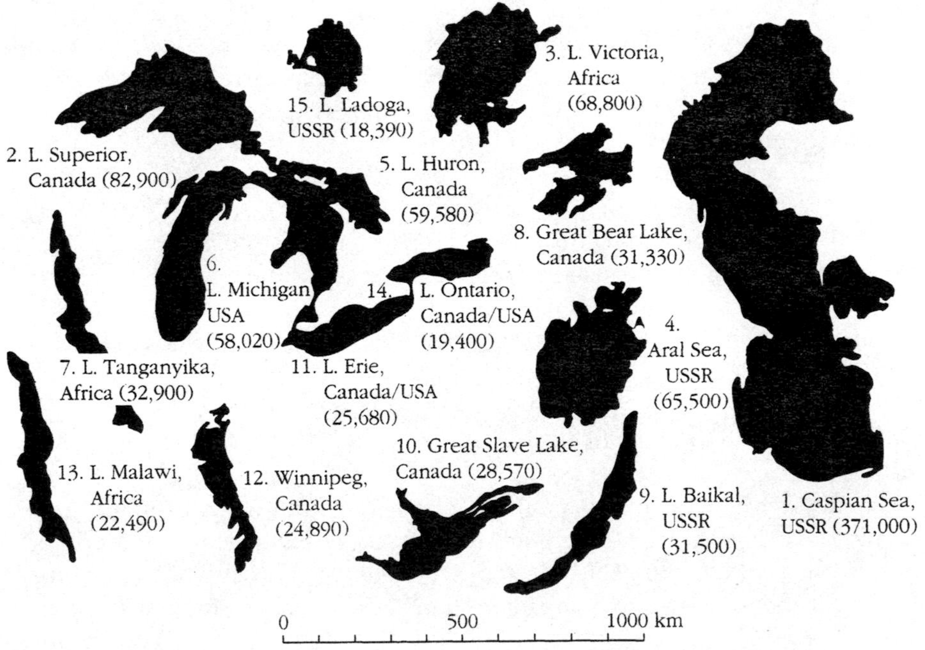

Figure 15-1 The 15 largest lakes on Earth all drawn to the same scale. The numbers indicate the rank in size and the figures in brackets the surface areas in km^2. (Redrawn, with permission, from Burgis and Morris 1987.)

tering streams. Nonetheless, the ecological distinctions between running and still water are considerable, as will be evident in the following discussion.

Lotic Ecosystems

Freshwater streams (springs, rivulets, creeks, brooks, etc.) and rivers change over their course from being narrow, shallow, and relatively rapid to become increasingly broad, deep, and slow moving. The degree of each feature depends on the streambed fall—that is, the vertical drop over its length. This shift in movement of the water is reflected in the substrate, which trends from being rocky and without sediment to the deep sediments found in deltas at the mouth of some great rivers like the Mississippi. Most streams are characterized by a repeating sequence of rapids and pools that decrease in frequency downstream (Figure 15-2). Waterfalls are not common features of lotic ecosystems (Figure 15-3).

As would be expected, only organisms capable of adhering to an exposed surface (rocks, limbs, etc.) are found in the upper reaches of a stream. Such adhering organisms are termed *periphyton* or *aufwuchs.* The periphyton includes attached clumped and filamentous green and blue-green algae and various sessile

Figure 15-2 Lower Ocqueoc Falls, Michigan, a typical lotic ecosystem of rapids and pools. (Photo by author.)

Figure 15-3 Akaka Falls, Island of Hawaii. A typical waterfall in a lotic ecosystem. (Photo by author.)

invertebrates, including larvae of insects such as blackflies and other midges, flattened mayflies and stonefly nymphs, and planarians. Farther downstream floating and emergent vegetation may be found, along with sessile invertebrates and those that burrow in the softer substrate, such as clams and burrowing mayfly nymphs. In both environments, we might find crayfish, fish of various sizes, from darters to trout, and salmon that must expend considerable energy, like Alice in Wonderland, to stay in the same place. Farther downstream we also find warm-water fish, such as catfish and carp, replacing the cold-water trout.

Chemically the upper reaches of lentic environments are rich in oxygen; as the water moves downstream and becomes more sluggish, the oxygen level tends to drop. Because of the continual addition of nutrients and detritus en route, nutrient levels tend to be higher downstream. In small streams in which producers are limited or absent, the major source of nutrients is from external ecosystems, that is allocthanous.

Lentic Ecosystems

Lentic ecosystems (pools, ponds, some swamps, bogs, lakes, etc.) vary considerably in physical, chemical, and biological characteristics. In general, they can be considered to have three zones—littoral, limnetic, and profundal.

The *littoral zone* extends from the shoreline to the innermost rooted plants, successively passing from rooted emergent forms, such as reeds and cattail, rooted species with floating leaves, such as water lilies, and in deeper waters to various submerged but rooted species (Figure 15-4; also see Figure 13-1b). This zone is populated by frogs, snakes, snails, clams, and a considerable variety of adult and larval insects. The depth and extent of the littoral zone, comprising as it does macrophytic plants, is determined by factors controlling the underwater light, its penetration and absorption. Among these factors is the effect of mixing caused by waves carrying varing amounts of sediment (Spence 1982).

Figure 15-4 Twin Ponds, northwest Pennsylvania; an example of a lentic ecosystem. (Photo by author.)

The *limnetic zone* is the open water down to the depth of light penetration; in shallow lentic environments the light may penetrate to the bottom. This zone contains *phytoplankton* that consist of diatoms, green and blue-green algae, and a variety of *zooplankton* from protozoans to microarthropods. It is also the zone for a variety of larger, swimming organisms, the *nekton*, including fish, amphibians, and larger insects.

The *profundal zone* occurs below the limnetic zone and in deep lakes, such as Lake Superior; this zone may constitute the largest water volume of a lake. The major food source in the profundal zone comes from a detritus rain from the limnetic (and thus photosynthetic) zone. The bottom consists mostly of decomposers. The nekton vary with the temperature and nutrient conditions; lake trout are found in the profundal zone of cold, nutrient-poor lakes whereas bass, pike, pickerel, and perch are found in this zone in warmer, nutrient-rich lakes.

Although classified as freshwater lakes, saline lakes are found throughout much of the world and particularly in arid climates. Sodium is typically the dominant cation and is often distributed in gradient fashion, which determines the distribution of both plant and animal species with different tolerances of the salts. Blinn's (1993) study of diatom communities in 63 saline lakes of western North America found the greatest diversity of these communities in physicochemical settings similar to marine environments.

The thermal properties of lakes and their role in the dynamics of lentic ecosystems were discussed in Chapter 4 (see also Figure 4-9 and accompanying discussion).

Limnology in general. The study of the physical, chemical, and biological properties and features of fresh waters is referred to as *limnology*, literally the study of marshes. For the most part, the foregoing generalized descriptions were based on lentic and lotic ecosystems in temperate climates inasmuch as it is in such regions that most limnological work has taken place.

There is a growing, yet still limited, body of research on tropical limnology. Lewis (1987), in reviewing the literature, notes that a basis exists for understanding the primary causes, effects, and interactions of factors controlling latitudinal trends in aquatic ecosystems. For example, the supply of energy and nutrients to autotrophs, as well as some important physical and chemical regulatory mechanisms, vary with latitude though they do not conform to notions that might be derived from analogies with terrestrial ecosystems as discussed earlier. Tropical limnology is definitely understudied and warrants much more effort than has been the case.

Another area that has had considerable study is the historical development of lakes through the study of fossil pollen and plant fragments. These paleolimnological studies can provide considerable insight into the dynamics of a contemporary aquatic ecosystem (Binford et al. 1983).

The literature on limnology is considerable and beyond the scope of this book. However, the following are recommended: Burgis and Morris (1987), Frey (1963), Hutchinson (1957, 1967, 1975, 1993), Likens (1985), Ruttner (1966), Taub (1984), and Whitton (1975).

MARINE ECOSYSTEMS

Oceans and Seas

Covering more than 70 percent of the Earth's surface, with an average depth of 3,750 m and a salinity averaging 3.5 percent (27 percent of which is sodium chloride), marine ecosystems are of singular ecological significance. The major oceans of the world (the Pacific, Atlantic, Indian, and Arctic) account for more than 90 percent of the total water surface of the Earth (Table 15-1).

Unlike land and freshwater ecosystems, the sea is continuous and is in continual circulation via the major surface currents (Figure 15-5), upwelling, seasonal turnover in the upper waters (Figure 4-10), waves, and tidal action. In waves, water oscillates back and forth without actually changing position. Tides consist of a rise and fall of water level caused by the gravitational effects of the Moon and Sun. When Earth, Moon, and Sun are, as at the time of a new and a full Moon, aligned, the gravitational pull is strongest, resulting in the highest, or *spring tides.* When the Moon and Sun are at right angles to each other, as at the time of the first and last quarters, their gravitational fields counteract, resulting in the smallest, or *neap tides.* Each of these tides occurs monthly. Tidal range varies from less than one-third of a meter to 15 meters in some enclosed bays.

In the absence of vertical mixing, the deep waters, which constitute 60 percent of the ocean, stay permanently near 3°C.

The concentration of nutrients in the ocean is low. Nitrates, phosphates, and other nutrients are measured in parts per billion in contrast to salts, such as sodium chloride, which are measured in parts per thousand. After hydrogen and oxygen, which constitute about 96 percent of the ocean's elements, the major elements (those constituting 0.3 to 1.9 percent) in descending order are chlorine, sodium, magnesium, sulfur, calcium, and potassium (Klemm et al. 1990). Minor elements (.00013 to .0067 percent), also in descending order, are bromine, carbon, nitrogen, strontium, boron, silicon, and fluorine. There is a wide spectrum of trace elements (less than .000063 percent) from argon to zirconium, some 45 in all.

TABLE 15-1 MAJOR OCEANS AND SEAS OF THE WORLD

	Area ($10^3 km^2$)	Greatest depth (m)	Average depth (m)	Volume ($10^3 km^3$)
Pacific Ocean	165,250	10,925	4,280	707,600
Atlantic Ocean	82,440	8,605	3,930	324,600
Indian Ocean	73,440	7,125	3,960	291,000
Arctic Ocean	14,090	5,450	1,205	17,000
South China Sea	3,685	5,015	1,060	3,905
Caribbean Sea	2,755	7,535	2,490	9,585
Mediterranean Sea	2,515	5,150	1,495	4,250
Bering Sea	2,270	4,800	1,440	3,300
Gulf of Mexico	1,555	3,735	1,640	2,330

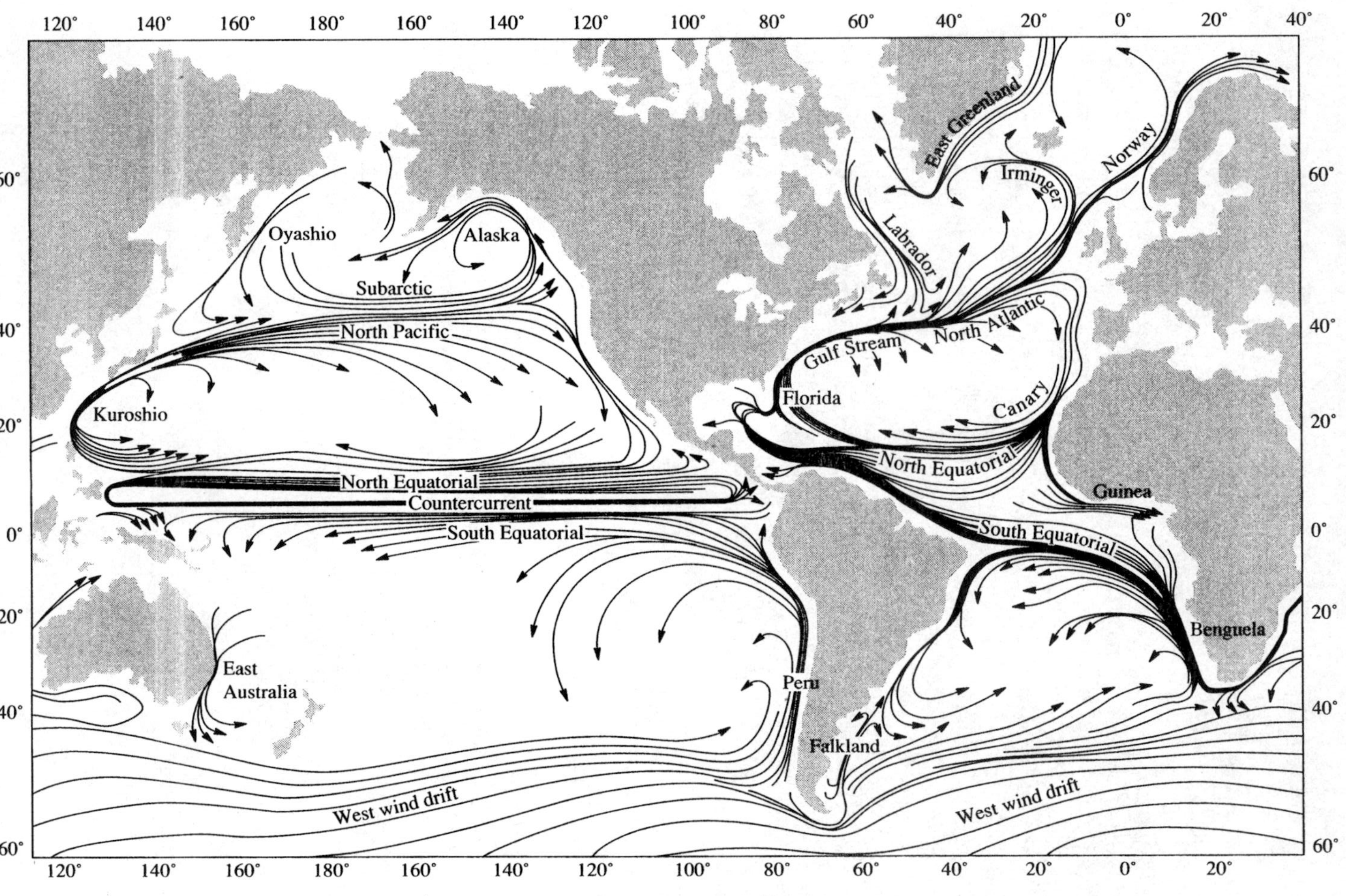

Figure 15-5 The major marine currents.

The major zones in marine ecosystems are the littoral, neritic, pelagic, and benthic (Figure 15-6). The *littoral,* or *intertidal zone* is the shoreline between land and open sea. It is subject to the physical violence of waves and tides and to fluctuations, sometimes extreme, of temperature, moisture, and light intensity. Along rocky shores (Figure 15-7), we find more sessile organisms (algae, barnacles, starfish, etc.) than in any other type of ecosystem (Sebens 1985). Along sandy shores, organisms are adapted by burrowing in or adhering to sand (e.g., ghost crabs, sand dollars,

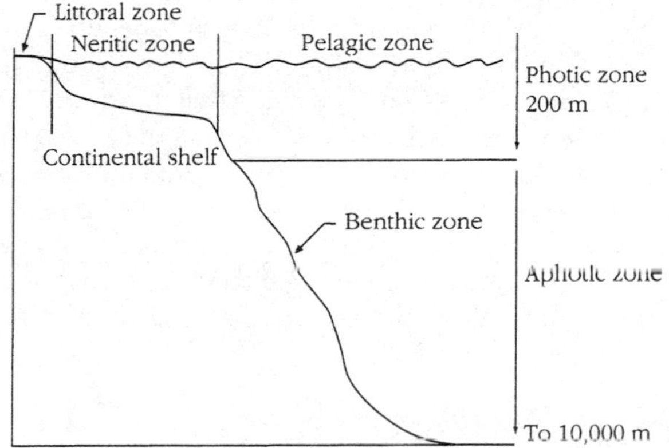

Figure 15-6 Zonation of the ocean.

Figure 15-7 Rocky shoreline at Acadia National Park, Maine. (Photo by author.)

polychaetes) (Figure 15-8). Mud flats, which occur in bays, harbor algae on the surface and often photosynthetic bacteria beneath the algae, along with an abundance of clams, worms, and crustacea. Coral reefs, fringes of coral formed around islands by colonial coelenterates, also may be regarded as littoral ecosystems. Because of their generally higher temperatures, depth of light penetration, and richer nutrient levels, coral reefs are among the most highly productive of all ecosystems (Goreau et al. 1979; Dubinsky 1990). Salt marshes are also intriguing ecosystems (Bertness 1992), as are wetlands because they are havens of biodiversity (Kusler et al. 1994).

The *neritic zone* is by a continental shelf, extending outward to its edge to a water depth of about 200 m. This zone constitutes about 7.5 percent of the total ocean area and is relatively rich in species and high in productivity owing to the depth of light penetration and the presence of nutrients washed from the land. This richness and productivity decrease gradually toward the deeper waters of the zone. Extensive algal communities of giant kelps, as well as of smaller uni- and multicellular forms, along with clams, snails, worms, and echinoderms dominate the bottom. Phytoplankton and zooplankton are relatively abundant and support some of the greatest fishing grounds in the world. Continental shelves are discussed extensively by Postma and Zijlistra (1988).

Figure 15-8 Sandy shoreline at Kaimu, or Black Sand Beach, Island of Hawaii. The black sand is of volcanic origin. Two years after this picture was taken a new lava flow covered the beach, and a new black sand beach was developed about 1 kilometer further into the ocean. (Photo by author.)

The *pelagic zone* is the open sea, constituting about 90 percent of the total ocean surface. Photosynthesis ocurs in the surface pelagic zone primarily by planktonic diatoms and dinoflagellates. Copepods and arrowworms are the major types of zooplankton, along with active swimming forms, such as shrimp, jellyfish, and ctenophores. Although the largest in size, this zone is low in nutrients and hence in both primary and secondary productivity. Nonetheless, summer blooms, which are absent in tropical waters but do occur in the Antarctic, are sufficient to support fin and blue whales; the latter are the largest animals alive, reaching lengths of 33 m and a mass of 13.6×10^4 kg.

As Daly and Smith (1993) have noted, since most plankton live in the sunlit waters between the surface and 75 to 200 m in depth, most of the production occurs in about 5 percent of the ocean, by total volume of water. Phytoplankton account for over 95 percent of marine photosynthesis, and marine primary production may account for more than 30 percent of the global total.

Also, in spite of its relatively low nutrient content, the top millimeter of the ocean plays a significant role in the global regulation of carbon dioxide (MacIntyre 1974). The organisms in the pelagic zone below the level of light penetration are completely dependent on the rain of detritus from the upper regions and are wholly heterotrophic. Carnivorous copepods and other crustacea are abundant in the upper waters of this nonphotosynthetic zone. In deeper waters many animals have reduced vision, others, including many fish, are bioluminescent, and some deep-water fish have light-producing organs.

The *benthic zone* extends from the edge of the continental shelf to the deepest ocean trenches. Organisms here are heterotrophic, many being anchored in the muddy ooze made up largely of siliceous and calcareous shells of foraminifera, snails, radiolarians, and diatoms. Among these "rooted" animals are sea lilies, sea fans, sponges, and brachiopods. Snails and clams embed in the mud while starfish, sea cucumbers, and sea urchins move on its surface.

Encompassing about 94 percent of the ocean bottom, the benthic zone is remarkably homogenous and stable in its physical and chemical parameters (Sanders and Hessler 1969). However, volcanic areas, like that south of the island of Hawaii in the Hawaiian archipelago, that are subject to underwater volcanic activity and the hot springs that are often associated with such areas, nourish intriguing forms of life, as well as laying down significant deposits of various ores (Edmond and Von Damm 1983).

The general stabilizing effect of the ocean on climate and weather, described in Chapter 4, can be altered by violent tidal waves and tsunamis and impacted by high winds of hurricanes and typhoons. There are also oscillating changes that occur in the ocean with regularity, such as El Niño.

The literature on the oceans is overwhelming, but for delightful and easy reading a classic by Ricketts and Calvin (1952 or earlier editions) and another by Russell and Yonge (1963) are recommended. Also see *Scientific American* (1969) and Sherman (1991).

Estuaries

An estuary is an ecosystem in which freshwater from rivers meets ocean water, the two being mixed by action of the tides. Estuaries include drowned river beds such as Chesapeake Bay, fjords (U-shaped basins gouged by glaciers) such as those found in Norway and Alaska, geological faults such as San Francisco Bay, river deltas such as the Mississippi and Nile, and shallow basins enclosed by a chain of offshore bars or barrier islands. The degree of mixing of salt- and freshwater depends in part on the morphology of the estuary basin, the rate and volume of freshwater flow, and the amount of tidal inflow. The biota of estuaries is, in turn, regulated by the degree of mixing; however, the major components are species restricted to the estuarine stiuation, such as oysters and crabs, and those that come from the sea, such as shrimp. Very few species are derived from freshwater and only those that are capable of osmoregulation in the saltier environment. Estuaries are highly productive ecosystems and characteristically are more productive than the adjacent river or sea. This high productivity results from estuaries being nutrient traps for both physical (based on the degree of mixing) and biological (rapid recycling) reasons. Also, estuarine producers, which include seaweeds and marsh grasses as well as benthic algae and phytoplankton, are capable of nearly year-round photosynthesis. Furthermore, the higher the fluctuation of water level, the higher is the productivity, the tides serving to remove wastes and to transport food and nutrients.

A still timely review of estuaries is that by Lauff (1967); also see Kennish (1990), Ketchum (1983), and McLusky (1989).

SUMMARY

VOCABULARY

benthic zone
estuary
lentic ecosystem
limnetic zone
limnology
littoral/intertidal zone
lotic ecosystem
neap tide
nekton
neritic zone
pelagic zone
periphyton/aufwuchs
phytoplankton
profundal zone
spring tide
zooplankton

KEY POINTS

- Freshwater, marine, and estuarine ecosystems are distinguished on the basis of differences in salt content; collectively aquatic systems encompass more than 70 percent of the Earth's surface, all but 2 percent being ocean and estuaries.

- Freshwater systems are either lentic (still water like lakes and ponds) or lotic (running water like streams and rivers). Each has a characteristic fauna and flora associated with differences in their physical and chemical makeup zonationally in lentic systems and along its length in lotic systems. Some lakes are more saline than the ocean and have a unique flora and fauna.
- Oceans account for more than 90 percent of the total water surface of Earth, are characterized by more or less distinct zones from the shore outward to open waters, and play a significant role in global regulation of carbon dioxide and the stabilization of climate.
- As an interface between fresh and marine ecosystems, estuaries are a nutrient trap and are therefore more productive than their adjacent waters.

PART VIII

Human Ecology

CHAPTER 16

The Nature of Human Ecology

If there is a lesson in the concepts presented in earlier chapters, it is that ecosystems are enormously complex structures. They have a multitude of intra- and intersystem relationships, with capacities for adjustment through negative-feedback channels and with tendencies to achieve dynamic homeostasis. Humans are but one of the interacting entities in ecosystems. They are contributors, participants, and receivers in ecological processes, just another species functioning in the flow of energy and cycling of nutrients, and whose population is subject to growth as well as regulation.

Humans are also the most powerful species in effecting alteration in ecosystems and ecological processes. Yet, like all species, *Homo sapiens* is immune neither to those processes nor the negative consequences of any detrimental ecological activity. In this sense, there is no ecosystem that can be called a human ecosystem as if it were quantitatively and/or qualitatively different in its structure-function relations from other ecosystems. Nor does a human ecosystem have unique and special properties exempting it from the inexorable operating rules of ecosystems, from interrelationships with other ecosystems, or from the cybernetic properties of its own system. Natural and human ecosystems are one and the same.

No ecosystem is unaffected by other ecosystems, and none is unaffected by humans, directly or indirectly. In this sense, Earth is the human ecosystem and, given the unparalleled capacity of humans to modify their environment, the fiduciary responsibility for true stewardship and guardianship becomes paramount (Kormondy 1970, 1974, 1981). In this sense, human ecology is not different in kind from any other kind of ecology, but rather differs only in the degree to which humans serve in their stewardship role.

Importantly, humans have no more prerogative to have dominion over the land than any other species. The environment serves not only humankind but other species as well. Rather, because of their capacity for cognition, reflection, and prediction, humans have responsibility for the environment far surpassing the responsibility of other species. That responsibility is one of being a steward and guardian. In contrast to almost all other species, humans can voluntarily control, or at least regulate, their environment and their population—as well as their behavior, genetic heritage, and evolutionary future. Exercise of this control has been particularly lax with regard to the size of the human population, as well as in the consumption of both renewable and nonrenewable natural resources and in the management of wastes, the consequences and byproducts of that consumption.

Human ecology, properly treated, is well beyond the scope and purpose of this book; several of the many good books and articles on aspects of this subject are cited in the References at the end of this book. Nonetheless, we will first explore in more detail the nature of human ecology and in subsequent chapters examine some limited aspects of the human population; pollution of air, water and land; desertification; fire; and anthropological ecology.

ORIGINS OF HUMAN ECOLOGY

The formal history of what can be called *human ecology* dates at least to 1921 as a subdiscipline of the social sciences, although a less tightly defined history can be traced back to the classical Greek philosophers (Young 1974, 1983; Hargrove 1989). Paul Sears (1956) antedates the Greek philosophers as being the first to consider human ecology when he noted that human ecology's "roots lie in the intuitive and empirical knowledge of ancient man who lived, perforce, in intimate relations with the natural world."

Several diverse and distinct intellectual traditions can be recognized in the relationship of humans and their environment: imperialist, arcadian, and scientific.

The Imperialist Tradition

This perspective of the human-nature relationship holds that humans have dominion over nature and derives from a very literal interpretation of the *Old Testament*, (White 1967).

> Let us make man in our image, after our likeness: and let them have dominion over the fish of the sea, and over the fowl of the air, and over the cattle and over all the earth, and over every creeping thing that creepeth upon the earth. Gen. 1:26

> and God said unto them, Be fruitful, and multiply, and replenish the earth and subdue it; and have dominion over the fish of the sea, and over the fowl of the air, and over every living thing that moveth upon the earth. Gen. 1:28

The *imperialist tradition* is reflective of the scientific tradition espoused by English philosopher Francis Bacon (1561–1626) that ascribed human dominion over nature through the exercise of reason. It is epitomized in the writing of the "father of systematics," Carl Linnaeus (1707–78): "All the treasures of nature, so artfully contrived, so wonderfully propagated, so providentially supported ... seem intended by the Creator for the sake of man" (*Discovery of Nature*, 1749). For Linnaeus, humanity had the right to manage nature for its profit. Charles Darwin (1809–1882) continued the imperialist tradition in recognizing the role of competitive interactions in ecological relationships, with each species holding a specific place or niche in the economy of nature.

The imperialist tradition was exercised and applied in the founding and expansion of the United States, as well as in Europe (Crosby 1986) and other countries (Nash 1967; Crosby 1986). This was succinctly encapsulated in the term Manifest Destiny, the nineteenth century doctrine that the United States had the duty and right to expand its territory and influence throughout North America, a reflection of a growing spirit of confidence in an age of population increase and westward movement. Wilderness stood in the way of providing food and was an imminent threat to personal and group security, and hence to survival. The result was wanton pillage, exploitation, and decimation of natural resources to an unprecedented extent in human history.

The Arcadian Tradition

Arcadia was a region of ancient Greece in the mid-Peloponnesus inhabited by pastoral people of idyllic satisfaction. The *arcadian tradition* that follows is one of contentment, encompassing a reverence for nature because of its beneficence. The embodiment of the arcadian view is found in the writings of the English parson naturalist Gilbert White (1720–93). His *The Natural History of Selbourne* (1798) epitomizes an Arcadia of idyllic contentment reaching back to Greek animism and pantheism, the very quintessence of harmony of humans and nature. In large measure this tradition developed as a reaction to the birth and growth of industrialism and a protest against the mechanistic analysis espoused by French philosopher Rene Descartes (1596–1650) that isolated scientists from society and its moral fabric.

The American transcendentalist Henry David Thoreau (1817–62) carried forward the arcadian tradition. Thoreau's interpretation of nature was an extension of the self, nature being best understood and appreciated by acknowledging correspondence and kinship with it. However, as Wooster (1977) notes, nature was, for Thoreau, not a flawless Newtonian machine but a "maimed and imperfect nature," largely the result of destruction of over 0.5 million acres of northeastern woodlands between 1750 and 1850.

In some large measure, Eastern religions and/or philosophies such as Buddhism, Taoism, and Confucianism, as well as many Amerindian beliefs, bear kinship to the arcadian view (Martin 1981, 1992; Whitehill 1981). None of these, however, contributed in any significant way to molding the emergent Western traditions of arcadianism, and certainly not of imperialism nor the scientific tradition described next.

The Scientific Tradition

As was noted in Chapter 1, the *scientific tradition* of ecology is first evidenced in Ernst Haeckle's 1870 definition, "By ecology we mean the body of knowledge concerning the economy of nature—the investigations of the total relations of the animal both to its inorganic and to its organic environment...." Here there is no direct mention of humans, nor were humans included in later conceptions of ecology by the leading figures in the field. While this absence of mention might be interpreted to mean that humankind is something apart from nature in the imperialist sense, it actually implies that humans are an integral part of, or not different from, other organisms in the nonromanticized Arcadian sense.

Contemporary ecology's ecosystem focus is that of a hard-nosed positivism stressing interrelatedness in both competition and cooperation, underscoring the fact that all components—abiotic and biotic (including humans)—are integral parts of nature rather than objects apart. In this tradition, humankind is not independent of nature's ways, and thus there is no substantive distinction between human ecosystems and natural ecosystems.

Contemporary Human Ecology

The pure, unadulterated ecologist is Cartesian in philosophical perspective, gathering knowledge about ecosystems for its own sake and having no goals but the elucidation of parts and relationships, predictions and projections. By contrast, the human ecologist's goal is philosophically Baconian, knowledge being gained toward the end of improving the human condition.

In another terminology, human ecology is often, if not typically, applied in its orientation rather than basic or theoretical in focus. This is witnessed in the adjectives often used to qualify the term human ecology: legal, economic, political, social, and psychological, among others. Its focus is that such application should lead to the betterment of the human condition that is philosophically justifiable in a deontological and/or utilitarian sense.

Because of the protean complexity of human ecology, it is singularly difficult, perhaps intellectually impossible, to be like the renaissance individuals who were able to encompass the myriad dimensions of world knowledge as it then existed. Today, we can only carve out components of this complexity—which is the tack taken in the subsequent and final chapters of this book—to gain some understanding and insight.

Values and Ethics of the Human Environment

Jeremiah Conway (1980) has commented, "Modern man tends not to dwell upon the earth; he tends to drift across it." As humans drift, however, values and ethics are consciously or unconsciously involved in responding to the environment. In a thoughtful and provocative essay, Rolston (1981 and also see Rolston 1988) has identified several major values in nature:

Economic:	nature as a resource to be used for human gain, something we can arrange to our ends
Life Support:	life is tethered to the biosphere, to its air, water, and land
Recreational:	a place to play, climb, swim, walk or run, which gives pleasurable appreciation and allows for participation
Scientific:	a milieu for intellectual activity, recreation, and stimulation
Aesthetic:	a vehicle for nonutilitarian searching for organic beauty, for perfection and imperfection
Life Value:	based on evolutionary kinship, the life we value in people is advanced from the life in monkeys, perch, and dandelions.

Since values are culturally derived, at least in part, not every culture would place the above entries in a list of values, nor would all Westerners or Easterners, New Yorkers or Oregonians, wealthy or poor, necessarily concur with each element. Even if there were consensus about the values, it is unlikely that there would be consensus as to the priorities among them. Whatever set of values is identified, however, adherents to that set of values make choices according to them (Kormondy 1982, 1984). Thus, the life value in nature lies at the heart of the Endangered Species Act of 1977, economic values at the heart of resource recycling, recreational values at the heart of the national park movement, and so on.

In his play *Professional Foul*, American playwright Tom Stoppard (1978) has one of his characters state, "There would be no moral dilemmas if moral principles worked in straight lines and never crossed each other." For it is only when two or more ethical values or principles come into conflict that choice is involved and a dilemma arises as to which to choose.

The ethical values in human-nature relationships devolve from the traditions of imperialism (humans apart from nature) and Arcadianism (humans as part of nature) and thus often lead to dilemmas involving choices between humans and nonhumans, between rights and entitlements of individuals and of society, as well as those choices between present and future generations. The ethics of manipulating nature at will derives from a consideration of whether the knowledge of how to control nature at any cost is prized or whether humankind is regarded as one of many species, but of no higher rank in the biosphere.

In practice, however, there is not usually an easy solution. For example, in the late 1970s, the Tellico Dam project in the eastern United States, presumably of benefit to humans, was interrupted for an endangered species, a tiny fish known as the snail darter. In the early 1990s, logging in old-growth forests of the northwestern United Sates was sharply curtailed for another endangered species, the spotted owl. Who has more right—humans or fish? Humans or owls?

Likewise, the choice between yielding to individual rights by compromising societal rights, or more usually vice versa, necessitates careful assessment of the issues of beneficence, justice, and fair play. Are the individual's rights to an off-road vehicle drive to be compromised by society's rights to conserve a fragile ecosystem? Is the individual's right of procreation to be compromised by the right of society to control its population to assure that it is justly fed, clothed, and housed? Do those living now have rights to consume natural resources to destruction or exhaustion without concern for what some regard as the identical rights of our descendants? In his 1864 *Man and Nature, or Physical Geography as Modified by Human Action,* George Perkins Marsh put this point well: "Man has too long forgotten that the earth was given to him for usufruct alone, not for consumption, still less for profligate waste."

The choices we make, or that society makes through cultural force, legislative mandate, or judicial decree are ultimately based on a value system. Such a system derives from the ethical principles we espouse, principles such as autonomy (personal liberty, independence, self-reliance), nonmaleficence (doing no harm or injury), beneficence (contributing to the health and welfare of others), and justice (receiving what one deserves and can legitimately claim). These principles become manifest as we judge the outcome or ends of an action as justifying the means of getting there (a utilitarian ethic) or whether under no circumstances can any set of means justify even a desirable end (a deontological ethic) (Hardin 1993, Hargrove 1989, Westra 1994).*

Because of our rationality, we have the capacity to choose and direct this human environment, by applying both ecological and ethical principles. It is in this further sense that human ecology is truly an applied field. We can, to a considerable extent, manipulate the human ecosystem to determine its quality (*Scientific American* 1989). Because of our humanness, we have a moral obligation to do so. We are indeed nature's keeper—and it is ours to keep.

SUMMARY

VOCABULARY

arcadian tradition
human ecology
imperialist tradition
scientific tradition

KEY POINTS

- No ecosystem is unaffected by other ecosystems, and none is unaffected by humans, directly or indirectly; human ecology is not different from other kinds of ecology except in the degree to which humans affect their environment and an emphasis on application.
- The human-nature relationship can be described in terms of three major traditions—imperialist, arcadian, and scientific.
- Several major values have been attributed to nature including economic, life support, recreational, scientific, aesthetic, and evolutionary kinship.

* For in-depth discussions and debates on ethical issues concerning the environment, consult the journal *Environmental Ethics*, vol. 1, 1979, et seq.

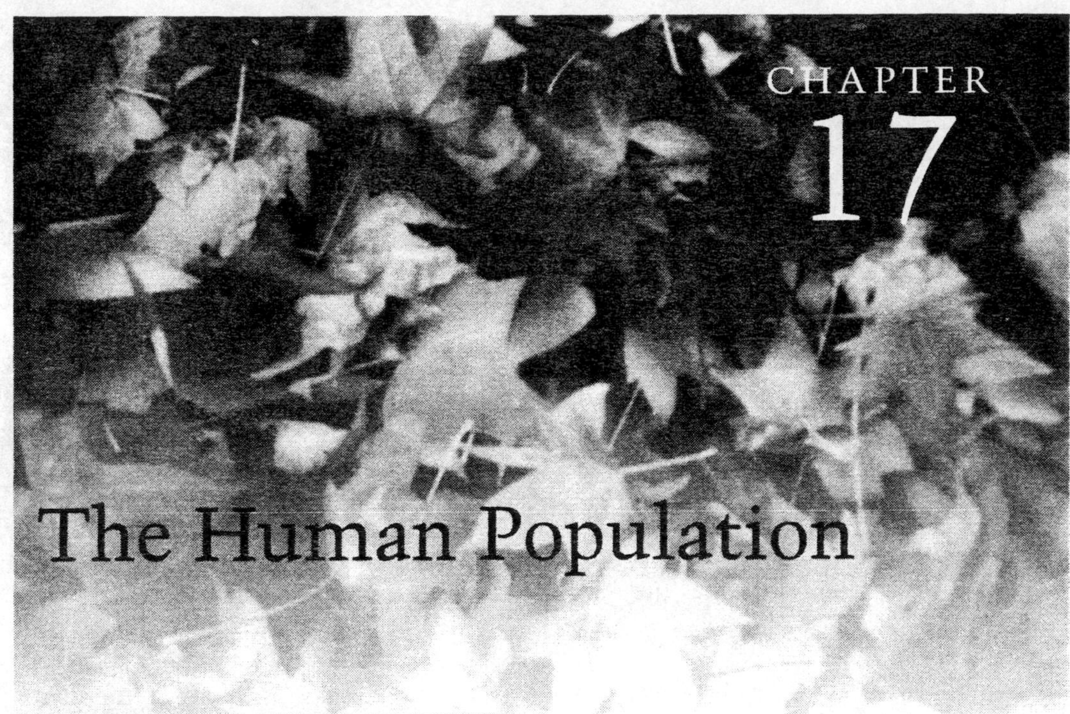

The Human Population

HUMAN POPULATION GROWTH

The first warning of the disparity between human population growth and its means of subsistence was given by English economist Thomas Robert Malthus (1766–1834). In his 1798 treatise, "An Essay on the Principle of Population, as It Affects the Future Improvement of Society," Malthus stated boldly and unequivocally, "I said that population, when unchecked, increased in a geometrical ratio, and subsistence for man in an arithmetical ratio." His concern was prompted by the rapid increase in population that began with the onset of the Industrial Revolution and his fear that such increase was counterproductive to progressing toward an ideal society, a perspective and ideal of the Age of Reason.

Malthus maintained that, because of their different growth characteristics, population would always outrun the food supply and therefore human numbers would have to be kept down by famine, disease, or war. In a second edition of his Essay, published in 1803, and in response to the storm of abuse the first edition engendered, he admitted that moral restraint (delayed marriage and sexual continence) might also counter the increase in population.

As students of evolution know, Malthus had a strong influence on Charles Darwin and Alfred Lord Wallace, whose theories of evolution by natural selection nevertheless would have been an anathema to Malthus.

In more recent times, Malthusian doom has been characterized in such expressions as "the population bomb" and the "population explosion," allusions to the rapidity of increase in the size of the human population. Some years ago, the metaphor "the mushroom crowd" was coined, an analog to the symmetry of an aboveground atomic blast. The phrase is apt in many respects: the rapidity of mushroom (and population) growth; the collective quality implied, many hyphae forming the fruiting body, many people forming the population body; and so on. But to a biologist, the mushroom metaphor has an alternative meaning—that of a largely temporary and purely reproductive structure and of organisms that are exclusively decompositional in activity. Human populations, in contrast, are more permanent, have a large amount of nonreproductive (as well as reproductive) activity, and are capable of creative output.

One thing learned from the study of natural populations is that many populations initially follow a sigmoidal pattern and subsequently show a variety of patterns—oscillations, fluctuations, and even a rather steady equilibrium. If the growth curve of the human population is, in fact, sigmoidal, is it still in the positive acceleration stage? Is it, in fact, to follow a J-shaped curve and be subject to ultimate if not imminent collapse? It is not apparent which of the several alternates may apply, but there is little question that world population is heading toward far greater numbers.

Yet another of the "lessons" that seems to be developing from the study of natural populations is that all populations are inherently capable of self-regulation in the face of environmental change. Whether that innate capacity exists in human populations and will be exercised fully or fast enough is another issue.

An expression of confidence in self-regulation should not be considered as being that of an assured and even benevolent Pollyanna ecologist. It is an assumption about human populations based on the study of other populations. But it does not necessarily assume the operation of any of the mechanisms employed by nonhuman populations to effect self-regulation, some of which are at once undesirable and unwanted. Yet a little additional reflection makes us wonder about some of the mechanisms that have, at least historically, contributed to less rapid human population growth—disease, pestilence, armed conflict, infanticide, and celibacy (Langer 1972, Dickeman 1975). The interpolation of conscious individual or family choice on procreation significantly alters the situation, as we shall see.

Global Human Population Growth

Because population growth is of global concern, it will be more appropriate at first to consider the world's population rather than to focus on that of a particular region or area. The growth of the world's population to date (Figure 17-1) allows for speculation regarding its growth curve in the future. Three different results can be predicted, assuming different periods by which the world's population would attain equilibrium—that is, with the number of births being equal to the number of deaths. If current trends continue, this equalization could occur by 2110, with a population of 10.5 billion or about two times the 1990 population of 5.3 billion. If the birthrate were to decrease at a faster pace, the halt in population growth could occur by 2040, with a population of 8 billion, not quite double that of 1982 and about 86 percent larger than that of 1990; and if the birthrate were to decline at a slower pace, the ultimate balance point would occur in 2130, with a population of 14.2 billion, slightly less than three times the present size.

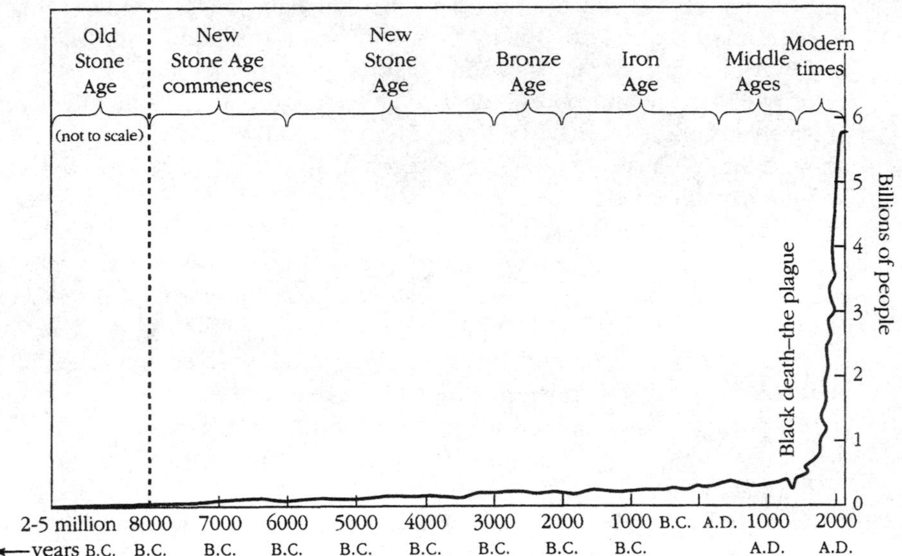

Figure 17-1. Growth of world population through human history. Data prior to 1800 are estimates; data from 1800 on from various sources.

The pattern of world population growth to date and the projections based on various growth trends allow us to put it into a variety of perspectives. We can see, for example, the dramatic reduction in the time that has been and that may be required for the population to increase in increments of 1 billion if present trends continue:

from 0 to 1 billion — 1800, or 2–5 million years
from 1 to 2 billion — 1930, or 130 years
from 2 to 3 billion — 1960, or 30 years
from 3 to 4 billion — 1975, or 15 years
from 4 to 5 billion — 1987, or 12 years
from 5 to 6 billion — 1996, or 9 years

Doubling time. *Doubling time* indicates the number of years in which a population would double in size if present growth rates were to continue. For example, a population of 1 million with a constant rate of growth and a 10 year doubling time would reach 2 million in 10 years, 4 million in 20 years, 8 million in 30 years, and so on.

The method of determining doubling time is derived directly from the exponential formula for continuous compounding as is used, for example, in computing money left on deposit in a savings bank. Doubling time for a population is calculated by dividing 69.3 (or 70 in round numbers) by the population's growth rate expressed as a percent. Thus, if the growth rate is 2 percent per year, the doubling time would be approximately 35 years ($70 \div 2 = 35$, or more precisely, $69.3 \div 2 = 34.7$). Likewise, if the interest rate on your deposit were 10 percent per year, the initial deposit would double in 6.9 years.

It is important to note that projections of doubling time in the future are just that, "projections." Whether such projections on population growth are realized will be dependent on a host of factors, such as changing birth and death rates, age structure, and migration. Nonetheless, the use of doubling time gives a quick approximation to the future or a retrospective on the past. For example, applying this concept to the human population, we can see the dramatic reduction in the time it has taken to about double the human population in the past (actual numbers and years) and the projected doubling time based on trends since 1965:

from 0.75 to 1.6 billion — 1750–1900, or 150 years (actual)
from 1.6 to 3.3 billion — 1900–1965, or 65 years (actual)
from 3.3 to 7.0 billion — 1965–2005, or 40 years (projection)

In 1982, when the population was 4.5 billion and the growth rate 1.7 percent a year, the world's population was projected to double in 41 years. In 1994, when the world's population stood at 5.6 billion (estimated) and the growth rate 1.6 percent, the doubling time was projected to be 43 years. At that rate, and given no other demographic changes, the world's population would be 11.2 billion in 2037, 22.4 billion in 2080, and about 31 billion in 2100.

The growth rate of the human population peaked in 1965 at 2 percent, the doubling time then being 35 years (Table 17-1). It has been dropping ever since, and the United Nations projects that it may drop to 1.5 percent in 2000 for a doubling time of 47 years and to 0.93 percent in the period 2020–2025 for a doubling time of 75 years. Given those projections, the world's population would be about 12 billion in 2047 and 16 billion in 2100. It is pertinent to note, however, that the growth rate decline leveled off in the 1980s (Horiuchi 1992).

Regional Human Population Growth

Perhaps it may clarify the matter for the reader from the northeastern part of the United States to state that the tristate area bounded by Trenton, New Jersey, Poughkeepsie, New York, and New Haven, Connecticut, which in the late 1960s had

TABLE 17-1 ANNUAL RATES OF WORLD POPULATION GROWTH (%)

Region	1950–1955	1960–1965	1975–1980	1990
World	1.77	1.99	1.81	1.80
More developed countries	1.28	1.19	0.67	0.5
Less developed countries	2.00	2.35	2.21	2.1

Data from various publications of the Population Reference Bureau, Inc.

about 18 million residents, is expected to have 30 million residents by 2000. Similarly, Shanghai, which had a population of 5.8 million in 1950 and of 14.3 million in 1980, is expected to have 23.7 million people by 2000. In contrast, London's population of 10.4 million in 1950 shrank to 10 million in 1980 and is expected to be less than 10 million in 2000.

But, as in the case of Shanghai and London, all regions of the world will probably not grow at the same rate over that period. In fact, the projections on a worldwide basis (Table 17-2) suggest that the 67 percent population increase in the tristate area by 2000 will be out of step with that in the developed regions, such as Europe (5 percent increase) or North America (14 percent), and even above that in the developing regions, such as Africa (59 percent).

The striking differences in projected growth of the developing regions of the world (e.g., Africa, South Asia, Latin America) as compared with the developed regions (e.g., Europe, North America, the former USSR) is dramatically evident in Figure 17-2. These growth patterns are consonant with the kind of age structure in each of these regions (Figure 17-3 and Figure 10-13) in which about 40 percent of the population in under 15 years of age (see also Table 17-3).

Based on United Nations data, the world's population from 1990 to 2025 will increase by a total of 3.2 billion people. Of the projected increase, less than 200 million (6 percent) will be in developed countries, and 3 billion (94 percent) will be in developing countries. Seventy percent of this increase is projected to occur in 20 less-developed countries as follows (increases in millions in parenthesis): India (592); China (357), Nigeria (188); Pakistan (144); Bangladesh (119); Brazil (95); Indonesia (83); Ethiopia (66); Iran (66); Zaire (64); Mexico (62); Tanzania (58); Kenya (53); Vietnam (51); Philippines (49); Egypt (40); Uganda (37); Sudan (34); Turkey (34); and South Africa (28). Given India's current population of 850 million and China's current population of 1.1 billion, their projected growth rates would yield populations of 1.4 and 1.5 billion, respectively, in 2025.

TABLE 17-2 POPULATION ESTIMATES FROM 1980 TO 2025 FOR MAJOR AREAS AND REGIONS OF THE WORLD

	Population (millions)							
Area	1980	1985	1990	1995	2000	2010	2020	2025
World	4,453	4,842	5,248	5,677	6,123	6,987	7,793	8,162
Africa	476	553	645	753	877	1,170	1,488	1,642
Latin America	362	406	453	501	550	647	742	787
North America	252	263	275	287	298	319	339	348
East Asia	1,182	1,252	1,317	1,390	1,470	1,589	1,662	1,696
South Asia	1,408	1,572	1,740	1,909	2,074	2,379	2,654	2,771
Europe	484	492	499	505	510	515	518	518
Oceania	23	25	27	29	30	34	38	40
USSR	265	279	291	303	314	334	352	361

Based on medium estimates and projections of the United Nations as assessed in 1982.

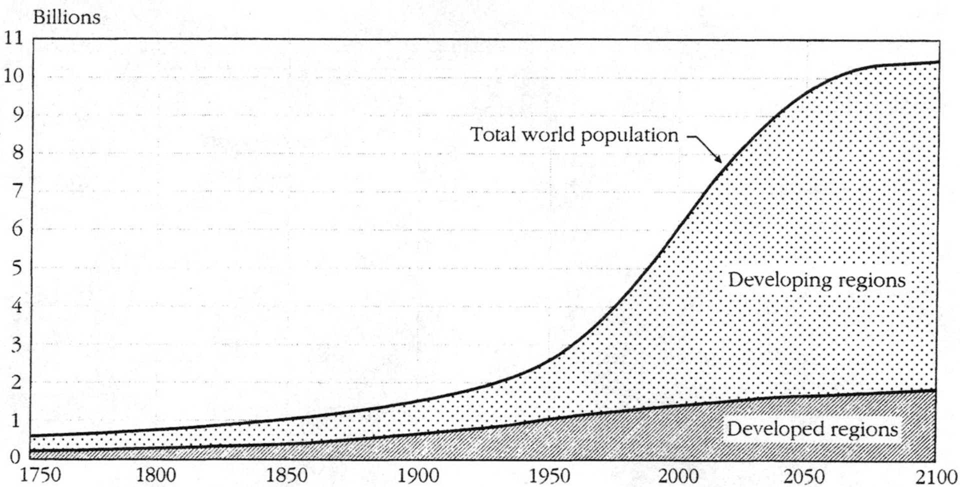

Figure 17-2. Human population growth in developed and developing countries, 1750–2100. Data beyond 1985 are projections. (Redrawn, by permission, from T. W. Merrick. 1986. *Population Bulletin* 41(2):4.)

Although the numerical and percentage increases are dramatic, they do not suggest the amount of space per person involved. We have already seen that density has much influence on regulatory mechanisms in natural populations. Thus it is the density of these various area and regional populations that may prove to be of considerable significance and to which some attention must be given (Table 17-3). Even the uninitiated should anticipate a different set of problems attendant to regions with low density but high growth rate (Africa, Latin America) as against those with low density and low growth rate (North America, the former USSR). A contemporary reading of the world's other ills of poverty, famine, and disease suggests correlation, if not causal connection, with regions of high growth rate regardless of their present density.

It is difficult for most people to grasp the significance of these projections and their implications for the next several decades. Numbers are numbers and the larger or more infinitesimal the number, the less is our the ability to visualize it. Reflect for a moment on a common measurement—the centimeter, 2.54 of which equal 1 inch. What conception do we really have of the diameter of a cell organelle measured in angstroms? (Each angstrom is 1/10,000 of a micron, the latter being 1/10,000 of a centimeter; hence the angstrom is 1/100,000,000 of a centimeter.) Or consider the distance of the Earth from its neighboring star, Alpha Centauri, measured as 4.35 light years or 25.6 trillion miles. So it is with speaking of billions or even millions of people. William Vogt put the situation aptly by suggesting that the reader count his pulse; if the pulse beat were normal, it would not quite keep up with the increase in world population, which was then about 50 million per year compared with about 38 million pulses per year. And even more telling, Vogt commented that the rate of increase at that time amounted to the equivalent of a city the size of

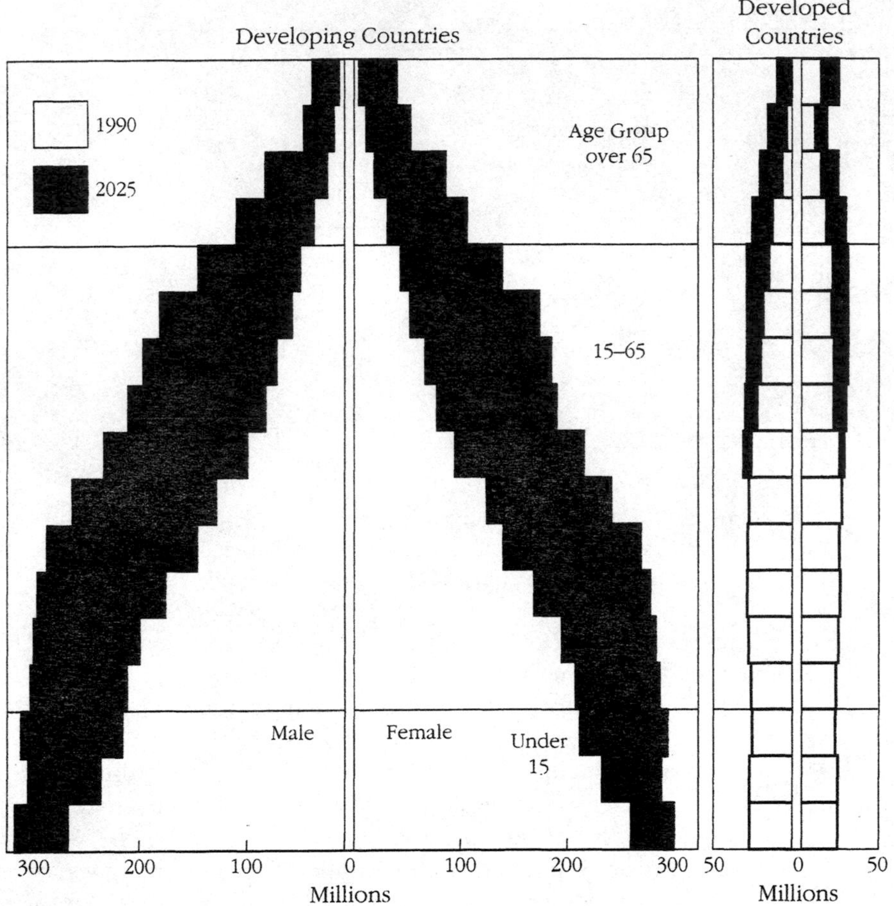

Figure 17-3. Age distribution of human populations in typical developing and developed countries in 1990 and projected to 2025. (Developed from data of the United Nations.)

Chicago—namely, about 4 million persons—each month. In 1990 the annual increase in world population was nearly 2.5 times the amount of normal annual pulses—namely, 90 million.

Natality and Mortality in Human Populations

With this background of past and projected human population growth, we can address the two forces that are behind these changes, *natality*, or birthrate, and *mortality*, or death rate. Both natality and mortality are calculated based on the number of instances per 1,000 individuals; thus, natality is the number of births per 1,000 people, and mortality is the number of deaths per 1,000 people.

BLE 17-3 BASIC DEMOGRAPHIC MEASURES OF THE MAJOR AREAS AND REGIONS OF THE WORLD, 1990

	Birth-rate	Death rate	Average annual rate of growth	Doubling time (years)	Total population (millions)[a]	Total area (10^6 km^2)	Density per km^2	Density per km^2 in 2000[b]
al World	27	10	1.8	39	5,321 (33)	135	39	45
ica	44	15	2.9	24	661 (45)	30	22	29
rth America	16	9	0.7	93	278 (22)	22	13	14
ope	13	10	0.3	266	501 (19)	5	100	102
a	27	9	1.9	37	3,116 (34)	27	115	131
in America	28	8	1.2	33	447 (38)	21	21	26
SR	19	10	0.9	80	291 (25)	22	13	14
eania	20	8	1.2	57	27 (27)	9	3	3

rcentage under age 15 in parentheses.
ased on United Nations estimate of moderate population growth.
ed, in part, on 1990 World Population Data Sheet, Population Reference Bureau, Inc.

As you have already surmised, the more important of these two opposing forces in the growth of human populations has been the reduction in mortality (Figure 17-4). Birthrates have dropped from about 40 to 15 per 1,000 since the middle of the nineteenth century in developed countries while remaining at the 40 births per 1,000 level until the early part of the twentieth century in developing countries. Birthrates in developing countries, while dropping, have not dropped as much as those in developed countries and stand twice as high. The death rate in developing countries, however, has dropped dramatically since the middle of the twentieth century and is now close to that of developed countries on a worldwide scale. With a continuing high birthrate but a dramatically reduced death rate, the gap between the two, which reflects the growth in population, has been increasing. This is the explanation behind the projected growth in developing countries seen in Figure 17-2.

Changes in Life Expectancy

About half of all the people alive in the world today have been born since 1945, the result not of a sustained birthrate but a reduction in death rate accomplished through improved application of new medical (drugs, surgery) and environmental (nutrition, sanitation, hygiene) technology. On the other hand, the reason this "born since 1945 group" constitutes no more than half the population is the decrease in the death rate of the "born before 1945 group" because of the same applied technology. For example, life expectancy worldwide increased from 61 years in 1980 (72 for developed countries and 57 for developing countries) to 64 years in 1990 (74 for developed countries and 61 for developing countries).

As used here, *life expectancy* is the estimated life span at the time of birth. For insurance purposes, life expectancy is estimated at any given age, typically the age

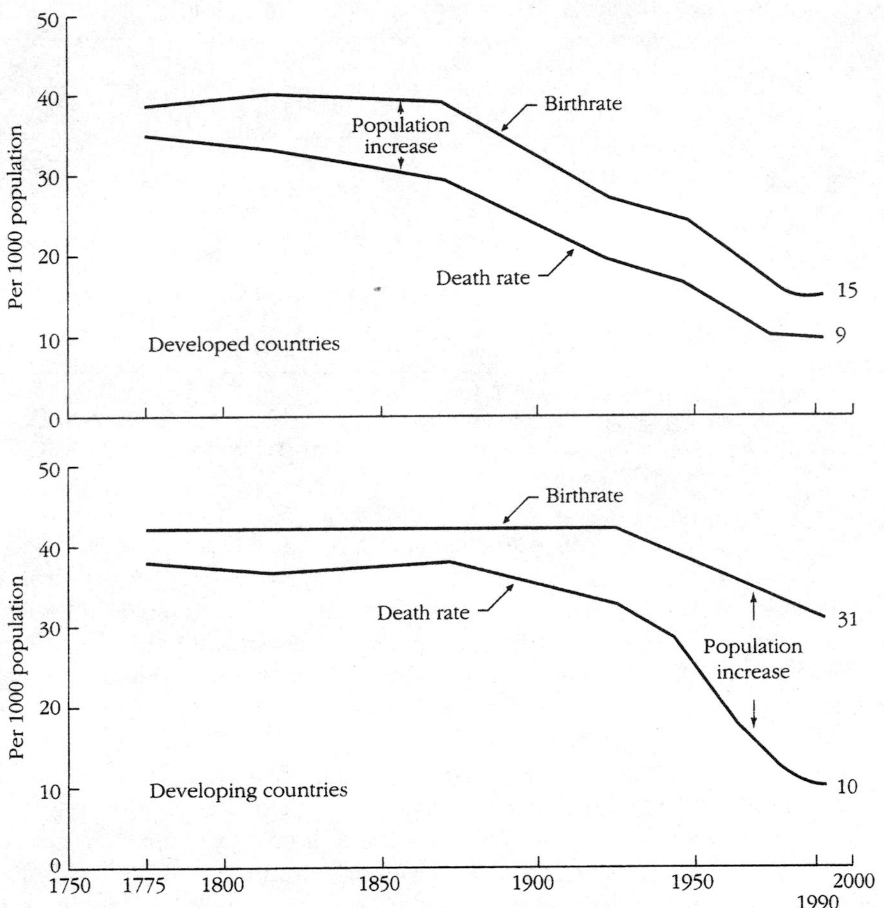

Figure 17-4. Birthrates and death rates in developed and developing countries from 1775 to 1990. (Based on United Nations and Population Reference Bureau, Inc. estimates.)

at which a policy is issued. Thus, at birth, life expectancy in the United States was 75 years in 1990, but at age 35 it drops to 41 years (meaning that at age 35, a person would be expected to live an additional 41 years), and at age 85 it drops to 6 years.

In the short time since the Pilgrims landed on Plymouth Rock, total life expectancy in the United States advanced from:

33 years to 47 years by 1900,
47 years to 60 years by 1930,
60 years to 71 years by 1971, and
71 years to 75 years by 1990.

There have been and continues to be significant differences in life expectancy by gender and ethnicity/race (Figure 17-5). For example, based on 1988 death certificates, life expectancy for Caucasians in the United States stood at 75.6 years (up from 75.3 in 1984) but at 69.2 years for African-Americans, the latter figure being down from 69.7 in 1984. The reversal for African-Americans seems to be related to the large increase in the number of early deaths caused by homicides, car accidents, drug abuse, and AIDS. But this group has higher death rates from many ailments, including cancer, heart disease, stroke, diabetes, liver trouble, and kidney failure suggesting that a combination, as well as interaction, of genetics, diet, and social factors contribute to the lower life expectancy.

Olshansky et al. (1990) suggest that for life expectancy at birth to increase from present levels to what has been referred to as the average biological limit to life (age 85), mortality rates from all causes of death would need to decline at all ages by 55 percent, and at ages 50 and over by 60 percent. Given that hypothetical cures for major degenerative diseases would reduce overall mortality by 75 percent, they conclude it to be highly unlikely that life expectancy at birth will exceed 85 years.

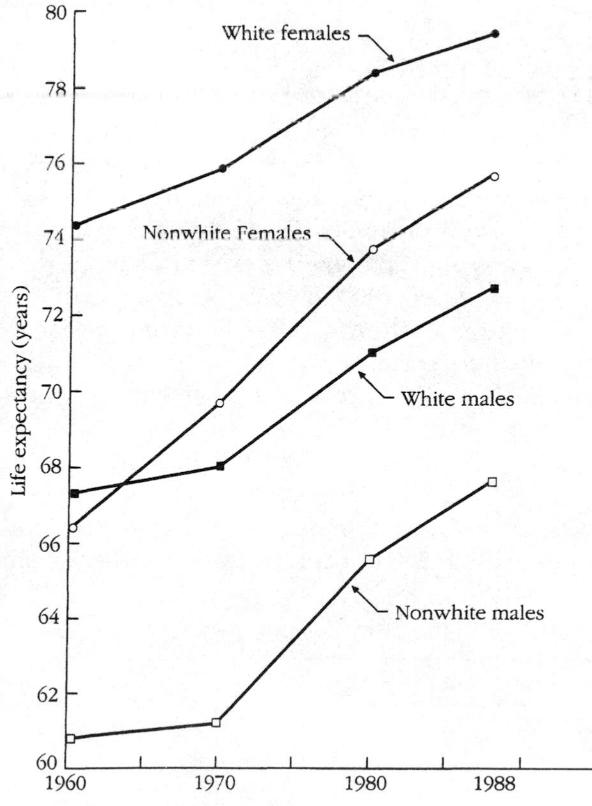

Figure 17-5. Life expectancy at birth in the United States, 1960–1988. (Data from *Statistical Abstracts of the United States* (1990).

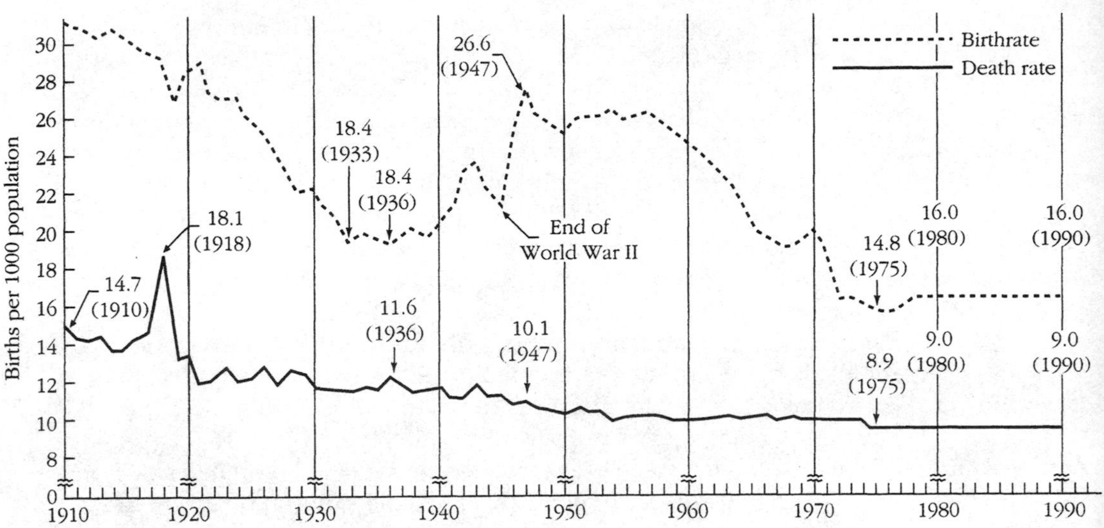

Figure 17-6 Birthrates and death rates in the United States, 1910–1990. (Based on data of Population Reference Bureau, Inc.)

Mortality in the Population of the United States

In the period 1910 to 1975, the total death rate in the United States, all ages included, dropped from 14.7 to 9.0 per 1,000 and has been sustained at that level ever since (Figure 17-6). (The dramatic rise in death rates in 1918 as seen in Figure 17-6 is attributed to a devastating flu that affected virtually the entire world.) During this period, a differential by gender and ethnicity/race nonetheless existed, as in the case of life expectancy. For example, in the period 1900 to 1988, the total death rate dropped from 17.2 to 8.8 but for nonwhite males the drop was to 9.9, for white males 9.6, for white females 8.6, and for nonwhite females 7.3.

The mortality curve for humans (Figure 17-7) reflects a relatively high infant mortality, a low death rate until about age 50, and then an increasing death rate. *Infant mortality* is defined as the number of deaths per 1,000 individuals between birth and 1 year of age. Again, as expected from the foregoing discussion of life expectancy and mortality in general, infant mortality also differs by gender and race/ethnicity; in general, the drop from 1900 to 1971 was by 90 percent, and from 1971 to 1987 by 50 percent:

	1900	1971	1987
White females	142.6	14.6	7.6
White males	175.9	19.1	9.6
Nonwhite females	299.5	26.2	13.6
Nonwhite males	369.3	32.3	16.9

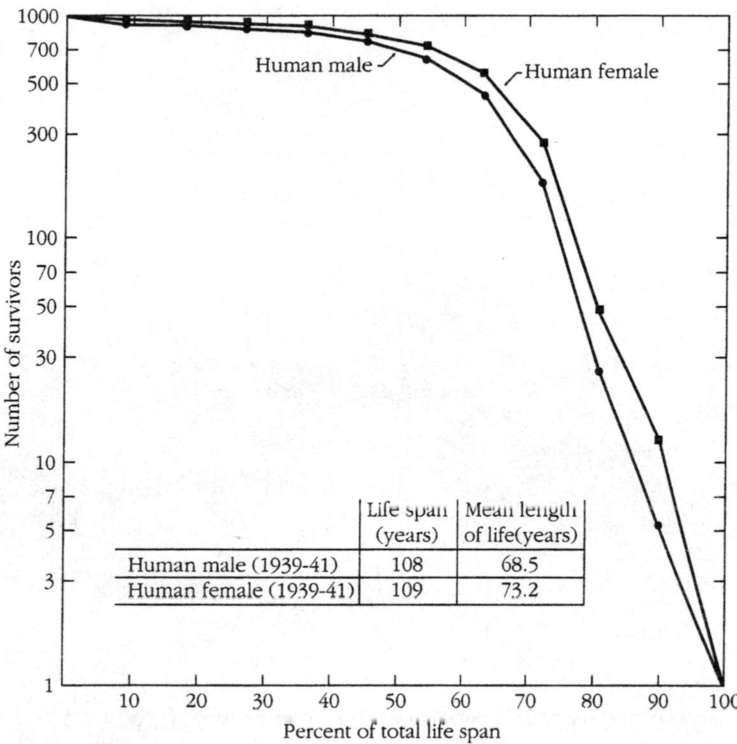

	Life span (years)	Mean length of life(years)
Human male (1939-41)	108	68.5
Human female (1939-41)	109	73.2

Figure 17-7 Differences in survivorship in human males and females at different stages of the life span, logarithmic plot. (Data on male and females, 1939–41, from L. I. Dublin, A. J. Lotka, and M. Spiegelman. 1949. *Length of Life.* New York: Roland Press.)

Mortality on a Global Level

These data again underscore the significant drop in mortality, in this case the infant mortality rate in the United States and, as seen in Table 17-4, throughout the world. Nonetheless, infant mortality in the United States in 1990 (9.7) tied with that of Ireland for 25th place, and was twice that of Japan. It varied from 7 deaths per 1,000 in affluent suburbs up to 60 deaths per 1,000 for migrant farm workers. In 1982, the United States ranked 14th in the world with an infant mortality rate of 12 deaths per 1,000, the same as Great Britain, Ireland, and Luxembourg. Given the leadership the United States has shown in so many areas, its infant mortality rate is not encouraging.

As seen in Table 17-4, the 1980 infant mortality rates vary widely between developed and developing countries: in Africa, from a low of 35 in Mauritius to a high of 217 in Gambia; in Asia, from a low of 8 in Japan to a high of 226 in Afghanistan; in Latin America and the Caribbean, from a low of 15 in Jamaica to a high of 168 in Bolivia. Comparison of the 1990 and 1980 rates demonstrates some dramatic drops

TABLE 17-4 THE 26 COUNTRIES WITH THE LOWEST AND HIGHEST INFANT MORTALITY RATES IN 1980 AND 1990

	Lowest			Highest	
	1990	1980		1990	1980
Japan	4.8	8	Afghanistan	182	205
Sweden	5.8	8	East Timor	166	175
Finland	5.9	9	Ethiopia	154	162
Iceland	6.2	11	Sierra Leone	154	208
Switzerland	6.8	10	Guinea	147	165
Singapore	6.9	12	Gambia	143	198
Canada	7.3	12	Central African Republic	143	190
Hong Kong	7.4	12	Mozambique	141	148
France	7.5	11	Angola	137	154
West Germany	7.5	15	Niger	135	200
Netherlands	7.6	10	Chad	132	165
Denmark	7.8	9	Guinea-Bissau	132	208
Malta	8.0	16	Somalia	132	177
Austria	8.1	15	Malawi	130	149
East Germany	8.1	13	Swaziland	130	168
Norway	8.4	9	Yemen, North	129	162
Australia	8.7	12	Bhutan	128	150
Luxembourg	8.7	11	Cambodia	128	212
Macao	9.0	78	Senegal	128	160
Netherlands Antilles	9.0	25	Mauritania	127	187
Spain	9.0	16	Cameroon	125	157
Belgium	9.2	12	Djibouti	122	—
Italy	9.5	18	Rwanda	122	127
United Kingdom	9.5	14	Equatorial Guinea	120	165
Ireland	9.7	16	Nigeria	121	157
United States	9.7	13	Bangladesh	120	153

Data from the 1980 and 1990 World Population Data Sheets, Population Reference Bureau, Inc.

in infant mortality. One of the most dramatic (not shown) is that of Liberia, which cut its 1980 rate of 154 to 83 by 1990!

Natality and Mortality on a Global Level

As these data indicate, there have been truly dramatic reductions in infant and adult mortality in the developed areas of the world but not so dramatic ones in the major developing areas. The subsequent effect on growth rate by reductions in mortality that are beginning to occur in undeveloped countries will be significant unless there is a corresponding reduction in natality. Changes in natality were rather summarily dismissed earlier as being of less importance than changes in mortality in the relatively recent changes in human population age structure and size. While this is the case, it is also imperative to recognize that even no upward change in birthrate, if accompanied by a decrease in death rate, will produce a burgeoning population. A high birthrate with a high mortality has quite different consequences than a high birthrate with a low mortality. The birthrate in developed countries, for instance, is generally about 15 to 20 per 1,000; in developing countries the rate is about 40. Given the present high infant mortality in developing countries, the situation of population size is markedly different from what it will be as mortality is reduced, especially if there is no downward change in natality. An assessment of the changes in natality that have occurred in the United States since 1910, their association with socioeconomic change, and their correlation with population growth (Figure 17-8) will be instructive on this point.

Natality and Fertility in the United States

It is immediately and dramatically evident that while the United States population burgeoned between 1950 and 1973, the birthrate did not show a corresponding change but rather one in the opposite direction (Figure 17-6). The increase was thus the result of reduced mortality. Had the high, peak natality rates of the immediate post-World War II years (1946–1948) been maintained and been coupled with the largely postwar reduction in mortality, the United States population would have reached some 370 million by the year 2000. Present estimates suggest about 270 million people.

The seeming lack of relevance of economic conditions to birthrate in the United States can be seen by looking at the periods of the two lowest birthrates, of economic depression on the one hand (the 1930s) and of unprecedented prosperity (1958–1966) on the other. In Sweden, however, there is a consistency between the period of economic depression (1930–1936) and a net reproduction rate at that time below the replacement level—namely, 0.81 (Davis 1967). Sweden is a nation that has matched reduction in mortality by reduction in natality.

The reader may speculate that although the United States birthrate has dropped, the fact that there are more units of 1,000 people, the baseline of comparison, would increase the total number of births. In fact, the total number of live births did show an increase from a low of about 2.3 million in 1933 to a peak of 4.3 in 1957. Since 1957, however, there was a steady decrease to about 3.2 million in 1973; in 1987 the number had risen to about 3.8 million. This shift is best explained as a function of *fertility*, the number of live births per 1,000 women between the ages of 15 and 49. In 1933 this rate was 76; it increased to 123 in 1957, dropped steadily to 97 in 1965, and to a low of 65.4 in 1984, hovering near that number since. Again, differences by race/ethnicity are evident: from 1960 to 1984, fertility dropped from 113 to 62 for white women and from 154 to 84 for nonwhite women. During the 1980s and 1990s the "fertile female" group increased to some 52 million, the result of the baby boom of the late 1940s to early 1960s. What effect this will have is, at the moment, problematic (Bouvier and DeVita 1991).

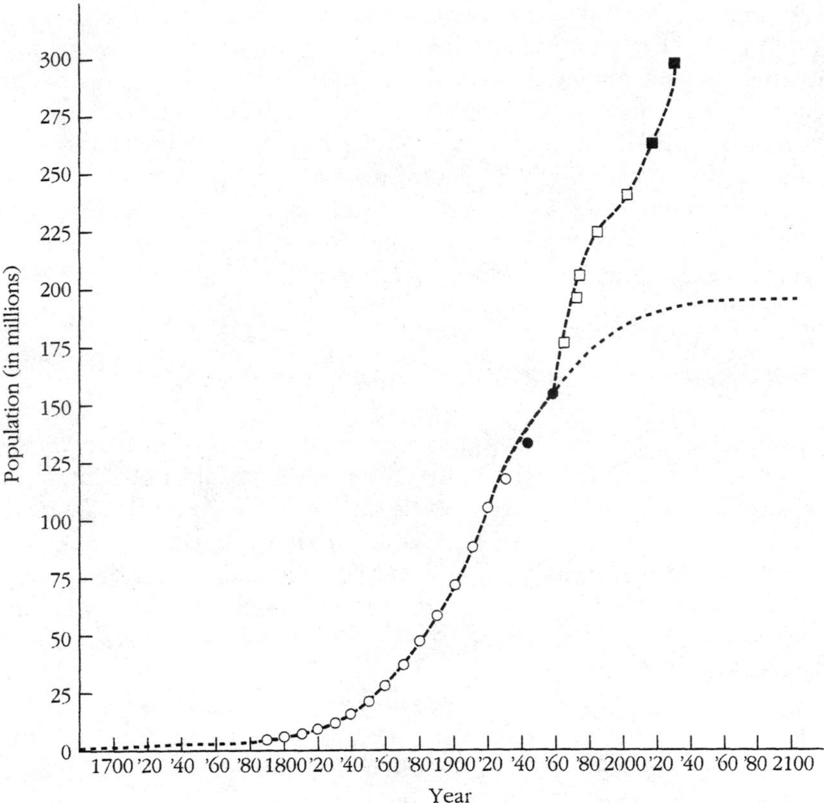

Figure 17-8 The census counts of the population of the United States from 1790 to 1940, inclusive (given by circles). The smooth curve is the logistic equation fitted to the census counts from 1790 to 1910 inclusive. The broken lines are extrapolations of the curve beyond the data to which it was fitted: the dashed portion from 1910 to 1940 is the part of the extrapolation that was tested by census counts (crossed circles) made since the logistic was originally fitted; the short dashed line shows the further extrapolation of the same curve; the squares and long dashed line represent the actual growth of population from 1950 through 1990; crossed squares and long dashed line represent projections at a moderate growth rate. (Basic graph by permission from R. Pearl, L. J. Reed, and J. F. Kish. 1940. *Science* 92:486–8. Copyright by the American Association for the Advancement of Science. Projections beyond 1990 are those of the United Nations.)

Many observers note, however, that certain conditions have tended to accelerate a decline in birthrate. A rapid economic development, with manufacturing rising faster than agriculture or construction (a capital-intensive rather than labor-intensive economy), has a consequence of increasing urbanization in limited space and a resulting relative disadvantage for large families. The downward trends of both fertility and birthrates in the United States strongly indicate there has been a self-imposed control on birthrate (Westoff 1986). Most interestingly, these trends were underway some 5 or more years before the contraceptive pill became widely available, the intrauterine device (IUD) widely used, or abortion widely and legally practiced. As most readers are doubtless aware, legalized abortion in Sweden and Japan, rather than widespread contraception, has resulted, particularly in the latter country, in dramatic reductions of birthrate. According to a study by Sklar and Berkov (1975), as a result of legislation in 1970 and 1971, the number of legal abortions in the United States rose from 200,000 in 1970 to 500,000 in 1971. The most dramatic effects were a reduction in the number of illegal abortions and deaths due to abortion as well as a decrease in the number of pregnancy-related marriages and illegitimate births. In 1980 the number of reported legal abortions was over 1.5 million (Cates 1982) and has remained at or above that level since.

PROBLEMS OF POPULATION GROWTH

Implicit in the foregoing account of human population growth, particularly in developing countries, is the notion that such growth is a problem (Keyfitz 1989). Rapid population growth is, in fact, followed by increased air and water pollution, increased demands for food and natural resources, such as minerals, forests, and water, increased urbanization and its concomitants of insufficient and inadequate housing, schools, and health services as well as loss of agricultural land and increased social and political unrest. There are, however, some observers who see population growth as beneficial to economic growth (Simon 1990, among others) and yet others who note that mortality data provide a gauge of economic deprivation (Sen 1993). The complexity of these environmental and economic problems is such that it is not possible to state categorically that population growth is the cause. It usually intensifies them, however. Certainly one of the problems exacerbated by population growth is that of feeding the world's population in order to avoid both starvation and malnutrition. Prior to exploring this problem, a little more reflection on regulating population growth is in order.

Controlling Growth

Although we can question the extent to which generalizations can be made about long-range predictions of population growth, to avoid predicting size and its probable consequences is to show no regard for future generations. Particularly troublesome in the matter of prediction is the future growth of developing nations. Here the sudden introduction of improvements in death control, developed gradually in Western countries, has brought dramatically rapid declines in mortality. Natality, however has not declined with sufficient rapidity to close the gap and thus those populations continue to burgeon (Figures 17-2, 17-4). Will a self-imposed natality, comparable to that which seems to have developed in the United States, Japan, and European countries, develop in other countries as their socioeconomic status improves? Or because many critical observers see Western-oriented contraceptive methods as impractical or improvident in non-Western cultures, does the regulation of these regions lie partly in the sanctioning of abortion, as in Japan and Sweden? There is no question but that improved medical and environmental technology will continue to depress mortality rates in all age groups, particularly in the early years of life, including the prenatal months. Then if growth rates are to be decelerated, the only mechanism of control is at the input side—natality.

It is precisely here, as Kingsley David (1967) has noted, that society runs amuck. Almost all proponents of population control have directed attention to family planning on the assumption that it is an effective means of regulating growth and have expended millions of dollars in the effort (Johnson 1987). There is no question but that population growth can be regulated by restraints on family size, but the paradox is that the principle of family planning stresses the right of parents to have the number of children they want, not the right of society to have the number of children it needs. This situation creates a dilemma; as long as families want more children than the number required for mere replacement, the population will continue to grow.

World fertility and replacement levels. Because many people remain childless and many die before reaching childbearing age, a given couple must replace more than themselves for the population as a whole to be maintained at *replacement level.* A country that achieves a fertility level of 2.1 (i.e., slightly more than two births per woman between 15 and 49) is at the point where births and deaths eventually balance out, that is, the replacement level. In 1990, *fertility rates* (the average number of births per female between 15 and 49 years of age) in 38 countries, almost the entire industrialized world, were at or below replacement level; by contrast, 20 nations had fertility rates of 7 or above, and another 27 had rates between 6.0 and 6.9 (Table 17-5). In these latter two groups of 47 nations, 34 are in Africa, and 12 are in the Middle East; of the African nations, those in sub-Saharan Africa have both the highest fertility and population growth rates in the world (Jacobson 1988; Caldwell and Caldwell 1990).

According to the World Fertility Survey and other surveys conducted largely in the 1970s to mid 1980s, the mean number of children desired ranged from a low of 2.1 in Hungary to 6.8 in Kenya (Kent and Larson 1982). On a regional basis, the ranges were: Africa, 6.1 (Sierra Leone) to 6.8 (Kenya); Middle East, 3.0 (Turkey) to 6.3 (Jordan); Asia and the Pacific, 2.2 (Japan) to 4.4 (Philippines); Latin America and the Caribbean, 3.6 (Haiti) to 5.1 (Paraguay); and in Europe, 2.1 (Hungary) to 2.8 (Spain). In the majority of the developing countries covered by the World Fertility Survey (12 in Asia and the Pacific, 10 in Latin America and the Caribbean, and 1 in Africa), women wanted between three and four children. However, in most of the countries, about 50 percent of the women wanted to avoid future births by the time they had three living children, a desire found to be particularly strong among women aged 35 and older (Robey et al. 1993). Less educated and rural women tended to

TABLE 17-5 COUNTRIES WITH THE LOWEST AND HIGHEST FERTILITY RATES[a]

Lowest			
1.5 or below	1.6 to 1.8	1.9 to 2.1	2.2 to 2.5
1.3 Italy	1.6 Belgium	1.9 Cuba	2.2 Guadaloupe
1.4 Austria	Denmark	2.0 Bulgaria	Ireland
Hong Kong	Japan	Malta	Puerto Rico
Luxembourg	Portugal	Mauritius	2.3 Bahamas
West Germany	Switzerland	Netherlands	China
1.5 Greece	South Korea	Antilles	Iceland
Netherlands	1.7 Antigua and Barbuda	New Zealand	Romania
Spain	Canada	Singapore	Sri Lanka
	East Germany	Sweden	2.4 Cyprus
	Finland	United States	Jamaica
	1.8 Australia	Yugoslavia	Reunion
	Barbados	2.1 Czechoslovakia	Uruguay
	France	Martinique	2.5 Chile
	Hungary	Poland	North Korea
	Norway		
	Taiwan		
Highest			
6.0 to 6.3	6.4 to 6.7	6.8 to 7.1	7.2 and above
6.0 Congo	6.4 Angola	6.8 Syria	7.2 Burkina Faso
6.1 Algeria	Gambia	7.0 Benin	Oman
Namibia	Liberia	Burundi	Saudi Arabia
Nepal	Mozambique	Gaza	Togo
6.2 Ethiopia	Senegal	Iraq	Zambia
Guinea	Sudan	South Yemen	7.4 Cote D'Ivoire
Swaziland	6.5 Mauritania	7.1 Afghanistan	Uganda
Zaire	Nigeria	Comoros	7.6 North Yemen
6.3 Iran	Sierra Leone	Niger	7.7 Malawi
Ghana	6.6 Djibouti	Tanzania	8.3 Rwanda
Solomon Islands	Madagascar		
	Maldives		
	Somalia		
	6.7 Kenya		
	Pakistan		

[a]Average number of births per female between 15 and 49 years of age. Data from 1990 World Population Data Sheet, Population Reference Bureau, Inc.

want, and have, larger families and, in some of the countries, religious or ethnic affiliations affected the average number of children desired. Finally, women who wanted no more children were much more likely to be using contraceptives than other women, an indication of trying to achieve their fertility goals.

Nevertheless, it is true that as long as the family is "free" to plan its size, there may well be increased population growth and, as death control expands to developing nations, there will be an even greater increase in population size.

Global family planning. Recent studies on world fertility indicate a drop from 1968 to 1990 of 4.6 to 3.5, primarily because of a rapid drop in fertility in less developed countries. Development, or socioeconomic progress—a factor associated with lower fertility and rate of population growth in developed countries—proved to have been too small to account for more than a relatively small part of the decline. Instead, contrary to the widely held view noted in the preceding paragraphs, organized family planning programs have been a major contributing factor. Begun in the early 1960s, these efforts have developed into a major, and successful, worldwide campaign involving international agencies, many national governments, and numerous foundations and private organizations (Donaldson and Tsui 1990; Jacobson 1988; Johnson 1987).

Countries such as Mexico and Thailand have devoted extensive resources to expanding services and supplying contraceptives, whereas in Brazil the efforts of private voluntary organizations have been the key to declining birth rates (Jacobson 1988). China's well-known national policy of the "one-child family" (Greenhalgh and Bongaarts 1987) has been successful in lowering its growth rate from 1.8 in 1970 to 1.3 in 1990 and its fertility rate from 3.8 in 1977 to 2.3 in 1989, the latter getting closer to the 2.1 replacement level that is now characteristic of the whole of East Asia (Tien et al. 1992; Feeney 1994). The rate of childbearing in China approximated 60 percent from 1970 to 1980 but was reversed by 1985–86 because of a boom in marriage that followed a relaxation of locally administered restrictions on marriage before the officially designated age (Coale et al. 1991).

One of the measures of this impact is a survey on the knowledge and use of contraceptives since the 1960s in 20 developing countries, including Mexico, Peru, Kenya, Pakistan, Bangladesh, and the Philippines (Mamlouk 1982). In 19 of the 20 countries (Nepal being the exception), three-quarters or more of the women who were or had been married know about contraception, with modern methods (e.g., the pill, condom, IUD) being more commonly known than traditional methods (e.g., abstinence, withdrawal, rhythm). Furthermore, modern methods were more frequently used than traditional methods, the most common being the pill, IUD, and sterilization. The highest use of contraception was by women in the midrange of their reproductive years; it was lower among younger and older women. Finally, usage increased with levels of educational attainment, paid employment, and urban residence.

In spite of the considerable progress in family planning in both developed and developing countries, there remain significant gaps between the number of women who are able to limit fertility and the number who express a desire to do so because of inadequate access to or knowledge of family planning methods. The World Fertility Survey demonstrated that 40 to 50 percent of women in 18 developing countries desire no more children but have no access to family planning; fertility rates could be reduced by 30 percent in these countries if unwanted births were prevented (United Nations Department of International Economic and Social Affairs 1987).

Given comparable socioeconomic settings, the percentage of couples who use contraceptives is higher in countries that have stronger family planning programs than in countries with weaker programs (Keyfitz 1989). Those with the weakest programs include the sub-Saharan African nations where there is very low contraceptive use—and the highest fertility rates (Table 17-5). Until recently, most African governments opposed family planning programs because they believed a curtailment of population growth would limit economic potential (Caldwell and Caldwell 1990). While the number of women using family planning has increased sixfold to over 400 million since 1960 (Donaldson and Tsui 1990), there are still areas where family planning concepts have not been effective. Fewer than one-fifth of Nigerian women have ever heard of a modern method of birth control, and, in Kenya, less than 40 percent of women familiar with at least one modern contraceptive method knew of a supply source; fewer than half of these women could reach the source on a 30-minute walk (Jacobson 1988).

An aging world. With the significant inroads being made on fertility and mortality, there is a consequence of major magnitude beginning to be felt worldwide, namely the aging of the human population. The number of elderly (those over 65 years of age) is growing at the rate of 2.4 percent a year, for a doubling time of 29 years; in 1985 there were 290 million elderly and a projection of 410 million by the year 2000 (Bureau of the Census 1987). Sweden is the "oldest country" in the sense of having the highest percentage of people over 65 years of age (17 percent), and Japan, currently with 12 percent elderly, is projected to reach 24 percent by 2025. Fifty-four percent of the elderly live in developing countries, a proportion that is projected to reach sixty-nine percent by 2025. Based on estimates by the United Nations in 1988 (Martin 1991), current and projected percentages of the elderly in selected locations are:

	1990	2025
United States	12.6	19.6
Japan	11.7	23.7
South Korea	4.7	13.9
China	5.8	13.0
Singapore	5.6	19.1
Malaysia	3.8	9.3

China's rapid transition to low fertility (i.e., the one-child family policy) and mortality result in a potential 4-2-1 population pyramid of four grandparents, two parents, and one child. With mortality at an already low level, natality will have a greater influence on population aging in China (Tien et al. 1992). This can be seen in the to-date and projected increase in the percentage of elderly: 1982, 4.9 percent; 1990, 5.9 percent; 2000, 7 percent; and 2025, 13 percent. Numerically, these increases translate to: 1980, 50 million; 1990, 66 million (the combined populations of California, New York, Texas, and Florida, the four largest U.S. states); 2020, 167 million. This increase has already placed a strain on the retirement pension system in China, which currently affects a relatively small portion of the population, and those are largely in urban areas.

Public policies dealing with aging range from attempts in Singapore to reverse it by encouraging more births, to efforts in Japan to accommodate it by increasing employment opportunities for older workers, thereby decreasing the impact on economic growth and public expenditures (Martin 1991). As noted earlier, there is some reason to believe the Japanese government is encouraging procreation to offset this aging (Jitsukawa and Djerassi 1994).

To date, the United States, whose elderly population is increasing more slowly than that in East Asia because of its longer postwar baby boom, has not developed any substantial policies for this inevitability. During the latter part of this century, the large United States baby boom population (those born between 1945 and 1965) will contribute mightily to the economy. But this same population in a few years will put a major strain on health care and the Social Security System, the major burden passing on to the post–baby boom (sometimes called "baby-bust") generation, which, because of a sharp drop in fertility rates, is much smaller. Perhaps there will be time to learn from our East Asian neighbors.

Cultures respond differently to their aging members. In the United States, there has been, in general, a negative connotation to aging and to the elderly. This view is markedly different from that common in many nonindustrialized cultures (Simmons 1970) and in Japan, where the elderly usually are respected (Palmore 1975). Generally speaking, there appears to be much diversity in the status of and attitudes toward the elderly within many cultural groups (Moore et al. 1980), with status being related to many variables, including economics, family structure, dependency, health, and so on. Another factor that may weigh more heavily in the future is the sheer number of elderly in the society, a relatively new phenomenon to which cultures must adapt. It remains to be seen how different cultures will, in fact, adapt to the likely substantial increases of the elderly in their midst.

FEEDING THE HUMAN POPULATION

These foregoing studies on reduction in fertility, even against dropping mortality, bode well for a number of developing countries to be at or near the replacement level (2.1) by 2000 if family planning continues. Were replacement levels to be reached on a worldwide basis by 2000, the resulting population would increase from 5.3 billion in 1990 to be more or less stabilized at slightly under 9 billion people by the end of the twenty-first century. Given the reality of present trends, however, the likelihood of this occurring is remote, since the world fertility stood at 3.5 in 1990, for a doubling time of 39 years. The size of the world's population at the end of the twenty-first century depends on when the replacement level is reached: if not until 2040, a population of 11 billion, slightly more than twice that of 1990; if not until 2080, a population of about 14 billion, slightly less than three times that of 1990.

Whichever scenario develops, the world's population will continue to grow and that means more people to feed. The primary issue becomes one of food—enough of it and well enough distributed to preclude starvation and malnutrition. This issue deserves further discussion.

Extent of Starvation and Malnutrition

The World Bank (1986) has estimated than more than 700 million people, or about 13 percent of the world's population, lack enough food for an active and healthy life, with the largest concentrations in Asia and sub-Saharan Africa. More recent estimates place the number at 950 million people (that is 1 in 5 people!) being chronically malnourished, that is too hungry to lead active, productive lives (McAfee 1990). Other estimates suggest that the number of hungry people increased five times faster in the 1980s than in the previous decade (World Food Council 1989). Others have estimated that perhaps a quarter of the world's population goes hungry during at least some season of the year (Clark 1989). UNICEF (1990) estimates that over 150 million children under 5 years old suffer from malnutrition; again most of these children live in Asia and sub-Saharan Africa. Malnutrition and its related diseases are responsible for the death of as many as 15 million children a year (McAfee 1990).

In 1986 the World Bank estimated that there were between 340 million and 730 million people in developing countries who lacked the income to ensure an adequate intake of dietary energy (Hendry 1988). Another way of stating this is that in 1980, 34 percent of the people in 87 developing countries did not obtain enough calories for an active working life, and 16 percent did not obtain enough calories to prevent stunted growth and serious health risks. About 80 percent of those who are undernourished live in countries with very low average incomes, two-thirds of the total residing in Asia and one-fifth in sub-Saharan Africa.

Malnutrition results from lack of adequate food, a deficiency of one or more essential nutrients (e.g., protein, calcium, vitamin C), or genetic or environmental illness in which there is interference with digestion, absorption, or metabolism. Too much food or too much of one type of nutrient is also regarded as a form of malnutrition and occurs primarily in the affluent parts of Western cultures. The extent and demography of starvation and malnutrition suggest that the answer lies in both expanded production of food, particularly highly nutritious kinds, and more equitable distribution of that food.

Earth's Carrying Capacity

Theoretical considerations. An initial perspective on feeding the world's population can be derived by considering carrying capacity, based on discussions of productivity in Chapter 6. *Carrying capacity* is equated to the parameter K of the logistic equation (see Chapter 10) and is generally regarded in terms defined by Odum (1953) as the "upper bound beyond which no major increase can occur (assuming no major changes in the environment)." Pulliam and Haddad (1994) have recently provided an extensive review of carrying capacity as applied to human population.

From the data on primary production (Figure 8-5) the estimated total annual net productivity is 2.5×10^{17} C for the oceans and 5.2×10^{17}C for the land, for a total of 7.7×10^{17} C per year. If the present human population of approximately 5.5 billion were exclusively herbivorous, and each person required an average of 2,200 C per day (the United Nations define adequate intake as 3,000 calories/day for a man and 2,200 calories/day for a woman), or 8×10^5 C per year, the total energy requirement would be 4.4×10^{15} C. This is about 0.4 percent of current net energy capture. We might jump to the happy thought of a population 150 times larger than at present—namely, 750 billion—supported by a current net production. The situation is complicated, however, by several factors: humans are not exclusively herbivores; there are numerous other herbivores; much plant biomass is not digestible by humans (e.g., trees); assimilation is not 100 percent efficient; and one-third of total plant production is in the ocean—the major untapped, unmanaged, and, so far, largely unpalatable plant production resource on the plant.

To try another perspective, assume that humans are completely carnivorous. The interjection of another trophic level, the herbivore, implies a substantial reduction in energy that would then be available. Based on the earlier discussion of this trophic-level energy transfer, we can expect an energy loss on the order of 85 to 90 percent from the net production of herbivores. Thus the 7.7×10^{17} C of net plant production would amount to 7.7×10^{16} C if 10 percent is actually transferred. At this level of energy use, the carnivore need (4.4×10^{15} C) would allow for a maximum total human population of about 75 billion. But in this calculation we have assumed no other carnivores and 100 percent energy transfer, neither of which is consistent with natural processes. And in both manipulations we have ignored the fact that food is not equally distributed throughout the world.

Realistic considerations. From a more realistic consideration of the Earth's carrying capacity, the United Nations Food and Agriculture Organization (FAO) in 1984 analyzed soil and climate data from 117 developing countries in order to estimate the food production potential for 15 major crops at three different levels of farming practice: (1) low input, or subsistence agriculture; (2) intermediate, with the introduction of some modern technology, conservation measures, and improved cropping patterns; and (3) high level, the equivalent of farming practices in industrialized countries (Hendry 1988). Even though there are many limitations to the FAO study, it did show that from a global perspective, subsistence-level agriculture would have been capable of supporting more than double the 1975 population of 2 billion in the 177 countries. However, by the end of the century, the margin would have narrowed significantly. By contrast, intermediate levels of agriculture could theoretically support four times the population projected for the year 2000, and high-level inputs and practices nine times that population.

At the national level, however, the FAO study found that of 117 countries, 54 would have been unable to feed their 1975 population (more than 1 billion people) from subsistence-level agriculture. Intermediate-level agriculture would have reduced the number of countries at risk to 24 (these countries having 4 percent of the total population); high technology agriculture would have led to only 13 countries being in critical condition (these have but 1 percent of the total population). The situation becomes exacerbated, according to this FAO study, for the projected population in the year 2000 when the number of critical countries would be increased to 64 at low-level agriculture, 28 at the intermediate level, and 17 at high levels.

Capacity of the land. Table 17-6 shows the amount of arable and grazing land in the world; it amounts to about half the total land surface, exclusive of the ice-covered portions of Antarctica and Greenland. Less than half of all arable land and more than half of all grazing land are currently in use. Of the soil that is suitable for agriculture, only 11 percent has no limitations, that is, it is sufficiently rich in nutrients and moisture so as to not require fertilizer or irrigation (Hendry 1988). Twenty-eight percent of potentially arable land would require irrigation and 23 percent would require fertilization.

These limitations notwithstanding, there was a 179 percent increase in croplands between 1950 and 1990 (from 538×10^6 hectares to $1{,}501 \times 10^6$) and a one percent reduction in grasslands and pastureland over the same period (Repetto 1987). On a regional basis, the increase in cropland ranged from a low of −4 percent in Europe to a high of 841 percent in Pacific developed countries. In Latin American and African nations, the substantial increase in cropland and pastureland was largely at the expense of the forests, a topic that will be treated in Chapter 19.

On a worldwide basis, 1 hectare of land currently supports about 3 persons compared to highly productive farmland, as in the United States Midwest, which can support 24 persons. Complicating the problem is the fact that much arable land is devoted to inedible crops, such as cotton, tobacco, rubber, and coffee, or the production of food for poultry and livestock. About one-half to two-thirds of all currently used arable land, barring fertilization and irrigation, must lie fallow or is used for pasturage.

TABLE 17-6 TOTAL ARABLE AND GRAZING LAND IN THE WORLD

Land surface	Hectares
Total land surface (exclusive of ice-covered portions of Antartica and Greeenland	13.0×10^9
Unusable (deserts, mountains, tundra, sandy or lateritic soils)	6.6×10^9
Arable	
Currently in use	1.4×10^9
Currently not used	1.8×10^9
Grazing	
Currently in use	1.8×10^9
Currently not used	1.4×10^9

World Food Production

Food production needs. The current rate of growth of agricultural output in many developed countries is about 2 percent a year; to meet the needs of populations whose growth rates are about 2 percent a year, agricultural output must increase to 3–4.5 percent (Wortman 1980). It is obvious that agricultural output must increase considerably more for those populations with higher growth rates. According to Crosson and Rosenberg (1989), world production of food is growing faster than population, a finding that would cause Malthus to turn in his grave. In the 1960s production of cereals increased annually by 3.7 percent whereas population increased by 2 percent; in the 1970s the respective increases were 2.5 and 1.8 percent, and in the 1980s, 2.1 and 1.6 percent. If the food supply were to increase at the latter pace, Crosson and Rosenberg project that there would be enough food for a stable world population of 10 billion in 100 years.

Prospects for future food production. Non-Malthusian optimists, those who do not see human population growth as being limited by physical resources available for agriculture, look to increased use of irrigation, fertilizers, and Western agricultural technology to increase productivity. They also look to a more enlightened and sophisticated social policy to ensure more equitable distribution of food. Roger Revelle (1983) theorizes that the amount of presently arable land (1.3 million hectares) could be increased more than threefold to about 4.2 billion hectares with irrigation. This would result in 0.62 hectare per person in the year 2000 compared to 0.37 in 1970. Coupled with increased use of fertilizers and pesticides (as much as 20 percent of all annual crops are destroyed by pests), flood control, land leveling and mechanization, crop yields could be substantially increased.

Also of promise is the creation of new crop varieties by genetic manipulation. For example, photosynthesis in rice and wheat follows the Calvin-Benson cycle or 3 carbon pathway and hence these grains are referred to as C3 plants. In contrast, maize and sorghum follow an alternative route in the dark reaction in which four carbon acids are the first products of photosynthesis; these are called C4 plants. These C4 plants are more efficient in photosynthesis and also grow particularly well at higher temperatures, as in the tropics. Some desert plants have a metabolic capacity to take up carbon dioxide at night when their stomates are open, the gas being stored for photosynthesis during the day. These desert plants drastically reduce water loss during the day through closing of their stomates. Genetic manipulation that would combine this "desert" capacity into C4 plants or that would combine rice and wheat with the more efficient C4 plants would greatly enhance photosynthetic efficiency and hence crop production. Gene transfer that would allow nitrogen-fixing bacteria, such as currently live in root nodules of legumes, to grow on grain plants would greatly increase productivity while reducing the need for nitrogen fertilizer. Progress in genetic engineering bodes well for such eventualities.

Genetic manipulation, however, runs the risk of eroding the genetic diversity of crop plants, the diversity that is the genetic basis for evolutionary adaptation to changing environments. This concern has led to the establishment of "gene banks," repositories for the gene pool of crop plants (Plucknett et al. 1987). These linchpins in the global effort to conserve this invaluable genetic heritage are especially important when we consider that the tropics contain the richest reservoirs of plant resources while the bulk of the capital and technology to develop these resources exists in the major industrial countries.

Major United States crops have a large genetic potential for yield that is unrealized because of the need for better adaptation of the plants to the environments in which they are grown. Native populations demonstrate that high productivity can occur, and genotypic selection for adaptation to such environments is already demonstrating an important role in improved productivity (Boyer 1982). Similarly, research on improving the interception of solar radiation by crop foliage and the efficiency of energy conversion in photosynthesis offers promise for increased productivity (Gifford et al. 1984).

Agricultural research on marginal lands that are ill-suited for food production in developing countries has been enhanced considerably by the International Agricultural Research Centers (now banded together under the title of Consultative Group on International Agricultural Research (CGIAR)), which, among other efforts, is emphasizing how varieties of plants are being tailored for introduction into the marginal areas (Plucknett and Smith 1982).

As noted above, arid lands constitute 28 percent of those that are potentially suitable for agriculture. In the United States alone there are substantial hectares of arid and semiarid rangeland, and there are even larger areas in Africa, Australia, and South America (Hinman 1984). Significantly, 90 percent of our food comes from a dozen crops, none of which is capable of being grown in arid conditions. There are xerophytic plants, however, and some of them have already shown economic, if not food, value. Among them are rubber from guayule, seed oil from jojoba, buffalo gourd and bladderpod, and resin from gumweed; of real economic value is the fact that after these products have been removed from the plants, the remaining material can be used for fuel, a valuable commodity in arid lands.

Limitations on future food production. These rather optimistic projections do carry, however, a substantial price tag for irrigation, fertilizers, mechanization, energy, and applied research (Reganold et al. 1990; Bongoarts 1994a). Brady (1982) has noted that much of the unprecedented increase in food production in developing countries has been due to chemical-based technologies and to the use of agricultural chemicals, that is fertilizers. Moreover, other observers are not so sanguine about these prospects even with significant capital investments. Lester Brown (1981), for example, notes that instead of increases in the use of presently arable land, the trend is in the opposite direction. Nearly 400,000 hectares of prime cropland in the United States were converted to nonfarm use each year between 1967 and 1977. Agricultural land equal to the size of the State of Ohio (about 65,000 km^2) has been covered by highways in the United States in the last 40 to 50 years. West Germany lost 2.5 percent of its agricultural land between 1960 and 1970, whereas France and Great Britain lost about 2 percent in the same time period.

As important as land available for cultivation is the loss of topsoil, which is subject to excessive erosion under intense cultivation. The State of Iowa loses 260 million tons of topsoil per year from its 146,000 square kilometers. Where wind erosion is particularly severe, as in the Great Plains, losses can average as much as 6 tons/ha in a year. Water erosion removes 2 billion tons of topsoil annually in the United States, whereas only about 1 billion tons are formed each year. In 1977 the United Nations reported that one-fifth of the world's cropland was being degraded by loss of topsoil.

Irrigation also has the potential disadvantage of waterlogging the soil. Estimates place the amount of irrigated soil that is waterlogged at 10 percent, and soil on which productivity has fallen at 20 percent. Irrigation in areas with high evaporation rates results in salinization of the soil. Finally, increasing demands for water in our cities and for industry not only diverts it from agricultural use but also puts severe strains on nonrechargeable water supplies of large aquifers. The Great Ogallala Aquifer, which extends from Nebraska to Texas and whose water capacity equals that of Lake Huron (Table 17-1), is one such victim. At present use rates its depletion will result in the loss of highly productive, but irrigated, farmland equivalent in area to the Commonwealth of Massachusetts.

Altered food distribution might also be considered, facilitated by some kind of world food bank. Between 1951 and 1971, for example, world production of cereal grains, the staple of the world food supply, doubled for an increase of about 40 percent per person. More than half the increase was absorbed by the 30 percent of the world's population in the developed countries.

Altered food production carried a warning derived from the rapid improvement in crop production during the 1960s and 1970s known as the Green Revolution. Although dramatic gains were experienced in the yields of many grain crops, many such yields proved to be protein deficient. A "good" protein food has a carbon-to-nitrogen ratio of 3.2:1. Ratios of 17:1 indicate protein deficiency. The high-yield corns and wheats of the Green Revolution had ratios of 30:1 and 26:1, respectively. In contrast, soybeans have a carbon-to-nitrogen ratio of 6:1 and many legumes average about 10:1. Peculiarly, nitrogen fertilizers tend to increase the carbon-to-nitrogen ratio because of the production of more carbohydrate and less protein.

Altered food habits would also help support a larger population. It has been estimated, for example, that the same amount of food that feeds America's population would feed seven times that number of Chinese on the average Chinese diet! However, as China's population grows and per capita incomes increase, the diet diversifies from a monotonous fare in which a starchy staple such as rice supplies 70 percent or more of calories to more meat, milk, and eggs, the latter creating a surge in grain demand. These cultural changes, coupled with a massive conversion of cropland to nonfarm use, raises the specter that China will soon no longer be able to feed itself, its demand for food then overwhelming the export capacity of the United States and other producers (Brown 1994).

A word on rice. The first Green Revolution affecting rice probably began with neolithic people who began the domestication of grasses belonging to the genus *Oryza*, the ancestor of today's rice. Rice is important, if not unique, among grain crops for two major reasons: it produces more calories per unit of cultivated area, corn being next; and it, unlike other grains, is grown almost entirely for direct human consumption. In the 1980s, 40 percent of the world's population depended on rice for at least half their diet, yet rice is grown on only 11 percent of the world's arable land (Van Slyke 1988).

Over 95 percent of the world's rice is grown in East, Southeast, and South Asia, China being the largest producer of all, with some 36 percent of the total (Figure 17-9). Within China, the Yangtze Valley accounts for 75 percent of the nation's output or about 25 percent of the world's total (Van Slyke 1988). Historically, although some rice was shipped on the Yangtze River system, about 90 percent was consumed in the producing area.

Although the pattern and technology of rice production has apparently changed little over several thousand years—the rice paddy, intensive cultivation, enormous amounts of hand labor—the acreage involved was pushed to the limits, aided and abetted by irrigation and terracing. This permitted much greater yields per hectare than less intensive methods of agriculture. About 1000 A.D. improved strains of rice were introduced, probably from Cambodia; the most significant of

Figure 17-9 Rice paddies near Mount Agung, Bali, Indonesia. (Photo by author.)

these new strains were early ripeners, which made two crops a year possible on the same land. Today, improved strains—the continuing Green Revolution of rice more resistant to disease, wilting, early dropping of grains, etc.—continues to develop through the International Rice Institute in the Philippines. It is largely the work of this organization that has helped make much of Asia self-sufficient in rice.

Some alternatives. By further judicious control and selection of the herbivores on which humans could come to depend as intermediaries in energy flow, there is a potential for increased efficiency and production at that level. This is an important point, for meeting the world's food needs is not only a quantitative but also a qualitative matter, with protein needs being the most difficult to meet. For in spite of efforts to meet caloric requirements based on agricultural output, the potential would still exist for serious protein deficiency. To obviate this condition, humans may need to adjust cultural attitudes to adopt some methods and sources of meeting protein needs that at present are considered unorthodox (Pirie 1967). Among unconventional food sources that have been suggested and on which research is being conducted are the freshwater manatee, the marine dugong, the capybara, a large South American rodent—all of which feed on aquatic weeds that contribute little to human nutrition—and the eland, an African antelope that is adapted to grazing on marginal lands not suited to agriculture. Alternately, the protein-rich residue left when oil is removed from soya, cottonseed, and sunflower, which is now used as animal feed or fertilizer, shows considerable potential as a protein source through improved extraction procedures. Still another unorthodox source, besides the use of algae and synthetic foods produced in the laboratory, is the use of microorganisms as food, microbes that grow on such substrates as sawdust, petroleum, and coal, substrates that are otherwise unusable directly in meeting our food needs (Rose 1981).

Modernization, or "Westernization," of agriculture has, among other limitations, the compelling one of being extremely energy intensive with large amounts of energy being used in growing crops (machines, fertilizers), in processing and packaging, in auto-powered shopping, and in refrigeration (used by the consumer). David Pimental and his associates at Cornell University (1976) have shown (Table 17-7) that a tremendous amount of energy is required in Western agricultural methodology compared with manual farming. In noninudstrialized cultures each calorie of energy invested in

TABLE 17-7 ENERGY COSTS IN FOOD PRODUCTION (C/ha)

	United States		
	1945	1970	Mexico
Total input	2,379,200	6,644,220	53,108
Corn yield	7,504,640	17,881,600	6,842,880
Return/input	3.15	2.69	128.8

Data from D. Pimenthal et al. 1976. *American Biology Teacher*, October, pp. 402–6.

agriculture produces 5 to 50 calories of food; in industrialized countries it takes 5 to 10 calories to get 1 food calorie. If all countries followed the latter pattern, the world would use 80 percent of its annual energy consumption just to produce food.

Concluding comment. These last few paragraphs have been directed toward only one major problem associated with population growth—quantitatively and qualitatively feeding an expanding population. The use or rate of use or misuse of natural resources—of space, air, and water, of renewable and nonrenewable resources—has not been discussed, neither has the impending as well as present disbalance between sources of supply and demand, essential and presumed. Too prevalent is the myth of the cornucopia, of unlimited supplies of food, space, and materials to meet an unlimited growth. The Earth is ultimately finite in its resources; they are ultimately of limited quantity. Demand must come into balance with supply and that portion of the supply that can be recycled must be recycled if humans are to survive as a species. No species can exist beyond its resources. There is a limit to growth, a limit to the size of population that can be sustained by a finite supply of resources. Human population size will need to be regulated and demand for supplies will need to be reduced.

In this context, John Bongaarts (1994b), of the Population Council, explores in depth three broad policy options to slow this population explosion. These include: (1) reducing unwanted pregnancies by strengthening family planning programs (the unmet need for contraception is nearly 25 percent in sub-Saharan Africa) through knowledge of and access to family planning services; (2) reducing the demand for large families through investments in human development including: raising educational levels; improving the economic, social and legal status of women; and de creasing infant and child mortality; and (3) interrupting population momentum, that is, the tendency of population size to increase for some time after fertility reaches a level consistent with long-range population stability. This can be achieved by additional declines in lifetime fertility to below the replacement level and raising the average age of women at childbearing.

These are formidable objectives fraught with political, economic, social, and often religious overtones. Nonetheless, they do warrant careful consideration in view of the likely untoward consequences of the current burgeoning rate of population growth, particularly in developing regions of the world.

SUMMARY

VOCABULARY

carrying capacity
doubling time
fertility
fertility rate
infant mortality
life expectancy
malnutrition
mortality
natality
replacement level

KEY POINTS

- The time it takes the world's population to increase by 1 billion has dropped from 2–5 million years from the origins of humans until 1800, to 9 years today, with a current doubling time of slightly more than 40 years.
- Not all regions of the world show the same rate of growth, developed countries being stable to declining, developing countries, in general, being yet on an accelerated scale.
- Although human mortality in both developing and developed countries is now about the same, birthrate (births per 1,000 people), which has declined from about 40 to 15 in developed countries over the last 150 years, remains at the level of about 40 in developing countries. Burgeoning populations are due both to the foregoing factors as well as the significant increase in life expectancy, resulting in some countries having an aging population.
- Infant mortality rates vary widely between, as well as within, developed and developing countries.
- In spite of some successes in family planning, fertility rates and replacement levels remain high in most developing countries.
- Starvation and malnutrition are consequences of overpopulation and inadequate production and/or distribution of food and affect some 13 percent of the world's population.
- Optimists and pessimists debate the carrying capacity and food production/distribution capability of Earth, proposing various strategies for meeting ever-increasing food demand.

CHAPTER 18

Anthropogenic Impact on Aquatic Ecosystems

This and the next two chapters will deal with the *anthropogenic* impact, that is human impact, on the three environmental components—water, land, and air—by focusing on some major environmental problems and issues. These problems and issues are tremendously broad, technical, and complex, and thus cannot receive exhaustive treatment here. However, like the topic of human population, the attention of students must be drawn to these issues since environmental problems will continue to affect their lives and welfare, both now and in the future.

In large measure, anthropogenic interaction with water, land, and air has largely resulted in pollution of those environmental components. It is important, therefore, to understand what pollution is before discussing the topics at hand.

POLLUTION

"To make or render unclean; to defile; desecrate; profane"—this is the meaning of *pollute* according to *Webster's New Collegiate Dictionary*. These are strong words, crisp and lucid. They are more direct but less informative than the following definition adopted by the Environmental Pollution Panel, of the President's Science Advisory Committee, in its report, "Restoring the Quality of Our Environment," in November 1965:

> Environmental pollution is the unfavorable alteration of our surroundings, wholly or largely as a by-product of man's actions, through direct or indirect effects of changes in energy patterns, radiation levels, chemical and physical constitution and abundances of organisms. These changes may affect man directly, or through his supplies of water and of agricultural and other biological products, his physical objects or possessions, or his opportunities for recreation and appreciation of nature.

The key phrase here, "unfavorable alteration," is one that we shall return to subsequently and consider at length. First, let us consider in more detail the nature of the pollutant itself.

The production of pollutants comes as the "by-product of man's actions"—they are the residues of things humans make, use, and throw away—their cans and bottles, metal and plastic caps, waste rock and mill tailings, pesticides and herbicides, automobile exhausts, and industrial discharges. They are concomitants of a technological society with a high standard of living. They increase both because of population increase and because of an increasing expectation for higher living standards. More is made, used, and thrown away. But as the Committee on Pollution of the National Academy of Science noted in 1966, "As the earth becomes more crowded, there is no longer an 'away.' One person's trash basket in another's living space." Pollution, then, can be regarded as primarily a result of human behavior that largely disregards consequences, one that unwisely uses and discards resources. But pollutants also include the "natural" byproducts of human metabolic activity and that of the organisms on which humans depend for food, their biological waste products, and farm animal excreta.

Seen in this perspective, a pollutant is not something apart from humans but is inherent in their very biology and culture; it is a result of their peculiar adaptations and attributes. In this framework, then, why should a "natural" byproduct be considered "unnatural"—that is, polluting? This is an important point to bear in mind. The so-called pollutants, a value- and emotion-laden word in today's society, are, in fact, normal byproducts of people as purely biological organisms and as creative social beings. They are the inorganic and organic wastes of metabolic and digestive processes and of creativity in protecting and augmenting the production of crops, in warming homes, in clothing the body, and in harnessing the atom. The problem is not in the natural elaboration of byproducts. It is in the disposition of them. As the National Academy of Sciences Committee aptly stated, the problem is a case of "a resource out of place"—too much of a resource in one system, not enough in another. The problem is a resource being present in a system that is not adapted to it and thus constituting an unaccustomed stimulus, stress, or "insult" to that system. These are stimulants that may terminate some or initiate other biological processes, alter efficiency, affect species composition and structure, and, in general, thereby alter the dynamics and development of an ecosystem.

Like the adage about the poor, byproducts will always be with us. They will increase as technology and living standards increase; they will become exacerbated as urbanization proceeds and more people live in smaller areas. Solutions do not and cannot lie solely in removing the cause because as long as humanity exists, it will have byproducts. Rather, answers lie in intelligent management of that production through regulating the "unfavorable alteration of our surroundings." Given that there are ultimate limits to the amount of these resources, it becomes an expression of wisdom to plan ahead for a steady-state system, not for an ever-expanding one.

WATER POLLUTION AND THE GREAT LAKES

The greatest reservoir of freshwater on the Earth's surface—a shoreline of 10,500 km, an area of 245,000 km^2, and a volume of 25,000 km^3, with a waterway extending 2,575 kilometers from Minnesota to the Atlantic Ocean—this describes the St. Lawrence Great Lakes, one of Earth's greatest natural resources for water, transportation, hydroelectric power, food, and recreation (Table 18-1). Nearly 20 percent of United States citizens and a third of Canada's population live in the Great Lakes basin. About one-fifth of United States industries and about half of Canadian industries lie within the Great Lakes basin. The lakes contain 95 percent of the United States' supply of freshwater and 20 percent of the world's. They supply water for 10 percent of all United States residents. They are also subject to the biological and cultural byproducts of these people.

What, in fact, can be said about the environmental changes wrought by the presence of humans in the region of the Great Lakes in no small degree describes human impact on the environment elsewhere: a reversal of succession in terrestrial systems and acceleration of it in aquatic ones.

Eutrophication

Before we discuss these matters, however, it is important to be reminded of a major principle about ecosystems—they age. They undergo succession. Freshwater lakes proceed inexorably, predictably, and ultimately to a semiterrestrial or fully terrestrial state. They initiate typically as nutrient-deficient, hence unproductive and detritus-free systems and develop into ones with increasing amounts of nutrients, thus to productive systems with considerable deposits of organic materials. This aging process from low production, or oligotrophy, to high production, or eutrophy, as a result of enrichment is referred to as *eutrophication*. Successional change, including the eutrophication of lakes, is, then, as natural a process as is individual development

TABLE 18-1 MORPHOMETRIC DATA FOR THE ST. LAWRENCE GREAT LAKES

Lake	Area (km^2)	Maximum depth (m)	Mean depth (m)	Volume (km^3)	Shoreline length (km)	Mean elevation above sea level (m)
Superior	83,300	307	145	12,000	3,000	183
Huron	59,510	223	76	4,600	2,700	177
Michigan	57,850	265	99	5,760	2,210	177
Erie	25,820	60	21	540	1,200	174
Ontario	18,760	225	91	1,720	1,380	75

Data on area, depth, volume, and shoreline, used by permission, from G. E. Hutchinson. 1957. *Treatise on Limnology*. New York: John Wiley and Sons. Data on elevation: U.S. Army Corps of Engineers, U.S. Lake Survey Chart, 1955.

or population growth. When the process is the result of anthropogenic activity, it can be referred to as *cultural eutrophication.*

As noted in Chapter 13, succession is accompanied by significant changes in structure and function, both biotic and abiotic. Some nutrients increase or become more readily available, and others are depleted (e.g., dissolved oxygen tends to be decreased, especially in deeper water). Chemical and physical properties are altered (e.g., electrical conductivity and thermal properties are changed). There are, of course, corresponding changes in the biota, with the two major, or at least most obvious, changes involving plankton and fish. Plankton, composed of microscopic or near-microscopic plants and animals of open water, are sparse in oligotrophic lakes. Their absence is responsible for the characteristic deep blue of such lakes, for there is nothing to reflect the wavelengths of "green light" and other portions of the visible spectrum back to the surface.

As enrichment proceeds through the import of nutrients from surrounding watersheds and by wind-blown dust coupled with internal mineral-fixation processes, the original dominant phytoplankters, the desmids, are replaced by diatoms, then by flagellates and other green algae, and, finally, by blue-green algae. Major population blooms of the blue-green algae create most of the problems in water works: they clog filters, release obnoxious and objectionable aromas and flavors, and turn away even the most inveterate swimmer from the peasoup consistency of a favorite swimming hole. Thus from a utilitarian point of view, eutrophic water, at least in this condition, is unusable without major and typically expensive remedial treatment. Almost everyone has observed such eutrophication if only in a puddle persisting in a rut made by an automobile tire on a front lawn.

Oligotrophic, or nutrient-poor, lakes are the source of many most-favored fish—trout, char, chub or lake herring, and whitefish. As eutrophication proceeds, these gourmet delights are replaced by progressively less relished forms, such as bass, perch, and pike, and still later by generally much less favored types, such as carp and sunfish.

The eutrophication process is, as is true of ecological succession in general, one that is measured on a geological time scale. The amount of natural eutrophication in a moderate-sized lake during a human lifetime is virtually imperceptible. The changes are slow but still inexorable, the result of dynamic interactions between the components of the ecosystems, and subject to the influences of surrounding ecosystems and the major environmental gradients.

In the Great Lakes, as in many other bodies of water, the natural process of eutrophication has been accelerated to a considerable degree by pollution, by misplacement of natural resources, the result being cultural eutrophication. Analysis of records from as far back as 1850 indicate that the effect has largely been since about 1910 and that it has been greatest in those lakes (Erie, Ontario, and Michigan) that have had the largest population growth within their drainage areas.

Eutrophication in the lower Great Lakes (Michigan, Erie, and Ontario) resulted from nutrient enrichment associated with early settlement and forest clearance that increased the amount of phosphorus, among other nutrients, discharged into one or more of the Lakes (Beeton 1965, Williams 1992). The introduction of phosphorus into phosphorus-limited waters results in depletion of silica because of increased production of diatoms that require silica for their skeletal framework. By analyzing silica and diatoms in sediment cores, Schelske et al. (1983) showed that diatom production peaked in Lake Ontario from 1820 to 1850, about 1880 in Lake Erie and not until 1970 in Lake Michigan. The explanation for the later time in Lake Michigan is tied to the use of phosphate detergents, which expanded tremendously beginning about 1940.

LAKE ERIE: A CASE STUDY IN POLLUTION AND POLLUTION CONTROL

Lake Erie is the shallowest and smallest of the Great Lakes (Table 18-1). Extending 240 miles from Toledo to Buffalo, its major source of water (about 80 percent) is Lake Huron, via a river system that services industrial Detroit (Figure 18-1). Additional water is derived from numerous rivers that drain agricultural land and also serve such major industrial cities as Toledo, Cleveland, Erie, and Buffalo.

The earliest records indicate that Lake Erie provided cool and cold-water habitats of low fertility. Primary production was low, and oxygen was present in deep waters throughout the year. Wetlands, marshes, bays, and rivers were of considerable extent, providing an abundance of shallow-water habitats conducive to fish spawning. Many fish species were characterized by large individuals; perhaps as much as 50 percent of the total biomass of all fish consisted of individuals over 5 kg in weight. With the influx of people came not only the addition of human and industrial wastes but also a vast reduction in wetlands and marshes, which were reclaimed for ports, buildings, and recreational beaches.

Eutrophic Changes in Lake Erie

Nutrients. Over the last 100 years or so there has been a significant increase in the amount of calcium, sodium, potassium, chlorides, and sulfates in Lake Erie (Figure 18-2). There has also been a corresponding increase in the total dissolved solids (soluble inorganic substances) from about 140 ppm in 1900 to nearly 190 in 1960. In both instances, the increase in these chemicals is most striking since about 1900, the time at which population growth in the region accelerated markedly from 3 million to 14 million in 1980. The fact that chloride and sulfate are important constituents of industrial and human sewage readily accounts for the significant increase in these chemicals in the lake. (Of the other Great Lakes, only Ontario shows comparable increases.) A threefold increase in phosphates from the 1930s to the 1970s is attributable to two major sources, some 60 to 80 percent came from household sewage— largely from laundry detergents—and some 20 to 40 percent from

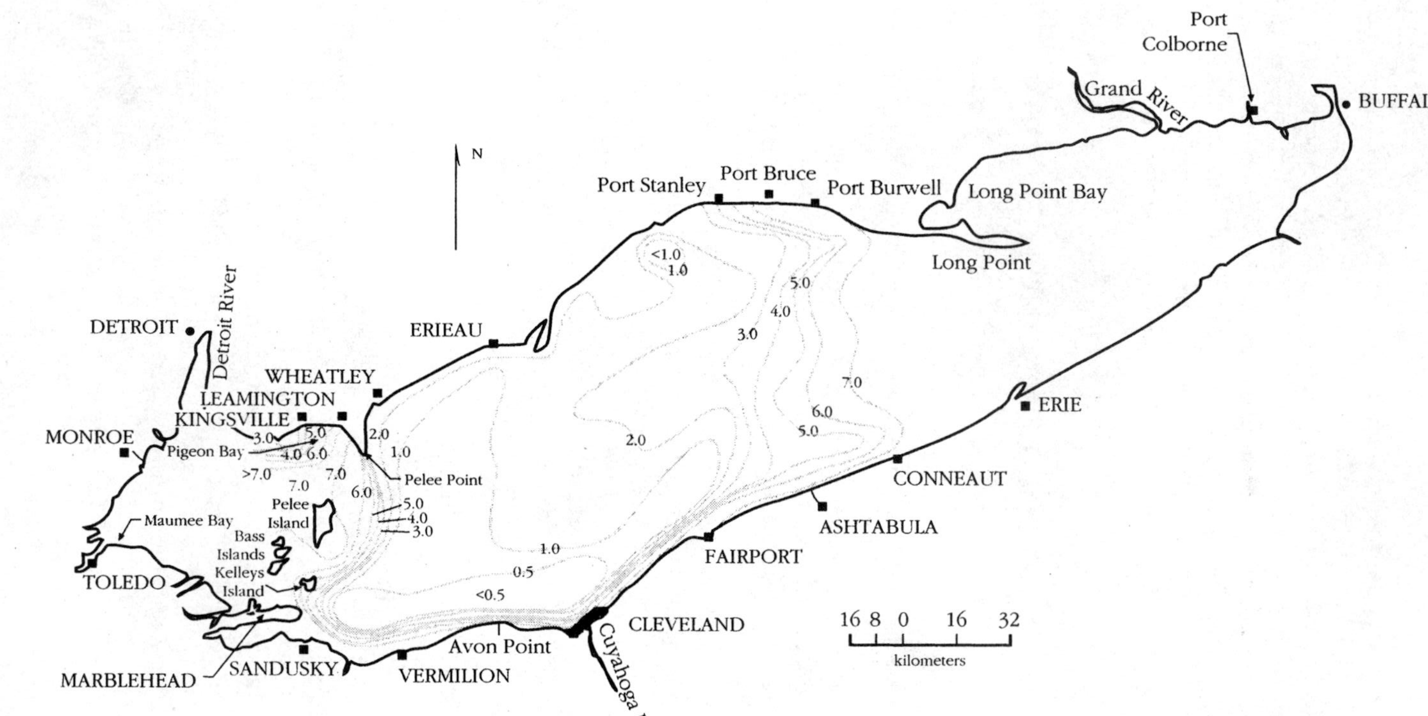

Figure 18-1 Distribution of dissolved oxygen (ppm) in the bottom waters of Lake Erie, 1959. (From A. M. Beeton. 1963. *Great Lakes Fisheries Commission Technical Report No. 6.*)

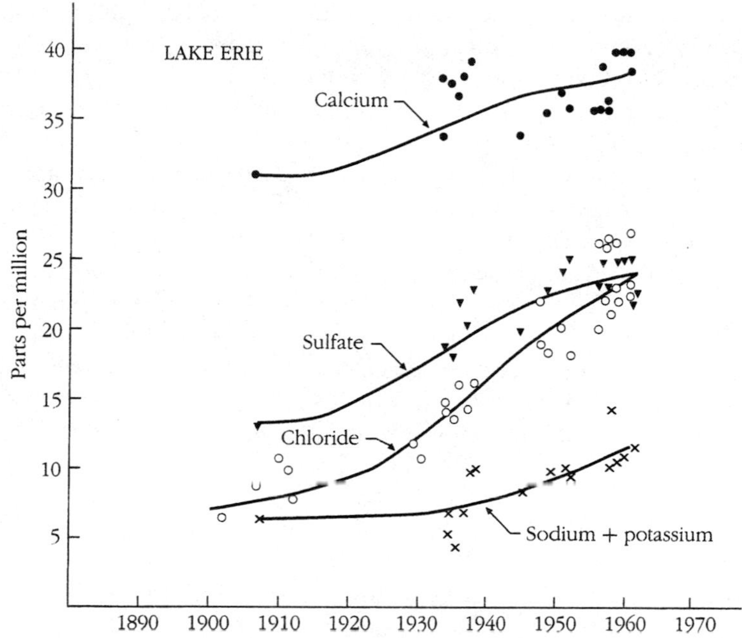

Figure 18-2 Chemical changes in Lake Erie, 1890–1960. (Redrawn by permission from A. M. Beeton. 1965. *Limnology and Oceanography* 10:240–54.)

fertilizers washed off the agricultural land of the region. Phosphate levels have since dropped markedly, from 15,260 tons/year in 1972 to 2,449 tons/year in 1985, as a result of provincial and state laws limiting the amount of phosphate in laundry detergents (Williams 1992).

Dissolved oxygen. Substantial changes in the amount of dissolved oxygen in the western and eastern portions of Lake Erie do not appear to have occurred since about 1930. In general, the shallow western end of the lake (average depth, 7.3 m, constituting 13 percent of the surface area and 5 percent of the volume of the lake) is subject to sufficient mixing of surface waters so that the hypolimnion is generally 80 percent saturated. Occasional stratification occurs and has resulted in a rather quick depletion of oxygen in the hypolimnion. The deep eastern end (maximum depth, 60 m; average depth, 24 m, and constituting 64 percent of both the surface area and volume of the lake) averages 60 to 70 percent oxygen saturation. The central basin (average depth, 18 m and constituting 24 percent of the area and 32 percent of the volume of the lake), however, is subject to severe oxygen depletion involving many hundreds of square miles, affecting about 70 percent of the bottom water (Figure 18-1). The frequency of these low oxygen periods is yet uncertain, but evidence over a 40-year period indicates that depletion is more frequent, extends over a longer period of time, and involves a greater area than in the past.

Plankton. On the biological side, evidence of eutrophic changes can be found in the effects on bacteria, plankton, bottom fauna, and fish. Unfortunately, the published studies on plankton do not allow the kind of valid comparison that was made for chemical changes in the lake. The data, however, that do exist provide evidence of recent eutrophication by changes in the composition and abundance of both phytoplankton and zooplankton, particularly in the western and central basin. From the records of the water filtration plant of Cleveland, for example, Charles C. Davis (1964) found an average increase in phytoplankton counts from 81 per ml in 1929 to 2,423 in 1962. Since 1919 there has been a nearly 20-fold increase in the annual algal biomass with a qualitative, and typically eutrophic, shift from dominance in fall blooms of diatoms (*Asterionella, Synedra*) to blue-green algae (*Anabaena, Microcystis, Aphanizomenon*). A particular nuisance is the alga *Cladophora*, shoreline accumulations of which sometimes reach 50 feet in width and 2 feet in depth. When it rots, it makes beaches completely useless for recreational purposes.

Bacteria. The bacterial situation is even more markedly different. Between 1913 and 1946, when the population on the area of the western end of Lake Erie grew from 1 million to 3.5 million, the coliform bacteria (*Escherichia coli*) count increased nearly threefold, from 175.2 per 100 ml to 448.9. The eastern end of the lake experienced no change in the same period. In the summers of the late 1960s as many as three-quarters of the swimming beaches along the southern shoreline of the lake were closed because the coliform bacteria count exceeded the safety level for public health. In 1967 out of 83 beaches in Michigan, Ohio, Pennsylvania, and New York, 27 were unsafe for swimming for the whole season, and 28 only periodically safe.

It will be of interest to some to note that *Escherichia coli*, the index organism for aquatic pollution, is itself a nonpathogenic bacterium of the intestine. However, *E. coli* is often pathogenic outside the hind gut, it being one of the most common causative agents for human urinary tract infections. Pathogenic bacteria are difficult to detect but are often intestinal associates of *E. coli.* Because the latter is discharged with body wastes and dies off outside the body more slowly than pathogens in general and can be detected by a fairly simple technique, it is used as an indirect assessment of the presence of pathogenic bacteria. The operating assumption is that as long as *E. coli* is present, there is a chance that some pathogenic bacteria are also present.

Bottom fauna. Significant changes in the distribution and abundance of the bottom fauna of western Lake Erie were detected in the period from 1930 to 1961 (Carr and Hiltunen 1965; Beeton 1961) (Table 18-2). In 1961 the tubificid worm, an oligochaete typical of low oxygen conditions and hence an empirical criterion of pollution, constituted 84 percent of the total bottom-dwelling organisms collected, a ninefold increase over a period of 30 years. Not so for the burrowing mayfly, *Hexagenia*, once so abundant that its synchronized emergence constituted a major nuisance. Highways were made slick by thousands of their bodies and shovels were required to remove them from sidewalks. The mayfly population has been reduced almost to extinction, a direct result of its inability to withstand low oxygen conditions. These direct comparisons with the most striking changes in tubificids, midges,

TABLE 18-2 AVERAGE DENSITY (number/m^2) OF BOTTOM ORGANISMS IN WESTERN LAKE ERIE IN 1930 AND 1961

Group	Number of stations compared	1930	1961	Ratio of numbers 1961-1930
Oligochaeta	33	677	5,949	8.79
Hexagenia sp. (Mayflies)	33	139	1	0.01
Tendipedidae (midges)	23	73	322	4.41
Sphaeriidae (fingernail clams)	23	221	438	1.98
Gastropoda (snails)	16	40	221	5.52

Reproduced by permission from J. F. Carr and J. K. Hiltunen. 1965. *Limnology and Oceanography* 10:551–69.

and mayflies indicate a sharp expansion of oxygen-demanding wastes between 1930 and 1961. In 1930, 263 km^2 or 26 percent of the area was classed as polluted. In 1961 the entire area (1,020 km^2) was so classed: 26 percent lightly, 51 percent moderately, and 23 percent heavily. The area of heavy pollution increased ninefold, from 26 km^2 in 1930 to 263 km^2 in 1961 (Figure 18-3); the relationships of this area to major centers of habitation and industry are obvious.

Commercial and sport fish. Marked changes have also occurred in the composition and abundance of both commercial and sport fish over the past 30 to 60 years (Applegate and Van Meter 1970; Parsons 1973; Smith 1972; Wolfert 1981, among others). Lake herring, sauger, blue pike, whitefish, walleye—all commercially significant and gastronomically satisfying—have declined dramatically (Figure 18-4): lake herring from a high of 22.2 million kg in 1918 to nil in 1965–69; sauger from a high of 2.8 million kg in 1916 to nil in 1965–69; blue pike from a high of 12.2 million kg in 1936 to nil in 1965–69; whitefish from a high of 3.2 million kg in 1949 to nil in 1965–69; walleye from a high of 7 million kg in 1956 to 0.4 million kg in 1965–69. Although the total fish production in the lake continues to be about 23 million kg, a drop from a peak of 35 million kg in 1915, it is of the less table-favored forms—more than half in yellow perch, the rest of carp, freshwater drum, white bass, channel catfish, and suckers, in that order.

Anthropogenic Factors in the Eutrophication of Lake Erie

A first reaction to this account of environmental alteration, particularly as it affects fish, would tend to indict a single cause—human pollution. As is the case for all ecosystems, however, a multiplicity of factors have been operating. In their review

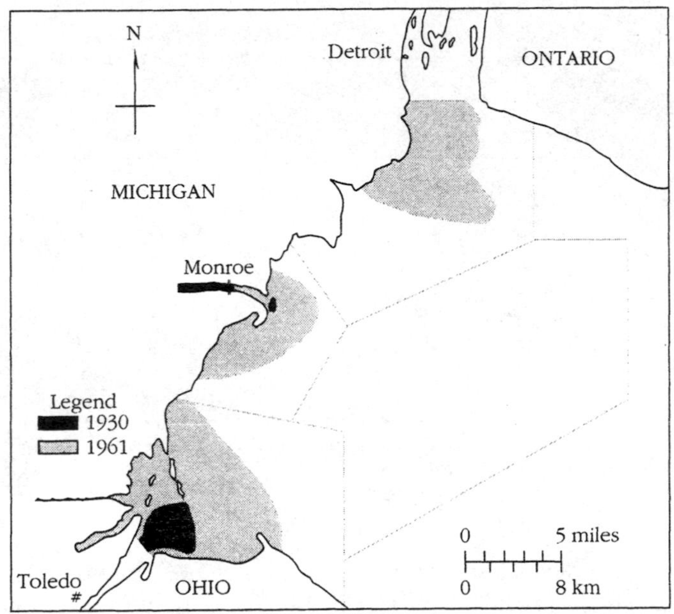

Figure 18-3 Zones of heavy pollution in western Lake Erie in 1930 and 1961. (Redrawn by permission from J. F. Carr and J. K. Hiltunen. 1965. *Limnology and Oceanography* 10:551–69.)

of Lake Erie's fish community over a 150-year period, Regier and Hartman (1973) concluded that the major cultural stresses, in decreasing order of their net effect, are:

1. one hundred fifty years of opportunistic and uncontrolled commercial fishing (a predatory process);
2. cultural eutrophication—that is, nutrient enrichment from sewage, fertilizer runoff, etc;
3. the introduction and/or invasion of nonindigenous species (e.g., predation by smelt that invaded the lake about 1931, cannibalism among blue pike—the sea lamprey that invaded Lake Erie in the late 1910s was never a serious parasite);
4. stream destruction and shoreline restructuring (e.g., the drainage of the great swamps at the western end of the lake), mill dam construction with consequent restrictions on suitable spawning, resting, and feeding habitats;
5. the release of toxic materials (e.g., mercury, polychlorinated biphenyls) from industrial sources and ships and the unintended introduction of such biocides as DDT.

This brief synopsis of changes in certain major physical, chemical, and biotic aspects of the Lake Erie ecosystem indicate that this is a system in which there are "resources out of place," as a result of which the system is experiencing an "unfavorable alteration." But we must also remember the other significant point made at

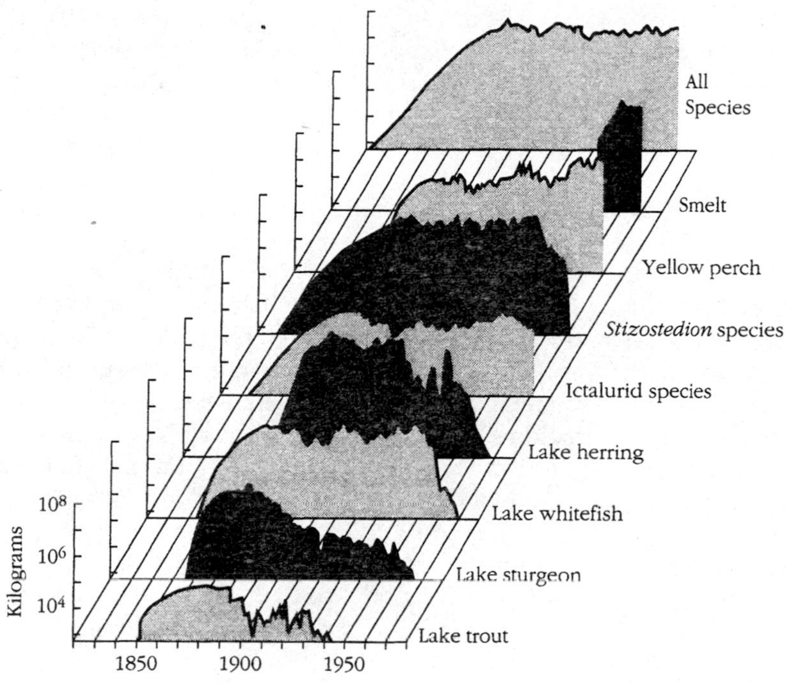

Figure 18-4 Annual catches of selected species and of all species combined by commercial fisheries in Lake Erie, 1820–1970. The vertical scale is logarithmic. *Stizostedion* species include walleye, blue pike, and saugers; Ictalurid species include channel catfish and bullhead. (Redrawn by permission from H. A. Regier and W. L. Hartman. 1973. *Science* 180:1248–55. Copyright 1973 by the American Association for the Advancement of Science.)

the beginning of this discussion on water pollution, that there is the natural process of aging, or eutrophication, which is a consequence of the dynamics of freshwater ecosystems. The problem in Lake Erie may be put in the words of the Federal Water Pollution Control Administration, in its statement before a Subcommittee on Government Operations of the House of Representatives of the 89th Congress: "The main body of the lake is deteriorating in quality at a rate greater than that of normal aging due to inputs of pollution resulting from the activities of man."

Restoring or Rehabilitating Lake Erie

Given the many changes that have occurred in Lake Erie over the past 150 or so years, is it possible to restore it? If so, to what point of restoration? If the goal were to restore the lake to its original, pristine condition as an oligotrophic lake with a rich mosaic of shallow-water habitats and large fish dominating the fauna, much of the positive changes that have ensued would have to be foregone. For example, all the intervening shoreline development for industry and residential and/or recreational use would have to be undone; popular introduced fish such as Pacific salmon as well as brown and rainbow trout would have to be exterminated. Even if restoration were feasible, given the positive changes in the interim, it is unlikely to be desirable. Instead agreement on some less drastic changes to accommodate a host of factors would lead to some degree of modification of the present situation, a rehabilitation rather than restoration of the Lake (Magnuson et al. 1980; Maugh 1979).

In 1972, the United States and Canadian governments signed the first Great Lakes Water Quality agreement, the focus of which was on conventional pollutants—phosphorus, oil, and visible solid wastes. With the expenditure of over $10 billion within the first 5 years of the rehabilitation program, extensive programs to reduce nutrient loading from municipal sewage and detergents were initiated. The primary objective of these programs has been to reduce phytoplankton production to a level that does not reduce the oxygen level to nil in the central basin of the lake. Because phosphate is the significant limiting factor in production, and because some 70 percent of all phosphorus generated in the United States comes from detergents, the major thrust of pollution abatement has been on phosphate removal from sewage. But as noted by Henry Reuss of the U.S. House of Representatives, the point of control is more direct: three major manufacturers of phosphate-containing detergents versus 200 million manufacturers (in the United States) of the other phosphates in sewage.

And if pollution were to be discontinued, how long would it take to rid the system of the pollutants currently present? According to Robert Rainey (1967), at the natural flow of water, it would take about 20 years for 90 percent of the wastes to be cleared from Lakes Erie and Ontario and hundreds of years from Lakes Superior and Michigan.

The average flushing time (or retention time) in Lake Erie is 2.6 years, in Lake Ontario 7.8 years, in Lake Michigan 30.8 years, and in Lake Superior 189 years (Rainey 1967). But, oligotrophication, a reversal of eutrophication, in no way assumes restoration of a previous, let alone pristine, state. In the case of fish, for example, the existing large populations of freshwater drum, carp, and goldfish, together with predation by rainbow smelt, will probably resist reintroduction of northern pike, lake whitefish, and lake herring. Perhaps even more significantly, inshore spawning grounds have been silted beyond restoration as a result of altered shorelines and disturbed or destroyed streams. Thus the likelihood of restoration of Lake Erie to anything similar to its condition 100 years ago is nil. But the lake can be rehabilitated by vigilant regulation of the addition of chemicals, by preventing additional modifications of shorelines and fish-breeding grounds, and by intelligent management of fish stocks, including the reintroduction of native species and control or elimination of invading species. Although Lake Erie may never again be the same as it once was, it can be better by far than it was in the 1970s before major rehabilitation efforts got underway.

Additional Pollution Problems in Lake Erie

Although the efforts to reduce phosphorus, oil, and visible solid wastes have been effective in Lakes Erie and Ontario, their curtailment brought to attention previously overlooked pollutants and sources of pollutants. These pollutants included a number with serious health implications such as mercury, asbestos, mirex (an insecticide and fire retardant), and chlorinated hydrocarbons, such as the dioxins, DDT, and polychlorinated biphenyls (PCBs). The previously unrecognized sources of pollutants included runoff from farmland and city streets, industrial discharges, airborne toxic substances, contaminated groundwater, and polluted sediments (Hileman 1988; Williams 1992). This recognition led to a 1978 United States and Canadian agreement, and a subsequent strengthening amendment in 1987, the goal of which was the virtual elimination of chemical discharge. The result of these agreements, and the corrective surgery they implemented, is that Lake Erie (as well as the other Great Lakes) is fair to excellent with respect to phosphorus and mixed with respect to toxic chemicals, although lead, PCBs, and DDT, all of which have been restricted or banned, have declined dramatically. However, nitrogen levels have been rising in all five Lakes.

Disturbingly, it is now believed that half of the inputs of toxic substances to Lake Superior come from atmospheric sources (Hileman 1988); since water eventually drains from Lake Superior to Lake Erie, the pollution problems of the latter will not just simply go away. Further, the sources of the airborne toxics may be very distant from the Great Lakes Basin and include disposal of contaminated sewage sludge on land, the open lagoon treatment of toxic wastes (especially by aeration), municipal waste incineration, sewage sludge incineration, and forced removal of volatile contaminants from groundwater.

A brief consideration of some of these additional pollutants is in order.

Mercury. Methyl mercury is a highly toxic substance that causes neurological damage, produces chromosomal aberrations, and results in congenital birth defects (Hammond 1971). Substantial mercury pollution in the Great Lakes became apparent in March 1970, when it was detected in pickerel shipped from Canada. Its source was identified as industrial waste in the form of inorganic mercury or phenyl mercury from which it is converted by anaerobic bacteria in the sediments into a methylated form.

Methyl mercury accumulates in fish both by way of the food chain and by direct absorption. The reason for concern was the episodes of mercury contamination in Sweden and Iraq (Bakir et al. 1973), as well as near Minamata, Japan, where 111 persons died or suffered serious neurological damage between 1953 and 1960 as a result of eating fish and shellfish from mercury-contaminated waters. Among the 111 were 19 congenitally defective babies born of mothers who had eaten mercury-contaminated fish and shellfish (Medati and Kormondy 1989; Totsuka 1989). Fortunately, typical mercury concentrations in Lake Erie fish were considerably below that in Minamata fish; furthermore, fish form much less a significant part of the American diet. Nonetheless, the "mercury scare" was significant in curtailing commercial fishing and in increasing regulation of mercury discharge in industrial wastes.

Although there has been control of point-source mercury pollution (industrial discharges) in the Great Lakes, including Lake Erie, its airborne input has not been brought under control. Mercury is an unusual pollutant because, according to the National Academy of Sciences, natural sources emit more than 1,000 metric tons annually whereas anthropogenic sources account for about 650 metric tons, the major sources of the latter being coal combustion and by municipal waste incineration (Hileman 1988).

Asbestos. Inhalation of asbestos fibers is acknowledged as a serious occupational and health problem because of its association with excessive rates of gastrointestinal and peritoneal cancer. In the early 1970s asbestos fibers were detected in substantial quantities in municipal water supplies from western Lake Superior water, including those for Duluth, Minnesota (Cook et al. 1974). The source of the asbestos was the large taconite tailings that had been discharged at Silver Bay, Minnesota, since the mid-1950s. Duluth residents had to resort to imported, bottled water to avoid the health hazards associated with asbestos fibers.

Chlorinated hydrocarbons. *Dioxins*, the most toxic substances known to humans and a byproduct impurity in pesticides (Madati and Kormondy 1989; Smith 1978) have been discovered in streams emptying into the Great Lakes. In spite of the early 1970s ban on DDT, which before the ban was responsible for the near extermination of cormorants and other birds such as herring gulls and terns that feed on Great Lakes fish, residues of this powerful pesticide persist in the environment, though at low levels (Williams 1992). The environmental effects of DDT are succinctly discussed by Buckley (1986).

Polychlorinated biphenyls (*PCBs*), before restrictions were placed on them, had been widely used in plastics, electrical insulation, and carbonless printing paper. They are among the most widespread contaminants of marine and freshwater environments and have been subject to control measures worldwide (Kormondy 1989; Quensen et al. 1988). In spite of controls, PCB levels are higher in the western regions of Lakes Erie and Ontario, as well as in Lake Michigan, than allowed by the United States/Canada agreements (Williams 1992).

The concern with PCBs lies in their adverse health effects on humans who eat Great Lakes fish. In the aquatic food chain, PCBs occurring in phytoplankton and zooplankton are progressively concentrated, or undergo *biomagnification*, as they move along the food chain such that relatively low levels in the plankton become magnified as much as 500 times in fish and 50,000 times in herring gull eggs. Women who feed on Great Lakes fish give birth to children with smaller birth sizes, lower gestational age, and neonatal behavioral defects (Hileman 1988).

Other pesticides such as atrazine, the most heavily used pesticide in the United States, which acts by inhibiting photosynthesis, have been found in streams entering the Great Lakes (deNoyelles et al. 1982). Oil spills have occurred in the Lakes, though not with the consequences for marine environments such as the Valdez disaster in Alaska in 1989 and elsewhere (Jackson et al. 1989; Kormondy 1989).

Concluding Comment

This case study of Lake Erie (and others of the Great Lakes) can be replicated many times over throughout the world. For example, see Lehman (1986) on Lake Washington in Seattle; Dunnette (1992) on the Willamette River in Oregon; Pavoni et al. (1992) on the lagoons of Venice; Cooper and Brush (1991) on a 2000-year history of anoxia in Chesapeake Bay; Pringle et al. (1993) on the Danube Delta; and Anderson (1994) on the toxic red tides affecting coastal waters throughout the world. Although discussion of pollution problems in stream, river, and marine ecosystems would be highly instructive, limitations of space preclude such an exploration. Nonetheless, it can be noted that in a more in-depth study we would find similarities of cause and effect when compared to the situation described in the case study of Lake Erie; however, we would also find differences arising from the specific physical, chemical, and biological characteristics of those ecosystems in general and of given specific ecosystems. The discharge of industrial chemicals and inadequately treated human sewage, runoff of agricultural fertilizers and pesticides, oil spills (Blumer and Sass 1972), and superheated water (Axtman 1975; Cairns 1972; Gibbons and Sharitz 1974) have occurred singly or in combination in many aquatic—as well as marine—ecosystems. Each occurrence serves as a reminder of the risks to human health as well as to economics and general amenities when aquatic ecosystems are converted to sewers and cesspools with the byproducts of anthropogenic activity. By the same token, the rectifying of pollution has required creative interaction between the scientific community and the political arena (local, regional, state, federal, and, increasingly, international governments) and continual diligence in the enforcement of regulations. To paraphrase American patriot John Philpot Curran's comment in 1790 that "the condition upon which God hath given liberty to man is eternal vigilance" (which is commonly quoted as "eternal vigilance is the price of liberty"), we can aptly state that "eternal vigilance is the price of a decent environment."

SUMMARY

VOCABULARY

anthropogenic
biomagnification
cultural eutrophication
dioxin
eutrophication
oligotrophic
PCBs
pollute/pollution

KEY POINTS

- Pollutants, "resources out of place," are the byproducts of human behavior and metabolism and the metabolism of organisms on which humans depend for food.
- Lake Erie has progressed from an oligotrophic to an eutrophic ecosystem at a pace accelerated by anthropogenic activity in the region resulting in increases in nutrients, reduction in dissolved oxygen, changes in the composition and abundance of plankton, increases in bacteria associated with human excrement, reduction in bottom fauna, and dramatic reductions in favored commercial fish. The Lake has also been subject to pollution from mercury, chlorinated hydrocarbons, and polychlorinated biphenyls, among other substances. Restoration is unlikely, and rehabilitation will be costly in time, money, and changed human behavior.

CHAPTER 19

Anthropogenic Impact on Terrestrial Ecosystems

Not all pollution, that is, "unfavorable alterations" resulting from "resources out of place," accelerate natural successional processes as in the instance of eutrophication of aquatic ecosystems. Some environmental insults to an ecosystem result in regression rather than progression. Snow "in its place," as in melting, adds moisture to the soil; snow "out of place," as in an avalanche, reduces the landscape to an earlier stage of its ecological and geological history. Strontium "in its place" as a stable isotope is harmless to organisms; "out of place," as a radioactive element, it can induce leukemia and perhaps result in death. Pesticides, properly used to control brush and tree growth, can maintain an agricultural ecosystem at a given stage of succession; improperly used and discharged, pesticides can destabilize other ecosystems by destroying key species near or at the top of the food chain. Fire, properly induced by human activity, can maintain a forest or prairie ecosystem, but fire can also revert such ecosystems to much earlier, less productive, stages of succession.

In this chapter, we will discuss the impact of anthropogenic activity on terrestrial ecosystems with respect to ionizing radiation and radioactive elements, pesticides, and the destruction of forests.

IONIZING RADIATION AND RADIOISOTOPES

Ionizing radiation "in its place," as in the sun and mineral deposits in the lithosphere, constitutes the background radiation of about 0.5 roentgen per year, which geneticists believe to be at least partically responsible for spontaneous mutation, the very tool of evolutionary adaptation. Ionizing radiation "out of place" in the form of atomic and hydrogen blasts, and from nuclear reactors and nuclear wastes, can not only directly cause death, but, more significantly, also increase rates of and kinds of mutation that are more likely to be detrimental to a species adapted to its current environment. But because humans live increasingly in an age in which nuclear energy assumes more dominance in heating homes, preserving food, powering industry, fueling transportation, and preparing for war, a discussion of some ecological impacts of ionizing radiation and radioisotopes as environmental pollutants is in order.

Ionizing Radiation

The disastrous ecological results of an atomic explosion were publicly recognized much later than were the immediate medical and military byproducts. It was not until the late 1950s that sources of ionizing radiation became available for the study of natural systems. The most significant studies of the effects of ionizing radiation on ecosystems were conducted by George Woodwell and his associates (Woodwell 1962; Woodwell and Sparrow 1963) at the Brookhaven National Laboratory on Long Island, New York; by Robert Platt (1965) of Emory University and his students at the Dawsonville Lockheed reactor in northern Georgia; by Frank McCormick and Frank Golley (1966) at the Savannah River Laboratory in South Carolina; and by Howard T. Odum (1970) and associates in tropical forests in Puerto Rico.

The Brookhaven experiments. At the Brookhaven Laboratory a 9,500-curie source of gamma irradiation, cesium-137, was centrally located in a stand of scarlet and black oaks and pitch pine, a part of the eastern pine-oak biome. The oaks were about 40 to 45 years old, with none over 64; some of the pines were as much as 100 years old. Exposure to the cesium source ranged from several thousand roentgens (R) per day a few meters away to 1.5 R at 125 m and to background levels at greater distances. Lowering the source into the lead shield enabled investigators to work in the forest with impunity.

Within 6 months of the beginning of the radiation exposure, five vegetation zones were apparent (Figure 19-1). Each widened appreciably over a 3-year period. Nearest the source, where the exposure was greater than 200 R/day, was a zone in which all woody plants and most herbaceous plants were killed within the first year. For perspective, the lethal dosage of radiation required to kill 50 out of 100 persons, the so-called LD_{50}, is about 400 R. The sedge, *Carex pensylvanica*, survived in this same zone, to within 18 m of the source, where it received 200 to 300 R/day, and there was some succession of herbaceous plants during the 3 years. Shielded in the "shadow" of large trees, *Carex* survived much nearer the source. Away from this "devastated zone," as Woodwell referred to it, was a zone dominated by *Carex pensylvanica* with other low herbs. This was followed by a shrub zone dominated by ericaceous shrubs, which had exposures from 40 to 150 R/day, an oak forest with exposures of 12 to 40 R/day, and, finally, a pine-oak forest where exposure was less than 12 R/day in 1962, 7 R/day in 1963, and 4.2 R/day in 1964.

Figure 19-1 The experimental forest at Brookhaven National Laboratory showing the results of 6 month's exposure to gamma radiation. (Photograph courtesy of Brookhaven National Laboratory.)

It is immediately evident that the zonation pattern was related to stratification of the vegetation, trees being more sensitive than the shrubs, pines more sensitive than oaks, and the ground layer being least sensitive. This effect, according to an hypothesis of Arnold Sparrow (1962) of the Brookhaven Laboratory, is directly related to the average volume of chromosomes (i.e., the larger the volume, the more sensitive the plant—namely, it presents a bigger target) and to the amount of bud shielding afforded by the plant itself (i.e., the larger the tree and/or the buds, the more shielding). Herbaceous species have less "unshielded biomass" aboveground and so are more resistant than trees. It is not unexpected that one of the corollaries of the observed zonation is a reduction in species diversity with increasing dosage (Figure 19-2). The untoward consequences of reduced diversity on community stability were discussed in Chapter 13. This zonation pattern and its concomitant reduction in species diversity, although not necessarily in productivity, have been shown in other studies of radiation effects and raise intriguing questions as we consider the parallel to the gradation of life forms and species diversity from tundra to tropics. It would appear, as Woodwell and Rebuck state, "that the gross patterns of radiation effects on vegetation are by no means unique but have abundant precedent in nature. . . . This means that evolution of radiosensitivity in all of its various ramifications has been controlled not by radiation but by other factors, or combinations of factors, that have had the same selective influence that ionizing radiation might have had" (Woodwell and Rebuck 1967).

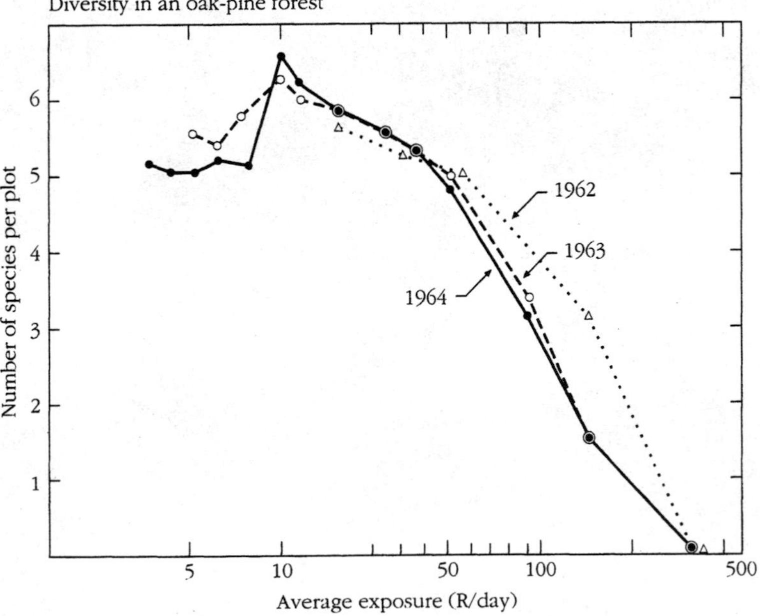

Figure 19-2 Effect of chronic irradiation on species diversity in an oak-pine forest. (Redrawn by permission from G. M. Woodwell and A. L. Rebuck. 1967. *Ecological Monographs* 37:53–69.)

Implications of the Brookhaven studies. Although Woodwell's study is a significant one, its limitation, from an applied sense, is that it is a study of chronic rather than acute ionizing radiation and that it is not a model for radiation effects from any known source except natural background radiation. Moreover, it deals only with effects of gamma radiation and, as increasing numbers of investigators have shown, the effects of beta emission are very important in both plant and animal communities. Nonetheless, this study serves as an index of the effects of any long-term, or chronic, stresses, such as the release of oxides from sulfur in smelting operations, the persistent use of herbicides, and such "natural" stresses as salt spray, reduced water availability, and elevational climatic gradients. All these continuous stresses result in a marked and systematic destruction or alteration of a forest with strata being removed layer by layer. In turn, this situation results in a reduction in species diversity, primary production, total respiration, and the nutrient pool.

Ionizing Radiation and Public Health

Humans are exposed throughout life to low-level ionizing radiation from cosmic rays, radioactive elements in the Earth's crust (e.g., radium, uranium), and emissions from certain radioisotopes occurring naturally in the body (e.g., ^{40}K and ^{14}C). This natural exposure is almost doubled by radiation from anthropogenic sources, mostly medical and dental in the form of x-rays, but also from radioactive substances incorporated in building materials and phosphate fertilizers, consumer products, such as television sets and smoke detectors, fallout from atomic weapons (Schull et al. 1981), and leakage from nuclear reactors. Although such radiation can induce a variety of cancers, cause chromosomal aberrations and hence genetic defects, and produce infertility and deficiencies in growth among other health effects, compared with other agents, the hazard from low-level ionizing radiation is relatively slight (Upton 1982; Beck and Krey 1983; Beebe 1982; Cobb 1989). Smoking, alcohol, and motor vehicles contribute, respectively, to 150,000, 100,000, and 50,000 deaths annually in the United States compared to 2,300 deaths caused by x-rays and 100 by nuclear power. Nonetheless, because of their mutagenic and unfavorable health consequences and their long-term activity, considerable anxiety exists over the storage of radioactive wastes, the largest single source of which are nuclear power plants (Cramer et al. 1988; Slovic et al. 1991; Jackson 1992). Radioactive wastes, if accessible to natural ecosystems, enter them through biogeochemical pathways and thus can affect the health not only of the ecosystem but also of humans dependent on such ecosystems. This topic will be discussed further in the next section.

Radioisotopes

A number of elements occur as different isotopes, that is forms of the element that have the same or very similar chemical properties and the same atomic number but different atomic weights. Thus, uranium exists as three isotopes—^{235}U, ^{238}U, and ^{239}U—the isotope numbers representing the number of neutrons in the nucleus. Although the isotopes of a given element are chemically identical, they do not necessarily behave the same. For example, ^{235}U undergoes *fission*, or splitting into two smaller atoms, whereas ^{238}U does not. Fission not only results in two smaller atoms but also the release of energy and two or three free neutrons. The lighter elements that are produced in the fission reaction are generally unstable isotopes of their respective elements; this includes isotopes such as iodine-131 (^{131}I), cobalt-60 (^{60}Co), and cesium-134 (^{134}Ce), which are not only unstable but release ionizing radiation in the form of alpha, beta, or gamma particles or rays. The unstable isotopes that release ionizing radiation are referred to as radioactive elements or *radioisotopes*. Parenthetically, fission is the basis for both the atomic bomb and nuclear weaponry, as well as the energy release from nuclear power plants; on the other hand, *fusion*, or the joining of elements as occurs on the Sun in the combining of hydrogen to form helium with the release of energy as a byproduct, is the basis of the thermonuclear or hydrogen bomb.

Radioisotopes and biogeochemical pathways. Fission byproducts, both of nuclear detonation and water-cooled atomic power reactors, do indeed constitute more of a potential hazard than direct ionizing radiation because they follow biogeochemical pathways. Radioactive iodine (^{131}I), which is produced as a byproduct of nuclear detonation, for example, is carried aloft into the atmosphere and through normal meteorological processes is transported for some distance and deposited on the vegetation and soil as "fallout." Because the decay rate (i.e., half-life) of iodine is fast (8 days), the amount that falls on the soil is unimportant—it decays before normal edaphic processes can get it into the plants. What *does* fall on the leaves is important, however, for plants incorporate it and, through grazing, can pass it in concentrated form on to cows and subsequently to humans in more concentrated form through the cow's milk. Behaving in typical iodine fashion, the radioisotope then concentrates in thyroid tissue, where its radiations can disrupt and damage the normal metabolism and growth of this endocrine gland, thyroid cancer being a typical result. An unhappy example was seen in the results of the 1954 atomic bomb test on Bikini, at the western end of the Marshall Islands. An unexpected wind shift carried fallout back onto the eastern islands, notably on Rongelap. As of 1968, 17 of 19 children on Rongelap who were under 10 years of age at the time of the test had developed thyroid abnormalities, many cancerous; most victims showed retarded growth. Similarly, atmospheric nuclear weapons tests in the People's Republic of China in September and November 1976 resulted in the presence of iodine-131 in milk and milk products in the United States, the heaviest concentrations occurring in the eastern half of the country (Smith et al. 1978). Fortunately, the potential health effects were judged insignificant: over a 45-year period only four more cases of thyroid cancer were predicted out of a total of 350,000 cases arising from other causes.

Radioisotopes and Lapland ecosystems. The actual fate of different radioactive elements has been demonstrated in a number of organisms. Our concern here is not only with one organism but also with ecosystems. The disposition of cesium (cesium-134) in white oak discussed earlier (Figure 9-2) is particularly instructive in this connection because of the eventual losses from the plant to the ecosystem through the normal processes of leaf fall and rain leaching. This amount, about 19 percent of the original input, is subject to recycling through normal biogeochemical processes back into the tree or into other vegetation and through grazing and detritus feeding to other components of the food web.

On a much larger scale, studies in the relatively simple tundra ecosystems of Alaska and Finland are particularly revealing. Reports by J. K. Miettinen (1969) of the University of Helsinki indicate that in 1965 Laplanders had reached the highest radiation exposure of any human population through dietary uptake of such radioactive elements as strontium, cesium, polonium, and iron. Although the total fallout of radioactive material in Lapland is about half that of Helsinki, the greater radiation exposure, which is on the order of 55 times, is a direct result of ecological processes. Radioactive elements constitute a source of enrichment for the nutrient-deficient soils and waters of Lapland. Some of these elements are synthesized into protoplasm in lieu of the normal element that it simulates chemically (e.g., strontium-90 in place of stable calcium, cesium-137 in place of potassium), and others become physically adsorbed onto surface tissues (e.g., zinc-65). As an aside, this latter point, physical adsorption of nutrients, is of considerable significance. Numerous studies, particularly in aquatic ecosystems, indicate that major uptake of radioactive elements—and hence probably of stable forms as well—is not by way of metabolic incorporation but by way of these nonbiological physical and chemical processes (Kormondy 1965).

For Laplanders, however, radiation exposure is directly dietary in origin. They depend heavily on reindeer meat and the reindeer, in turn, depend heavily on the abundant and widespread lichen known as reindeer moss (*Cladonia*) (see Figure 2-1). Reindeer moss concentrates the fallout radioisotopes both biologically and physically, and these radioisotopes are, in turn, transferred to reindeer as they graze. The directness of this transfer is shown by the fact that the concentration of particular radioisotopes increased by four times in both lichens and reindeer meat between 1961 and 1964. At this point, as much as 99 percent of the strontium-90 goes no farther than the reindeer. It becomes trapped in bone. Not so, however, with other elements, which get passed directly to humans as a result of their carnivorous and milk-drinking habits. If these elements are not metabolized or excreted, they become concentrated to levels considerably above the base level measured at the ground surface. For example, in a study of cesium-137 in the lichen-caribou-Eskimo food chain, it was shown that humans had twice the concentration of the caribou and that wolves and caribou had as much as three times the concentration of the lichen.

The high concentration ability of some organisms, some of which are on the order of thousands of times above the environmental level, has led to the suggestion that such species be used to decontaminate a system. They could be introduced, allowed to concentrate the pollutant, and be periodically harvested and disposed of in some appropriate fashion. But in discussing the situation in Lapland, another clue for the potential management of these contaminants presented istelf—namely, the nutrient status of the soil and water. The nutrient-deficient soils and waters of Lapland take up these introduced isotopes like long-lost friends. But a number of studies have shown that the uptake of given elements can be regulated by the mere presence of other elements. The biological uptake of strontium-90, for instance, bears an indirect relationship to the amount of calcium present. In a stiuation in which the ratio of calcium to strontium-90 is above a certain level, plants will preferentially take up the calcium even though the actual abundance of the strontium is considerable. This is a promising line of resource management but is far from being well understood or practiced.

Radiation Accidents and Other Exposures

Since the dawn of the nuclear age in 1945, scientists, engineers, and technicians have sought valiantly to contain the potential hazards of nuclear-bomb testing, nuclear-plant meltdowns, and environmental contamination from disposal of nuclear wastes (Madati and Kormondy 1989). Even before 1945, excessive medical and dental use of x-rays and other radiation treatments (e.g., x-rays were once used for fitting shoes!) had been brought under control as knowledge of adverse effects became better understood. Unfortunately, however, accidents involving radiation and radioactive elements do occur, most often as a result of human negligence; infrequent or not, such incidents heighten public concern for health and safety.

The nuclear age's dawn, the bombing of Nagasaki and Hiroshima in Japan and the later bomb testing on Bikini Atoll, have left permanent scars on people and the environment (Schull et al. 1981). According to a report in the quarterly *Pravda* on the secret history of the Soviet nuclear establishment, half the workers at the Cheliabinsk site in the Ural Mountains east of Moscow, the site of postwar production of nuclear materials, were routinely receiving 100 rem per year, 20 times the maximum annual dose a worker is allowed to receive in the United States (Marshall 1990). (A rem is a unit of radiation that reflects the type of radiation and its relative penetrating power). By contrast to this extremely heavy dosage per year in the Soviet Union, it has been calculated that the lifetime dose from 1944 through the early 1980s in the United States was 3 rem and about 11 rem in Great Britain.

Meltdowns. There were serious *meltdowns*, the melting of the core of a nuclear reactor, of nuclear reactor power plants at Three Mile Island in the United States (near Pittsburgh) in March 1979 and at Chernobyl in the USSR in April 1986 (Anspaugh et al. 1988; Goldman 1987; Davidson et al. 1987; Read 1993; Travis 1994), resulting in both loss of life and serious and extensive radiation damage to people. Radioactive gases released from the Three Mile Island meltdown on March 28, 1979, traveled as far as Albany, New York, a distance of about 500 km, in 18 to 24 hours resulting in an exposure to xenon-133 of about 0.004 percent of the annual whole-body dose of radiation from natural sources (Whalen et al. 1980).

The fallout from Chernobyl was the equivalent of 10 Hiroshima bombs and produced geiger counter readings of at least 250 roentgens an hour, millions of times above normal; it has permanently contaminated agricultural land the size of Holland, necessitating the evacuation of about 200,000 people, although nearly 5 million people still live in contaminated areas (Dobbs 1991). The official death toll remains at 31, which was the number of Chernobyl workers and firefighters killed in the immediate aftermath of the explosion; unoffical estimates run between 5,000 and 10,000 deaths.

There is still lingering concern as to whether all facts concerning the disaster, including the number of immediate and longer-term casualties, have been made known (Lemonick 1989; Dobbs 1991; Medvedev 1991). Not surprisingly, given atmospheric movements, fallout from Chernobyl was found in the form of cesium-134 and cesium-137 just a few months later in Greenland after these radioactive elements had been transported by polar winds three-quarters of the way around the Earth (Figure 19-3).

Goiania and cesium-13. Goiania lies about 120 miles southwest of Brazil's capital of Brazilia and has a population of 1 million. In mid-September 1987, a junkyard dealer pried open a lead cylinder, which, unknown to him, contained a capsule of radioactive cesium-137; the cylinder had been used and then disposed of as scrap by a medical clinic that had used the isotope in the treatment of cancer (Roberts 1987 and Gorman 1987). During the few days following the cesium removal, more than 244 people were exposed to the capsule and at least one, the 6-year-old niece of the junk dealer, ate some of the contaminant in a sandwich before the problem was identified. Within 3 months, 4 people died, 2 were in guarded condition, and about 15 others remained hospitalized. The 4 who died had received radiation dosages in the equivalent of 4,000 chest x-rays; one of the 4 was the 6-year-old.

Although Goiania would appear to be an unusual incident—actually in 1984 there was an eerily similar occurrence in Ciudad Juarez in Mexico (Stengel 1984)—constituting, with Chernobyl, the worst radiation accident in history and second in severity only to the wartime bombings of Nagasaki and Hiroshima, it is a reminder of the vigilance that must surround the use of a resource that can get "out of place." It is a reminder of the extent to which humans can be their own worst enemy when they contaminate their own nest.

The Disposal of Radioactive Wastes

Nuclear power, heralded in the 1950s and 1960s as the most promising alternative to fossil fuels in meeting industrial and domestic energy needs, has nonetheless been confronted by two major concerns: potential leakage, if not meltdowns, that would release radioactive materials to the environment (as in the Three Mile Island and Chernobyl incidents); and storage of radioactive wastes. The latter concern has confronted governments not only in the United States but also around the world and, in turn, challenged scientists to develop not only safe disposal methods but ones that are acceptable to the public as well (Carter 1977a, 1977b; Hafele 1990; Krauskopf 1990; LaPorte 1978).

There are two distinct problems in the management of nuclear wastes: high-level, thermally-hot radioactive wastes that decay fairly rapidly; and long-lived radioisotopes whose decay rates are very low (Angino 1977). Recycling of radioactive wastes still does not appear to be technologically feasible, thus storage or containment is the focus of attention. Interim storage of radioactive wastes for 50 to 100 years (Figure 19-4) allows for a sufficient amount of decay so that the heat problems for long-term storage are greatly reduced (Harrison 1984). Also, since nuclear wastes are liquid and bulky and require large containment vessels, the first

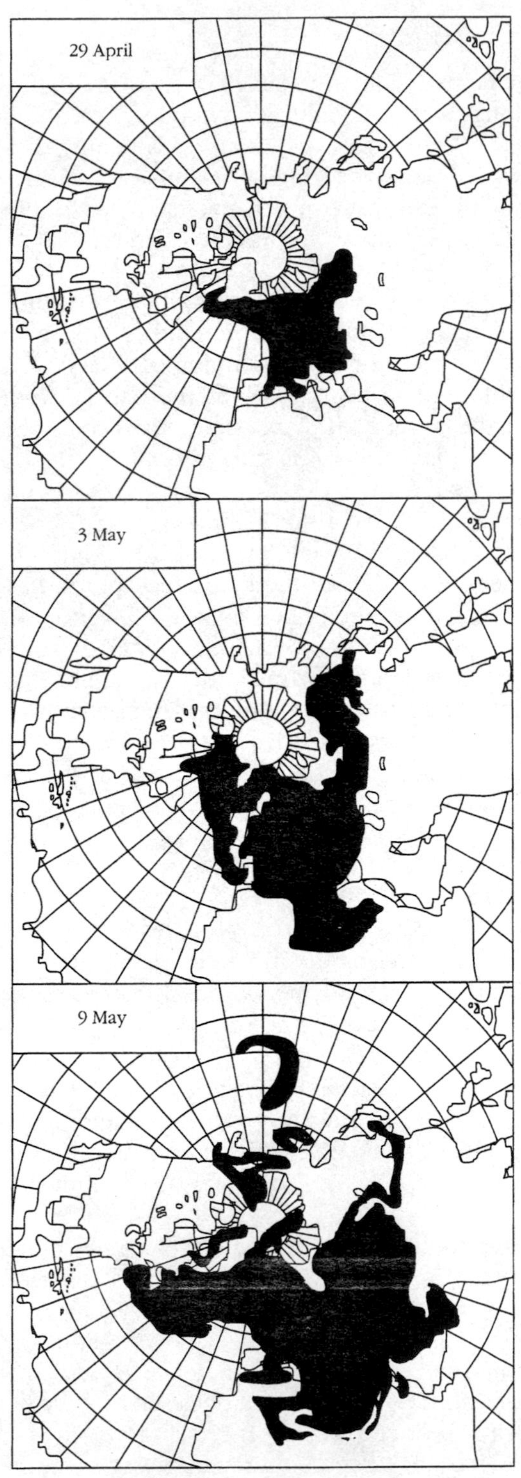

Figure 19-3 The spread of cesium-137 emitted from Chernobyl. (Reproduced with permission from C. I. Davidson et al. 1987. *Science* 237:633–4. Copyright 1987 by the American Association for the Advancement of Science.)

Figure 19-4 Six 2.9 million liter (1.0 million gallon) double-walled carbon-steel tanks for storage of high level radioactive wastes shown under construction at the Department of Energy's Hanford Plant in Richland, Washington. At completion, the outer steel wall was encased in reinforced concrete, and the entire tank was backfilled with 2 to 3 meters of soil over the top cover. (Photograph courtesy of the U.S. Department of Energy.)

efforts to reduce the liquid by a factor of 9 or 10 were made in the 1960s (Carter 1977b) and are still a critical step in providing for long-term storage.

Although there have been proposals to store long-lived radioactive wastes in the subseabed (Hollister et al. 1981) or literally to dump such wastes in the ocean (Norman 1982), the most promising proposals are for long-term storage in waste packages (Figure 19-5) in thick, unsaturated, geological strata deep underground (Figure 19-6) in rock salt and rocks of igneous origin, such as granite and basalt (Winograd 1981; Witherspoon et al. 1981; Bredehoeft and Maini 1982; *American Scientist* 1982). A substantial number of sites throughout the United States are being studied for such storage, including rock salt sites in the Great Lakes and Gulf Coast Basins, basalt exposures in the northwest (notably around the Hanford Reservation), and granite in many parts of the country (Figure 19-7). Because of the potential for leaching by

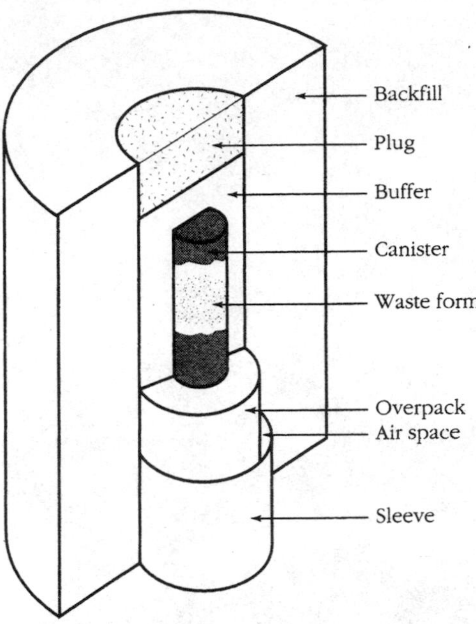

Figure 19-5 A model of a package of radioactive wastes prepared for long-term underground storage. (Redrawn by permission from C. Klingsberg and J. Duguid. 1982. *American Scientist* 70:182–90, journal of Sigma Xi, The Scientific Research Society.)

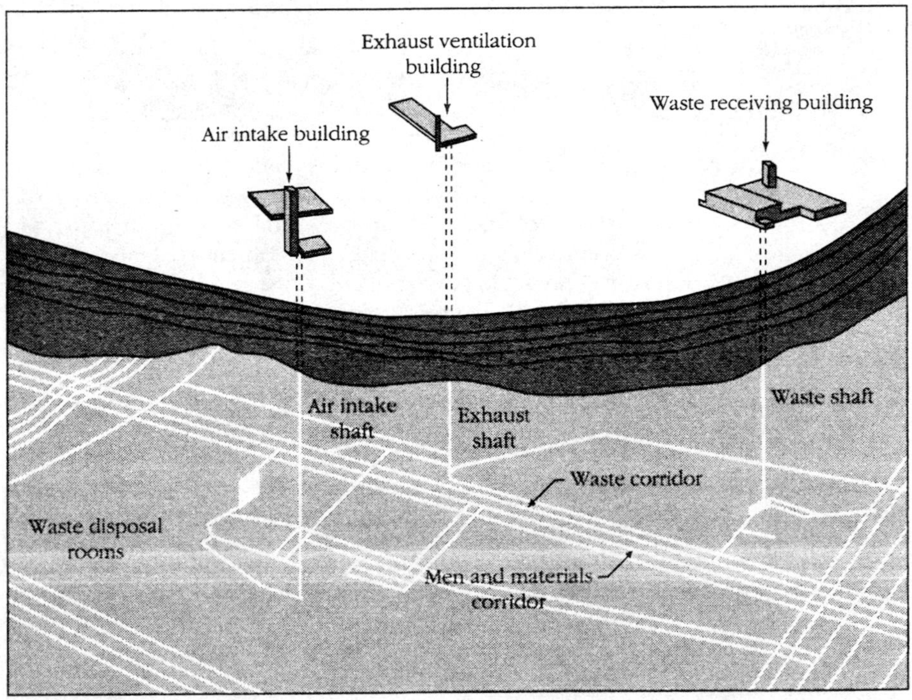

Figure 19-6 A model of a geological repository for solidified radioactive wastes. (Redrawn by permission from S. Gonzales. 1982. *American Scientist* 70:191–200, journal of Sigma Xi, The Scientific Research Society.)

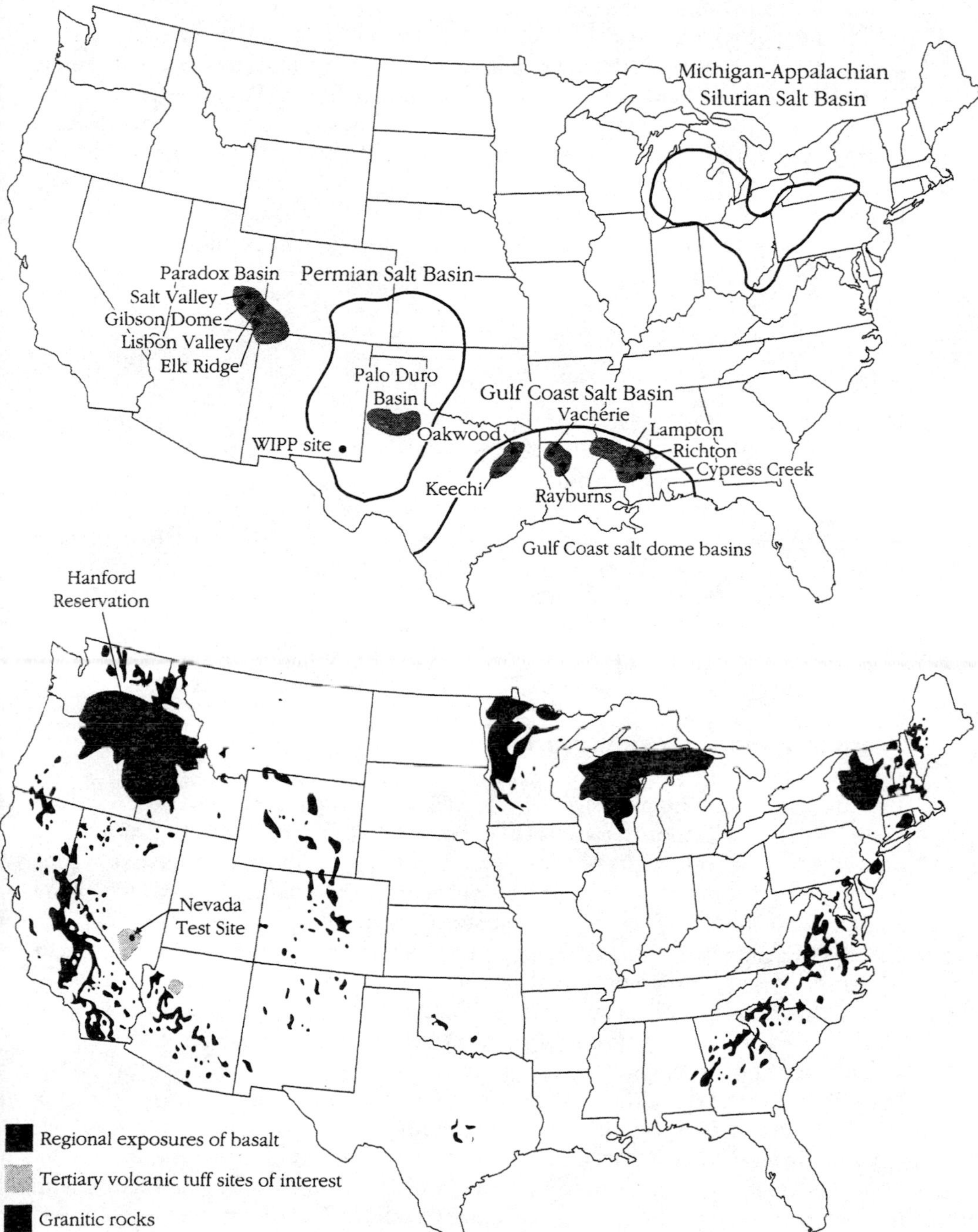

Figure 19-7 Rock salt and granitic rock sites in the United States that are under consideration as potential locations for long-term storage of radioactive wastes.

circulating groundwater in all but deep underground strata (and even there), considerable research is being conducted on immobilizing the radioactive wastes in ceramics made, in part, of titanium, hence called titanate ceramics (Ringwood 1982).

The U.S. Nuclear Regulatory Commission and the U.S. Environmental Protection Agency have established regulations and performance standards for containment for up to 10,000 years after disposal. Obviously none of the developers of the regulations will be around then, thus the effect of full compliance with these standards can now only be assessed through the development of computer scenarios based on the probability of various events occurring in the future (Campbell and Cranwell 1988).

PESTICIDES

Pesticides encompass a broad array of chemicals that are used to control plant growth (herbicides), insects (insecticides), worms (vermicides), bacteria (bacteriocides), fungi (fungicides), and other organisms. Almost without exception, pesticides are used to benefit humans by the control of unwanted species of plants (e.g., weeds in a lawn), disease-causing bacteria (e.g., *Pneumococcus*, the causative agent of pneumonia), disease-carrying insects (e.g., malarial or yellow fever mosquitoes) or in controlling destruction of agricultural products by disease-causing organisms (e.g., corn-borers, the virus of anthurium blight). Although they are by and large mostly beneficial, pesticides used both properly and improperly are also "resources out of place," that is pollutants, that are subject to normal ecological processes.

Pesticides and Ecological Processes

By virtue of biogeochemical cycling, the DDT that until recently was used in large quantities to control mosquitoes has a potential and often an actual regulatory effect on other biotic components of ecosystems. Some pesticides are subject to normal biological degradation by soil and water bacteria. The commonly used weedkiller 2,4,-D is one of them—it is quite quickly destroyed in the soil. This particular compound is a synthesized growth regulator and simulates many of the naturally occurring ones. It is not surprising, then, that such natural, or pseudonatural, products are ecologically accommodated in normal ecosystem dynamics. On the other hand, other plant growth regulators and insecticides are very resistant to biological decomposition. If such nonbiodegradable compounds should become "fixed" by the soil by being bound to the clay or humus fractions, they are of less consequence to an ecosystem. Dieldrin and DDT, both insecticides, tend to be fixed by some types of soils, particularly those rich in clay, and thus are less available under given conditions.

It is significant to note, however, that a substantial proportion of insecticides are not permanently fixed by the soil. Also, they collect on plant surfaces and can thus be passed directly along a food chain. And, importantly, because of differential metabolic and retention capacities of organisms, residue levels tend to increase with higher trophic levels, popularly spoken of as *biological magnification*. For example, a study was made of DDT concentrations in organisms along the south shore of Long Island, where this insecticide was used to control mosquitoes for some 20 years. The level of 0.00005 ppm in water was concentrated to 26.4 ppm in cormorants, a bird that eats large fish, for a 500,000-fold increase!

In the following section, DDT has been selected for a case study because its chemical and ecological behavior and effects have been well studied. As with any case study, however, there is the implication that all cases, in this instance all pesticides, behave similarly. However, pesticides are not all, like DDT, chlorinated hydrocarbons; there are organophosphates, organochlorines, carbamates, and natural organic pesticides, among others. Their activity, persistence, and effect, that is their ecotoxicology, differ from that of DDT to varying degrees, a topic well beyond this text.

DDT, a Case Study

Although first synthesized in 1874, the insecticide properties of *DDT* (*d*ichloro *d*iphenyl*t*richloroethane) were not recognized until 1939 after which it was widely used against many insect pests. Its use throughout the world is credited with saving millions of lives from such diseases as typhus and malaria, which are carried by insect vectors, and with the control of countless agricultural insect pests (Buckley 1986). Unfortunately, DDT had unforeseen negative effects, as will be noted below, mostly deriving from its persistence, a characteristic that allowed its being transported from its initial site of application so that almost no part of the Earth's surface was free of it (Woodwell et al. 1971).

Public concern, at least in the United States, became widespread in 1962 with the publication of Rachel Carson's *Silent Spring*, which chronicled the adverse effects of this insecticide. This book had phenomenal impact and resulted in curtailment of many uses of DDT within only a few years; in 1972, the U.S. Environmental Protection Agency announced the cancellation of all remaining uses of this powerful insecticide. Worldwide, however, DDT is still being used, and thus is another "resource out of place" that has not just gone away (Kormondy 1989; Moriarity 1988).

Because it serves as a reference point to the potential unwanted consequences of other pesticides, a brief review of DDT's effects will be instructive.

DDT and the Bermuda petrel. Among the world's rarest birds, with a population of only about 100, the Bermuda petrel (*Pterodroma cahow*) feeds at sea and visits land only to breed—and breeds only on Bermuda. Yet Charles F. Wurster (1968) reported finding significant levels of DDT (6.44 ppm) in unhatched eggs and dead chicks. The only source of the insecticide was its oceanic food chain and the only access to the latter was by runoff from the mainland more than 1,000 km away. In itself, this distance from the direct source of contamination is striking. But more significant is the fact that various lines of evidence implicated DDT as the probable major cause of declining reproduction in the petrel, at a rate at which reproduction would have failed completely by 1978.

DDT and diminished reproduction in other birds. The correlation between DDT and diminishing reproduction is based on rather considerable and diversified evidence. Taking one example out of several that might be cited, healthy osprey produce 2.2 to 2.5 young per nest, but a Maryland colony containing a DDT level of 3 ppm in its eggs produced only 1.1 young per nest. There is some evidence to suggest that DDT interferes with calcium metabolism, resulting in symptoms of calcium deficiency and creating a hormonal disturbance that results in delayed ovulation and inhibition of gonad development. These disturbances may prove to be interrelated and to be related to unsuccessful reproduction. It is important to note that a statistically significant decrease in calcium content of eggshells occurred between 1946 and 1950 in several species of carnivorous birds (peregrine falcon, golden eagle, sparrow hawk)—a period in which there was widespread introduction of DDT into the environment. It is also significant to note that subsequent to the ban on DDT in December 1972, reproduction in a number of birds has increased. For example, bald eagles in northwestern Ontario declined from 1.26 young per breeding area in 1966 to a low of 0.46 in 1974. Since then the rate increased to 1.12 in 1981 (Grier 1982).

DDT and impact on humans. The story for humans is no less alarming. According to analyses by the Food and Drug Administration in the 1970s, residues of pesticide chemicals were found in about half of the thousands of samples of food examined each year. About 3 percent of the samples contained levels in excess of, or not authorized by, legal tolerances. In a study based on the period June 1964 through April 1966, residues of chlorinated organic pesticides (DDT, dieldrin, lindane, etc.) were commonly found in all diet samples and all kinds of food except beverages. Meat, fish, poultry, and dairy products accounted for more than half the intake even though there is little direct application of pesticides to these products—they got there by the food chain. Moreover, DDT accounted for one-third of the total. Yet as the study noted, the producing area for the food was typically outside the purchasing or consuming area—a situation not unlike the petrel and its DDT intake. These DDT levels (about 0.0014 mg/kg body weight) were well below the acceptable daily intake level (0.01 mg/kg body weight) established in 1965 by the World Health Organization Expert Committee on Pesticide Residues and the Food and Agriculture Organization of the United Nations.

Direct effects on the nervous system and hormone balance have yet to be demonstrated in humans, although there are known effects on other vertebrates. These findings coupled with the high frequency of contamination (more than half the samples over a period of years), the then accelerating use of DDT and other pesticides, and the biological magnification phenomenon created sufficient public pressure that the use of DDT was quite stringently restricted by the federal government.

Various interests periodically press for elimination of restrictions and sometimes succeed. In one such instance in the late 1970s in eastern Washington, DDT was sprayed over 430,000 acres of Douglas-fir forests to control the tussock moth, even though the evidence before spraying, and corroborated after, was that the moth population was dropping as a result of natural causes. The untoward consequence was an extensive contamination of food sources of cattle and sheep and the appearance of high concentrations of DDT in the milk and meat of these animals.

Other Pesticides—Kepone, Mirex, and Dioxin

Some brief comments on several other pesticides expand the story of DDT, demonstrating that regulations on their use have too often waited for serious incidents. In other instances questions may still exist about unfavorable effects, but answers are not necessarily easily forthcoming.

Kepone. The pesticide *kepone* was not banned until 1976, and only after a number of workers in the kepone-manufacturing Life Sciences Products Company on the James River in Virginia fell ill. More than 70 workers were identified as victims of kepone poisoning with such symptoms as tremors, liver damage, and even sterility (Madati and Kormondy 1989). The children of employees of the Company were reported as being contaminated through contact with their fathers. In addition to banning the manufacture and use of kepone, large portions of the James River and its seafood-rich mouth of Chesapeake Bay were contaminated and declared out-of-bounds for fishing. Although kepone is a persistent pesticide like DDT, it was rapidly buried in James River sediments; however, disturbance by hurricanes or dredging can return the highly contaminated sediments to the surface (Cutshall, Larsen and Nichols 1981).

Mirex and fire ants. Fire ants (*Solenopsis invicta* and *Solenopsis richteri*), accidentally imported from South America in the early part of the twentieth century, had spread to nine southern states by the late 1950s. Although they do not do significant damage to crops or livestock, fire ants bite people with venomous stings (a single ant can deliver 20 bites) comparable to those of wasps or bees. An individual who is sensitive or allergic to the venom can succumb, and the bites alone can be most unpleasant and painful.

To the rescue came *mirex*, a chlorinated hydrocarbon closely related to DDT, kepone, and other pesticides; in fact, about 2 to 5 percent of mirex degrades to kepone (Carlson et al. 1976). It was regarded as an excellent replacement for other insecticides because of its selectiveness and effectiveness—ants drag home the mirex-tainted soybean oil and corncob bait and wipe out a whole colony (Holden 1976). However, by 1975 evidence had begun to accumulate demonstrating that mirex was carcinogenic (cancer-causing or cancer-inducing) in rats and mice and that 25 percent of the population living in sprayed areas had mirex stored in their bodies. These findings were compounded by the kepone incident on the James River (described above) and the recognition that kepone is not only closely related to mirex but is a degradation product of mirex as well. The end result: mirex has joined the list of banned chemical insecticides.

Dioxin. *Dioxin* (TCDD, or 2,3,7,8-tetrachlorodibenzo-*p*-dioxin) is a highly toxic compound found as a contaminant in certain herbicides (e.g., 2,4,5-T), chlorophenols, and other related chemicals. It is, in fact, an impurity that can be formed during the manufacture of 2,4,5,-T (Moriarty 1988) and other phenoxy herbicides. It is very persistent in soil except when exposed to light (McConnell et al. 1984). This pesticide has been shown to cause death, endometriosis, and spontaneous abortions in human females (Gibbons 1993), and cancer in guinea pigs and rats, and is suspected of doing so in humans as well (Roberts 1991a).

Dioxin is a known contaminant in "Agent Orange," an herbicide widely used in the 1960s by the U.S. military to defoliate Viet Cong sanctuaries during the Vietnam War. Although a number of Vietnam veterans have developed various cancers and fathered seriously handicapped children, direct linkage to Agent Orange (actually its dioxin) has been a matter of continual contention among Vietnam veterans, Dow Chemical Company (the major manufacturer of Agent Orange), and the political establishment (Magnuson 1990). Nonetheless, discovery of dioxin in the soil of Times Beach, Missouri, led to the evacuation of the town and a $33 million government offer to buy out all property owners; and, a modern state office building in Binghampton, New York, was abandoned and sealed after a transformer fire in 1981 spewed dioxin-tainted soot throughout the 18-story structure (Rempel and Randolph 1983).

In addition to these incidents, one that occurred in Seveso, Italy, in July 1976 has direct bearing. An explosion at the ICMESA Chemical Plant, which was manufacturing the herbicide TCP, released a cloud of dioxin that resulted in the death of pets and other animals within a few days (Madati and Kormondy 1989). Humans developed skin rashes, itching, vomiting, and diarrhea; long-term effects are still being studied. Unfortunately, a mass evacuation of the 700 inhabitants did not occur for several days as company and political officials argued, unnecessarily exposing the population to chronic dosages longer than should have been the case (Walsh 1977). The full details and effects of this disaster are as yet not clear; nor are the linkages to dioxin as uncontrovertible as scientists would like (Stone 1994).

Concluding Comment

As Pimental (1992) points out, nearly one-half of all potential world food supply is lost to pests, 35 percent of potential crop production and an additional 20 percent of postharvest production. But although pesticides save about 10 percent of the world food supply, they cause serious environmental and public health problems as noted above. These environmental and health side effects justify government regulation, a subject of continuing societal debate (Zilberman et al. 1991). But bans on pesticide use often carry economic impacts in lower production levels and consequent higher prices. This dilemma demands continued surveillance of pesticides and pesticide levels that are of concern to human health and safety, while simultaneously identifying alternative chemicals (Moffat 1993) and noneconomic regulatory strategies. Investigations are also needed on the effect of pesticides in all kinds of populations, including the human one, to preserve the biodiversity that is imperative to the biosphere as a whole. Perhaps the singular failure in pest population control, however, has been the slow growth of investigating mechanisms of natural control by parasites and predators of the pest itself.

DESTRUCTION OF FOREST AND GRASSLAND ECOSYSTEMS

Of all the anthropogenic impacts on ecosystems, perhaps few are as awesome and on as grand a scale as the major and systematic destruction of large sized ecosystems, such as forests—boreal, temperate, and tropical—and grasslands. Boreal and temperate forests have been subject to acid precipitation (see Chapter 20), largely a consequence of industrial activity; tropical forests have been subject to clearing for agricultural and urban/suburban development; and grasslands have been subject to *desertification*, the shift to desertlike conditions resulting from human misuse of the land. We will first consider desertification followed by discussion of the destruction of tropical forests; the impact of acid precipitation will be considered in the next chapter.

Desertification

Deserts of the world. Deserts cover about one-third of the Earth's landmass (Figure 19-8) and support about one-sixth of the world's population. They range from hyperarid (e.g., the Sahara in Africa and the Sonoran and Mojave in North America) to arid (Sahel in Africa, Chihuahuan in Mexico, Gobi in China) to semiarid (Great Plains in the United States, Kalahari in Africa) (Gore 1979).

Topographic deserts are deficient in rainfall because they are situated near the centers of continents, far from the oceans (e.g., the Gobi), or because they are cut off from rain-bearing winds by high mountains (e.g., the Mojave). Tropical deserts are those found in zones from 5° to 30° north and south of the equator and largely result from global wind patterns. At the equator, in the rainy tropics, hot air rises, shedding its moisture as it cools. The cool air begins to subside and warm up again between 15° and 30° north of the equator. The subsiding air is too dry for clouds and rain to form, the high pressure belt parching the Earth from the Sahara (which contains half the desert surface of the world) through the Arabian Desert in the Middle East and the Great Indian Desert in northwest India and Pakistan. South of the equator, a similar belt leads to the great Kalahari and Namib Deserts of Africa, as well as the Atacama Desert of Chile, the Peruvian Desert, and the barren outback of Australia.

Extent of desertification. Although deserts have replaced agricultural land before and since the advent of civilization (e.g., the Tigris–Euphrates Valley, where agriculture began, is now salt desert), the past 150 or so years have witnessed an alarming expansion of deserts. Desertification is a major global concern. According to the United Nations, each year the rate of expansion is about 80,000 km^2, or about

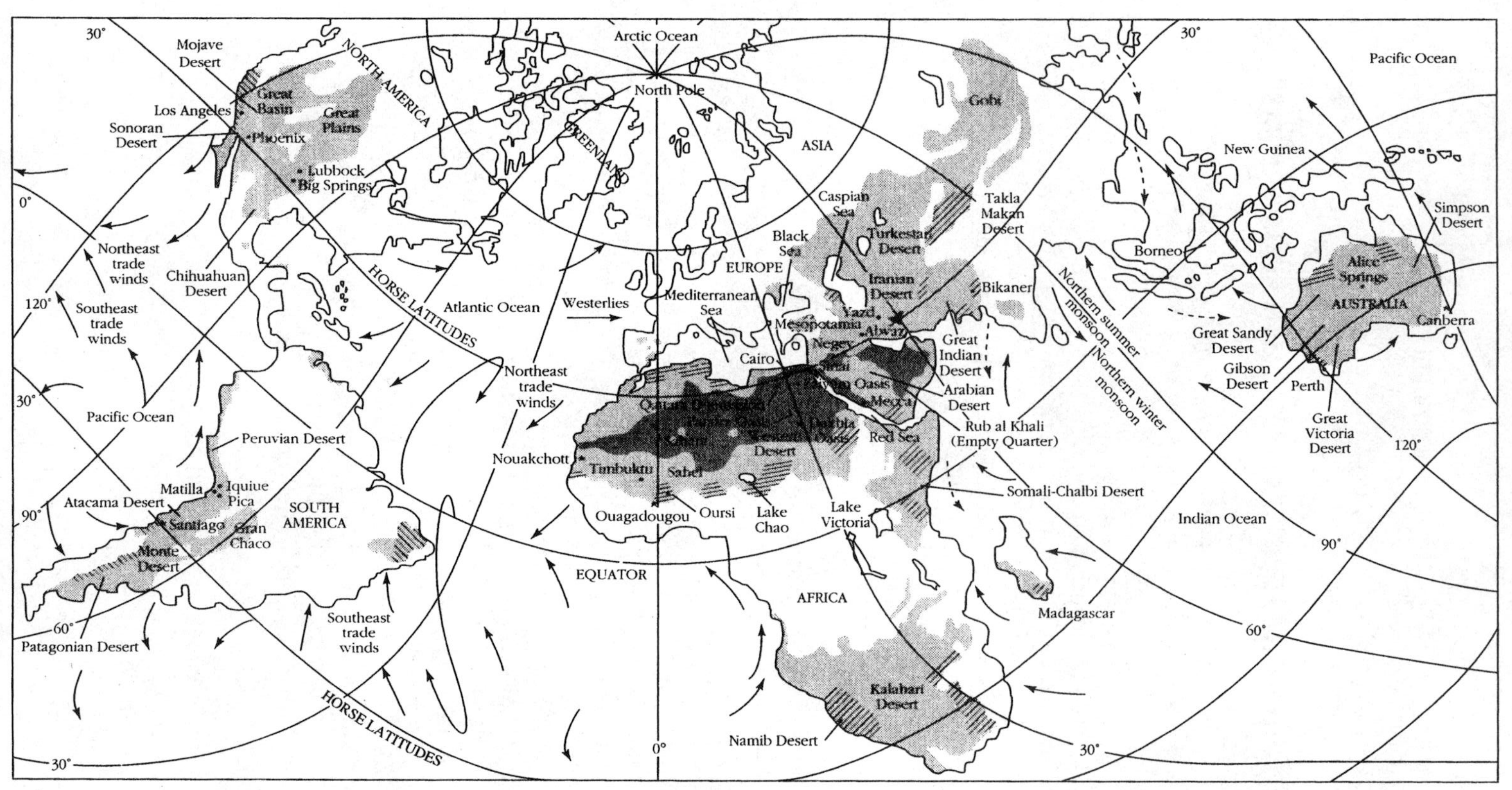

Figure 19-8 The deserts and dry areas of the world.

the size of Maine. The total area threatened by future desertification is some 39,000,000 km^2, an area equal to that of the United States, the former USSR, and Australia combined. In the United States alone the annual impact of desertification is estimated at $1 billion in terms of loss of income in Arizona, New Mexico, and western Texas from declining production per acre as well as loss of productive land. Gupta (1988) has estimated that desertification of the world has affected collectively an area larger than Brazil (Brazil is larger than the contiguous 48 United States). Desertification is ongoing in some 22 African countries, and in the 7 countries in the Sahel zone, where the rate of deforestation is seven times the Third World average, desertification is rampant (Brown and Flavin 1988). According to the World Bank, desertification in Mali alone has expanded the Sahara southward by 350 kilometers in 20 years.

Impact of desertification. Those most affected by desertification are pastoral nomads, who account for about 6 percent of the world's population living in dry-land environments. Prior to 1968, for example, 65 percent of Mauritania's population were nomads (Erbsen 1979). Then a severe drought struck the Sahel, an arid region bordering the southern edge of the Sahara (Beskit and Mubarek 1984). As many as 250,000 people and millions of animals died over a 6-year period. The nomads who survived migrated in large numbers to the capital city of Nouakchott, whose population grew from 12,300 in 1964 to about 200,000 in 1980. Nouakchott, which means "windy city," is itself threatened by desert takeover, the Sahel's sands drifting relentlessly and unceasingly through the city's streets. Mauritania's desertification has been exacerbated by recurrent droughts (Glantz 1987), but even when and if sufficient rains fall, the Sahel area in general will continue to feel the pressure of increasing population and increasing rural to urban migration (Walsh 1988; Beshir and Mubarek 1989).

The situation in China is no less severe, where deserts make up more than 13 percent of the land area (Walker 1982). Although various methods are being used to convert deserts to farmland, the rate of desertification continues to accelerate. From 1949 to the mid-1980s, soil desertification areas increased by 65,000 km^2 with degenerated grasslands covering one-fifth of that total (Jin and Cheng 1989). Coupled with forest destruction that has reduced China's percentage of the world's forest to 3 percent (Ryan and Flavin 1995), expansion of agricultural areas, with its attendant overgrazing and water loss, has caused soil erosion estimated to be 5 billion tons annually, equivalent to the country's annual fertilizer output.

Causal factors of desertification. Desertificaton in Mauritania is characteristic of the process in general: an overgrowth of livestock that could not be sustained during drought because of the reduction of vegetation by overgrazing. Also complicating the situation was the fact that forests were cut for fuel, thereby removing the natural protection of wind breaks and addition of humus to the soil. Cultivation and irrigation, overgrazing, deforestation, mining, recreation, and urbanization are all factors involved in desertification. Cultivation and irrigation contribute to desertification by wind erosion of improperly cultivated soil, consumptive use of water supplies, salinization of soils, and even abandonment of agricultural lands to barren wastelands.

Overgrazing, however, is often the common denominator of desertification in arid and semiarid regions: in one part of the southwestern United States ecologists calculated that the land could support 16,000 sheep and instead 140,000 sheep were grazed. Since 1957 the number of livestock has increased sixfold in this region. In western India, from 1950 to 1961, the area available exclusively for grazing dropped from 13 to 11 million hectares, whereas the population of goats, sheep, and cattle increased from 9.4 to 14.4 million.

Semiarid grasslands are essentially homogenous ecosystems with respect to water, nitrogen, and other soil resources—that is, they are distributed with considerable regularity. Long-term grazing, however, leads to an increase of heterogeneity of these resources, promoting invasion by desert shrubs (Schlesinger et al. 1990). In turn, this invasion leads to further localization of soil resources under shrub canopies and loss of soil fertility by erosion and gaseous emisson in the barren areas between shrubs. Such a situation is a positive feedback that leads to desertification of formerly productive land in southern New Mexico and in other regions, such as the Sahel.

Deforestation, however, is not regarded as important a desertification agent in the United States as it is in Africa, Asia, and Latin America, where large numbers of people have no other energy resource. Mining can create local deserts through the removed overburden, plus processing wastes that create unproductive areas. Finally, urbanization and recreation, such as the use of off-road vehicles that results in compaction of soil and thereby increases runoff and erosion, are also significant agents in desertification in some areas (Iverson et al. 1981).

Dust storms. Because of their lifting and carrying capacity, winds can transport particulate and other materials for considerable distances. Winds blowing across desert, desertified, or even parched prairie erode the top soil, carrying it windward. For example, the harmattan winds that blow across the Sahara carry large amounts of particulate dust in the air into Nigeria (and elsewhere), the dust load having increased in recent years because of the increase in transport vehicles plying untarred roads and increased desertification (Egunjobi 1989).

In the United States the most spectacular and devastating instances of desertification were the dust stroms of the 1930s. Airborne dust in clouds several miles high carried hundreds of millions of tons of dust for thousands of miles. This dust was the valuable and productive topsoil of what became known as the Dust Bowl, an area that included parts of Texas, New Mexico, Colorado, Oklahoma, and Kansas (Figure 19-9). In 1937 the Soil Conservation Service estimated that 43 percent of a 16-million-acre area in the heart of the Dust Bowl has been seriously damaged by wind erosion (Lockeretz 1978).

In the area of the Dust Bowl in the mid-1800s native vegetation was shortgrass (buffalo grass and blue grama) that was conducive to grazing animals, such as bison and subsequently cattle, notably longhorn breeds. Overgrazing, among other factors, ended the open-range cattle industry, and the grasslands were converted to cul-

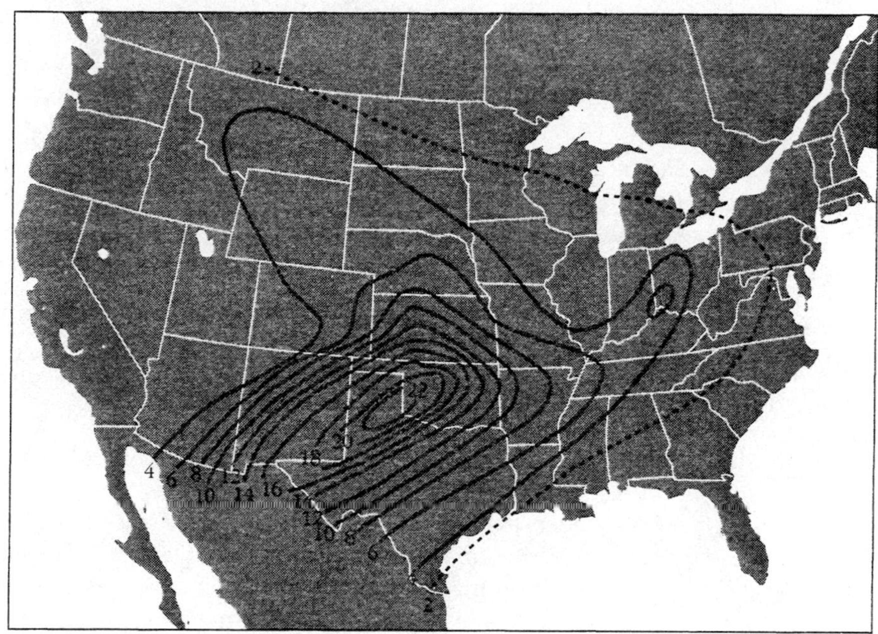

Figure 19-9 Dust storms during March 1936. The contours indicate the number of days with dust storms or dusty conditions. (R. J. Martin. 1936. *Monthly Weather Review* 63:87–8.)

tivated crops. Droughts led to wind erosion and frequent dust storms, culminating ultimately in the 1931 storms.

Despite steps toward erosion control, additional dust storms have occurred since then: the region tends to be subject to major droughts at intervals of about 20 to 22 years. More land was damaged annually by wind erosion in the Great Plains in the 1950s than in the 1930s, and soil loss in the 1970s was comparable to that of the 1930s. Pimentel et al. (1976) has estimated that in the last 200 years at least one-third of the topsoil of United States croplands has been destroyed, involving about 81 million hectares and that 850 tons of soil per year were moved by the wind in the western region of the United States alone. Pimentel further estimates that 1 billion tons per year are removed from the entire United States. Combined with 4 billion tons washed from the land, the gross soil erosion is equivalent to about 18 cm of soil from 2 million hectares.

To convey further the distance effect of dust storms and top-soil erosion, dust fallout across the Atlantic Ocean from the African continent measured at Barbados in the West Indies, a distance of 4,700 kilometers west of the Sahel Desert, increased from 8 micrograms/m^3 in 1967–1968 to 15 micrograms in 1972 and to 24 micrograms in 1973 (Brown and Flavin 1988). Erosion of top soil is, of course, a serious long-term threat to crop production (Larson, Pierce, and Dowdy 1983).

Land management and desertification. Halting or decelerating the rate of desertification requires intelligent land management. This process includes using arid-adapted vegetation to provide an adequate cover that not only reduces wind speed but also traps dust-blown air. Creating barriers, such as rows of trees to cut down wind speed, is also a deterrent, as is maintaining a rough, cloddy soil surface that can result from returning organic matter to the soil. Yet certain soils are more susceptible to erosion and, unfortunately, the sandy soils of the Sahara, among other places, fall in this category. But ultimately climate is the significant and still largely uncontrollable factor. Nonetheless, avoiding or mitigating the known human activities that contribute to desertification can halt the advance of *Deserts on the March*, the title of a prophetic and foreboding book by ecologist Paul Sears (1935). The choice is ours.

Tropical Forest Destruction

Forests have probably been subject to anthropogenic destruction, or *deforestation*, since the earliest days of *Homo sapiens* and its predecessor species. The initial needs for fuelwood and construction materials continued to increase as populations expanded and as industrial construction, discovery of mineral resources in the forested areas, and eventually roadways and highways, and new and successively larger settlements increased the demand for forest products. Crosby (1986) describes destruction of Greenland's forest by Norsemen and of Madeira's forest by Portuguese settlers in the fourteenth century, as well as elsewhere, as European imperialistic activity spread throughout the world. Incidentally, Madeira, meaning "wood," received its name because of its great forests.

Causes of deforestation. Tropical rainforests (see Figure 14-1 for their worldwide distribution and Chapter 14 for a description of them) have been particularly subject to destruction since the 1950s due to increased export of tropical hardwoods to Europe, Japan, and the United States (Aldhous 1993; Gupta 1988; Skole and Tucker 1993). Tropical hardwoods have a relatively homogenous surface and large dimensions making it possible to produce long, wide planks of uniformly high quality, prized in the timber industry (Vincent 1992). Harvesting of a forest itself takes a toll during construction of the necessary access roads.

In drier areas, tropical rainforests are being exploited for the collection of fuelwood that, along with charcoal, is the main energy source for a large number of the poor in the Third World. Every form of fuelwood biomass, including wood, twigs, crop residues and grass, is especially important in Nepal, India, China, Kenya, Zimbabwe, Brazil and Egypt, among other countries (Gupta 1988), and in many of them, it is short supply (Figure 19-10). Even developed countries are increasingly turning to obtaining energy from plants to bypass the use of fossil fuels (Lawson, Callaghan and Scott 1984).

Agricultural practices have also taken their toll on tropical forests. Poor landless peasants tend to move into the edges of the forests seeking farming areas; while

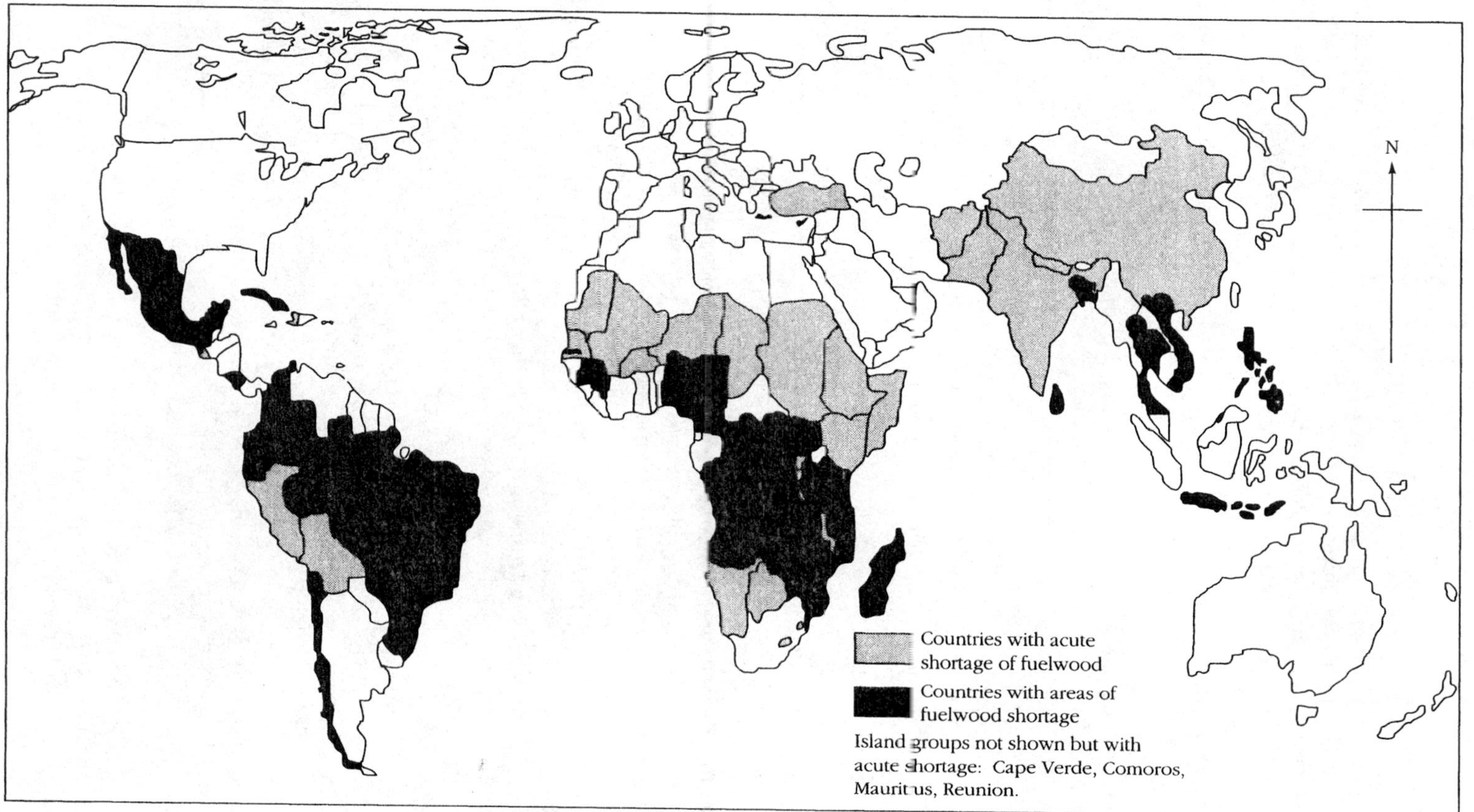

Figure 19-10 The status of fuelwood in Third World countries. (Redrawn with permission from A. Gupta. 1988. *Ecology and Development in the Third World.* London: Routledge.)

large areas are cleared for plantations (e.g., rubber and palm oil groves have replaced the rainforest on the lower slopes of Malaysia) or for cattle ranches as in Central America and the Amazon Basin, which produce lean meat used for the hamburger, hot dog, and canned meat industries in the United States (Gupta 1988). In many tropical countries, government policies that encourage exploitation by excessive logging and clearing for ranches and farms are largely to blame for the accelerating destruction of tropical forests (Repetto 1990).

Consequences of deforestation. One of the immediate effects of deforestation is an acceleration of soil erosion with the consequences of decreased fertility, and hence productivity, because of the removal of nutrients and organic matter (Ewel et al. 1981; Jordan 1986; Whitford 1992). Soil erosion also converts relatively homogenous land surfaces to a series of gullies, reducing access, as well as increasing the sediment load in streams, which reduces their productivity. The increased sediment load can also lead to choking of a stream resulting in flooding of adjacent land and increased leaching of valuable nutrients from that soil. Logging that results in substantial increases of downed woody debris can dramatically increase susceptibility to fire (Uhl and Kauffman 1990).

Tropical forests are sources of many valuable products, such as oils, gums, rubber, fibers, dyes, tannin, resins, and turpentine, as well as a wide variety of fruits and ornamental plants. They are also the source of ingredients for drugs used in the treatment of leukemia and Hodgkin's disease and for contraceptive pills, as well as the source for other pharmaceuticals, such as strychnine, reserpine, curare, and quinine (Gupta 1988). More significant perhaps is the fact that many tropical rainforest species have not been fully explored for their health-related and other beneficial products. For example, during the Green Revolution (see Chapter 17), tropical rain forest species were used extensively in hybridization of cereals. Finally, these rain forests are inhabited by a large number of highly diversified animals and plants, providing a worldwide gene pool resource for evolutionary adaptation. Obviously, destruction of tropical rain forests imposes a considerable negative impact on biodiversity (Wilson 1988).

Among the less well-understood and hence highly debated consequences of tropical rain forest destruction are the long-term and global effects. It is known that deforestation decreases moisture in the soil, evapotranspiration, and precipitation and that it increases surface temperature and reflection of greater amounts of solar energy back to the atmosphere. All of these changes influence regional climate (Shukla et al. 1990) as well as changes in rainfall and atmospheric heat flux on a global scale. That is, the consequence of localized disruptions may have effects many miles distant, as is true of other ecological disturbances, as we have already pointed out.

The greatest impact on global climate, however, may well be an exacerbation of the Greenhouse Effect, which was discussed in Chapter 8 and will be treated further in Chapter 20. Deforestation on a large scale reduces the uptake of carbon dioxide in photosynthesis, and it is that unused and enlarging pool of carbon dioxide that affects the Earth's ambient temperature (Woodwell et al. 1983).

Extent of deforestation. Although there is disagreement on the exact rates of deforestation, partly because of the difficulty of defining what 'deforestation' is, there is little disagreement that at present levels of destruction much of the world's tropical rain forests will have disappeared by the end of the twentieth century (Jordan 1985). The World Resources Institute (1990) estimates that 12 million hectares (slightly larger than the state of Kentucky) of tropical forests were lost in eight countries in 1987, more than four times the rate calculated by the United Nations's Food and Agriculture Orgânization in 1980. Further, satellite sensing shows that tropical forest are vanishing at a rate of 16 to 20 million hectares a year (about the size of Missouri), much faster than the U. N. estimate of 11 million. And Brazil alone accounts for about 65 percent of the total. Because of its devastating impact on the world, Mitchell (1991) has termed deforestation as one of the "Four Horsemen of the Apocalypse."

By 1985, Madagascar, one of the richest biological areas of the world and whose diverse flora and fauna are most endangered, had only 50 percent of the 7.6 million hectares of rainforests it had in 1950 and only 30 percent of the estimated original extent of 11.2 million hectares (Green and Sussman 1990). For comparison, the original forested area of 11.2 million hectares is about the size of Kentucky, whereas that area in 1985 was about the size of Massachusetts and Connecticut combined. The average annual rate of deforestation in the 35 year period from 1950 to 1985 was 111,000 hectares (that is about one-third the size of Rhode Island) and was most severe in areas of low topography and high population density. If this pace of cutting continues, only forests on the steepest slopes will be surviving in the year 2020.

In Central America, substantial amounts of tropical forest were lost between 1940 and the mid-1980s, and El Salvador has been completely deforested (Figure 19-11). With a total of 12.6 million hectares of forested land in the mid-1980s, destruction was proceeding at an annual rate of over 400,000 hectares (slightly larger than the size of Rhode Island), mostly because of cattle ranching and to a lesser extent to colonization and logging. In Panama, more than 70 percent of the tropical rain forests has vanished, leading to two untoward consequences on the strategically important Panama Canal: (1) silting of Madden Lake, which is a backup source of water for the Canal's operation; and, (2) a decrease of about 10 percent in annual rainfall (which is critical to the Canal's need for 52 million gallons of water for the passage of each ship) since the turn of the century (Borrell 1987).

In Brazil's Amazon Basin, the World Bank estimated that as much as 12 percent, or 60 million hectares of the rainforest (equivalent to the size of Indiana, Iowa, Michigan, Ohio, and Wisconsin combined) had been destroyed between 1978 and 1989 (Golden 1989). A subsequent estimate based on satellite remote sensing reduced this figure to about 5 percent, or 25 million hectares (equivalent to the size of Michigan and Ohio combined). The discrepancy is partly due to the fact that the base in one case is on the total land area of Amazonia, including savanna and wetlands, rather than just rainforest. Whether the reduction is 5 percent or 12 percent, the rate and amount of destruction are substantial. Given the impact of tropical forests on world climate described above, there is grave and worldwide concern. And, although there is climatic evidence that the Amazon rainforest is well adapted to a number of natural disturbances, the enigma is whether it has the resilience to tolerate anthropogenic exploitation (Colinvaux 1989; Hollaway 1993).

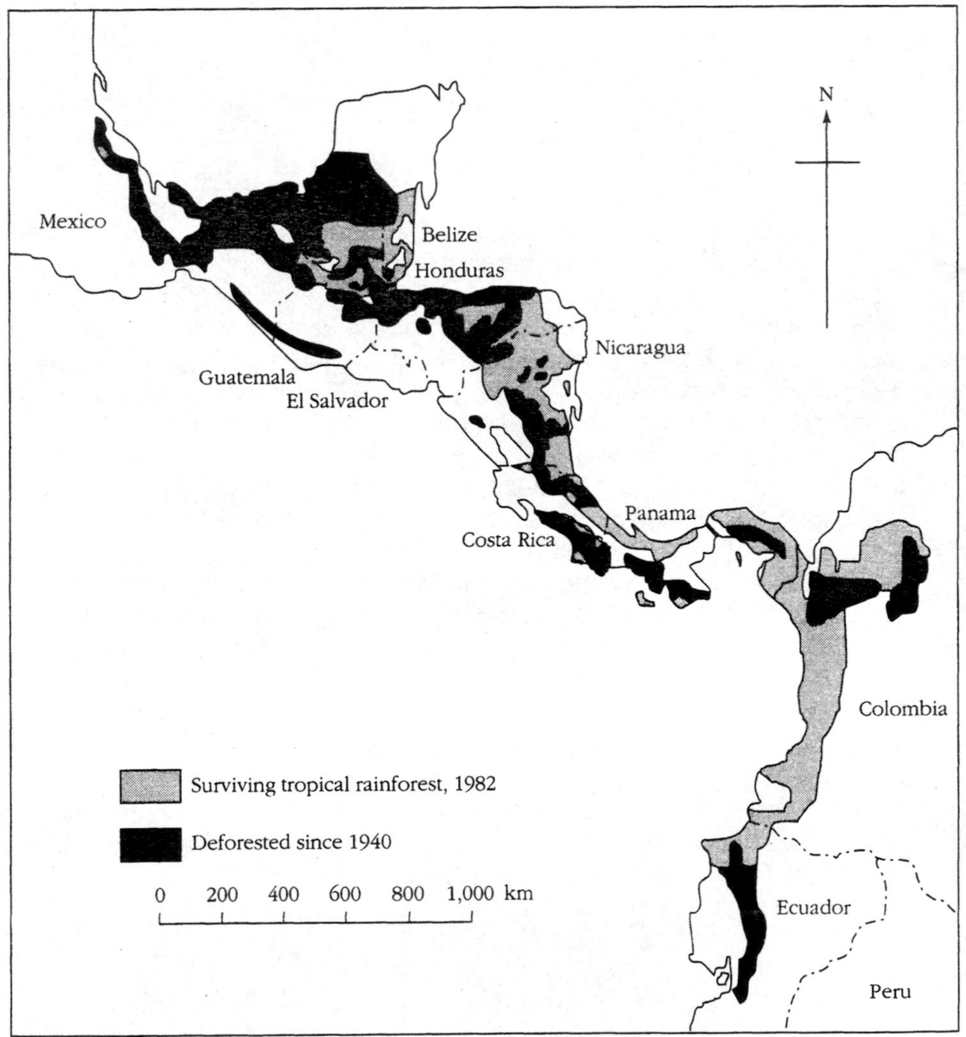

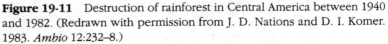

Figure 19-11 Destruction of rainforest in Central America between 1940 and 1982. (Redrawn with permission from J. D. Nations and D. I. Komer. 1983. *Ambio* 12:232–8.)

SUMMARY

VOCABULARY

biological magnification
DDT
deforestation
desertification
dioxin
fission
fusion
kepone
meltdown
mirex
radioisotopes

KEY POINTS

- Ionizing radiation, particularly acute rather than chronic, may have significant adverse mutagenic as well as public health effects; fallout from nuclear bomb testing, meltdowns of nuclear reactor plants, and accidental contamination have had serious consequences; disposal of nuclear wastes is yet to be safely provided.
- Although all pesticides do not behave the same, some are subject to biological magnification, thus rendering higher trophic levels more subject to adverse effects from those that are toxic.
- The rate of desertification continues to accelerate as more and more humans require more space for living, more fuel for cooking and heating, and more land for agriculture; dust storms are a possible consequence of desertification in some areas.
- Tropical forests have been increasingly subject to deforestation since the 1950s, their wood being ideal for timber purposes and fuel and the land being used for agriculture. Long-term effects on climate are the subject of considerable debate as is the reduction in biodiversity.

CHAPTER 20

Anthropogenic Impact on the Atmosphere

"Resources out of place" in the atmosphere have significant impacts on human health, the environment, and climate. Since the atmosphere is the medium through which respiratory gases are exchanged, organisms that meet their respiratory needs through this vehicle can have their health affected by airborne contaminants. Precipitation in its many forms transports atmospheric pollutants to terrestrial and aquatic ecosystems; some of these pollutants enrich whereas others adversely affect such systems. Other atmospheric pollutants contribute to destruction of the protective ozone layer, thus affecting not only human health but also contributing to a warming of the Earth. Excesses of carbon dioxide, a normally small component of the atmosphere, likewise can affect climate.

In this chapter, we will first discuss the nature of particulates, the form in which many, but not all, atmospheric pollutants occur and then proceed to discussions of five major areas of contemporary concern about the atmosphere: ozone depletion, global climate modification, nuclear winter, heavy metal pollution, and acid rain.

PARTICULATES

Particulate matter, or aerosols, in the air is the result of both natural and human processes, the former contributing globally about nine times more than the latter, or about 2.3 billion tons annually versus 296 million tons (Fennelly 1976). Particulates may be minute solid particles or liquid droplets as small as a cluster of several molecules or as large as a visible dust particle. The very fine particles behave more like a gas and are subject to Brownian motion, coagulation, and condensation; large particulates are more like solid matter and are influenced by gravity.

Large, or primary, particulates range in size from 1 to 100 micrometers, the modal point being around 10 micrometers, and are injected directly into the atmosphere from such human sources as industrial smokestacks, gravel crushers, and blast furnaces and from such natural sources as forest fires and ocean spray (Table 20-1) (Fergusson 1992). They also arise from oil fires ignited by purposeful action, as occurred during the 1991 Gulf War in Kuwait (Linden 1991; Hobbs and

TABLE 20-1 HUMAN SOURCES OF PARTICULATE POLLUTION IN THE UNITED STATES

Source	Emissions (millions of tons per year)	Total
Natural dusts		63
Forest fires		56.3
Wildfire	37	
Controlled fire		
Slash burning	6	
Accumulated litter	11	
Agricultural burning	2.4	
Transportation		1.2
Motor vehicles		
Gasoline	0.420	
Diesel	0.260	
Aircraft	0.030	
Railroads	0.220	
Water transport	0.150	
Nonhighway use		
Agriculture	0.079	
Commercial	0.012	
Construction	0.003	
Other	0.026	
Incineration		0.931
Municipal incineration	0.098	
On-site incineration	0.185	
Wigwam burners (excluding forest products disposal)	0.035	
Open dump	0.613	
Other minor sources		1.284
Rubber from tires	0.300	
Cigarette smoke	0.230	
Aerosols from spray cans	0.390	
Ocean salt spray	0.340	
Total		122.715

Data from Midwest Research Institute. 1971. *Particulate Pollutant Svstem Studv.* Vol. I: Mass Emissions.

Radke 1992) The smaller, or secondary, particles, range in size from less than .01 micrometer to about 1 micrometer, the modal point being around 0.1 micrometer, and result largely from chemical reactions in the atmosphere on both natural and anthropogenic components. The secondary acidic particles, including sulfates and nitrates, are responsible for reducing visibility and causing erosive damage to materials in the form of "acid rain," which will be discussed later, and are derived almost exclusively from the combustion of fossil fuels (Shaw 1987). The smaller of the secondary particulates result from photochemical reactions, the larger by coagulation or condensation of photochemically generated particles. Larger primary and secondary particulates settle out of the atmosphere by gravity; the smaller particulates tend to be washed out by precipitation.

Long-Range Transport of Particulates

As noted above, primary particulates are more subject to the influence of gravity than are secondary particulates, which can be transported considerable distances in relatively short periods of time (see Figure 19-3). Even remote parts of the Arctic, which most scientists had assumed to be pristine (Kerr 1979), have been found to be contaminated by sulfates, vanadium, and other trace elements during the winter and spring when pollutants travel 5,000 km and more from industrial areas in Europe, Asia, and perhaps North America (Kerr 1981b). Arctic pollution levels are not as heavy as those near most cities, but are still high: New York City's air contains 10 times more soot than that of Berkeley.

Carbon, in the form of soot, organic matter, and carbonates is a major constituent of particulates in the atmosphere. Unlike most other aerosols that only scatter light, soot carbon is able to absorb significant levels of visible light. This leads to increased retention of solar energy and therefore to a warming effect augmenting the Greenhouse Effect, which is due largely to carbon dioxide. In a study in which atmospheric soot carbon was sampled over the Atlantic Ocean from Hamburg, West Germany, to Montevideo, Uruguay, Andreae (1983) found significant amounts over large, remote areas, often similar to concentrations in rural continental areas. Biomass burning in tropical regions was identified in this study as an important source of the soot carbon; it has been indicted again in subsequent studies (Crutzen and Andreae 1990).

Although long-term transport of particulates has been amply evidenced, local or nearby sources of pollution are not insignificant as will be seen in the case of the Meuse Valley, London, and Donora incidents described below. Considerable controversy has existed as to the contributions of midwestern industrial and domestic sources of pollutants on the northeast, a matter we will explore further in a discussion of acid precipitation. Rahn and Lowenthal (1985) showed that midwestern sources dominated northeastern sources in Vermont in July 1982, during a major pollution episode and during other episodic incidents. However, nearer northeastern sources are comparably important in the long-term, providing a persistent base level upon which midwestern aerosols are superimposed every few days.

Effects on Human Health

Particulate matter larger than 10 micrometers tends to be retained in the upper respiratory tract and thus can be disposed of readily. Secondary particulate matter, however, easily travels to the alveoli becoming an irritant and as well as progressively more dangerous to health. The adverse effects range from such common ailments as hayfever and other allergic reactions to serious illness and death, particularly from heavy concentrations of particulates whose effects become exacerbated by *atmospheric inversions*, situations in which an upper layer of warm air prevents vertical mixing from below. Atmospheric inversions can lead to a rapid buildup of pollutants. A brief description of several major air pollution incidents derived from Madati and Kormondy (1989) will exemplify these points.

Meuse Valley, Belgium—1930. The highly industrial Meuse Valley is 24.0 by 2.4 km and surrounded by hills up to 100 m high. In addition to domestic air-pollution sources such as coal burners, motor vehicles, and coal-burning railways, the Valley was populated by: steel plants with coking installations, rolling mills, blast furnaces, boilers and welding facilities; metallurgical and foundry works and coking plants; coal-burning, electric-power generating plants; glass, ceramic and brick works; coal and gas furnaces and lime kilns; zinc plants with reduction furnaces, retorts, and drying ovens; a sulfuric acid plant with roasting ovens; and a concentrated fertilizer plant with ovens. Among the possible air pollutants from such installations were: ammonia, arsine, carbon dioxide, carbon monoxide, cement and lime dust, drops of tar, phenol and naphthalene, formaldehyde, hydrocarbons, hydrofluoric acid, hydrogen, hydrogen chloride, methyl and ethyl alcohol, nitrogen dioxide, silica, soot and metal dust, sulfur dioxide, and sulfur trioxide.

The topography of the Valley and the generation of now known air pollutants presented a situation waiting for disaster to happen. And that occurred between December 1 and 5, 1930, when there was a convergence of anticyclones characterized by high atmospheric pressures, mild winds of a general easterly direction, a worsening of fog, and cooling at night—the conditions of atmospheric inversion. Typical signs and symptoms of a respiratory irritant, including chest pains, cough, shortness of breath, and irritation of the mucous membranes and eyes were the result. Six thousand people became seriously ill, and sixty died. Upon investigation, it was concluded that the sulfur oxides were the most likely culprits. Control measures were finally directed at the major polluting sources.

Donora, Pennsylvania—1948. Situated 32 km (20 miles) south of Pittsburgh in the steel-belt of Pennsylvania, Donora's industries included: steel plants with blast furnaces, looping rod mills, wire drawing and finishing; a zinc plant with reduction furnaces, retorts, and drying ovens; a sulfuric acid plant; glass works and a coal-burning electricity-generating power plant. Atmospheric conditions coupled with the presence of such probable air pollutants as arsine, cadmium, carbon dioxide and carbon monoxide, chlorides, fluoride, hydrogen, iron, carbonyl, lead, manganese, nitrogen oxides, sulfur dioxide, and zinc, in addition to large particulates, caused serious illness among 6,000 people and the deaths of 20 others. The reduction in air pollution in the general Pittsburgh area since this catastrophe has been nothing less than remarkable.

Poza Rica, Mexico—1952. A low-altitude inversion coupled with a high concentration of haze and a very light wind compounded an overflow of amine solution in a sulfur removal plant. The overflow blocked the gas lines to the pilot lights of the flares so that, instead of being burned, hydrogen sulfide escaped unburned into the atmosphere. Hydrogen sulfide poisoning claimed 22 lives and hospitalized 320 people.

London, England—1952. Between December 5 and 13, 1952, almost 4,000 people died when a low temperature inversion and almost a complete absence of wind movement intensified London's infamous, periodic fog (Logan 1953). The height of the inversion was determined to be between 60 to 90 m, with a haze layer extending to 600 m. Preceding the disaster, particulate matter averaged 0.50 mg/m^3 and, during the disaster period, it averaged 4.46 mg/m^3, a ninefold increase. The sulfur dioxide levels rose to 1.33 parts per million, also a ninefold increase over the average of the preceding days. As in other incidents, those individuals with a history of heart and respiratory ailments were most subject to adverse effects of air pollution; in the London incident, more than 80 percent of the deaths occurred in people with such predisposing conditions.

Bhopal, India—1984. In 1980 India started manufacturing the toxic gas methyl isocyanate (MIC), which is used to make Sevin (carbaryl), a pesticide that can kill more than 100 types of insects attacking 100 different crops. For reasons unknown, water entered one of three underground storage tanks that held some 40 tons of MIC, building up pressure and causing a leak on December 3. For unknown reasons, the safety scrubber chimneys, filled with the detoxifying sodium hydroxide solution through which the gas should have escaped, failed to function. The outcome was that the noxious MIC gas started seeping from the tank, and the northwesterly winds spread it over the densely populated city of Bhopal. Within days and with no known treatment or antidote, MIC maimed more than 100,000 people and killed more than 2,500, making Bhopal the worst industrial accident ever.

Effects of long-term exposure. The several foregoing incidents were acute, or short-term, rather than chronic, or long-term, in nature. Except for the Bhopal incident and the Seveso and Chernobyl incidents described in Chapter 19, individuals living in the Meuse Valley, Donora, and London had been exposed for years to the various air pollutants. The short-term duration of the London incident, for example, would not have promoted the onset of the bronchitis, coronary diseases, pneumonia, cancer of the lung, and respiratory tuberculosis that accounted for the majority of the 3,000 deaths; however, in all cases it decidedly exacerbated already-diseased conditions. Long-term exposure to airborne particulates has been tied to a variety of respiratory diseases, such as bronchitis, bronchial asthma, pulmonary edema, cardiovascular diseases, cancer of the skin and lungs, dermatitis, and hyperpigmentation of the skin.

Particulates and Materials

Besides the unaesthetic quality of sedimenting soot and dirt, the corrosive effects of particulates are insidious. Metals, paints, and textiles, as well as the masonry of buildings and statuary, are not immune to the effects of particulate air pollution. Annual costs and losses are considerable. One estimate suggests that soiling of personal property in a pollution-prone city adds $85 per year to a person's cleaning costs. The destruction of imposing and important artifacts of historical civilizations, such as those from ancient Athens, is another sorry consequence of aerosol pollution.

Particulates and Air Quality Improvement

Given the adverse impact of particulate pollution on human health and materials, improvement in air quality is a desirable goal. But controlling pollution has attendant costs. A thorough discussion of cost benefits of any particular pollution containment is beyond the scope of this book; however, some economists have determined that the costs of many new pollution controls in the nation as a whole and in the Los Angeles metropolitan area in particular exceed the benefits, perhaps by a considerable margin (Krupnick and Portney 1991).

Particulates and Climate

Particulates can affect climate by both absorbing and scattering sunlight, thus reducing the amount of sunlight reaching the Earth (see Figure 4-4) and lowering the temperature. Sulfate aerosol particles are among such particulates, directly scattering short-wavelength solar radiation and modifying the shortwave reflective properties of clouds (Charlson et al. 1992). But particulates can also absorb and reradiate energy in the infrared part of the spectrum, creating the Greenhouse Effect (described in Chapter 8 and later in this chapter), and thereby warming the Earth. This *quid pro quo* of different particulates impacting climate will be discussed more fully later in this chapter.

Volcanoes and Climate

In addition to volcanic activity being a possible source of global significance for a number of elements in the form of particulate material (Mroz and Zoller 1975), the most studied effects of particulate intrusions on climate have been those resulting from volcanic eruptions. Perhaps the first to suggest that volcanoes modify climate and weather was Benjamin Franklin who tied the 1783 eruption of Laki Volcano in Iceland (which, incidentally, produced the greatest lava flood in recorded history) to an abnormally severe winter in 1783–84 in Europe (Decker and Decker 1989). The reason, Franklin speculated, was that fine ash and gases from the eruption may have blocked out the sun's rays.

In 1816 snow fell in June and killing frosts continued through August in New England in what has been referred to as "the year without a summer" (Stommel and Stommel 1979) caused by a major volcanic explosion of Mount Tambora in Indonesia. The eruption of Krakatau in Indonesia on August 27, 1883, collapsed the 800-meter high peak to 300 meters below sea level and generated a tsunami (or tidal wave) that killed 36,000 people in nearby Java and Sumatra and eventually reached the shores of England. Solar radiation decreased 10 percent, and temperatures in Europe were below normal for the next 3 years (Decker and Decker 1989). The major volcanic eruption of Mount Agung in Bali, Indonesia, in 1963, resulted in 2,000 deaths and a year-long decrease in starlight measured worldwide; in addition, stratospheric temperatures increased as much as 6°C and world temperature dropped 0.4°C for 3 years after the eruption (Decker and Decker 1989). Despite the magnitude of the 1980 violent eruption of Mount St. Helens, in the State of Washington, little or no worldwide atmospheric or surface-cooling effects were detected (Kerr 1981a; Decker and Decker 1989). By contrast, the 1982 eruption of long dormant El Chichon Volcano in southern Mexico created a stratospheric haze 100 times more opaque than that of Mount St. Helens and that persisted for nearly 3 years. Nonetheless, no significant global cooling was detected.

The eruption of Mount Pinatubo in the Philippines in June 1991, provided scientists with the first unambiguous, direct measurements of large-scale volcanic "radiative forcing," that is a cooling impact on climate (Minnis et al. 1993; Brasseur and Granier 1992). Some 25 to 30 million tons of sulfur dioxide gas were lofted into the atmosphere causing a strong cooling effect almost immediately through a direct increase in albedo over mostly clear areas and adding to the albedo in cloudy areas. Based on his computer climate model, James Hansen of the National Aeronautics and Space Administration's Goddard Institute for Space Studies had predicted that Pinatubo's debris shade would lower global temperature by one-half degree centigrade; the year-end temperature reports for 1992 dramatically confirmed this prediction (Kerr 1993)! Parenthetically, the aerosol sulfur compounds from Mount Pinatubo created some of the most spectacularly colorful sunsets throughout the globe for well over a year.

Decker and Decker (1989) conclude that volcanic eruptions can inject large amounts of fine ash and great quantities of sulfur dioxide gas into the stratosphere and thereby significantly affect the solar radiation reaching the Earth's surface. If these eruptions occur in equatorial regions, average global temperatures may be lowered by as much as 1°C. Although climate is affected by a number of factors (see Chapter 4), volcanic activity is perhaps the most significant historically.

OZONE, POSITIVE AND NEGATIVE

Ozone, or trimolecular oxygen (O_3), can be a positive or a negative influence on human health, as well as that of other organisms, depending on whether it is present in the troposphere or stratosphere. In the latter case, ozone is a positive, serving as a shield in screening out most of the damaging portions of ultraviolet radiation (wavelengths between 290 and 320 nanometers) (Cicerone 1987). In the troposphere, ozone is a negative as it appears as a component of smog with a number of deleterious effects, as will be discussed below.

Stratospheric Ozone

The chemistry of ozone. Ozone is formed in the stratosphere by the action of solar ultraviolet radiation on molecular oxygen (O_2) to form atomic oxygen (O) followed by recombination of O atoms with O_2 to form O_3 (Molina and Molina 1992):

1. $O_2 \rightarrow O + O$
2. $O + O_2 + M \rightarrow O_3 + M$,

where M is any third body, such as N_2 or O_2, that removes the energy of the reaction and stabilizes O_3. Photolysis of ozone results in the production of O_2,

3. $O_3 \rightarrow O + O_2$
4. $O + O_3 \rightarrow O_2 + O_2$,

this fourth reaction providing the oxygen sink for the production of ozone by reactions 1 and 2.

Much of recent stratospheric research has been concerned with gaining an understanding of the additional pathways by which the "odd oxygen" (that is, O and O_3) can be removed (McElroy and Salawitch 1989). The major culprits are compounds with halogen radicals and, most notably, the families of chlorofluorocarbons, hydrofluorocarbons, and hydrochlorofluorocarbons (Rowland 1989).

The hypothesis indicting the chlorofluorocarbons in the destruction of stratospheric ozone was first advanced by Molina and Rowland (1974) who postulated that solar photodissociation of chlorofluorocarbons (or more precisely chlorofluoromethanes) in the stratosphere produces significant amounts of chlorine atoms that lead to the destruction of atmospheric ozone. Thus, trichlorofluoromethane ($CFCl_3$) acts in a catalytic chain of reactions as follows:

1. $CFCl_3$ + ultraviolet light → Cl
2. $Cl + O_3 \rightarrow ClO + O_2$
3. $CLO + O \rightarrow Cl + O_2$,

with the chlorine descending through the ozone layer for ultimate deposition in precipitation as hydrochloric acid (HCl) (Rowland 1989).

Among other chemical properties of the chloroflurocarbons is their stability: the atmospheric lifetimes for $CFCL_3$ and CF_2Cl_2 (dichlorodifluoromethane) are estimated at 75 years and 111 years respectively (Cunnold et al. 1986). Even if all emissions of chlorofluorocarbons were to be universally stopped, the discharges to date would be experienced for some time to come.

Uses of chlorofluorocarbons. *Chlorofluorocarbons*, known collectively as CFCs, enjoy technical popularity because of their inertness and volatility, $CFCl_3$ being employed as an aerosol propellant and CF_2Cl_2 as a refrigerant (commercially known as Freon). Other CFCs are used as blowing agents for soft polyurethane foams, such as that used in seat cushions, and as agents for metal cleaning and drying, industrial sterilization of medical equipment, and fast freezing of foods. Although the use of $CFCl_3$ in aerosol spray cans was banned in the United States in 1977, few acceptable alternatives have been found for other uses of the CFCs (Manzer 1990).

Impact of ozone destruction. A 16 percent reduction in stratospheric ozone will produce a 44 percent increase in the amount of damaging ultraviolet radiation reaching the Earth's surface, and a 30 percent reduction will double such radiation (Maugh 1980). Studies by the National Academy of Sciences note that increases in the ultraviolet radiation would increase cases of skin cancer, both nonmelanoma and the more severe melanoma that is fatal nearly one-third of the time, reduce the productivity of agricultural crops, destroy larval forms of some marine life, and produce a slight warming of the Earth's atmosphere (Maugh 1980, 1982, 1984; MacKenzie and El-Ashry 1988; Roberts 1989b). Several groups of scientists working independently have estimated a reduction in photosynthesis in the order of 6 to 16 percent by marine phytoplankton because of increased ultraviolet radiation (Smith et al. 1992; Cullen et al. 1992) It has also been recognized that ultraviolet radiation affects the immune system by decreasing the viability of circulating lymphocytes.

Given present levels of discharge of CFCs, the National Academy estimates a 2 to 5 percent increase in the incidence of skin cancer for every 1 percent decrease in ozone concentration (Maugh 1982). Recent estimates note that even if the production of CFCs remains limited according to the protocol signed in Montreal in 1987 for the protection of the ozone layer (Crawford 1987), the abundance of active chlorine (2 parts per billion by volume in the early 1980s) is expected to reach 6 to 7 parts per billion by 2050 (Brasseur and Hitchman 1988), or a 300 percent increase!

The ozone holes. Between 1977 and 1984, the springtime amounts of ozone over Halley Bay, Antarctica, decreased by more than 40 percent (Figure 20-1), and later research showed that the region of ozone depletion spanned over several million square miles, wider than the continent itself (Figure 20-2), and extended from 12 to 24 kilometers in altitude (Farman et al. 1985; Stolarski 1988). Within the "ozone hole," as it has come to be known, the abundance of ozone is about half that of what it was a decade before. The 1990 ozone hole (140 Dobson units, see Figure 20-2) marked the third year out of four in which the depletion was of this order (121 Dobson units in 1987 and 125 units in 1989) (Kerr 1990c). In 1992, a record low of global ozone was reached, 2 to 3 percent lower than any earlier year,

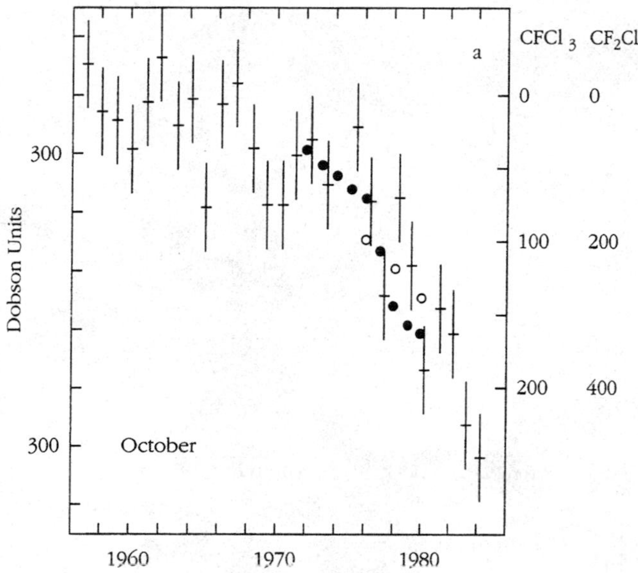

Figure 20-1 The decline of ozone in the Antarctic in springtime from 1956 to 1985 as measured in Dobson units at Halley Bay (vertical line indicates mean and standard deviation). The Dobson unit, named for the measuring instrument used, is equal to one part per billion by volume. Also shown are the volumes of $CFCl_3$ (dark circles in parts per thousand) and CF_2Cl_2 (open circles in parts per thousand). Note that the scales for the CFCs (right side of graph) are inverted from those of ozone concentration; thus, their concentration increases down the figure. Put alternately, as the CFCs increase, ozone decreases. (Redrawn with permission from J. C. Farman, B. G. Gardiner, and J. D. Shanklin. 1985. *Nature* 315:207–10).

the largest decreases being in the regions from 10°S to 20°S and 10°N to 60°N (Gleason et al. 1993). The eruption of Mount Pinatubo contributed to this depletion through the release of chlorine gases (Tabazadeh and Turco 1993).

The Antarctic ozone hole is the consequence of chlorine chemistry coupled with the special conditions of Antarctic meteorology (Anderson et al. 1991; Brune et al. 1991; Kerr 1987a, 1987b; Rowland 1989; Tolbert, et al. 1988). Briefly, the long polar night (summer in the Northern Hemisphere) produces low temperatures that allow the formation of stratospheric clouds on the surface on which hydrochloric acid and chlorine nitrate are converted to molecular chlorine (Toon and Turco 1991). Spring (August–September) sunlight releases the atomic chlorine triggering the ozone-destroying reactions described above, a set of chain reactions that runs unhindered for 5 or 6 weeks into early spring (October), causing ozone depletion of 95 percent or more in the stratosphere and as much as 60 percent in total ozone over all altitudes (Frederick and Snell 1988; Kerr 1989a; Schoerberl and Hartmann 1991).

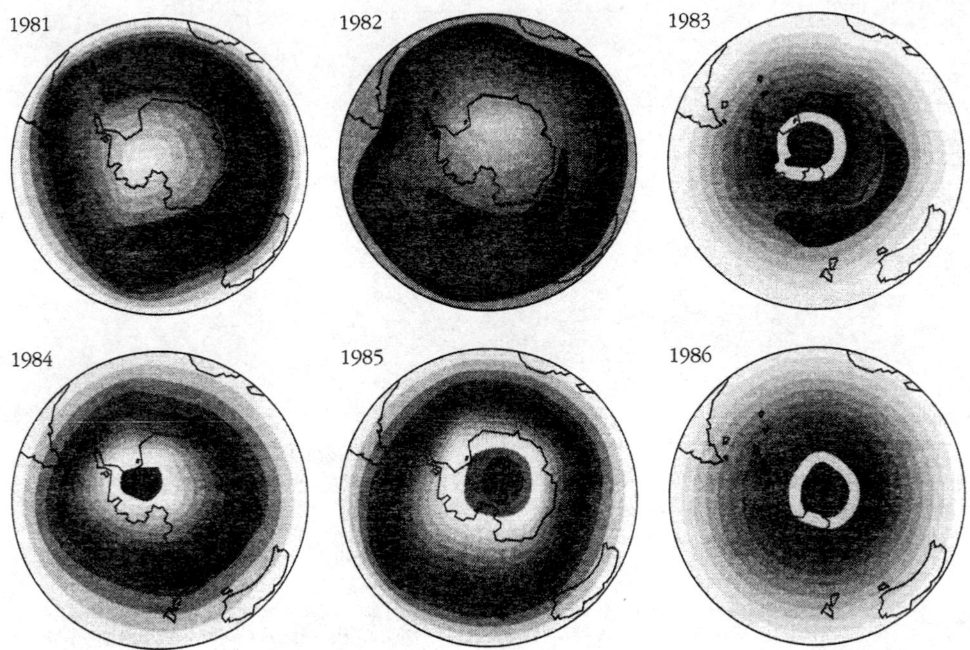

Figure 20-2 The degree and extent of ozone depletion over Antarctica from 1981 to 1986, with the clear central area showing the least amount of ozone, the progressively grayer areas showing increasing amounts of ozone, and the darkest crescent-shaped band showing an ozone-rich region. (Based on data from the U.S. National Aeronautics and Space Administration.)

Unfortunately, the decreases in total ozone have not been confined to the Antarctic spring season but are global in extent, with the largest decreases occurring in the middle and high latitudes in all seasons of the year (Bowman 1988; Kerr 1988b; Lemonick 1992; Stolarski et al. 1992). Studies of the Arctic stratosphere, for example, have demonstrated that the same cloud-mediated chemistry that leads to the Antarctic ozone hole appears over the Arctic (Brune et al. 1988; Kerr 1989b; Mount et al. 1988; Solomon et al. 1988). As observed so far, the Arctic ozone losses are not as severe as those in the Antarctic: Total ozone destruction did not exceed a few percent, and the hardest hit layers lost only 15 to 20 percent of their ozone (Kerr 1990a).

By contrast, during the 13-year period from 1957 through 1970, the global distribution of total ozone showed an upward trend of about 7.5 percent per decade in the Northern Hemisphere and about 2.5 percent in the Southern Hemisphere (London and Kelly 1974). It should be noted, however, that these studies did not include observations over either of the poles and occurred during the time of remarkable increases in the use of CFCs.

A cure for ozone destruction? Based on a numerical model, Cicerone et al. (1991) have suggested that the injection of 50,000 tons of the alkanes ethane or propane into the Antarctic stratosphere could suppress ozone loss. In other scenarios of the model where smaller amounts of the alkanes were injected, ozone depletion increased. In addition to a number of yet poorly defined physical and chemical processes and other uncertainties about ozone production and destruction, the authors acknowledge there would be major difficulties in delivering and distributing the needed alkanes.

Other proposals include a kind of planetary engineering by lofting 10 to 20 balloon-borne platforms carrying sheets, wires, or curtains of a metal whose electrons can easily be dislodged by ultraviolet light, the electrons converting the reactive chlorine atoms into inactive chloride ions (Amato 1994). Viggiano et al. (1995), however, have shown subsequently that the abundance of chlorine-containing ions is so small that using the proposed negative ion chemistry to mitigate ozone depletion is not feasible.

Tropospheric Ozone

Chemistry of tropospheric ozone. The basic mechanism of photochemical air pollution, or *smog*, was first described by A. J. Haagen-Smit (1952) who established that the photooxidation of hydrocarbons in the presence of nitrogen oxides produces ozone. The chemistry is similar to that which occurs in the stratosphere, but the specific agents are typically the nitrogen oxides and hydrocarbons, both byproducts of industrial as well as domestic fossil fuel combustion, primarily by automobiles (Cleveland and Graedel 1979; Seinfeld 1989):

1. NO_2 + sunlight $\rightarrow$ $NO + O$
2. $O_2 + O \rightarrow O_3$
3. $NO + O_3 \rightarrow NO_2 + O_2$

The first reaction indicates that ozone formation is a daytime phenomenon and demonstrates the role of nitrous oxides in ozone production, whereas the third reaction shows their role in ozone destruction. Since NO is the principal oxide of nitrogen emitted from combustion sources, ozone is destroyed near strong sources of combustion and formed away from such sources.

Hydrocarbons apparently affect ozone concentrations by oxidizing NO to NO_2. Since trees emit significant quantities of hydrocarbons, these natural emissions can substantially affect urban ozone levels especially where nitrous oxide levels are concentrated (Chameides et al. 1988). This is yet another reason to shift away from gasoline dependence to ethanol- or methanol-fueled vehicles and to compromise on other hydrocarbon and/or nitrous oxide anthropogenic sources (Russell 1988).

Impact of tropospheric ozone. Although ozone is only one of the undesirable components of smog, it has been used by the U.S. Environmental Protection

Agency as the indicator of photochemical smog, set initially at 80 parts per billion and raised to 120 parts per billion in 1979 (Cleveland and Graedel 1979). Whereas ozone in the stratosphere is a protective agent, it is otherwise in the troposphere (McKee 1994). For humans, ozone is the predominant air pollutant causing irritation of the mucous membranes. It has been indicted as a significant factor in dieback of vegetation downwind from heavily polluted cities and as having negative effects on the production of such crops as corn, wheat, soybeans, and peanuts (MacKenzie and El-Ashry 1988). Tinhgey et al. (1994) indicate that approximately 90 percent of air pollution–induced crop loss in the United States is from tropospheric ozone. Responses can include changes in metabolism, reduced photosynthesis, leaf necrosis (one of the earliest and most obvious manifestations of ozone injury to crops), leaf drop, and altered growth and quality. Nonetheless, because of its effectiveness in killing microorganisms, ozone has the potential for being used as a disinfectant in sewage treatment and has, in addition, the side benefit of breaking down into oxygen, which can improve the water quality of the purified sewage.

CARBON AND WORLD CLIMATE

As has been noted several times already, humans have been altering their environment over large geographic areas since the domestication of fire, plants, and animals. The progression from hunter to farmer to technologist has increased the variety and pace more than the geographic extent of human impact on the environment, resulting in significant climatic changes closely related in time to anthropogenic environmental changes in many regions on Earth (Sagan et al. 1979). Perhaps the most compelling and at the same time contentious of these changes has been the impact of increased release of carbon dioxide on global climate. This legitmate concern about global climate has resulted in an outpouring of publications, the best of which are: *Annual Review of Ecology and Systematics* 1992; Charlson and Wigley 1994; Dunnette and O'Brien 1992; Gates 1993; Hays and Smith 1993; Vitousek 1994; Woodward 1992; Wyman 1991.

The Greenhouse Effect

Carbon dioxide plays a critical role in controlling the Earth's climate because as an aerosol it absorbs and reflects or scatters incoming radiation on the one hand and absorbs and reradiates outgoing infrared radiation on the other. This latter phenomenon results in what is popularly called the *Greenhouse Effect*, an analogy to what happens in a greenhouse.

The greenhouse analogy, however, is not perfect. Greenhouse glass and Greenhouse gases, like carbon dioxide, allow the passage of sunlight to the Earth's surface and then keep the heat within. The difference is that in a greenhouse all the heat is trapped, except that which is allowed to escape by regulating the glass louvers; by contrast, the Greenhouse Effect gases trap only the heat in infrared radiation. However, as less heat escapes through the glass louvers of a greenhouse, the temperature within the greenhouse rises, and similarly as the level of Greenhouse gases rises, Earth's climate warms up.

Sources of Greenhouse carbon dioxide. Although most organisms release carbon dioxide as a byproduct of cellular metabolism, the sources responsible for the increase in carbon dioxide in the atmosphere are anthropogenic, and these are largely from the burning of fossil fuels and the destruction of the world's forests.

Fossil fuels such as coal and gas do not yield the same amounts of carbon dioxide per unit of energy. Coal releases one-third more carbon dioxide per unit of energy than does oil and almost twice as much as natural gas. Coal is largely carbon, and its combustion leads to most of its energy being released as carbon dioxide. In contrast, oil contains more hydrogen and natural gas even more; their combustion yields less carbon dioxide and more water than does coal. Today, oil supplies 38 percent of global energy needs, coal 20 percent, and natural gas 20 percent (Oppenheimer and Boyle 1990). Nearly 6 billion tons of carbon were emitted into the atmosphere in 1989 due to fossil-fuel burning, and emissions continue to grow by 2 percent a year.

As was noted in an earlier discussion, atmospheric buildup of carbon dioxide is one of the consequences of destruction of tropical forests (Chapter 19) (Woodwell et al. 1983). This is the result of less carbon dioxide being taken up by photosynthesis and excess carbon dioxide being released by burning of the trees (Detweiler and Hall 1988). For example, a hectare of trees stores about 50 tons of carbon. Prior to 900 A.D., forests covered 90 percent of Europe; by 1900 forests covered only 20 percent. In the western llanos or plains of Venezuela, 33 percent of the forests were reduced between 1950 and 1975 (Figure 20-3). In 1987, slash-and-burn operations in Brazil covered an area almost the size of Maine (Oppenheimer and Boyle 1990).

Broecker et al (1979) take the opposite position, arguing that the regrowth of previously cut forests and the enhancement of forest growth resulting from excess carbon dioxide in the atmosphere have about balanced the rate of forest destruction in the past few years.

Changes in carbon dioxide levels. As was noted in Chapter 8, there is some disagreement as to the exact amount of carbon dioxide that has been added to the atmosphere by anthropogenic activity, but that there has been an increase is not debated. Various estimates, as noted earlier, suggest a 20 to 30 percent increase since the mid-nineteenth century level of about 260 to 275 ppm. Studies like those on the slopes of the still active volcano, Mauna Loa, in Hawaii (Figure 20-4) and numerous others have established incontrovertibly that the increases are real even if the percentage increases and actual amounts are debatable (National Research Council 1983a; Revelle 1983).

The fate of carbon dioxide discharged to the atmosphere is also subject to some disagreement. Stuiver (1978) estimates that 34 percent is stored in surface and thermocline waters and shallow-water sediments and 13 percent in the deep seas. This would leave the airborne fraction at 53 percent. Some portion of this is consumed

Figure 20-3 Slash-and-burn agriculture near Punta de Mata, east of Caracas, Venezuela. The land is cleared of forest and then burned; bananas are grown on the then available soil for 2 or 3 years, and the plot is then abandoned as another slash-and-burn operation takes place on another piece of land. (Photo by author.)

in photosynthesis, but whether that portion results in a net balance to the atmosphere or to the biota, as noted in the previous paragraph, is a matter of contention. Regardless of this latter point, the actual amount of carbon dioxide in the atmosphere has been increasing since 1850 and is projected to continue to do so (Figure 20-5).

Direct observations as well as computer modeling indicate that global warming due to the Greenhouse Effect is indeed occurring (Hileman 1989; Houghton and Woodwell 1989; Jones and Wigley 1990; Kerr 1990b, 1991; LaBrecque 1989, among many others). A warming trend that began about 1880 has continued (Figure 20-6), and Jones and Wigley (1990) suggest that temperatures will rise between 1°C and 4°C by the year 2040 to 2065.

Although there have been dissenters on the warming phenomenon (Kerr 1989c; Roberts 1989b), a report by the International Panel on Climate Exchange in mid-summer 1990 brought virtual unanimity among some 170 scientists from 25 different countries that a warming is on the way, and, in fact, that the average global temperature has increased between 0.3°C and 0.6°C since the late nineteenth century (Kerr 1990b). According to the Panel's estimate, anthropogenic enhancement of carbon dioxide since 1965 will be doubled during the next 35 years if no significant steps are taken to reduce the yearly increases of carbon dioxide and other Greenhouse gases.

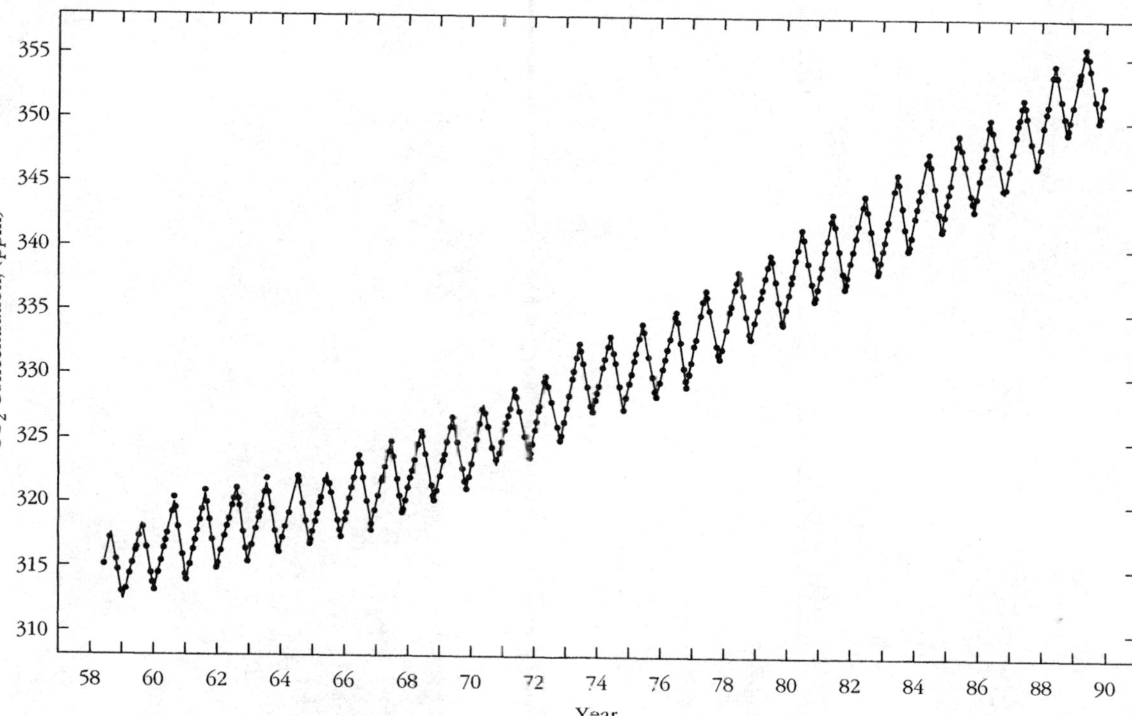

Figure 20-4 Monthly variations in the concentration of atmospheric CO_2 content at Mauna Loa Observatory, Hawaii, observed with a continuously recording nondispersive infrared gas analyzer. The yearly oscillations reflect seasonal variations in the amount of photosynthesis; the dots indicate monthly average concentration. (Data courtesy of Climate Monitoring and Diagnostics Laboratory, Environmental Research Laboratory, NOAA, Boulder, CO, 1991.)

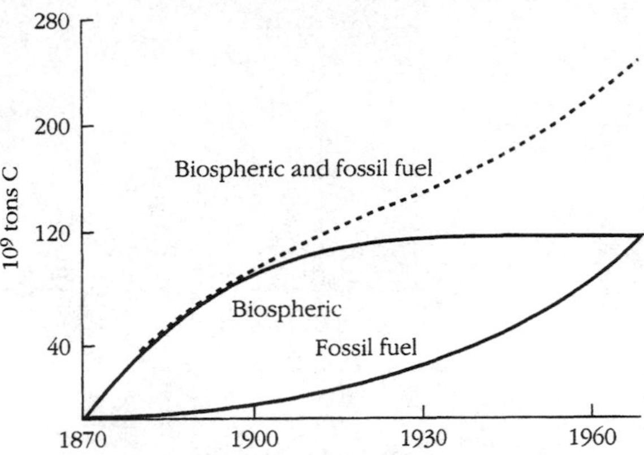

Figure 20-5 Cumulative CO_2 release to the atmosphere from the combustion of fossil and biospheric (wood, humus) fuels. (Redrawn by permission from M. Stuiver. 1978. *Science* 199:253–8. Copyright 1978 by the American Association for the Advancement of Science.)

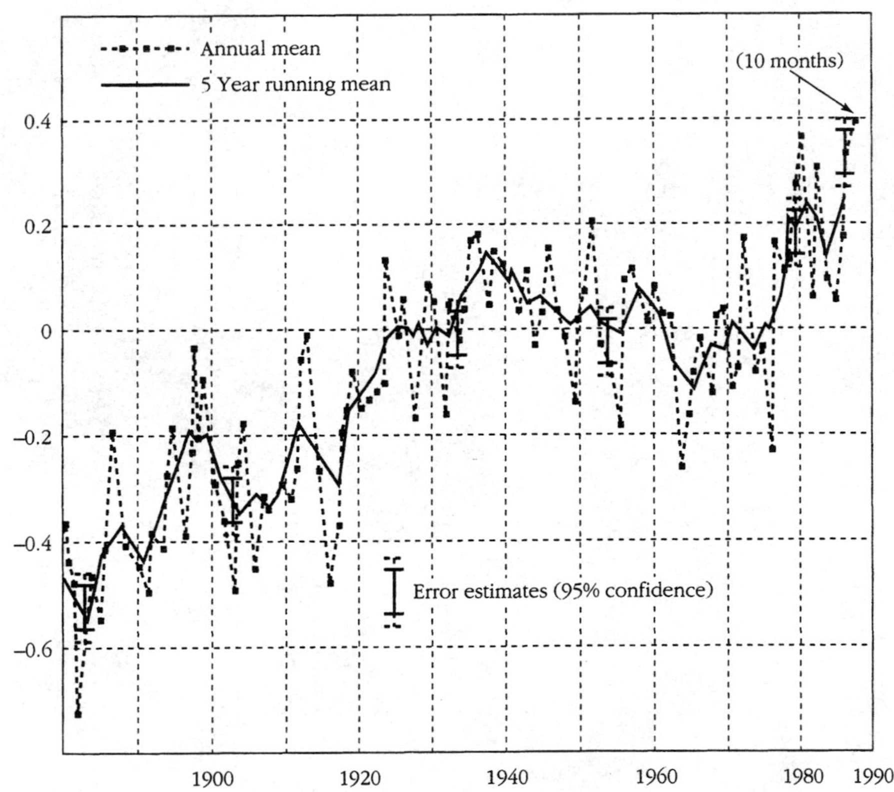

Figure 20-6 Global warming from 1880 to 1990 as shown by deviations from the 1950–1980 mean temperature. Except for the period 1940 to 1975 when the trend inexplicably reversed itself, the increase has been consistently upward. (Redrawn with permission of James Hansen from M. La Brecque. 1989. *Mosaic* 20(4):1–17.)

Mitigation of the Greenhouse Effect. Except for admonitions about reducing the consumption of fossil fuels, strategies to reduce the impact of the Greenhouse Effect have received relatively little sophisticated consideration (Rubin et al. 1992). In the previous section it was shown that aerosols, both natural and anthropogenic, have a mitigating impact by increasing the albedo and thus reflecting and scattering more heat back into the atmosphere (also see Kiehl and Briegleb 1993; Penner et al. 1992). However, the increase of airborne particulates, as we have seen, can have untoward consequences on human health. John Martin, of the Moss Landing Marine Laboratory, proposed a quick fix of the Greenhouse Effect by injecting iron into the Southern Ocean near Antarctica to trigger a massive bloom of phytoplankton; although the idea was considered wild (later even by Martin himself), a small scale test has been proposed by the American Society of Limnology and Oceanography (Roberts 1991b). However, the study would not be so much directed at mitigating global climate change as to try to determine why the phytoplankton of the Southern Ocean grow so poorly even though the waters are rich in nutrients such as phosphorus and nitrogen.

Consequences of the Greenhouse Effect. Based on an extensive review of vast quantities of evidence, the United Nations Intergovernmental Panel on Climate Change (IPCC 1990) used a five-star rating system to describe the probability of various scenarios for global warming (Table 20-2). Among other projections: global warming by 2.5°C (or as high as 4°C) with a consequent rise in sea level between 8 and 29 centimeters by 2030, and continental interiors—the bread baskets of North America and Eurasia—might be dry during the summer (Karl et al. 1991).

The severity of the impact would depend on the level of atmospheric carbon dioxide, the amount and kind of particulates, solar activity (which in fact varies) (Friis-Christensen and Lassen 1991), the Earth's wobble, and still other factors that interact to produce the complex environmental component known as climate. Already, however, there is evidence of: widespread warming of the permafrost by 2 to 4°C in northernmost Alaska (Lachenbruch and Marshall 1986); an increase of 2°C in air and lake temperature and a 3-week increase in ice-free season on lakes of the central boreal forest (Schindler et al. 1990); large-scale changes in precipitation over the Northern Hemisphere over the last 30 to 40 years (Bradley et al. 1987); and, an increase in coastal upwelling (Bakun 1990) but only a slight rise in sea-level, on the order of 1 millimeter per year (Peltier and Tushingham 1989); and a reduction from 88 to 30 hectares of pine (*Pinus elliottii* var. *densa*) in the Florida Keys over a 70-year period, the result of higher sea level that brought increased salinity to the soil and groundwater (Ross et al. 1994).

Some scientists have projected a positive side of the Greenhouse Effect by way of growth enhancement of natural vegetation in the presence of abundant carbon dioxide (Eamus and Jarvis 1989; Esser 1992; LaMarch et al. 1984; Lindroth et al. 1993). In contrast, others are less sanguine, largely because of our lack of sufficient knowledge of the many factors impacting global climate (Ausubel 1991; Bazzaz and Fajer 1992; Korner and Arnone III 1992; Mooney 1991b; Paine 1993) Others have suggested a world with an abundance of male alligators, migrating species of trees, and a plethora of parasites (Roberts 1988).

TABLE 20-2 THE PROBABLE MAJOR ENVIRONMENTAL CHANGES RESULTING FROM A DOUBLING OF CO_2 AS DEDUCED FROM MODELS[a]

Degree of confidence[a]	Predicted change
Temperature	
*****	The lower atmosphere and Earth's surface warm
*****	The stratosphere cools
***	Near the Earth's surface, the global average warming lies between +1.5°C and +4.5°C, with a "best guess" of 2.5°C
***	The surface warming at high latitudes is greater than the global average in winter but smaller than in summer. (In time-dependent simulations with a deep ocean, there is little warming over the high-latitude ocean.)
***	The surface warming and its seasonal variation are least in the tropics
Precipitation	
****	The global average increases (as does that of evaporation); the larger the warming, the larger the increase
***	Increases at high latitudes throughout the year
***	Increases globally by 3 to 15 percent (as does evaporation)
**	Increases at mid-latitudes in winter
**	The zonal mean value increases in the tropics, although there are areas of decrease. Shifts in the main tropical rain bands differ from model to model, so there is little consistency between models in simulated regional changes
**	Changes little in subtropical arid areas
Soil Moisture	
***	Increases in high latitudes in winter
**	Decreases over northern midlatitude continents in summer
Snow and Sea Ice	
****	The area of sea ice and seasonal snow cover diminish

[a]The number of stars indicates the degree of confidence, five stars indicating virtual certainty and one star low confidence in the prediction. From IPCC. 1990. *Climate Change: The IPCC Scientific Assessment.* Eds. J. T. Houghton, G. J. Jenkins, and J. J. Ephraums. Cambridge University Press. Reprinted by permission.

Whether the uncertainties about these estimates and projections are large enough to suggest delaying the development of new policies and regulations on Greenhouse gases becomes not a scientific question but a value judgment (Schneider 1989).

A geological perspective on the Greenhouse Effect. Quite naturally, we of the latter part of the twentieth century are increasingly concerned about the consequences of the Greenhouse Effect on our lives and livelihoods. Geologically, however, it appears the Earth has experienced a number of "greenhouse" episodes followed by "icehouse" episodes (Figure 20-7) (Berner 1990; Berner and Lasaga 1989; COHOP Members 1988; Kasting and Ackerman 1986; McGowran 1990, among others). During the Mesozoic period, 100 million years B.P., dinosaurs roamed in a climate that was warm and maritime with narrow temperature gradients across latitudes

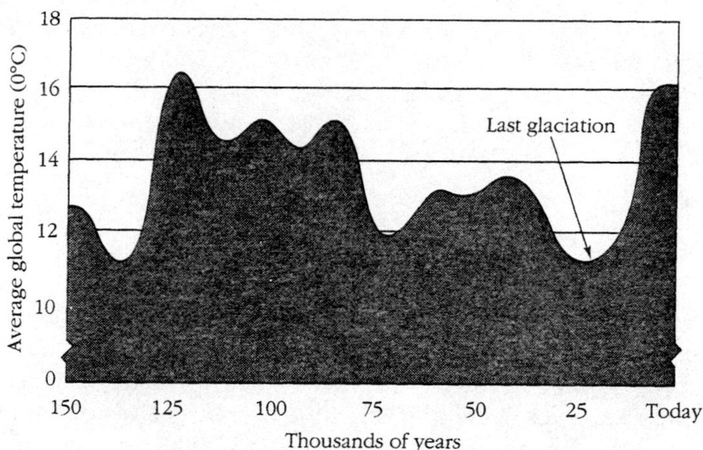

Figure 20-7 Greenhouse and icehouse episodes over the last 150,000 years. Note that global warming of 2 to 6°C would make the Earth the warmest it has been in 150,000 years. (Redrawn with permission from B. Hileman. 1989. *Chemical & Engineering News* 67:25–44. Copyright 1989 American Chemical Society.)

and within the oceans; polar icecaps were small or even nonexistent, and global sea level was high. During the succeeding Eocene "icehouse" period, temperature gradients were steeper, climate was continental with stronger seasonal variation, and ocean circulation was more vigorous (McGowran 1990). Speculation is that natural geochemical processes resulted in the slow build-up of atmospheric carbon dioxide and caused the consequential Greenhouse Effect (Berner and Lasaga 1989; Raynaud et al. 1993).

Other Greenhouse gases. Although most of the concern about the Greenhouse Effect has been about carbon dioxide in the atmosphere, other gases such as the chlorofluorocarbons, nitrous oxide, carbon monoxide (CO), and methane are also among the culprits (Mooney, et al. 1987). Chlorofluorocarbons, for which the United States leads world production, are very potent agents since one molecule of a CFC can trap as many as 10,000 molecules of carbon dioxide. Carbon monoxide from tropical burning rivals transportation and industry as sources of carbon monoxide, which continues to increase in the northern hemisphere (Kahlil and Rasmussen 1984; Newell et al. 1989); however, more recent studies from 27 locations between 71°N and 41°S show that CO levels decreased between 1990 and 1993 (Novelli et al. 1994). Methane (CH_4) makes substantial contributions to the Greenhouse Effect (Blake and Rowland 1988; Brauman et al. 1992). Besides its elimination as a waste product from termite and ruminant guts, substantial amounts of methane are formed in anaerobic situations such as wetlands (Bubier et al. 1993), swamps (Pulliam 1993), rice paddies, and sanitary landfills. India's methane emission from rice growing and livestock is second only to that emitted from the United States (World Resources Institute 1990). Warming of the soil increases the rate at which anaerobic bacteria breakdown organic matter into methane. Since methane is a Greenhouse gas, the more that is produced, the more the Earth warms; the more the Earth warms, the more methane is released, which in turn leads to more warming of the Earth. A positive-feedback system in operation.

Carbon Dioxide—A Lethal Gas

Although generally regarded as beneficial with respect to human health, carbon dioxide can cause asphyxiation and death preceded by an anesthetic effect that can produce sensory hallucinations. On August 21, 1986, a sudden release of carbon dioxide from Lake Nyos, a crater lake, in the northwest area of Cameroon, West Africa, caused the deaths of at least 1,700 people and 3,000 cattle within a 10 km area (Kling et al. 1987). Birds, insects, and small mammals were not seen for 48 hours after the event, but plant life was essentially unaffected. Two days after the disaster, the lake surface was calm but littered with floating mats of vegetation and had turned from its normal clear blue color to a rusty red. A water surge had washed up the southern shore to a height of about 25 meters, and a fountain of water and froth had splashed over an 80 meter-high promontory on the southwestern shore.

The gas was presumably derived from magma beneath the Lake in this basaltic volcano country and was stored in the hypolimnion. Its release was triggered not by volcanic activity but presumably because the volume of gas reached its limits of containment.

An earlier release of lethal gas, probably also carbon dioxide, from Lake Monoun, also a crater lake, caused the deaths of 37 people. These two instances are the only recorded events in which gas released from lakes has caused the loss of human life. Kling et al. (1987) note that the nature of the gas released suggests that hazardous lakes may be identified and monitored to reduce the danger of such incidents in the future.

NUCLEAR WINTER

One of the enigmas of paleontology has been explaining the extinction of the dinosaurs and about 75 percent of the species alive at the time, the boundary between the Cretaceous and Tertiary geological periods. Although there is still some disagreement, most scientists now accept the hypothesis advanced by father and son scientists Walter and Louis Alvarez and their associates (Alvarez et al. 1980; Alvarez and Asaro 1990). The relationship of this hypothesis to the topic of nuclear winter will be evident soon. An alternate interpretation of dinosaur extinction focuses on the role of cataclysmic volcanism (Courtillot 1990; Glen 1990).

The Alvarez postulate is that a giant meteorite, or an asteroid or comet, smashed into Earth causing a rapid heating or a rapid cooling that led to the extinctions. Rapid heating might have occurred because of the heat generated by the meteor itself as it traveled through the atmosphere; rapid cooling would have followed from the attenuation of sunlight as dust clouds rose from the impact, the attenuation also leading to a reduction or even an absence of photosynthesis, which in turn would have led to extinction by starvation of herbivores and subsequently carnivores.

The Alvarez hypothesis was predicated on the finding of a thin worldwide layer of iridium-enriched sediment deposited at the time of the Cretaceous/Tertiary boundary. Iridium is a scarce element on Earth but is relatively abundant in meteorites, which form craters on impact (Grieve 1990).

Climatic Consequences of Nuclear War

From the discussion in Chapter 19, we are aware of the effects of radioactive fallout and burning due to the heat intensity of a nuclear bomb; their severity depends on the megatonnage of the bomb. Studies have suggested that a large-scale nuclear war (5,000 to 10,000 megatons) would result in 750,000 immediate deaths from the blast alone, a total of about 1.1 billion deaths from the combined effects of blast, fire, and radiation, and another 1.1 billion injuries (Ehrlich et al. 1983). Most of these casualties would occur in the midlatitude region (30° to 70°) of the Northern Hemisphere and especially in the United States, the former U.S.S.R., Europe, and Japan. Although the human devastation wrought by a nuclear war would be almost beyond comprehension and warrants deep concern, the discussion here will be confined to the potential impact of nuclear war on climate, since that would impact the remaining 2 to 3 billion survivors in the rest of the world and possibly lead to more casualties globally than the direct effect of nuclear war.

The Alvarez hypothesis encouraged reconsiderations of the effects of nuclear war on climate. One of the first of such studies was that by Crutzen and Birks (1983), which suggested that massive fires ignited by nuclear explosions could generate quantities of sooty smoke that would attenuate sunlight and perturb climate. In turn, this led to one of the seminal studies of the potential impact of nuclear war on climate, the TTAPS report, an acronym based on the first initial of the authors' names (Turco, Toon, Ackerman, Pollack, and Sagan 1983). A subsequent study by the National Research Council (1985) and further appraisals by TTAPS (Turco et al. 1990) have largely concurred in a number of possible consequences, summed up in the term "nuclear winter," as follows:

- generation of sufficient smoke to decrease average solar intensities by 50 percent or more on a hemispheric scale and up to 100 percent in regions covered by the initial hemispheric smoke load;
- average land cooling could reach 10° to 20°C, and the interior of continents could cool by up to 35°C with subzero temperatures possible in summer in some regions;
- upper air layers could be heated by as much as 100°C thereby promoting interhemispheric transport and horizontal mixing of smoke clouds;

- nitrogen oxides resulting from the nuclear detonations would be deposited in the stratosphere causing severe depletions of the ozone layer in the Northern Hemisphere;
- failure of the summer Asian monsoon season with consequent drought in the affected areas.

Choices

The foregoing discussions of health problems caused by increases in tropospheric ozone, ozone depletion by nitrogen oxides and CFCS, the consequences of global warming by increased levels of carbon dioxide and other Greenhouse gases, and global cooling by nuclear war pose the conundrum, "will Armageddon come through freezing or burning to death?" The cloud of pessimism such a question elicits is reminiscent of Oscar Wilde's definition of a pessimist, namely "one who, when he has the choice of two evils, chooses both." As was pointed out in Chapter 16, such a dilemma must be resolved. Neither alternative is desirable, but it is patent that we do have some critically significant choices to make for our own sake, as well as for generations to come.

HEAVY METAL POLLUTION

Activities associated with modern society—industry, power generation, transportation—have subjected the environment to many novel chemical materials over the past 200 years; PCBs and CFCs are notorious among these newcomers. These activities have also increased the airborne levels of naturally occurring heavy metals above the base level that results from geological weathering processes on Earth's crust or release from ocean waters. The chemical composition of atmospheric particles collected near sea level over the North Atlantic, for example, indicates that such metals as aluminum, manganese, iron, cobalt, chromium, magnesium, and calcium as well as zinc, copper, lead, and selenium are derived from either crustal weathering or the ocean (Duce et al. 1975). As already noted, volcanoes constitute a significant source of heavy metals (Decker and Decker 1989). The explosion of Mount St. Helens in the State of Washington resulted in heavy ash deposition both near and at great distances from the volcano (Siegel and Siegel 1982). Nonetheless, natural sources of heavy metals appearing in the atmosphere are greatly enhanced by anthropogenic activity (Kormondy 1989).

The Nature of Heavy Metals

There is no precise definition of a heavy metal, and common usage has tended to further obscure the term. In general, however, *heavy metals* are those that exist as cations under biologically significant conditions. Typically the heavy metals occur as salts and other combined forms rather than as elements; mercury is an exception.

Some 80 of the more than 100 elements may be classified as metals, and of them about three-quarters can be regarded as heavy metals. The most biologically significant and the most studied of these heavy metals include cadmium, mercury, lead, zinc, chromium, and copper.

As particulates, heavy metals, in one form or another, have airborne residency times of up to one month or more. Such particles, as noted in the discussion of particulates, are deposited not only close to the source of emission but also on ecosystems some considerable distances away. Smelters, roads, urban areas, sewage incinerators, and mine waste heaps tend to have the greatest rates of deposition nearest them, especially of the heavier particles, including asbestos from rock quarries (Carter 1977c), lead from fires (Davidson et al. 1981), and mercury from geothermal power plants (Robertson et al. 1977). Smaller and lighter particles tend to be more widely dispersed. Also affecting dispersal are the morphology of the deposition surface, the age of the aerosol, and wind speed, as well as the topography of an ecosystem (e.g., valley versus open plain) and floristic composition.

Unlike PCBs, whose toxicity depends on the structure of the whole molecule, a metal such as copper is toxic in its own right, although the toxicity can vary greatly with the molecular form (Moriarity 1988). Once released into the environment, the potential for pollution by copper does not disappear with a change of chemical form.

Ecological Effects of Heavy Metals

Ecologically, the interactions of heavy metals and ecosystems range from that in which the ecosystem is an unaffected sink for the pollutant, to that in which the intermediate dosage produces sublethal, adverse effects in individuals and populations, or to high dosage conditions resulting in lethal effects. The last effects, of course, are most disruptive of ecosystem diversity and stability. Because of the interaction of environmental factors, the initial disruption may be by some other agent (e.g., fire, acid precipitation) and the longer-term effect may be the result of heavy metal contamination. Atmospheric deposition during the growing season contributes one-third or more of the estimated total flux of lead, zinc, and cadmium from the forest canopy to soils beneath an oak stand in the Tennessee Valley; but interactions between these metal particles deposited on dry leaf surfaces and subsequent acid precipitation can result in metal concentrations on leaves that are considerably higher than in rain alone (Lindberg et al. 1982).

Excessive soil burdens of heavy metals do not necessarily have adverse effects on plant growth and productivity. Moriarity (1988) cites a number of instances in which there is not only metal resistance in plants but where growth is stimualted by the presence of certain heavy metals. For example, seeds of the grass *Agrostis tenuis* growing on contaminated soils adjacent to copper mines had greater percentages of germination and growth of survivors when sown experimentally on contaminated soils than did adjacent populations on uncontaminated soils. In other cases, adverse effects do occur, including reduced productivity through stunted growth and chlorosis.

Because of uptake through leaves or roots or by physical adsorption on plant surfaces, particularly in aquatic situations, heavy metals can not only become incorporated in plant material but can also be passed on to humans as was noted earlier in the case of mercury contamination in Minimata, Japan. Humans can also be directly contaminated by inhalation or ingestion of dusts and dirt. A study of lead contamination in the Toronto area showed that between 13 and 30 percent of the children living in the contaminated area had absorbed excessive amounts of lead, the major route having been by direct ingestion (Roberts et al. 1974). Metabolic changes were found in the majority of 21 children with excessive lead burdens; 10 to 15 percent of this group showed subtle neurological dysfunction and minor psychomotor abnormalities.

Legislation and regulation to reduce atmospheric heavy metal contamination has made some progress internationally but there is yet a considerable gap between what is known and what is needed to be done in the way of control (Kormondy 1989). Federal and state clean air acts have moved dramatically to reduce airborne particulates, including the heavy metals, but resistance, largely because of the costs for controls and reductive technologies, is found in many quarters (Mitchell 1991). Monitoring techniques have improved considerably, and many newspapers daily report various pollutant levels. Among the more intriguing monitoring techniques is the use of honeybees, which have been found to be accurate in detecting concentrations of arsenic, cadmium, and fluoride in Puget Sound, as well as in other situations (Bromenshenk et al. 1985).

ACID PRECIPITATION

An Historical Perspective

The spiritual father of the term "acid rain" is considered to be the English scientist Robert Angus Smith whose 1872 book, *Air and Rain—The Beginnings of a Chemical Climatology*, provided detailed discussions of many of the characteristics of this now well-recognized, though not necessarily well-understood, phenomenon (Brandt 1987). Although a considerable amount of study continued well into the twentieth century, discussion of acid precipitation was not revived until the mid-1950s by Scandinavian scientists. Since that time there has been both increased recognition of and research on the phenomenon and a resulting refinement of its definition (Kormondy 1989; Likens et al. 1979; Moriarity 1988; National Research Council 1981, 1983b; among many others).

Definition, Causal Factors, and Dispersal

Acid precipitation defined. Although there remains some understandable differences of opinion, the generally accepted definition of *acid precipitation* is the increase in hydrogen ion content of precipitation water—rain, snow, sleet, fog, etc.—

due to anthropogenic sources. The standard of a pH value of 5.6 for "pure rain" was defined for many years based on the equilibrium reaction of deionized water with a 0.03 percent volume of carbon dioxide. Recent information from extensive rainwater analyses in isolated areas shows that a background pH between 4.5 and 5.5 and with a mean of about 4.9 is more reasonable (Brandt 1987).

For some comparisons with the standard, the average pH of rainfall in Toronto in February 1979 was 3.5; in the fall of 1981, fog in Los Angeles had a pH as low as 2.2. The most acidic rainfall in the United States appears to have been 1.4 in Wheeling, West Virginia; this is not much above the acidity of battery acid, which has a pH of 1. Precipitation that fell before the Industrial Revolution and that was preserved in glaciers and continental ice generally has a pH above 5.

Causal factors. Two strong acids are the major agents in acid precipitation—sulfuric and nitric. Approximately 70 percent of the acids derive from oxidation and hydrolysis of sulfur dioxide and hydrogen sulfide and 30 percent from various nitrous oxides and other compounds, all attributed largely to the burning of fossil fuels (Figure 20-8). In 1975, the 28.5 million tons of sulfur emissions in the United States were derived as follows: 65 percent from electric utilities, 25 percent from industrial sources, and 10 percent from nonferrous smelters (Ember 1981). In that same year, the 24.4 million tons of nitrogen emissions were derived as follows: 45 percent from transportation (mostly automobiles), 28 percent from electric utilities, and 27 percent from other sources.

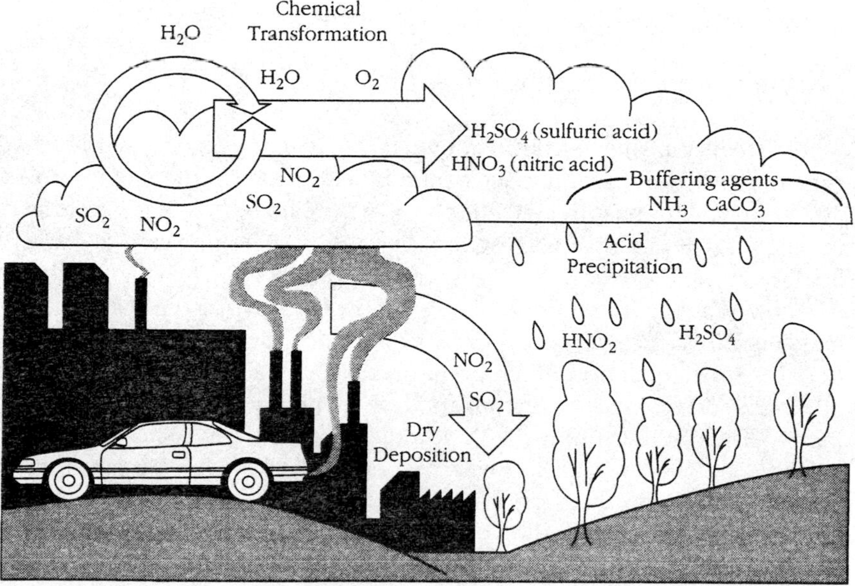

Figure 20-8 The chemical formation of acid precipitation.

Like the rise in atmospheric carbon dioxide, the increased levels of sulfur and nitrogen in the atmosphere have occurred in the last 150 or so years during which fossil fuel consumption has increased markedly. This is even experienced in a nation as undeveloped as China, as well as in areas remote from large populations or industrial enterprises in Australia, Canada, and even the United States (Brezonik et al. 1980; Galloway et al. 1987; Lewis and Grant 1980; Nriagu et al. 1987). It is not always the case, however, that natural and anthropogenic components can be distinguished (Galloway et al. 1984; Hoffman, 1990).

Dispersal. Because of their aerosol nature, sulfur and nitrogen oxides are capable of being transported considerable distances before being precipitated out. Despite considerable controversy, most of the acid rain in northeastern North America is believed to be derived from the Midwest, particularly from those areas in which emission sources are both concentrated and high. Eighty percent of the sulfur pollution in North America is generated east of the Mississippi River, mainly in pockets concentrated around Sudbury, Ontario, and the Ohio River Valley (Ember 1981). There is considerable evidence, however, that acid rain has been spreading southward and westward (Figure 20-9). In 1955–1956 acid rain with a pH of 4.5 extended west to central Ohio and south to southern Virginia; in 1972–1973 it had extended west to central Missouri and south to northern Louisiana, Mississippi, and Georgia.

An important factor is that acid rain is a transboundary problem that involves not only the United States and Canada but also every nation in Europe whether industrialized or not (Kormondy 1989). Air knows no boundaries. Such is also true for multiboundary lakes like Lake Erie and rivers like the Danube that traverse more than one nation. Pollution is seldom self-contained.

Effects of Acid Precipitation

Effects on biogeochemical cycles. Given the composition of acids in acidic precipitation, the major impact would be expected on the sulfur and nitrogen cycles with not inconsequential effects on other cycles affected by acidity or by direct interaction with sulfur or nitrogen compounds. The nitrogen cycle, as noted earlier, involves pH changes and is, in part, regulated by pH levels and changes. Thus, it may well be that the nitrogen cycle is more sensitive to acidification than the other major nutrient cycles. For example, denitrification is completely suppressed at a pH of 3.5, volatilization of ammonia increases with increased pH, extremes of pH affect the survival of *Rhizobium*, and free-living nitrogen-fixing organisms are even more pH sensitive (Sprent 1987). Rudd et al. (1988) have shown that experimental acidification of two small soft-water lakes in northwestern Ontario inhibited nitrifying bacteria at pH values of 5.4 to 5.7 with a resulting accumulation of ammonium; when the pH level was allowed to recover above 5.7, nitrification resumed after 1 year. Since forests are a major nitrogen-containing component of the biosphere, any forest destruction by acid precipitation (or other activities) will result in major perturbations of the nitrogen cycle. Finally, although the total global nitrogen pool

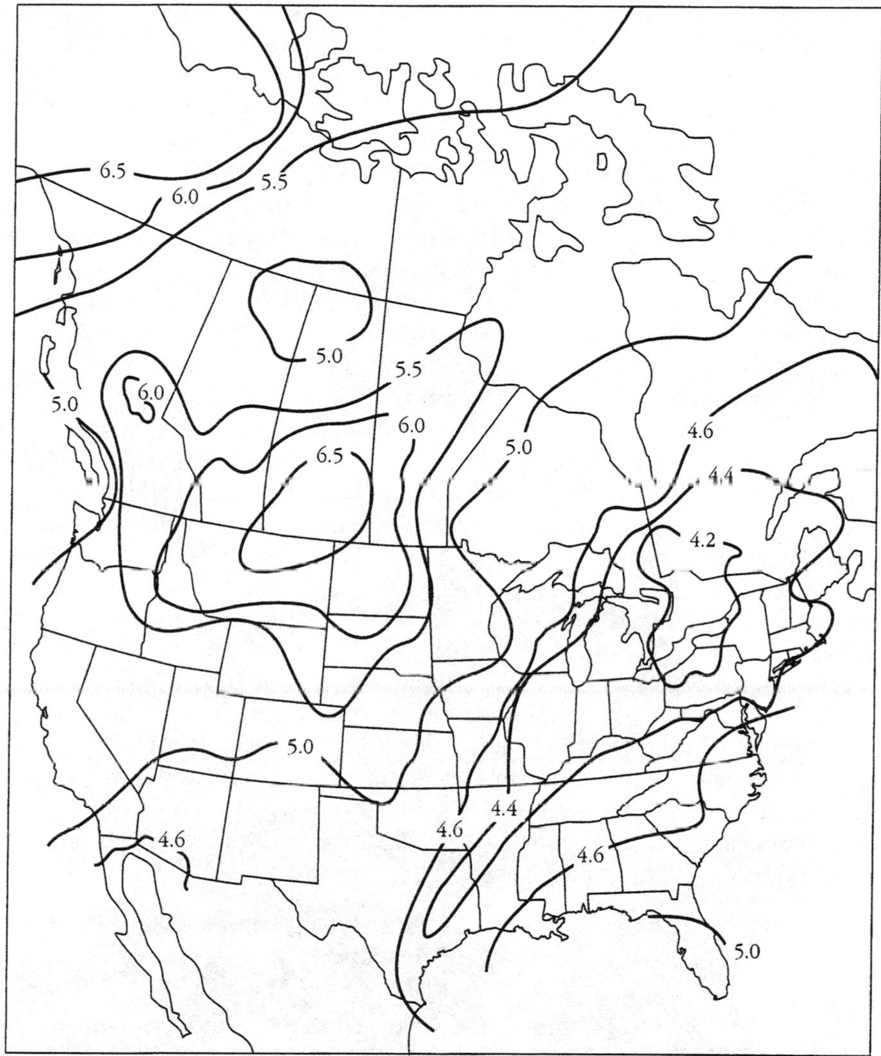

Figure 20-9 Average pH of annual precipitation in 1979. (National Atmospheric Deposition Program.)

is probably not affected by acid precipitation, there is likely an increased proportion of the total in biologically active form.

Effects on aquatic ecosystems. Acidification of aquatic ecosystems occurs naturally and anthropogenically. Naturally occurring acid lakes consist of: (1) those that form in areas of igneous rocks or where the substrate is largely sand; typically small with weak carbonate-bicarbonate buffering capacity, these lakes often have a pH of 5.6 to 5.7; (2) those that occupy geological substrates similar to acid lakes in category one but have accumulated considerable organic matter or humus and contain peat and bog plants; they generally have a pH of about 4.5, but some have a pH as low as 3.4 (Patrick et al. 1981). The humus in the latter lakes chelates the heavy metals that are soluble under low pH conditions so they do not become toxic to many forms of aquatic life. In both types of acid lakes, microbial, plant, and animal species are found that have adapted to the acidic conditions. Although many of these acid lakes have been exposed to long-term (i.e., 10,000 years or more) acidification processes (Ford 1990), many have experienced an increase in acidity from anthropogenic sources, and most notably from the combustion of fossil fuels. Atmospheric deposition was the dominant source of acid ions in 75 percent of the 1,180 lakes and 47 percent of the 4,670 acidic streams analyzed by the National Surface Water Survey (Baker et al. 1991).

In contrast to naturally occurring acid lakes, many freshwater ecosystems in northeastern America (and elsewhere) are soft-water lakes with poor buffering capacity, that is, the ability to neutralize additions of acids or alkalis. In fact, any ecosystem whose bedrock is highly siliceous (granite, some gneisses, quartzite, quartz sandstone) falls in this category. The extent of areas that are geologically vulnerable to acid precipitation include not only the long-recognized northeastern United States, but large areas of Minnesota, Wisconsin, upper Michigan, several southeastern states, and many of the mountainous areas of the West (Schindler 1988). Half of the 700,000 lakes in the six eastern provinces of Canada south of 52°N are extremely acid-sensitive. The Netherlands, Belgium, Denmark, Switzerland, Italy, West Germany, and Ireland, along with Scandinavia and the United Kingdom, and vast areas of Precambrian and Cambrian geology in Asia, Africa, and South America are also acid-sensitive.

Specific effects of increased acidification of aquatic ecosystems are demonstrated by the following:

- fish populations in southern Norway have been adversely affected by acidification of lakes in a 33,000 km^2 area, and all lakes in an area of 13,000 km^2 are fishless;
- in 1975 82 of 214 high-altitude lakes in the Adirondacks were devoid of fish, and 51 percent of the lakes had a pH of 5 or less;
- in the late 1970s nine rivers in Nova Scotia had a pH below 4.7 and were barren whereas formerly they had been teeming with salmon (Ember 1981);
- in 1980 some 140 acidified lakes in Ontario were devoid of fish, and thousands of lakes were experiencing biological damage;
- in 1984, of 107 lakes surveyed near Helsinki, Finland, half were either severely acidified or projected to lose all fish;
- in the mid-1980s in Sweden all bodies of freshwater were acidified with 15,000 too acidified to support sensitive aquatic life and an additional 1,800 nearly lifeless (Starke 1988).

In addition to the direct effect of acid on fish, the acid mobilizes such metals as aluminum from the surrounding soil, which then enters lakes by runoff. When combined with high acidity, aluminum is toxic to fish (Moriarity 1988).

Many lakes and streams in acid-sensitive areas show some resistance to acidification through natural sources of acid-neutralizing reactions (Johnson 1979: Schindler et al. 1986; Schindler 1988). In large watersheds with well-developed soils, terrestrial watersheds supply most of the acid-neutralizing capacity, and the degree of soil development controls the extent of the acidification, at least in Adirondack lakes. In-lake generation of alkalinity is attributed mostly to microbial reduction of sulfate and nitrate; recent work has shown that most of the sulfate reduction occurs not in the anoxic hypolimnion but in the epilimnion.

In areas of eastern Canada, lakes have recovered after the acidity of precipitation has been reduced through a decrease in sulfate deposition (Schindler 1988). For example, in the Sudbury, Ontario area, a combination of smelter closures and sulfur dioxide controls has reduced sulfate emissions by about two-thirds of their early 1970s value. This decrease in sulfate deposition has resulted in an increased alkalinity and pH in nearby lakes along with a decrease in concentrations of sulfates, aluminum, and toxic trace metals. Schindler (1988) estimates that sulfate reductions would need to be decreased from their present values of 20 to 50 kg/ha/yr in most of eastern North America and western Europe to somewhere between 9 to 14 kg/ha/yr to protect the most sensitive aquatic ecosystems.

Zooplankton food webs in 46 Precambrian shield lakes near Sudbury, Ontario, showed decreases in species richness and numbers of predatory and competitive links, among other relationships, when the pH dropped below 5.0 in the 1970s (Locke and Sprules 1994). By 1990, when the pH had increased by two units, species richness and numbers of predatory and competitive links had all increased relative to the 1970s values.

Liming (i.e., the addition of limestone, $CaCO_3$) has been proposed as an antidote to low or lowered pH levels (Gloss et al. 1988) and has met with some success, but only on a temporary basis. Attempts to lime a watershed as an option for mitigating lake acidity do not appear very promising on a long-term basis either (Mackum et al. 1994).

Effects on soils. Soil acidity affects a wide range of ecological processes, including the solubility and exchange reactions of inorganic nutrients and toxic metals, the activities of soil animals and microorganisms, and the weathering of soil minerals. Changes in soil acidity result from an array of interacting processes that produce and consume hydrogen ions, including cation exchange, mineral weathering, anion adsorption, and nutrient uptake from the deep soil, many of which, in turn, are induced by changes in land use and consequent vegetative succession (Brandt 1987; Krug and Frink 1983; Mohnen 1988). Precipitation is only one source of acidity, though in some situations it is very significant. For example, the pH of soil near the Copper Hill ore smelter in Tennessee (Figure 20-10) was reduced to the range of 3.2 to 3.8 compared with 4.5 to 5.0 in similar soils farther from the smelter (Wolt and Lietzke 1982).

Figure 20-10 The effects of emissions from the copper smelter in Copper Hill, Tennessee. (Photo by author.)

At very low pHs and high precipitation rates over long periods, reduction in the decomposition of humus occurs, presumably because of a shift in the kinds and abundance of bacteria and fungi.

Assessment of the probable impacts of acid deposition must include an accounting of both the buffering capacity of ecosystems and the natural production of hydrogen ions within ecosystems. Short-term acceleration of soil acidification will occur if the soils are characterized by: low cation exchange capacity (i.e., low in clay and organic matter); moderate or high pH; low amounts of weatherable minerals and low rates of renewal by soil mixing or erosion; low sulfate adsorption capacity; shallow profile; and, atmospheric inputs of concentrated mineral acids (Binkley and Richter 1987).

Effects on terrestrial ecosystems. Natural and anthropogenic sources of acidic pollution are well known to affect adjacent terrestrial ecosystems, as in the case of volcanic eruptions, sulfur steam vents (Figure 20-11), and ore smelters (Figure 20-10). As an instance of natural acid pollution, Freedman et al. (1990) have described the impact of long-term emissions of sulfur dioxide and acidic aerosols on tundra plants (*Artemisia tilesii* and *Arctogrostis latifolia*) in the Smoking Hills of the Northwest Territories of Canada. The acids are derived from spontaneous combustion of sulfur-bearing bituminous shale caused by the heat that is produced as

Figure 20-11 Effects on vegetation of sulfur steam at Kilauea Crater, Volcanos National Park, Hawaii. The Kau Desert lies downwind of Kilauea Crater's sulfur-producing vents. (Photo by author.)

pyritic sulfur becomes exposed to oxygen and subsequently undergoes an exothermic oxidation to sulfate. This same reaction probably underlies the ignition of waste coal heaps and underground coal mines where pyrites are also exposed to atmospheric oxygen.

The impact of acidic precipitation on forest ecosystems is not so clear. Although an estimated 30,718 million hectares of forest (about 22 percent of total forest area) were reported as damaged in Europe in 1986 (Starke 1988), and well-documented declines in red spruce (*Picea rubens*), which range from the southern Appalachian Mountains to northeast United States and Canada, and of several southern pines (*Pinus taeda*, *P. elliottii*, and *P. echinata*) in the mountain, piedmont, and coastal regions of southeast United States have been noted, the causal connection between acid precipitation and the declines has been difficult to pinpoint (Brandt 1987; Mohnen 1988; Pitelka and Raynal 1989). The decline, or *Waldsterben* (forest death) as German scientists call it, is most dramatic in high-elevation coniferous forests; for example, more than 50 percent of the red spruce above 850 meters in the Adirondacks in New York, Green Mountains in Vermont, and the White Mountains of New Hampshire have died within the past 25 years (Mohnen 1988). In contrast, Kauppi et al. (1992) found a general increase in forest biomass in Austria, Finland, France, Germany, Sweden, and Switzerland as a result of the fertilizing effects of airborne pollutants.

Although there is considerable controversy concerning the causal connection of acidic precipitation in forest decline, there is increasing recognition of the multiplicity of factors that might well be involved, including acid precipitation that may act less as a lethal agent than as a stress factor exacerbating other conditions (McLaughlin and Norby 1991). Pitelka and Raynal (1989) note a number of factors that could be acid precipitation-related including: excess nitrogen interfering with winter hardening of buds and foliage; soil acidification and cation leaching leading to nutrient deficiency; aluminum or other heavy metal toxicity caused by soil acidification; and ozone-acid mist causing nutrient leaching from foliage. Acid precipitation could further stress plants already weakened by disease, insect and parasitic invasions, shortages of light, water or essential nutrients, and injury from floods, high winds, and ice storms (Mohnen 1988; Schulze 1989). And not to be discounted is the genetic determination of life spans and consequent natural diebacks (Mueller-Dombois 1988a, 1988b). All of these potential factors are compounded by the variations in the levels of precipitation (Goodess 1987).

Effects on humans and human artifacts. No adverse health effects have been directly attributed to acid waters although there is concern that acidified drinking water could leach toxic copper from pipes. In fact, in southwest Sweden, children have contracted diarrhea from high copper levels in drinking water. There is also speculation that concentrations of aluminum and other metals in drinking water derived from acidified water supplies may be related to incidences of mental disorders and that the breathing of acid sulfate aerosols may result in respiratory and cardiac problems.

Acid rain has caused direct damage to architectural structures by corroding such famous monuments as the six Caryatids that support the porch of the Acropolis in Athens, the Parthenon, the Taj Mahal, Rome's Colosseum, the Lincoln Memorial, and the Washington Monument. The U.S. Council on Environmental Quality estimated in 1979 that the annual cost in the United States alone in architectural damage due to acid rain was nearly $2 billion. The National Academy of Sciences put the total damage to materials, forests, agriculture, aquatic ecosystems, and health and water systems at $5 billion in 1978.

Concluding Comment

There is little question that acid precipitation and carbon dioxide buildup are two of the most significant current environmental issues. Both are directly related to the burning of fossil fuels, particularly high-sulfur coal and oil, and the latter is tied directly to industrial development. The costs involved in either shifting to alternate, nonpolluting energy sources and/or reducing the discharge of sulfur and nitrogen by such technology as the installation of scrubbers are considerable. They must be paid by a society geared to continuing industrial development either by direct assessment to industry or by passing the costs on to the consumer. This expense must be weighed against the environmental costs, some of which are unknown, of inadequately controlled emission. The problem is then both economic and environmental, as well as political.

As Woodwell (1989), among others, has noted, tight causal proofs are elusive in ecology as in most sciences. Although tying the knot of acidic precipitation to forest decline may be likened to a child first trying to tie shoestrings, the effects of acid precipitation on soil and water, along with the possible indirect if not direct effects on forest ecosystems, leave no doubt about the need to control its causes with environmentally and economically attractive solutions (Mohnen 1988; Levin 1992).

EPILOGUE

RETROSPECTIVE

From the material on human ecology presented in these last five chapters, it is relatively easy to be caught up in the emotion elicited by the term pollution. What is important, however, is to realize again that what we are dealing with are byproducts, the wastes of a natural ecological agent—*Homo sapiens.* We are, and our wastes are, as much a part of the living world as the microbe and its wastes. But humans have a peculiar and potential dominance over ecosystems even though we are in no way independent of them; when we insult an ecosystem, we can expect to be slapped back. As we put resources out of place, we can expect changes—undesirable concentrations of toxic substances, consequent reductions in the number and abundance of species, and a resultant and consequential community instability. Humans cannot, nor can any organism, exist without producing wastes. The problem, then, is intelligent and resourceful management. The quality of our environment will be a reflection of our capacity to manage these wastes. It will require much more understanding of ecological ramifications, as well as much more of the interplay of the myriad facets of a complex society, than we have presently before we can resolve these issues.

Solutions to the problems discussed in these chapters, as well as to the many not discussed—climate control, strip-mine wastes, other pollutants, open-pit mining—will certainly not come with ease. For many of them, we do not have the scientific wisdom to make reliable predictions and, if we did, we often lack the data with which to do so. These problems and others that have been discussed are not only ecological. They are sociological, economic, political, legal, psychological, and, in the final analysis, ethical as well. These problems will not disappear by attempting to return to a bygone day, to some Arcadia or Camelot, or to maintain the status quo. No responsible ecologist can argue for either. To do so would be contrary to the most basic of ecological principles—dynamic and adaptive change. The landscape of America was indeed changed after our forebears came to it, but it had changed before. Where I sat in northern Ohio writing the first edition of this book was formerly forest, but earlier it was also once ocean and then glacier; where I wrote this edition was ocean floor and subsequently covered by lava until it breached the ocean surface and continued building new land, as it does today. Change is the essence of nature, both the gradual change that transforms a pond to a forest, or a tundra to a taiga, and the devastating kind by which a tornadic wind denudes a landscape. Humanity's hopes and chances, however, lie in the regulation of the changes that they, as a natural and integral part of the landscape, induce. We are partners with other natural processes in the stewardship of resources and the management of cultural and biological wastes. The future depends on our intelligence and ability to develop a body of knowledge about such management and on the wisdom that few now possess to apply that knowledge.

PERSPECTIVE

We tend to get so caught up in today's issues and problems that too often we fail to listen to the lessons from less technologically complex times or present cultures in which environmental relationships were or still are the absolute determinants of human survival. A brief reconnaissance of three cultures may be instructive on this point.

North American Paleo-Indians

The Clovis Culture, named for a site in New Mexico, is dated to about 11,000 to 12,000 years B.P. and is perhaps the earliest Paleo-Indian culture pattern in North America (Hester 1970). This Great Plains culture was based on a big-game economy adapted to the utilization and exploitation of grassland ecosystems with their abundance of permanent water in small ponds and valley streams, the latter populated with juniper and oak. Large herds of giant bison and other, now extinct, Pleistocene animals, including mammoths, roamed the landscape. There appear to have been two basic big-game hunting patterns in this Great Plains culture: stalking and stampeding. For stalking bison and mammoths, campsites were situated so the animals could be observed as they came for water, at which time they were killed, usually in small numbers, as they were drinking. In stampeding, bison were driven into a canyon, stream, or over a cliff, where as many as several hundred animals were crushed or drowned. Although it is presumed that the Clovis Culture derived from emigrees from the Old World, the absence of an emphasis on hunting of horses and the absence of mammoth traps suggests hunting techniques specifically adapted to New World conditions (Wendorf and Hester 1962).

Although the big-game hunting economy established in the Great Plains is thought to have been practiced from the West Coast to the East Coast and south to the valley of Mexico, the evidence is more circumstantial than solid. Even though some big-game hunting took place, cultural adaptations to local environmental conditions occurred: a seed-grinding, small-game–hunting, shellfish-gathering culture in what is now Southern California; a salmon-fishing and bird- and mammal-hunting culture west of the Cascades in Oregon and Washington; a plant-collecting, along with a big-game–hunting, culture in the Great Basin; and in northeastern North America, a large-game–hunting (musk-ox, mammoth, mastodon, bison, moose, among others) culture complemented by marine mammal-fishing of whales, walrus, and seals, which occurred in the Great Lakes (Hester 1970).

The Valleys of Hawaii

Waimanu, Pololu, Honokane Nui, and Honokane Iki are the names of erosional valleys on the windward side of the Island of Hawaii, the largest of the islands in the Hawaiian archipelago. Formed over a million years ago as streams wore down the lava and ash deposits of this volcanic island, the valleys were subject to sea level rise and fall and continual wave action resulting in V-shaped profiles, sea cliffs as high as 400 m, steep lateral walls, and an extensive base level surface (Figure 20-12). Temperatures now average 22°C in the winter months and 24°C in the summer, and rainfall varies from 2.0 to 2.5 m. The combination of temperature and rainfall creates a humid regime that is believed not to have changed significantly for centuries.

The natural flora of the Hawaiian Islands is the product of adaptive radiation of perhaps less than 300 immigrant species introduced over the course of several million years (Terry 1988). Ninety-five percent of the almost 2,000 species native to

Figure 20-12 Waipio Valley, Island of Hawaii, showing the characteristic features of the windward V-shaped valleys. (Photo courtesy of Ronald Terry.)

Hawaii are endemic, and evolution has favored the loss of thorns, poisons, and secondary compounds that provide defense against invading plants and predators.

Polynesians, who came initially from the Marquesas and later from Tahiti beginning about 300 A.D., brought taro, sweet potatoes, yams, and bananas, as well as breadfruit (*'ulu*), mountain apple (*'ohi'a 'ai*), and coconut trees. Prior to Western contact in 1778 when Captain Cook arrived, subsistence farming in these Hawaiian valleys appears to have consisted of irrigated and nonirrigated taro patches, with intercropping of banana, sugar cane, and other crops and trees on the banks and terraces. A variety of plants used for fiber, medicines, dyes, ceremonial items, and timber were harvested in both wild and nonwild areas. Socially, this was a stratified society in which land-use rights were allocated through great chiefs (*ali'i*) to lesser chiefs, and finally to common people who farmed the land.

The arrival of Captain Cook in 1778 initiated the beginning of significant changes in vegetation, land use, and social structure. Among the first new tree crops were papaya and citrus, followed by mango, avocado, coffee, and guava. Human diseases, which decimated the native population, coupled with migration away from the rural areas contributed to a breakdown of social structure, including severing the tie of commoners to the land. The latter condition was exacerbated by the introduction of fee-simple land titles, a concept of land ownership that was completely unfamiliar to the Hawaiians and led to widespread dispossession of their land. The introduction of sugar cane, rice, and pineapple plantation agriculture with contract or hired labor, taxes, and a money economy were diametrically opposed to Hawaiian values and customs, as well as farmstead agriculture. The valley communities were gradually abandoned, and by 1930 none of these valleys retained any inhabitants except for an occasional hunter or camper. Forests now thrive where once there were human populations living off the land and the sea. In fact, Kikuchi (1976) argues that indigenous aquaculture influenced the development of social stratification in Hawaii.

Lest this brief scenario suggest that these Polynesian cultures lived in complete harmony and equilibrium with their environment, it must be noted that long-term destruction to ecosystems probably occurred in Hawaii prior to Captain Cook. This included deforestation and consequent soil erosion and other damage including the near-to-actual extinction of various species of birds whose feathers were highly prized for adornment.

Likewise, environmental degradation has been found on Melanesian islands and elsewhere on other Pacific islands. As Clark (1990) has noted,"... Islanders did what all peoples, especially pioneers, do: in their efforts to make a living they actively manipulated, modified, and, at times, degraded the ecosystems they lived in, producing environmental changes that in turn required ecological adaptations and social adjustments." Transformations of natural landscapes into cultural landscapes resulted in extinctions of flora and fauna, endangered other species, and degraded soil, vegetation, and reef. But, as Clarke further notes, they also developed sustained-yield systems of agriculture, agroforestry, and reef use that still operate productively today.

Altiplano Andeans and Energy Flow

The Andean altiplano is a high plateau surrounding Lake Titicaca, where few of the valleys are below elevations of 3,700 m above sea level, and habitable valleys are found up to 5,000 m above sea level. At these elevations the air is rarified and contains much less oxygen than at sea level; for example, air at 4,000 m contains about 60 percent of the oxygen found in air at sea level. Air temperatures are frequently below freezing and, because of the rarified air, solar radiation is intense. Ecologically, the vegetation zone is an essentially treeless alpine grassland referred to as subalpine *paramo*, or *puna*, consisting of bunchgrasses, herbaceous and hard-cushion plants, shrubs, mosses, and lichens (Figure 20-13). Wild animals include small mammals, some deer, and small herds of camel-related vicuna and guanaco.

A study by Thomas (1973; see also Little and Moran 1976) focused on the flow of energy in these altiplano Andean communities, particularly those 8,000 people living in the region of Nunoa, about 125 km northwest of Lake Titicaca. The farmer-pastoralists of the community, sociologically the *indigena* class, are usually native Andean Indians with little European admixture. The hacienda system of farms is owned privately by a social class known as *mestizo;* the dominant way of life for the *indigena,* who are landless tenants, is performing services for the hacienda in exchange for the privilege of living on the land. A hacienda's economy is based on herding llamas, alpacas, and sheep, which contribute hides and skins for bedding, wool for clothing and blankets, dung for fuel and fertilizer, and meat products for food. Haciendas below 4,300 m are used for herding and agriculture (the staple white potato, other tubers and native grains) (Figure 20-14); haciendas above 4,300 are used for herding only.

(a)

(b)

Figure 20-13 *Paramo* of the high Andes in Venezuela, comparable to that found in the Peruvian Andes. (a) El Negro, near Tovar; (b) Frailijones zone near Aguila a Punanjo. (Photos by author.)

Figure 20-14 Harvesting white potatoes in the high Andes village of Pueblo Llano, Venezuela. (Photo by author.)

From an energy flow perspective, the producers are the *puna* wild native grasses, primary consumers are alpacas, llamas, sheep, horses, and cattle, and second-order consumers are humans, with wild carnivores playing a relatively minor role in energy flow. The subsistence economy is an efficient one in that a reasonable degree of balance can be maintained within the ecosystem; that is, most clothing, food, fertilizer, and fuel needs are provided within the system.

With respect to energy expenditures involved in cultivation and herding, about 12 percent (51,800 C) is expended on the crops and 88 percent (384,200 C) on herding. Crops produce about 44 percent (595,000 C) of the energy in the system and herds about 56 percent (760,000 C); the latter figure consists of food produced (222,000 C) plus exports (538,000 C). From an efficiency perspective, cultivated crops provide about 11.5 times the food energy per unit of work energy expended in cultivating the crop (595,000 C ÷ 51,800 C = 11.5), whereas livestock provides about twice the food energy per unit work energy expended in herding activities (760,000 C ÷ 384,200 C = 2.0).

Since the exports (538,000 C) are traded for food products of higher caloric value (222,000 C + 2,442,000 C), the overall efficiency of herding can be considered to increase almost four-fold to 7.5 (2,886,000 C ÷ 384,200 C = 7.5). This calculation shifts the percentage of food produced by cultivation from 44 percent to 17 percent and the percentage of food produced directly and indirectly by herding to

83 percent. If the energy value of dung, which is used for both fuel and fertilizer, were incorporated into the energy flow model, the energetic efficiency of herding would increase even more.

It is apparent that the mixed economy of herding and agriculture provided a stable human ecosystem for the Altiplano Andeans: herding, the more stable resource, providing protein, clothing, fertilizer, and a basis of trade; agriculture, insufficient alone to provide all the food needed for survival, can be limited by crop failures.

Although this account suggests an equilibrium and harmonious use of the environment, there is evidence of pre-Conquest environmental degradation, including deforestation in this region, as well as elsewhere in South America. Because the vast *puna* grasslands of the antiplano regions of Peru were common at the time of Conquest, deforestation for construction and fuel purposes must have occurred much earlier (Thomas and Winterhalder 1976). Evidence of severe erosion due, in part, to overgrazing dates back 2,000 years, and in more recent times, shrubs and other woody vegetation have been removed from large, now-eroded areas to provide fuel for population centers and railroad steam engines.

Prospective

As can be seen from the preceding examples, the idea of past, or simpler, human societies being harmoniously at equilibirum with the environment may be more myth than reality. What is clear, however, is that the scale of disharmony and resulting environmental degradation has been changed radically as human population burgeoned and required more and different uses of the environment, placing tremendous pressures on Earth's life-support capacity. It is absolutely necessary that collectively we develop and apply the knowledge required to conserve and wisely manage Earth's resources. To this end, the foremost thinking in the ecological profession is for an initiative that focuses on the necessary role of ecological science in the wise management of Earth's resources and the maintenance of Earth's life-support systems; this thrust is proposed as the Sustainable Biosphere Initiative (SBI) (Lubchenco et al. 1991).

What is demanded of us today is a rethinking of our place in nature, a rethinking of attitudes about the total environment—in the words of Leopold (1949), the development of a new ethic for the land, and of Potter (1971), the development of an ethic of survival. The roots of the crisis in which we find ourselves are deep in the outlook Westerners, in particular, have had about the land—land as an adversary to be conquered, as a servant to be exploited for our own ends, as a possession of rightful and eminent domain, and, most importantly, land of unlimited capacity. These concepts must give way to an ecological conscience, to a love, respect, admiration, and understanding for the total ecosystem of which we are part and to an ethic that ensures the survival of the human species with quality, dignity, and integrity. Otherwise our course is one of collision, an inexorable Armageddon.

SUMMARY

VOCABULARY

acid precipitation
atmospheric inversion
chlorofluorocarbons
Greenhouse Effect
heavy metals
ozone
particulates
smog

KEY POINTS

- Depending on their size and form, particulates may be transported substantial distances from their points of origin and have significant adverse effects on public health, especially during atmospheric inversions, as well as on climate.
- Stratospheric ozone screens out most of the damaging portions of ultraviolet radiation whereas tropospheric ozone is a cause of skin cancer, among other public health effects.
- Stratospheric ozone depletion (ozone holes), resulting from the discharge of chlorofluorocarbons, is increasing, as is the amount of tropospheric ozone that largely results from the burning of fossil fuels.
- Although there is debate about the degree to which Earth is being subjected to a Greenhouse Effect (caused by acknowledged increases in the amount of atmospheric carbon dioxide and other Greenhouse gases since the beginning of the industrial revolution), most scientists agree there is global warming. Major impacts of global warming would be melting of polar ice caps with subsequent rise in sea level and an adverse shift in climate on the continents.
- In some circumstances, carbon dioxide, a necessity for photosynthesis and hence virtually all life, can be lethal.
- A likely consequence of nuclear war would be a nuclear winter, owing to particulate matter in the atmosphere preventing the penetration of sunlight.
- Some 60 elements (e.g., lead, asbestos, mercury, copper) form compounds that make them heavy metals, most of which have adverse effects on public health and property.
- Acid precipitation generally occurs some distance from the source of the acids and typically has adverse effects on ecosystems through acidification of water and soil.
- Precontact cultures did not necessarily exist in a harmonious equilibrium with their environment, nor do contemporary ones; but because of population growth, contemporary cultures are placing extreme pressure on Earth's life-support capacity.

References

ABRAMS, P. A. 1994. The fallacies of "ratio-dependent" predation. *Ecology* 75:1842–50.

ABRAMSKY, Z., and C. R. TRACY. 1979. Population biology of a "noncycling" population of prairie voles and a hypothesis on the role of migration in regulating microtine cycles. *Ecology* 60:349–61.

ALDHOUS, P. 1993. Tropical deforestation: Not just a problem in Amazonia. *Science* 259:1390.

ALLAN, T., and A. WARREN. 1993. *Deserts. The encroaching wilderness.* Oxford: Oxford University Press.

ALVAREZ, L. W., W. ALVAREZ, F. ASARO, and H. W. MICHAEL. 1980. Extraterrestrial cause for the Cretaceous-Tertiary extinction. *Science* 208:1095–1108.

ALVAREZ, W., and F. ASARO. 1990. An extraterrestrial impact. *Scientific American* 263(4):78–84.

AMATO, I. 1994. High-flying fix for ozone loss. *Science* 264:1401–2.

AMBROGGI, R. P. 1977. Underground reservoirs to control the water cycle. *Scientific American* 236 (5):21–27.

American Scientist. 1982. Several articles on storage of radioactive wastes. 70:180–207.

AMUNDSON, R. 1994. John T. Gulick and the active organism: adaptation, isolation, and the politics of evolution. In *Darwin in the Pacific,* P. Rehbock and R. MacLeod, eds., 110–39. Honolulu: University of Hawaii Press.

ANAGNOSTAKIS, S. L. 1982. Biological control of chestnut blight. *Science* 215:466–71.

ANDERSON, D. M. 1994. Red tides. *Scientific American* 271(2):62–8.

ANDERSON, J. G., D. W. TOOHEY, and W. H. BRUNE. 1991. Free radicals within the Antarctic vortex: The role of CFCs in Antarctic ozone loss. *Science* 251:39–46.

ANDREAE, M. O. 1983. Soot carbon and excess fine potassium: Long-range transport of combustion-derived aerosols. *Science* 220:1148–51.

ANDREWARTHA, H. G., and L. C. BIRCH. 1954. *The distribution and abundance of animals.* Chicago: University of Chicago Press.

ANDREWS, R. M. 1991. Population stability of a tropical lizard. *Ecology* 72:1204–17.

ANGINO, E. E. 1977. High-level and long-lived radioactive waste disposal. *Science* 198:885–90.

Annual Review of Ecology and Systematics. 1992. Special section on global environmental change. 23:1–235.

ANSPAUGH, L. R., R. J. CATLIN, and M. GOLDMAN. 1988. The global impact of the Chernobyl reactor accident. *Science* 242:1513–9.

ANTONOVICS, J., and D. A. LEVIN. 1980. The ecological and genetic consequences of density-dependent regulation in plants. *Annual Review of Ecology and Systematics* 11:411–52.

APPLEGATE, V. C., and H. D. VAN METER. 1970. A brief history of commercial fishing in Lake Erie. Fishery Leaflet 630. Washington, DC: U.S. Department of the Interior.

ARRIGO, K. R., and C. R. MCCLAIN. 1994. Spring phytoplankton production in the western Ross Sea. *Science* 266:261–2.

ATKINSON, D. 1994. Temperature and organism size—a biological law for ectotherms? In *Advances in ecological research,* M. Begon and A. H. Fritter, eds., vol. 25, 1–58. London: Academic Press

AULIE, R. P. 1971. Origin of the idea of the nitrogen cycle. *American Biology Teacher* 33:461–71.

AUMANN, G. D., and J. T. EMLEN. 1965. Relation of population density to sodium availability and sodium selection by Microtine rodents. *Nature* 208:198–99.

AUSUBEL, J. H. 1991. A second look at the impacts of climate change. *American Scientist* 79:210–21.

AVERILL, A. L., and R. J. PROKOPY. 1987. Intraspecific competition in the tephritid fruit fly *Rhagoletis pomonella. Ecology* 68:878–86.

Axler, R. P., G. W. Redfield, and C. R. GOLDMAN. 1981. The importance of regenerated nitrogen to phytoplankton productivity in a subalpine lake. *Ecology* 62:345–54.

AXTMANN, R. C. 1975. Environmental impact of a geothermal power plant. *Science* 187:795–803.

BAIRD, D., and R. E. ULANOWICZ. 1989. The seasonal dynamics of the Chesapeake Bay ecosystem. *Ecological Monographs* 59:329–64.

BAKER, L. A., A. T. HERLIHY, P. R. KAUFMANN, and J. M. EILERS. 1991. Acidic lakes and streams in the United States: The role of acidic deposition. *Science* 252:1151–54.

BAKER, R. G., A. E. SULLIVAN, G. R. HALLBERG, and D. G. HORTON. 1989. Vegetational changes in western Illinois during the onset of Late Wisconsin glaciation. *Ecology* 70:1363–76.

BAKIR, F., S. F. DAMLUJI, L. AMIN-ZAKI, M. MURTADHA, A. KHALIDI, N. Y. AL-RAWI, S. TIKRITI, H. I. DHAHIR, T. W. CLARKSON, J. C. SMITH, and R. A. DOHERTY. 1973. Methylmercury poisoning in Iraq. *Science* 181:230–41.

BAKUN, A. 1990. Global climate change and intensification of coastal ocean upwelling. *Science* 247:198–201.

BARBOUR, M. G., and W. D. BILLINGS. 1988. *North American terrestrial vegetation.* Cambridge: Cambridge University Press.

BARKALOW, F. S., JR., R. B. HAMILTON, and R. F. SOOTS, JR. 1970. The vital statistics of an unexploited gray squirrel population. *Journal of Wildlife Management* 34:489–500.

BARTHOLOMEW, B. 1970. Bare zone between California shrub and grassland communities: The role of animals. *Science* 170:1210–12.

BATTAN, L. J. 1974. *Weather.* Englewood Cliffs, NJ: Prentice-Hall.

BAZZAZ, F. A. 1990. The response of natural ecosystems to the rising global CO_2 levels. *Annual Review of Ecology and Systematics* 21:167–96.

BAZZAZ, F. A., and E. D. FAJER. 1992. Plant life in a CO_2-rich world. *Scientific American* 266(1):68–74.

BAZZAZ, F. A., and S. L. MIAO. 1993. Succession status, seed size, and responses of tree seedlings to CO_2, light and nutrients. *Ecology* 74:104–12.

BAZZAZ, F. A., and W. E. WILLIAMS. 1991. Atmospheric CO_2 concentrations within a mixed forest: Implications for seedling growth. *Ecology* 72:12–16.

BEARDSLEY, T. 1989. Gaia. *Scientific American* 261(6):35–36.

BECK, H. L., and P. W. KREY. 1983. Radiation exposures in Utah from Nevada nuclear tests. *Science* 220:18–24.

BEEBE, G. W. 1982. Ionizing radiation and health. *American Scientist* 70:35–44.

BEETON, A. M. 1961. Environmental changes in Lake Erie. *Transactions of the American Fisheries Society* 90:153–59.

BEETON, A. M. 1965. Eutrophication of the St. Lawrence Great Lakes. *Limnology and Oceanography* 10:240–54.

BEGON, M., and M. MORTIMER. 1986. *Population ecology.* 2d ed. Sunderland, MA: Sinauer Associates Publishers.

BEGON, M., J. L. HARPER, and C. R. TOWNSEND. 1986. *Ecology. Individuals, populations, and communities.* Sunderland, MA: Sinauer Associates Publishers.

BEHRENSMEYER, A. K., J. D. DAMUTH, W. A. DIMICHELE, R. POTTS, H.-D. SUES, and S. L WING, eds. 1992. *Terrestrial ecosystems through time: Evolutionary paleoecology of terrestrial plants and animals.* Chicago: University of Chicago Press.

BENEMANN, J. R. 1973. Nitrogen fixation in termites. *Science* 181:164–65.

BERG, B., and G. EKBOHM. 1983. Nitrogen immobilization in decomposing needle litter at variable carbon:nitrogen ratios. *Ecology* 64:63–67.

BERGERON, Y. 1991. The influence of island and mainland lakeshore landscapes on boreal forest fire regimes. *Ecology* 72:1980–92.

BERGERON, Y., and J. BRISSON. 1990. Fire regime in red pine stands at the northern limit of the species' range. *Ecology* 71:1352–64.

BERNER, R. A. 1990. Atmospheric carbon dioxide levels over phanerozoic time. *Science* 249:1382–86.

BERNER, R. A., and A. C. LASAGA. 1989. Modeling the geochemical carbon cycle. *Scientific American* 260(3):74–81.

BERRYMAN, A. 1992. The origins and evolution of predator-prey theory. *Ecology* 73:1530–35.

BERTNESS, M. D. 1992. The ecology of a New England salt marsh. *American Scientist* 80:260–68.

BERTNESS, M. D., and A. M. ELLISON. 1987. Determinants of pattern in a New England salt marsh plant community. *Ecological Monographs* 57:129–47.

BERTNESS, M. D., L. GOUGH, and S. W. SHUMWAY. 1992. Salt tolerances and the distribution of fugitive salt marsh plants. *Ecology* 73:1842–51.

BESHIR, M. E. H., and M. O. MUBAREK. 1989. Sudan. In *International handbook of pollution control,* E. J. Kormondy, ed., 297–307. Westport, CT: Greenwood Press.

BEYERS, R. J. 1966. The pattern of photosynthesis and respiration in laboratory microecosystems. In *Primary productivity in aquatic environments*, C. S. Goldman, ed., 61–64. Berkeley: University of California Press.

BILLINGS, W. D. 1988. Alpine vegetation. In *North American terrestrial vegetation,* M. C. Barbour and W. D. Billings, eds., 391–420. Cambridge: Cambridge University Press.

BINFORD, M. W., E. S. DEEVEY, and T. L. CRISMAN. 1983. Paleolimnology: An historical perspective on lacustrine ecosystems. *Annual Review of Ecology and Systematics* 14:255–86.

BINKLEY, D., and D. RICHTER. 1987. Nutrient cycles and H^+ budgets of forest ecosystems. *Advances in Ecological Research* 16:1–51.

BIRCH, L. C. 1957. The role of weather in determining the distribution and abundance of animals. *Cold Spring Harbor Symposium on Quantitative Biology* 22:203–15.

BJORKSTROM, A. 1979. A model of CO_2 interaction between atmosphere, oceans, and land biota. In *The global carbon cycle*, B. Bolin, E. T. Degens, S. Kempe, and P. Ketner, eds., 403–57. New York: John Wiley and Sons.

BLACKMAN, F. F. 1905. Optima and limiting factors. *Annals of Botany* 19:281–95.

BLAKE, D. R., and F. S. ROWLAND. 1988. Continuing worldwide increase in tropospheric methane, 1978 to 1987. *Science* 239:1129–31.

BLINN, D. W. 1993. Diatom community structure along physicochemical gradients in saline lakes. *Ecology* 74:1246–63.

BLISS, L. C. 1988. Arctic tundra and polar desert biome. In *North American terrestrial vegetation,* M. G. Barbour and W. D. Billings, eds., 1–32. Cambridge: Cambridge University Press.

BLUMENSCHINE, R. J., and J. A. CAVALLO. 1992. Scavenging and human evolution. *Scientific American* 267(4):90–96.

BLUMER, M., and J. SASS. 1972. Oil pollution: Persistence and degradation of spilled fuel oil. *Science* 176:1120–2.

BODENHEIMER, F. S. 1958. Animal ecology today. *Monographiae Biologicae* 6:1–276.

BOLIN, B. 1970. The carbon cycle. *Scientific American* 223(3):124–32.

BOLIN, B., E. T. DEGENS, S. KEMPE, and P. KETNER, eds. 1979. *The global carbon cycle.* Published on behalf of the Scientific Committee on Problems of the Environment (SCOPE) of the International Council of Scientific Unions (ICSU). New York: John Wiley and Sons.

BONAN, G. B., and H. H. SHUGART. 1989. Environmental factors and ecological processes in boreal forests. *Annual Review of Ecology and Systematics* 20:1–28.

BONGAARTS, J. 1994a. Population policy options in the developing world. *Science* 263:771–76.

BONGAARTS, J. 1944b. Can the growing human population feed itself? *Scientific American* 270(3):36–42.

BOOSE, E. R., D. R. FOSTER, and M. FLUET. 1994. Hurricane impacts to tropical and temperate forest landscapes. *Ecological Monographs* 64:369–400.

BORMANN, B. T., and J. C. GORDON. 1984. Stand density effects in young red alder plantations: Productivity, photosynthate partitioning, and nitrogen fixation. *Ecology* 65:394–402.

BORMANN, B. T., F. H. BORMANN, W. B. BOWDEN, R. S. PIERCE, S. P. HAMBURG, D. WANG, M. C. SNYDER, C. Y. LI, and R. C. INGERSOLL. 1993. Rapid N_2 fixation in pines, alder and locust; evidence from the sandbox ecosystem study. *Ecology* 74:583–98.

BORMANN, F. H., and G. E. LIKENS. 1967. Nutrient cycling. *Science* 155:424–28.

BORMANN, F. H., and G. E. LIKENS. 1979. *Pattern and process in a forested ecosystem.* New York: Springer-Verlag.

BORMANN, F. H., G. E. LIKENS, and J. M. MELILLO. 1977. Nitrogen budget for an aggrading northern hardwood forest ecosystem. *Science* 196:981–83.

BORRELL, J. 1987. Trouble ahead for the Canal? *TIME* (March 2):63.

BOULIER, R., ed. 1983. *Tropical savannas.* Amsterdam: Elsevier Science Publishers.

BOULIER, R., and M. HADLEY. 1970. The ecology of tropical savannas. *Annual Review of Ecology and Systematics* 1:125–52.

BOULTON, A. J., D. G. PETERSON, N. B. GRIMM, and S. G. FISHER. 1992. Stability of an aquatic macroinvertebrate community in a multiyear hydrologic disturbance regime. *Ecology* 73:2192–2207.

BOUVIER, L. F., and C. J. DEVITA. 1991. The baby boom—entering middle life. *Population Bulletin* 46(3):1–35.

BOVBJERG, R. V. 1970. Ecological isolation and competitive exclusion in two crayfish (*Orconectes virilis* and *Orconectes immunis*). *Ecology* 51:225–36.

BOWDEN, W. B. 1984. A nitrogen-15 isotope dilution study of ammonium production and consumption in a marsh sediment. *Limnology and Oceanography* 29:1004–15.

BOWDEN, W. B. 1986. Nitrification, nitrate reduction, and nitrogen immobilization in a tidal freshwater marsh sediment. *Ecology* 67:88–99.

BOWDEN, W. B., and F. H. BORMANN. 1986. Transport and loss of nitrous oxide in soil water after forest clear-cutting. *Science* 233:867–69.

BOWMAN, K. P. 1988. Global trends in total ozone. *Science* 239:48–50.

BOX, E. 1978. Geographical dimensions of terrestrial net and gross productivity. *Radiation and Environmental Biophysics* 15:305–22.

BOYER, J. S. 1982. Plant productivity and environment. *Science* 218:443–48.

BRADLEY, R. S., H. F. DIAZ, J. K. EISCHEID, P. D. JONES, P. M. KELLY, and C. M. GOODESS. 1987. Precipitation fluctuations over northern hemisphere land areas since the mid-19th century. *Science* 237:171–75.

BRADY, N. C. 1982. Chemistry and world food supplies. *Science* 218:847–53.

BRAMRYD, T. 1979. The effect of man on the biogeochemical cycle of carbon in terrestrial ecosystems. In *The global carbon cycle,* B. Bolin, E. T. Degens, S. Kempe, and P. Ketner, eds., 183–218. New York: John Wiley and Sons.

BRANDT, C. J., ed. 1987. *Acidic precipitation. Formation and impact on terrestrial ecosystems.* Kommission Reinhaltung der Luft. Dusseldorf, Germany: Verein Deutscher Ingenieure.

BRASSEUR, G., and C. GRANIER. 1992. Mount Pinatubo aerosols, chlorofluorocarbons, and ozone depletion. *Science* 257:1239–42.

BRASSEUR, G., and M. H. HITCHMAN. 1988. Stratospheric response to trace gas perturbations: Changes in ozone and temperature distributions. *Science* 240:634–37.

BRATTSTEN, L. B., C. W. HOLYOKE, JR., J. R. LEEPER, and K. F. RAFFA. 1986. Insecticide resistance: Challenge to pest management and basic research. *Science* 231:1255–60.

BRAUMAN, A. M., D. KANE, M. LABAT, and J. A. BREZNAK. 1992. Genesis of acetate and methane by gut bacteria of nutritionally diverse termites. *Science* 257:1384–87.

BRAUN-BLANQUET, J. 1932. *Plant sociology: The study of plant communities.* Translated, revised, and edited by G. D. Fuller and H. S. Conard. New York: McGraw-Hill, Inc.

BREDEHOEFT, J. D., and T. MAINI. 1981. Strategy for radioactive waste disposal in crystalline rocks. *Science* 213:293–96.

BREZONIK, P. L., E. S. EDGERTON, and C. D. HENDRY. 1980. Acid precipitation and sulfate deposition in Florida. *Science* 208:1027–29.

BRIAND, F., and J. E. COHEN. 1984. Community food webs have scale-invariant structure. *Nature* 307:264–67.

BRIAND, F., and J. E. COHEN. 1987. Environmental correlates of food chain length. *Science* 238:956–60.

BRIDGES, E. M. 1978. *World soils.* 2d ed. Cambridge: Cambridge University Press.

BRILL, W. J. 1979. Nitrogen fixation: Basic to applied. *American Scientist* 67:458–66.

BRINSON, M. M., A. E. Lugo, and S. Brown. 1981. Primary productivity, decomposition and consumer activity in freshwater wetlands. *Annual Review of Ecology and Systematics* 12:123–61.

BROECKER, W. S. 1973. Factors controlling CO_2 content in the oceans and atmosphere. In *Carbon and the biosphere*, G. M. Woodwell and E. V. Pecan, eds., 32-50. Washington, DC: U.S. Atomic Energy Commission.

BROECKER, W. S., T. TAKAHASHI, H. J. SIMPSON, and T.-H. PENG. 1979. Fate of fossil fuel carbon dioxide and the global carbon budget. *Science* 206:409–18.

BROMENSHENK, J. J., S. R. CARLSON, J. C. SIMPSON, and J. M. THOMAS. 1985. Pollution monitoring of Puget Sound with honey bees. *Science* 227:632–34.

BRONK, D. A., P. M. GLIBERT, and B. B. WARD. 1994. Nitrogen uptake, dissolved organic nitrogen release, and new production. *Science* 265:1843–46.

BRONSTEIN, J. L., P.-H. GOUYON, C. GLIDDON, F. KJELLBERG, and G. MICHALOUD. 1990. The ecological consequences of flowering asynchrony in monoecious figs: A simulation study. *Ecology* 71:2145–56.

BROOKS, J. M., M. C. KENNICUTT II, C. R. FISHER, S. A. MACKO, K. COLE, J. J. CHILDRESS, R. R. BIDIGARE, and R. D. VETTER. 1987. Deep-sea hydrocarbon seep communities: Evidence for energy and nutritional carbon sources. *Science* 238:1138–42.

BROUGHTON, W. J. 1981–86. *Nitrogen fixation*. vol. 1, *Ecology*, 1981; vol. 2, *Rhizobium*, 1982; vol. 3, *Legumes*, 1983; vol. 4, *Molecular Biology*, 1986. Oxford: Clarendon Press.

BROWN, J. H., and B. A. MAURER. 1987. Evolution of species assemblages: Effects of energetic constraints and species dynamics on the diversification of the North American avifauna. *American Naturalist* 130:1–17.

BROWN, L. R. 1981. World population growth, soil erosion, and food security. *Science* 214:995–1002.

BROWN, L. R. 1994. Facing food insecurity. In *State of the world 1994. A Worldwatch Institute report on progress for a sustainable society*, L. Starke, ed., 177–97. New York: W. W. Norton and Company.

BROWN, L. R., and C. FLAVIN. 1988. The Earth's vital signs. In *State of the world 1988. A Worldwatch Institute report on progress toward a sustainable society*, L. Starke, ed., 3–21. New York: W. W. Norton and Company.

BRUNE, W. H., J. G. ANDERSON, D. W. TOOHEY, D. W. FAHEY, S. R. KAWA, R. L. JONES, D. S. MCKENNA, and L. R. POOLE. 1991. The potential for ozone depletion in the Arctic polar stratosphere. *Science* 252:1260–66.

BRUNE, W. H., D. W. TOOHEY, J. G. ANDERSON, W. L. STARR, J. F. VEDDER, and E. F. DANIELSEN. 1988. *In situ* northern mid-latitude observation of ClO, O_3 and BrO in the wintertime lower stratosphere. *Science* 242:558–62.

BUBIER, J. L., T. R. MOORE, and N. T. ROULET. 1993. Methane emissions from wetlands in the midboreal region of northern Ontario, Canada. *Ecology* 74:2240–54.

BUCKLEY, J. 1986. Environmental effects of DDT. In *Ecological knowledge and environmental problem-solving,* Committee on the Applications of Ecological Theory to Environmental Problems, Commission on Life Sciences, National Research Council, 358–74. Washington, DC: National Academy Press.

Bulletin of the Torry Botanical Club. 1926. 53:7–26.

BULLOCK, T. H. 1955. Compensation for temperature in the metabolism and activity of poikilotherms. *Biological Reviews* 30:311–342.

BUREAU OF THE CENSUS. 1987. *An aging world.* International Population Reports Series P-95, No. 78. Washington DC: U.S. Department of Commerce.

BURGIS, M. J., and P. MORRIS. 1987. *The natural history of lakes.* Cambridge: Cambridge University Press.

BURKE, I. C. 1989. Control of nitrogen mineralization in a sagebrush steppe landscape. *Ecology* 70:1115–26.

BURNETT, W. C., M. J. BEERS, and K. K. ROE. 1982. Growth rates of phosphate nodules from the continental margin off Peru. *Science* 215:1616–18.

BUSH, G. S. 1982. An environmental analysis of forest patterns. *American Scientist* 70:18–25.

BUSH, M. B., D. R. PIPERNO, P. A. COLINVAUX, P. E. DEOLIVEIRA, L. A. KRISSEK, M. C. MILLER, and W. E. ROWE. 1992. A 14 300-yr paleoecological profile of a lowland tropical lake in Panama. *Ecology* 62:251–75.

BUTCHER, S. S., R. J. CHARLSON, G. H. ORIONS, and G. V. WOLFE, eds. 1992. *Global biogeochemical cycles.* London: Academic Press.

CAIRNS, J., JR. 1972. Coping with heated waste water discharges from steam-electric power plants. *BioScience* 22:411–19.

CALDWELL, J. C., and P. CALDWELL. 1990. High fertility in sub-Saharan Africa. *Scientific American* 262(2):118–25.

CALLOWAY, R. M., and F. W. DAVIS. 1993. Vegetation dynamics, fire, and the physical environment in coastal central California. *Ecology* 74:1567–78.

CAMPBELL, F. E., and R. M. CRANWELL. 1988. Performance assessment of radioactive waste repositories. *Science* 239:1389–92.

CANFIELD, D. E., and D. J. DES MARAIS. 1991. Aerobic sulfate reduction in microbial mats. *Science* 251:1471–73.

CANTLON, J. E. 1953. Vegetation and microclimates on north and south slopes of Cushetunk Mt., New Jersey. *Ecological Monographs* 23:241–70.

CAPONE, D. G., and E. J. CARPENTER. 1982. Nitrogen fixation in the marine environment. *Science* 217:1140–42.

CAREY, J. R., P. LIEDO, D. OROZCO, and J. W. VAUPEL. 1992. Slowing of mortality rates at older ages in large medfly cohorts. *Science* 258:457–61.

CARLISLE, A., A. H. F. BROWN, and E. J. WHITE. 1966. The organic matter and nutrient elements in the precipitation beneath sessile oak (*Quercus petraea*) canopy. *Journal of Ecology* 54 (1):87–98.

CARLSON, D. A., K. D. KONYHA, W. B. WHEELER, G. P. MARSHALL, and R. G. ZAYLSKIE. 1976. Mirex in the environment: Its degradation to kepone and related compounds. *Science* 191:939–41.

CARNIGAN, R., and J. KALFF. 1980. Phosphorus sources for aquatic weeds: Water or sediments. *Science* 207:987–89.

CARPENTER, E. J,. and K. ROMANS. 1991. Major role of the cyanobacterium *Trichodesmium* in nutrient cycling in the North Atlantic ocean. *Science* 254:1356–58.

CARR, J. F., and J. K. HILTUNEN. 1965. Changes in the bottom fauna of western Lake Erie from 1930 to 1961. *Limnology and Oceanography* 10:551–69.

CARRICK, R. 1963. Ecological significance of territory in the Australian magpie. *Proceedings of the XIII International Ornithological Congress* 2:740–53.

CARSON, W. P., and G. W. BARRETT. 1988. Succession in old-field plant communities: Effects of contrasting types of nutrient enrichment. *Ecology* 69:984–94.

Carter, L. J. 1977a. Radioactive wastes: Some urgent unfinished business. *Science* 195:661–66, 704.

CARTER, L. J. 1977b. Nuclear wastes: Popular antipathy narrows search for disposal sites. *Science* 197:1265–66.

CARTER, L. J. 1977c. Asbestos: Trouble in the air from Maryland rock quarry. *Science* 197:237–40.

CARY, S. C., C. R. FISHER, and H. FELBECK. 1988. Mussel growth supported by methane as sole carbon and energy source. *Science* 240:78–80.

CATES, W., JR. 1982. Legal abortion: The public health record. *Science* 215:1586–90.

CHAMEIDES, W. L., R. W. LINDSAY, J. RICHARDSON, and C. S. KIANG. 1988. The role of biogenic-hydrocarbons in urban photochemical smog: Atlanta as a case study. *Science* 241:1473–75.

CHAPIN, S. F., N. FLETCHER, K. KIELLAND, K. R. EVERETT, and A. E. LINKINS. 1988. Productivity and nutrient cycling of Alaskan tundra: Enhancement by flowing soil water. *Ecology* 69:693–702.

CHAPMAN, R. F. 1982. *The insects. Structure and function.* Cambridge, MA: Harvard University Press.

CHAPMAN, R. N. 1928. The quantitative analysis of environmental factors. *Ecology* 9:111–22.

CHARLSON, R. J., T. L. Anderson, and R. E. McDuff. 1992. The sulfur cycle. In *Global biogeochemical cycles*, S. S. Butcher, R. J. Charlson, G. H. Orions, and G. V. Wolfe, eds., 285–300. London: Academic Press.

CHARLSON, R. J., S. E. SCHWARTZ, J. M. HALES, R. D. CESS, J. A. CROAKLEY, JR., J. E. HANSEN, and D. J. HOFFMANN. 1992. Climate forcing by anthropogenic aerosols. *Science* 255:423–30.

CHARLSON, R. J., and T. M. L. WIGLEY. 1994. Sulfate aerosol and climatic change. *Scientific American* 270(2):48–57.

CHITTY, D. 1960. Population processes in the vole and their relevance to general theory. *Canadian Journal of Zoology* 38:99–113.

CHITTY, D. 1967. The natural selection of self-regulatory behavior in animal populations. *Proceedings of the Ecological Society of Australia* 2:51–78.

CHOI, G. H., and D. L. NUSS. 1992. Hypovirulence of chestnut blight fungus conferred by an infectious viral cDNA. *Science* 257:800–803.

CHRISTENSEN, N. L. 1988. Vegetation of the southeastern coastal plain. In *North American terrestrial vegetation,* M. G. Barbour and W. D. Billings, eds., 317–63. Cambridge: Cambridge University Press.

CHRISTIAN, J. J. 1961. Phenomena associated with population density. *Proceedings of the National Academy of Sciences* 47:428–49.

CICERONE, R. J. 1987. Changes in stratospheric ozone. *Science* 237:35–42.

CICERONE, R. J., S. ELIOTT, and R. P. TURCO. 1991. Reduced Antarctic ozone depletions in a model with hydrocarbon injections. *Science* 254:1191–94.

CLARK, J. S. 1986. Coastal forest tree populations in a changing environment, southeastern Long Island, New York. *Ecological Monographs* 56:259–77.

CLARK, J. S. 1990. Fire and climate change during the last 750 years in northwestern Minnesota. *Ecological Monographs* 60:135–59.

CLARK, W. C. 1989. Managing planet Earth. *Scientific American* 261(3):47–54.

CLARK, W. C. 1990. Learning from the past: Traditional knowledge and sustainable development. *The Contemporary Pacific* 2:233–53.

CLARKE, G. L. 1954. *Elements of ecology.* New York: John Wiley and Sons.

CLAUSEN, J., D. D. KECK, and W. M. HIESEY. 1948. Experimental studies on the nature of species. III. *Carnegie Institute of Washington Publication No. 581*:1–129.

CLEMENTS, F. E. 1905. *Research methods in ecology.* Lincoln, NE: University Publishing Company.

CLEMENTS, F. E. 1916. Plant succession: Analysis of the development of vegetation. *Carnegie Institute of Washington Publication No. 242.*

CLEVELAND, W. S., and T. E. GRAEDEL. 1979. Photochemical air pollution in the northeast United States. *Science* 204:1273–78.

CLOUDSLEY-THOMPSON, J. L. 1984. *Key environments. Sahara Desert.* New York: Pergamon Press.

COALE, A. J., W. FENG, N. E. RILEY, and L. F. DE. 1991. Recent trends in fertility and nuptiality in China. *Science* 251:389–93.

COBB, C. E., JR. 1989. Living with radiation. *National Geographic* 175:403–37.

COHOP (Cooperative Holocene Mapping Project). 1988. Climatic changes of the last 18,000 years: Observations and model simulations. *Science* 241:1043–52.

COLE, C. V., INNIS, G. S., and J. W. B. STEWART. 1977. Simulation of phosphorus cycling in semiarid grasslands. *Ecology* 58:1–15.

COLE, J. T., N. F. CARACO, G. W. KLING, and T. K. KRATZ. 1994. Carbon dioxide supersaturation in surface waters of lakes. *Science* 265:1568–70.

COLEMAN, D. C., C. P. P. REID, and C. V. COLE. 1983. Biological strategies of nutrient cycling in soil systems. *Advances in Ecological Research* 13:1–55.

COLINVAUX, P. A. 1989. The past and future Amazon. *Scientific American* 260:102–108.

COLLINS, M., ed. 1990. *The last rainforests. A world conservation atlas.* Oxford: Oxford University Press.

COLLINS, S. L., and L. L. WALLACE. 1990. Fire in North American tall grass prairies. Norman, OK: University of Oklahoma Press.

CONNELL, J. H. 1961. The influence of interspecific competition and other factors on the distribution of the barnacle *Chthamalus stellatus. Ecology* 42:710–23.

CONNELL, J. H., and R. O. SLAYTER. 1977. Mechanisms of succession in natural communities and their role in community stability and organization. *American Naturalist* 111:1119–44.

CONWAY, J. 1980. The energy crisis: Its ethical dimension. *Contemporary Philosophy* 8:7–8.

COOK, P. M., G. E. GLASS, and J. H. TUCKER. 1974. Asbestiform amphibole minerals: Detection and measurement of high concentrations in municipal water supplies. *Science* 185:853–55.

COOPER, S. R., and G. S. BRUSH. 1991. Long-term history of Chesapeake Bay anoxia. *Science* 254:992–96.

COPELAND, B. J. 1965. Evidence for regulation of community metabolism in a marine ecosystem. *Ecology* 46:563–64.

COUPLAND, R. T., ed. 1979. *Grassland ecosystems of the world: Analysis of grasslands and their uses.* London/New York: Cambridge University Press.

COURTILLOT, V. E. 1990. A volcanic eruption. *Scientific American* 263(4):85–92.

COURTNEY, S. P. 1986. The ecology of Pierid butterflies: Dynamics and interactions. *Advances in Ecological Research* 15:51–131.

COVINGTON, W. W. 1981. Changes in forest floor organic matter and nutrient content following clear cutting in northern hardwoods. *Ecology* 62:41–48.

COWLES, H. C. 1899. The ecological relations of the vegetation of the sand dunes of Lake Michigan. *Botanical Gazette* 27:95–117, 167–202, 281–308, 361–91.

COX, P. A., and M. J. BALICK. 1994. The ethnobotanical approach to drug discovery. *Scientific American* 270(6):82–87.

CRAMER, J., B. R. LEAVITT, and J. M. NASH. 1988. They lied to us. *TIME* (October 31):60–65.

CRAWFORD, M. 1987. Landmark ozone treaty negotiated. *Science* 237:1557.

CRONAN, C. S., W. A. REINERS, R. C. REYNOLDS, JR., and G. E. LANG. 1978. Forest floor leaching: Contributions from mineral, organic, and carbonic acids in New Hampshire subalpine forests. *Science* 200:309–11.

CROSBY, A. W. 1986. *Ecological imperialism. The biological expansion of Europe, 900–1900.* New York: Cambridge University Press.

CROSSON, P. R., and N. J. ROSENBERG. 1989. Strategies for agriculture. *Scientific American* 261(3):128–35.

CROWLEY, P. H., R. M. NISBET, W. S. C. GURNEY, and J. H. LAWTON. 1987. Population regulation in animals with complex life histories: Formulation and analysis of a damselfly model. *Advances in Ecological Research* 17:1–59.

CRUTZEN, P. J., and J. W. BIRKS. 1982. The atmosphere after a nuclear war: Twilight at noon. *Ambio* 11:114–25.

CRUTZEN, P. J., and M. O. ANDREAE. 1990. Biomass burning in the tropics: Impact on atmospheric chemistry and biogeochemical cycles. *Science* 250:1669–78.

CULLEN, J. J., R. J. NEALE, and M. P. LESSER. 1992. Biological weighting function for the inhibition of phytoplankton photosynthesis by ultraviolet radiation. *Science* 258:646–50.

CUNNOLD, D. M., R. G. PRINN, R. A. RASMUSSEN, P. G. SIMMONDS, F. N. ALYEA, C. A. CARDELINO, A. J. CRAWFORD, P. J. FRASER, and R. D. ROSEN. 1986. Atmospheric lifetime and annual release estimates for $CFCL_3$ and CF_2CL_2 from 5 years of ALE data. *Journal of Geophysical Research* 91:10.917–10.817.

CURTIS, J. T., and R. P. MCINTOSH. 1951. An upland forest continuum in the prairie-forest border region of Wisconsin. *Ecology* 32:476–96.

CUTSHALL, N. H., I. L. LARSEN, and M. M. NICHOLS. 1981. Man-made radionuclides confirm rapid burial of kepone in James River sediments. *Science* 213:440–42.

CWYNAR, L. C., and R. W. SPEAR. 1991. Reversion of forest to tundra in the central Yukon. *Ecology* 72:202–12.

DALY, K. L., and W. O. SMITH, JR. 1993. Physical-biological interactions influencing marine plankton production. *Annual Review of Ecology and Systematics* 24:555–85.

DANSEREAU, P. 1951. Description and recording of vegetation upon a structural basis. *Ecology* 32:172–229.

DARWIN, C. R. 1845. *Voyage of the Beagle.* 2d ed. New York: P. F. Collier and Son, 1937.

DARWIN, C. R. 1859. *On the origin of species by natural selection.* London: Murray.

DAUBENMIRE, R. F. 1968. *Plant communities: A textbook of plant synecology.* New York: Harper and Row.

DAVIDSON, C. I., T. C. GRIMM, and M. A. NASTA. 1981. Airborne lead and other elements derived from local fires in the Himalayas. *Science* 214:1344–46.

DAVIDSON, C. J., J. R. HARRINGTON, M. J. STEPHENSON, M. C. MONAGHAN, J. PUDYKIEWICZ, and W. R. SCHELL. 1987. Radioactive cesium from the Chernobyl accident in the Greenland ice sheet. *Science* 237:633–34.

DAVIDSON, E. A., S. C. HART, and M. K. FIRESTONE. 1992. Internal cycling of nitrate in soils of a mature coniferous forest. *Ecology* 73:1148–56.

DAVIDSON, J., and H. G. ANDREWARTHA. 1948. Annual trends in a natural population of *Thrips imaginis* (Thysanoptera). *Journal of Animal Ecology* 17:193–99.

DAVIS, C. C. 1964. Evidence for eutrophication of Lake Erie from phytoplankton records. *Limnology and Oceanography* 9:275–83.

DAVIS, K. 1967. Population policy: Will current programs succeed? *Science* 158:730–39.

DAVIS, M. B. 1969. Palynology and environmental history during the Quaternary Period. *American Scientist* 57:317–32.

DAVIS, M. B. 1976. Pleistocene biogeography of temperate deciduous forests. *Geoscience and Man* 13:13–26.

DAVIS, R. B. 1966. Spruce-fir forests of the coast of Maine. *Ecological Monographs* 36:79–94.

DAYTON, P. K., M. J. TEGNER, P. E. PARNELL, and P. B. EDWARDS. 1992. Temporal and spatial patterns of disturbance and recovery in a kelp forest community. *Ecological Monographs* 62:421–45.

DEANGELIS, D. L., W. M. POST, and C. C. TRAVIS. 1986. *Positive feedback in natural systems.* New York: Springer-Verlag.

DECKER, R., and B. DECKER. 1989. *Volcanoes.* New York: W. H. Freeman and Company.

DEEVEY, E. S., JR. 1947. Life tables for natural populations of animals. *Quarterly Review of Biology* 22:282–314.

DUBLIN, L. I., A. J. LOTKA, and M. SPIEGELMAN. 1949. *Length of life.* New York: Ronald Press.

DEL MORAL, R., and L. C. BLISS. 1993. Mechanisms of primary succession: Insights resulting from the eruption of Mount St. Helens. *Advances in Ecological Research* 24:1–66.

DELCOURT, H. R. 1979. Late Quaternary vegetation history of the eastern Highland Rim and adjacent Cumberland Plateau of Tennessee. *Ecological Monographs* 49:255–80.

DELWICHE, C. C. 1970. The nitrogen cycle. *Scientific American* 223(3):137–46.

DELWICHE, C. C. 1981. *Denitrification, nitrification, and atmospheric nitrous oxide.* New York: John Wiley and Sons.

DENOYELLES, F., W. D. KETTLE, and D. E. SINN. 1982. The responses of plankton communities in experimental ponds to atrazine, the most heavily used pesticide in the United States. *Ecology* 63:1285–93.

DETWEILER, R. P., and C. A. S. HALL. 1988. Tropical forests and the global carbon cycle. *Science* 239:42–47.

DIAMOND, J. M. 1978. Niche shifts and the rediscovery of interspecific competition. *American Scientist* 66:322–31.

DICKEMAN, M. 1975. Demographic consequences of infanticide in man. *Annual Review of Ecology and Systematics* 6:107–37.

DICKMAN, C. R. 1986. An experimental study of competition between two species of dasyurid marsupials. *Ecological Monographs* 56:221–41.

DIEHL, S. 1992. Fish predation and benthic community structure: The role of omnivory and habitat complexity. *Ecology* 73:1646–61.

DIXON, R. K., S. BROWN, R. A. HOUGHTON, A. M. SOLOMON, M. C. TREXLER, and J. WISNIEWSKI. 1994. Carbon pools and flux of global forest ecosystems. *Science* 263:185–90.

DOBBS, M. 1991. Chernobyl: Symbol of Soviet failure. *Washington Post* 114:142(April 26): 1, 38.

DONALDSON, P. J., and A. O. TSUI. 1990. The international family planning movement. *Population Bulletin* 45(3):1–47.

DUBINSKY, Z., ed. 1990. *Coral reefs.* Amsterdam: Elsevier Science Publishers.

DUBLIN, L. I., A. J. LOTKA, and M. SPEIGELMAN. 1949. *Length of life.* New York: Ronald Press.

DUCE, R. A., G. L. HOFFMAN, and W. H. ZOLLER. 1975. Atmospheric trace metals at remote Northern and Southern Hemisphere sites: Pollution or natural? *Science* 187:59–61.

DUDLEY, W. C., and M. LEE. 1988. *Tsunami.* Honolulu: University of Hawaii Press.

DUGDALE, V. A., and R. C. DUGDALE. 1962. Nitrogen metabolism in lakes. II. Role of nitrogen fixation in Sanctuary Lake, Pennsylvania. *Limnology and Oceanography* 7:170–77.

DUGGINS, D. O., C. A. SIMENSTAD, and J. A. ESTES. 1989. Magnification of secondary production by kelp detritus in coastal marine ecosystems. *Science* 245:170–73.

DUNNETTE, D. A. 1992. Assessing global river water quality: Case study using mechanistic approaches. In *The science of global change. The impact of human activities on the environment,* D. A. Dunnette and R. J. O'Brien, eds., 260–86. Washington, DC: American Chemical Society.

DUNNETTE, D. A., and R. J. O'BRIEN, eds. 1992. *The science of global change. The impact of human activities on the environment.* Washington, D.C: American Chemical Society.

DUNWIDDIE, P. W. 1986. A 6000-year record of forest history on Mount Rainier, Washington. *Ecology* 67:58–68.

DYKSTERHUIS, E. J., and E. M. SCHMUTZ. 1947. Natural mulches of litter of grasslands with kinds and amounts on a southern prairie. *Ecology* 28:163-79.

EAMUS, D., and P. G. JARVIS. 1989. The direct effects of increase in the global atmospheric CO_2 concentration on natural and commercial temperate trees and forests. *Advances in Ecological Research* 19:1–55.

EDMOND, J. M., and K. VON DAMM. 1983. Hot springs on the ocean floor. *Scientific American* 248(4):78–93.

EDWARDS, J. S., and P. SUGG. 1993. Arthropod fallout as a resource in the recolonization of Mount St. Helens. *Ecology* 74:954–58.

EGERTON, F. N. 1977. A bibliographic guide to the history of general ecology and population ecology. *History of Science* 15:189–215.

EGUNJOBI, J. K. 1989. Nigeria. In *International handbook of pollution control,* E. J. Kormondy, ed., 285–95. Westport, CT: Greenwood Press.

EHRENFELD, D. 1988. Why put a value on biodiversity? In *Biodiversity*, E.O. Wilson and F. M. Peter, eds., 212–16. Washington, DC: National Academy Press.

EHRLICH, P. R., and E. O. Wilson. 1991. Biodiversity studies: Science and policy. *Science* 253:758–62.

EHRLICH, P. R., J. HARTE, M. A. HARTWELL, P. H. RAVEN, C. SAGAN, G. M. WOODWELL, J. BERRY, E. S. AYENSU, A. H. EHRLICH, T. EISNER, S. J. GOULD, H. D. GROVER, R. HERRERA, R. M. MAY, E. MAYR, C. P. MCKAY, H. A. MOONEY, N. MYERS, D. PIMENTAL, and J. M. TEAL. 1983. Long-term biological consequences of nuclear war. *Science* 222:1293–1300.

ELLIOTT-FISK, D. L. 1988. The boreal forest. In *North American terrestrial vegetation*, M. G. Barbour and W. D. Billings, eds., 33–62. Cambridge: Cambridge University Press.

ELTON, C. 1927. *Animal ecology*. London: Sidgwick and Jackson.

EMBER, R. 1981. Acid pollutants: Hitchhikers ride the wind. *Chemical & Engineering News* 59(37):20–31.

ENDLER, J. A. 1993. The color of light in forests and its implications. *Ecological Monographs* 62:1–27.

ERBSEN, C. E. 1979. Mauritania losing life-death struggle against the sands of the Sahara. *The Bulletin* (November 22):53.

ESKEW, D. L., R. M. WELCH, and E. E. CARY. 1983. Nickle: An essential micronutrient for legumes and possibly all higher plants. *Science* 222:621–23.

ESSER, G. 1992. Implications of climate change for production and decomposition in grasslands and coniferous forests. *Ecological Applications* 2:47–54.

EVANS, A. C., and W. J. McL. GUILD. 1947. Studies on the relationships between earthworms and soil fertility. I. Biological studies in the field. *Annals of Applied Biology* 34:307–30.

EVANS, F., and F. SMITH. 1952. The intrinsic rate of natural increase for the human louse, *Pediculus humanus* L. *American Naturalist* 86:299–310.

EVENARI, M., I. NOY-MEIR, and D. W. GOODALL, eds. 1985. *Hot deserts and arid shrublands*. Amsterdam: Elsevier Science Publishers.

EVENARI, M., L. SHANAN, and N. TADMORE. 1971. *The Negev. The challenge of a desert*. Cambridge, MA: Harvard University Press.

EWEL, J., C. BERISH, B. BROWN, N. PRICE, and J. RAICH. 1981. Slash and burn impacts on a Costa Rican wet forest site. *Ecology* 62:816–29.

EYRE, S. R. 1963. *Vegetation and soils, a world picture*. Chicago: Aldine Publishing Company.

FAHEY, T. J. 1983. Nutrient dynamics of aboveground detritus in lodgepole pine (*Pinus contorta* ssp. *latifolia*) ecosystems, southeastern Wyoming. *Ecological Monographs* 53:51–72.

FALKOWSKI, P. G., Y. KIM, Z. KOLBER, C. WILSON, C. WIRICK, and R. CESS. 1992. Natural versus anthropogenic factors affecting low-level cloud albedo over the North Atlantic. *Science* 256:1311–13.

FARMAN, J. C., B. G. GARDINER and J. D. SHANKLIN. 1985. Large losses of total ozone in Antarctica reveal seasonal ClO_x/NO_x interaction. *Nature* 315:207–10.

FARRELL, T. M. 1991. Models and mechanisms of succession: An example from a rocky intertidal community. *Ecological Monographs* 61:95–113.

FAWCETT, M. H. 1984. Local and latitudinal variation in predation on an herbivorous marine snail. *Ecology* 65:1214–30.

FEENEY, G. 1994. Fertility decline in East Asia. *Science* 266:1518–23.

FENCHEL, T. 1988. Marine plankton food chains. *Annual Review of Ecology and Systematics* 19:19–38.

FENNELLY, P. F. 1976. The origin and influence of airborne particulates. *American Scientist* 64:46–56.

FERGUSSON, J. E. 1992. Dust in the environment: Elemental composition and sources. In *The science of global change. The impact of human activities on the environment,* D. A. Dunnette and R. J. O'Brien, eds., 117-33. Washington, DC: American Chemical Society.

FITZPATRICK, E. A. 1980. *Soils. Their formation, classification and distribution.* London: Longman Group Ltd.

FLUECK, W. T. 1994. Effect of trace elements on population dynamics: Selenium deficiency in free-ranging black-tailed deer. *Ecology* 75:807–12.

FOLT, C., and C. R. GOLDMAN. 1981. Allelopathy between zooplankton: A mechanism for interference competition. *Science* 213:1133–35.

FORD, M. S. (J.). 1990. A 10,000-yr history of natural ecosystem acidification. *Ecological Monographs* 60:57–89.

FOUKAL, P. V. 1990. The variable sun. *Scientific American* 262(2):34–41.

FOWLER, N. L. 1986. Density-dependent population regulation in a Texas grassland. *Ecology* 67:545–54.

FOX, G. A. 1989. Consequences of flowering-time variation in a desert annual: Adaptation and history. *Ecology* 71:1294–306.

FOX, L. R. 1975. Cannibalism in natural populations. *Annual Review of Ecology and Systematics* 6:87–196.

FRANCIS, C. M., M. H. RICHARDS, F. COOKE, and R. F. ROCKWELL. 1992. Long-term changes in survival rates of lesser snow geese. *Ecology* 73:1346–62.

FRANK, P. W., C. D. BOLL, and R. W. KELLY. 1957. Vital statistics of laboratory cultures of *Daphnia pulex* DeGeer as related to density. *Physiological Zoology* 30:287–305.

FREDERICK J. E., and H. E. SNELL. 1988. Ultraviolet radiation levels during the Antarctic spring. *Science* 241:438–40.

FREDRICKSON, A. G., and G. STEPHANOPOULOS. 1981. Microbial competition. *Science* 213:972–79.

FREED, L. A., S. CONANT, and R. C. FLEISCHER. 1987. Evolutionary ecology and radiation of Hawaiian passerine birds. *Trends in Ecology & Evolution* 2:196–203.

FREEDMAN, B., V. ZOBENS, T. C. HUTCHINSON, and W. I GIZYN. 1990. Intense, natural pollution affects arctic tundra vegetation at the Smoking Hills, Canada. *Ecology* 71:492–503.

FREY, D. G., ed. 1963. *Limnology in North America.* Madison: University of Wisconsin Press.

FRIEDEN, E. 1972. The chemical elements of life. *Scientific American* 227(1):52–60.

FRIEDERICHS, K. 1958. A definition of ecology and some thoughts about basic concepts. *Ecology* 39:154–59.

FRIEND, A. D., and F. I. WOODWARD. 1990. Evolutionary and ecophysiological responses of mountain plants to the growing season environment. *Advances in Ecological Research* 20:59–124.

FRISS-CHRISTENSEN, E., and K. LASSEN. 1991. Length of the solar cycle: An indicator of solar activity closely associated with climate. *Science* 254:698–700.

FRITTS, H. C. 1972. Tree rings and climate. *Scientific American* 226(5):93–100.

FRITTS, H. C., and T. W. SWETNAM. 1989. Dendroecology: A tool for evaluating variations in past and present forest environments. *Advances in Ecological Research* 19:111–88.

FULLER, M. 1991. *Forest fires. An introduction to wildland fire behavior, management, firefighting, and prevention.* New York: John Wiley and Sons.

GALLOWAY, J. N., G. E. LIKENS, and M. E. HAWLEY. 1984. Acid precipitation: Natural versus anthropogenic components. *Science* 226:829–31.

GALLOWAY, J. N., Z. DIANWU, X. JILING, and G. E. LIKENS. 1987. Acid rain: China, United States, and a remote area. *Science* 236:1559–62.

GANTER, P. F. 1984. The effects of crowding on terrestrial isopods. *Ecology* 65:438–45.

GARDNER, L. R. 1990. Simulation of the diagenesis of carbon, sulfur, and dissolved oxygen in salt marsh sediments. *Ecological Monographs* 60:91–111.

GARNER, W. W., and H. A. ALLARD. 1923. Further studies in photoperiodism, the response of the plant to relative length of day and night. *Journal of Agricultural Research* 23:871–920.

GARRELS, R. M., A. LERMAN, and F. R. MACKENZIE. 1976. Controls of atmospheric O_2 and CO_2: Past, present, and future. *American Scientist* 64:306–15.

GASCON, C. 1991. Population- and community-level analyses of species occurrences of central Amazonian rainforest tadpoles. *Ecology* 72:1731–46.

GATES, D. M. 1962. *Energy exchange in the biosphere.* New York: Harper and Row.

GATES, D. M. 1993. *Climate change and its biological consequences.* Sunderland, MA: Sinauer Associates Publishers.

GAUSE, G. F. 1934. *The struggle for existence.* Baltimore: Williams and Wilkins Company.

GEIGER, R. 1941. *The climate near the ground.* Cambridge, MA: Harvard University Press.

GEORGE, T. L., A. D. FOWLER, R. L. KNIGHT, and L. E. MCEWEN. 1992. Impacts of severe drought on grassland birds in western North Dakota. *Ecological Applications* 2:275–84.

GHOLZ, H. L., R. F. FISHER, and W. L. PRITCHETT. 1985. Nutrient dynamics in slash pine plantation ecosystems. *Ecology* 66:647–59.

GIBBONS, A. 1993. Dioxin tied to endometriosis. *Science* 262:1373.

GIBBONS, J. W., and R. R. SHARITZ. 1974. Thermal alteration of aquatic ecosystems. *American Scientist* 62:660–70.

GIFFORD, R. M., J. H. Thorne, W. D. HITZ, and R. T. GIAQUINTA. 1984. Crop productivity and photoassimilate partitioning. *Science* 225:801–808.

GILBERT, J. J. 1985. Competition between rotifers and *Daphnia. Ecology* 66:1943–50.

GILBERT, J. J., and R. S. STEMBERGER. 1985. Control of *Keratella* populations by interference competition from *Daphnia. Limnology and Oceanography* 30:180–88.

GLANTZ, M. H. 1987. Drought in Africa. *Scientific American* 256:34–40.

GLAZOVSKAYA, M. S. 1983. *Soils of the world.* 2 vols. New Delhi: Amerind Publishing Company. Translated from Russian.

GLEASON, H. A. 1926. The individualistic concept of the plant association. *Bulletin of the Torry Botanical Club* 53:7–26.

GLEASON, J. F., P. K. BHARTIA, J. R. HERMAN, R. MCPETERS, P. NEWMAN, R. S. STOLARSKI, L. FLYNN, G. LABOW, D. LARKO, C. SEFTOR, C. WELLEMEYER, W. D. KOMHYR, A. J. MILLER, and W. PLANET. 1993. Record low global ozone in 1992. *Science* 260:523–26.

GLEESON, S. K. 1994. Density dependence is better than ratio dependence. *Ecology* 75:1834–35.

GLEN, W. 1990. What killed the dinosaurs? *American Scientist* 78:354–70.

GLOSS, S P., C. L. SCHOFIELD, and R. E. SHERMAN. 1988. An evaluation of New York State liming data and the applications of models from Scandinavian lakes to Adirondack lakes. *Water, Air and Soil Pollution* 41(1–4):241–78.

GLYNN, P. W. 1988. El Niño–southern oscillation 1982–83: Nearshore population, community, and ecosystem responses. *Annual Review of Ecology and Systematics* 19:309–45.

GOERING, J., and V. DUGDALE. 1966. Estimate of the rate of denitrification in a subarctic lake. *Limnology and Oceanography* 11:113–17.

GOLDEN, F. 1989. A catbird's seat on Amazon destruction. *Science* 246:201–202.

GOLDMAN, M. 1987. Chernobyl: A radiobiological perspective. *Science* 238:622–23.

GOLDWASSER, L., and J. ROUGHGARDEN. 1993. Construction and analysis of a large Caribbean food web. *Ecology* 74:1216–33.

GOLLEY, F. B. 1960. Energy dynamics of a food chain of an old-field community. *Ecological Monographs* 30:187–206

GOLLEY, F. B. 1983. *Tropical rain forest ecosystems. Structure and function.* Amsterdam: Elsevier Science Publishers.

GOLLEY, F. B. 1993. *A history of the ecosystem concept in ecology.* New Haven, CT: Yale University Press.

GOLLEY, F. B., and J. B. GENTRY. 1965. A comparison of variety and standing crop of vegetation on a one-year and twelve-year abandoned field. *Oikos* 15:185–99.

GOMEZ-POMPA, A., C. VAZQUES-YANES, and S. GUEVARA. 1972. The tropical rain forest: A nonrenewable resource. *Science* 177:762–65.

GOODESS, C. M. 1987. Precipitation fluctuations over Northern Hemisphere land areas since the mid-19th century. *Science* 237:171–75.

GOODMAN, D. 1975. The theory of diversity-stablity relationships in ecology. *Quarterly Review of Biology* 50:237–66.

GORE, R. 1979. An age-old challenge grows. *National Geographic* 156:594–639.

GOREAU, T. F., N. I. GOREAU, and T. J. GOREAU. 1979. Corals and coral reefs. *Scientific American* 241(2):124–36.

GORMAN, C. 1987. A battle against deadly dust. *TIME* (November 16):66.

GOSZ, J. R., R. T. HOLMES, G. E. LIKENS, and F. HERBERT BORMANN. 1978. The flow of energy in a forest ecosystem. *Scientific American* 238(3):93–102.

GREEN, G. M., and R. W. SUSSMAN. 1990. Deforestation history of the eastern rainforests of Madagascar from satellite images. *Science* 248:212–15.

GREENHALGH, S., and J. BONGAARTS. 1987. Fertility policy in China: Future options. *Science* 235:1167–72.

GRELLER, A. M. 1988. Deciduous forest. In *North American terrestrial vegetation,* M. G. Barbour and W. D. Billings, eds., 287–316. Cambridge: Cambridge University Press.

GRIER, J. W. 1982. Ban of DDT and subsequent recovery of reproduction in bald eagles. *Science* 218:1232–35.

GRIEVE, R. A. F. 1990. Impact cratering on the earth. *Scientific American* 262(4):66–73.

GRIMM, E. C. 1984. Fire and other factors controlling the Big Woods vegetation of Minnesota in the mid-nineteenth century. *Ecological Monographs* 54:291–311.

GRIMM, N. B. 1987. Nitrogen dynamics during succession in a desert stream. *Ecology* 68:1157–70.

GROFFMAN, P. M., G. J. HOUSE, P. F. HENDRIX, D. E. SCOTT, and D. A. CROSSLEY, JR. 1986. Nitrogen cycling as affected by interaction of components in a Georgia piedmont agroecosystem. *Ecology* 67:80–87.

GROFFMAN, P. M., C. W. RICE, and J. M. TIEDJE. 1993. Denitrification in a tallgrass prairie landscape. *Ecology* 74:855–62.

GROFFMAN, P. M., D. R. ZAK, S. CHRISTENSEN, A. MOSIER, and J. M. TIEDJE. 1993. Early spring nitrogen dynamics in a temperate forest landscape. *Ecology* 74:1579–85.

GROOMBRIDGE, B., ed. 1992. *Global diversity: Status of the Earth's living resources.* London: Chapman and Hall.

GUPTA, A. 1988. *Ecology and development in the Third World.* London: Routledge.

HAAGEN-SMIT, A. J. 1952. The chemistry and physiology of Los Angeles smog. *Industrial and Engineering Chemistry* 44:1342–46.

HADLEY, J. L., and W. K. SMITH. 1986. Wind effects on needles of timberline conifers; seasonal influence on mortality. *Ecology* 67:12–19.

HAECKEL, E. 1870. *Generelle Morphologie der Organismen: Allgemeine Grundzuge der Organischen Formen-wissenschaft, Mechanisch Begrundet durch die von Charles Darwin Reformirte Descendenz-Theorie.* Berlin: Vorlag von Georg Reiner.

HAFELE, W. 1990. Energy from nuclear power. *Scientific American* 263(3):137–44.

HAIRSTON, N. G., J. D. ALLEN, R. K. COLWELL, D. J. FUTUYMA, J. HOWELL, M. D. LUBIN, J. MATHIAS, and J. H. VANDERMEER. 1968. The relationship between species diversity and stability: An experimental approach with protozoa and bacteria. *Ecology* 49:1092–101.

HALL, F. G., D. B. BOTKIN, D. E. STREBEL, K. D. WOODS, and S. J. GOETZ. 1991. Large-scale patterns of forest succession determined by remote sensing. *Ecology* 72:628–40.

HALL, S. J., and D. G. RAFAELLI. 1993. Food webs: Theory and reality. *Advances in Ecological Research* 24:187–239.

HALVORSON, J. J., E. H. FRANZ, J. L. SMITH, and R. A. BLACK. 1992. Nitrogenase activity, nitrogen fixation, and nitrogen inputs by lupines at Mount St. Helens. *Ecology* 73:87–98.

HAMMOND, A. L. 1971. Mercury in the environment: Natural and human factors. *Science* 171:788–89.

HANSON, H. C. 1958. Principles concerned in the formation and classification of commmunities. *Botanical Review* 24:65–125.

HARDIN, G. 1960. The competitive exclusion principle. *Science* 131:1292–97.

HARDIN, G. 1993. *Living within limits. Ecology, economics, and population taboos.* New York/Oxford: Oxford University Press.

HARGROVE, E. C. 1989. *Foundations of environmental ethics.* Englewood Cliffs, NJ: Prentice-Hall.

HARMON, M. E., J. F. FRANKLIN, F. J. SWANSON, P. SOLLINS, S. V. GREGORY, J. D. LATTIN, N. H. ANDERSON, S. P. CLINE, N. G. AUMEN, J. R. SEDELL, G. W. LIENKAEMPER, K. CROMACK, JR., and K. W. CUMMINS. 1986. Ecology of coarse woody debris in temperate ecosystems. *Advances in Ecological Research* 15:133–302.

HARPER, J. L. 1961. Approaches to the study of plant competition. *Symposia of the Society for Experimental Biology* 15:1–39.

HARRINGTON, G. N. 1991. Effects of soil moisture on shrub seedling survival in a semi-arid grassland. *Ecology* 72:1138–49.

HARRIS, J. A. 1967. Some competitive relationships between *Agropyron spicatum* and *Bromus tectorum*. *Ecological Monographs* 37:89–111.

HARRIS, W. F., P. SOLLINS, N. T. EDWARDS, B. E. DINGER, and H. H. SHUGART. 1975. Analysis of carbon flow and productivity in a temperate deciduous forest ecosystem. In *Productivity of world ecosystems,* D. E. Reichel, J. F. Franklin and D. W. Goodall, eds., 116–23. Washington, DC: National Academy of Sciences.

HARRISON, J. M. 1984. Disposal of radioactive wastes. *Science* 226:11–14.

HAYES, P., and K. SMITH, eds. 1993. *The global Greenhouse regime. Who pays?* Tokyo: United Nations University Press.

HEMOND, H. F. 1980. Biogeochemistry of Thoreau's Bog, Concord, Massachusetts. *Ecological Monographs* 50:507–26.

HEMOND, H. F. 1983. The nitrogen budget of Thoreau's Bog. *Ecology* 64:99–109.

HENDERSON, L. J. 1913. *The fitness of the environment.* New York: The Macmillan Company.

HENDRY, P. 1988. Food and population: Beyond five billion. *Population Bulletin* 43(2):1–40.

HESTBECK, J. B. 1986. Multiple regulation states in populations of the California vole (*Microtus californicus*). *Ecological Monographs* 56:161–81.

HESTER, J. J. 1970. Ecology of the North American paleo-Indian. *BioScience* 20:213–17.

HEWETT, S. W. 1988. Predation by *Didinium nasutum*: Effects of predator and prey size. *Ecology* 69:135–45.

HILEMAN, B. 1988. The Great Lakes cleanup effort. *Chemical and Engineering News* 66:22–39.

HILEMAN, B. 1989. Global warming. *Chemical and Engineering News* 67:25–44.

HINMAN, C. W. 1984. New crops for arid lands. *Science* 225:1445–48.

HIXON, M. A., and J. P. BEETS. 1993. Predation, prey refuges, and the structure of coral-reef fish assemblages. *Ecological Monographs* 63:77–101.

HOBBS, P. V., and L. F. RADKE. 1992. Airborne studies of the smoke from the Kuwait oil fires. *Science* 256:987–91.

HOBBS, R. J., and C. H. GIMINGHAM. 1987. Vegetation, fire and herbivore interactions in heathland. *Advances* in *Ecological Research* 16:87–173.

HOFFMAN, D. J. 1990. Increase in the stratospheric background sulfuric acid aerosol mass in the past 10 years. *Science* 248:996–1000.

HOLDEN, C. 1976. Mirex: Persistent pesticide on its way out. *Science* 194:301–303.

HOLLISTER, C. D., D. R. ANDERSON, and G. R. HEATH. 1981. Subseabed disposal of nuclear wastes. *Science* 213:1321–26.

HOLLOWAY, M. 1993. Sustaining the Amazon. *Scientific American* 269(1):90–99.

HOLMEN, K. 1992. The global carbon cycle. In *Global biogeochemical cycles,* S. S. Butcher, R. J. Charlson, G. H. Orions, and G. V. Wolfe, eds., 239–62. London: Academic Press.

HORIUCHI, S. 1992. Stagnation in the decline of the world population growth rate during the 1980s. *Science* 257:761–65.

HOUGHTON, R. A., and G. M. WOODWELL. 1989. Global climatic change. *Scientific American* 260(4):36–44.

HOUGHTON, R. A., J. E. HOBBIE, J. M. MELILLO, B. MOORE, B. J. PETERSON, G. R. SHAVER, and G. M. WOODWELL. 1983. Changes in the carbon content of terrestrial biota and soils between 1860 and 1980: A net release of CO_2 to the atmosphere. *Ecological Monographs* 53:235–62.

HOWARTH, R. W. 1988. Nutrient limitation of net primary production in marine ecoystems. *Annual Review of Ecology and Systematics* 19:89–110.

HOWARTH, R. W., and J. J. COLE 1985. Molybdenum availability, nitrogen limitation, and phytoplankton growth in natural waters. *Science* 229:653–55.

HOWARTH, R. W., J. R. FRUCI, and D. SHERMAN. 1991. Inputs of sediment and carbon to an estuarine ecosystem: Influence of land use. *Ecological Applications.* 1:27–39.

HUFFAKER, C. B., K. P. SHEA, and S. G. HERMAN. 1963. Experimental studies on predation. Complex dispersion and levels of food in an acarine predator-prey interaction. *Hilgardia* 34(9):305–30.

HUNT, H. W., E. R. INGHAM, D. C. COLEMAN, E. T. ELLIOTT, and C. P. P. REID. 1988. Nitrogen limitation of production and decomposition in prairie, mountain meadow, and pine forest. *Ecology* 69:1009–16.

HURD, L. E., and R. M. EISENBERG. 1990. Arthropod community responses to manipulation of a bitrophic predator guild. *Ecology* 71:2107–14.

HURD, S. C., G. E. NASON, D. D. MYNOLD, and D. A. PERRY. 1994. Dynamics of gross nitrogen transformations in an old-growth forest: The carbon connection. *Ecology* 75:880–91.

HUTCHINSON, G. E. 1957, 1967, 1975, 1993. *A treatise on limnology*, vols. I, II, III, IV. New York: John Wiley and Sons.

HUTCHINSON, G. E. 1959. Homage to Santa Rosalia, or why are there so many kinds of animals? *American Naturalist* 93:145–59.

HUTCHINSON, G. E. 1965. *The ecological theatre and the evolutionary play.* New Haven, CT: Yale University Press.

INGHAM, R. E., J. A. TROFYMOW, E. R. INGHAM, and D. C. COLEMAN. 1985. Interactions of bacteria, fungi, and their nematode grazers: Effects on nutrient cycling and plant growth. *Ecological Monographs* 55:119–40.

IPCC (INTERGOVERNMENTAL PANEL ON CLIMATE CHANGE). 1990. *Climate change: The IPCC scientific assessment*, J. T. Houghton, G. J. Jenkins, and J. J. Ephraums, eds. Cambridge: Cambridge University Press.

IVERSEN, J. 1960. Problems of early postglacial forest development in Denmark. *Danmarks Geologiske Undersogelse, Series IV* 4:1–32.

IVERSON, R. M., B. S. HINCKLEY, R. M. WEBB, and B. HALLET. 1981. Physical effects of vehicular disturbances on arid landscapes. *Science* 212:915–16.

IVES, J. D., and R. G. BARRY, eds. 1974. *Arctic and alpine environments.* London: Methuen and Company.

JACKSON, J. B. S., J. D. CUBIT, B. D. KELLER, V. BATISTA, K. BURNS, H. M. CAFFEY, R. L. CALDWELL, S. D. GARRITY, C. D. GETTER, C. GONZALEZ, H. M. GUZMAN, K. W. KAUFMANN, A. H. KNAPP, S. C. LEVINGS, M. J. MARSHALL, R. STEGER, R. C. THOMPSON, and E. WEIL. 1989. Ecological effects of a major oil spill on Panamanian coastal marine communities. *Science* 243:37–44.

JACKSON, J. O. 1992. Nuclear time bombs. *TIME* (December 7):44–5.

JACKSON, R. M., and F. RAW. 1966. *Life in the soil.* Studies in Biology, No. 2. New York: St. Martin's Press.

JACKSON, S. T., R. P. FUTYMA, and D. A. WILCOX. 1988. A paleoecological test of a classical hydrosere in the Lake Michigan dunes. *Ecology* 69:928–36.

JACOBS, M., and R. A. A. OLDEMAN. 1987. *The topical rain forest: A first encounter.* New York: Springer-Verlag.

JACOBSON, J. 1988. Planning the global family. In *State of the world 1988. A Worldwatch Institute report on progress toward a sustainable society,* L. Starke, ed., 151–69. New York: W. W. Norton and Company.

JAFFE, D. A. 1992. The nitrogen cycle. In *Global biogeochemical cycles,* S. S. Butcher, R. J. Charlson, G. H. Orions, and G. V. Wolfe, eds., 261–84. London: Academic Press.

JAHNKE, R. A. 1992. The phosphorus cycle. In *Global biogeochemical cycles,* S. S. Butcher, R. J. Charlson, G. H. Orions, and G. V. Wolfe, eds., 301–15. London: Academic Press.

JEANNE, R. L. 1979. A latitudinal gradient in rates of ant predation. *Ecology* 60:1211–24.

JEFFERY, D. 1989. Yellowstone. The great fires of 1988. *National Geographic* 175(2):255–73.

JENNY, H. 1941. Factors of soil formation. New York: McGraw-Hill Inc.

JENNY, H. 1980. *The soil resource: Origin and behavior.* New York: Springer-Verlag.

JIN, R. L., and Z. K. CHENG. 1989. People's Republic of China. In *International handbook of pollution control,* E. J. Kormondy, ed., 357–76. Westport, CT: Greenwood Press.

JITSUKAWA, M., and C. DJERASSI. 1994. Birth control in Japan: Realities and prognosis. *Science* 265:1048–51.

JOHNSON, D. M., P. H. CROWLEY, R. E. BOHANAN, C. N. WATSON, and T. H. MARTIN. 1985. Competition among larval dragonflies: A field enclosure experiment. *Ecology* 66:119–28.

JOHNSON, E. A., and S. L. GUTSELL. 1994. Fire frequency models, methods, and interpretation. *Advances in Ecological Research* 25:239–87.

JOHNSON, E. A., and C. P. S. LARSEN. 1991. Climatically induced changes in fire frequency in the southern Canadian Rockies. *Ecology* 72:194–201.

JOHNSON, N. M. 1979. Acid rain: Neutralization within the Hubbard Brook ecosystem and regional implications. *Science* 204:497–99.

JOHNSON, S. P. 1987. *World population and the United Nations. Challenge and response.* Cambridge: Cambridge University Press.

JONES, P. D., and T. M. L. WIGLEY. 1990. Global warming trends. *Scientific American* 263(2):84–91.

JORDAN, C. F. 1982. The nutrient balance of an Amazonian rain forest. *Ecology* 63:647–54.

JORDAN, C. F. 1985. *Nutrient cycling in tropical rainforest ecosystems: Principles and their application in management and conservation.* New York: John Wiley and Sons.

JORDAN, C. F. 1986. Ecological effects of clearcutting. In *Ecological knowledge and environmental problem-solving,* Committee on the Applications of Ecological Theory to Environmental Problems, Commission on Life Sciences, National Research Council, ed., 345–57. Washington, DC: National Academy Press.

JÖRGENSEN, B. B. 1990. A thiosulfate shunt in the sulfur cycle of marine sediments. *Science* 249:152–54.

JÖRGENSEN, B. B., M. F. ISAKSEN, and H. W. JANNASCH. 1992. Bacterial sulfate reduction above 100°C in deep-sea hydrothermal vent sediments. *Science* 258:1756–57.

JOSHI, M. M., and J. P. HOLLIS. 1977. Interaction of *Beggiatoa* and rice plant: Detoxification of hydrogen sulfide in the rice rhizosphere. *Science* 195:178–79.

JUDAY, C. 1940. The annual energy budget of an inland lake. *Ecology* 21:438–50.

KAHLIL, M. A. K., and R. A. RASMUSSEN. 1984. Carbon monoxide in the Earth's atmosphere: Increasing trend. *Science* 224:54–56.

KALKAT, G. S., R. H. DAVIDSON, and C. L. BRASS. 1961. The effect of controlled temperature and humidity on the residual life of certain insecticides. *Journal of Economic Entomology* 54:1186–90.

KANESHIRO, K. Y. 1988. Speciation in the Hawaiian *Drosophila*. *BioScience* 38:258–63.

KANESHIRO, K. Y., and C. R. B. BOAKE. 1987. Sexual selection and speciation: Issues raised by Hawaiian *Drosophila*. *Trends in Ecology and Evolution* 2:207–12.

KAREIVA, P., ed. 1994. Space: The final frontier for ecological theory. *Ecology* 75:1–47.

KARL, T. R., R. R. HEIM, JR., and R. G. QUAYLE. 1991. The Greenhouse Effect in central North America: If not now, when? *Science* 251:1058–61.

KASTING, J. F., and T. P. ACKERMAN. 1986. Climatic consequences of very high carbon dioxide levels in the Earth's early atmosphere. *Science* 234:1383–85.

KASTING, J. F., O. B. TOON, and J. B. POLLACK. 1988. How climate evolved on the terrestrial planets. *Scientific American* 258(2):90–97.

KAUFFMAN, J. B., R. L. SANFORD, JR., D. L. CUMMINGS, I. H. SALCEDO, and E. V. S. B. SAMPAIO. 1993. Biomass and nutrient dynamics associated with slash fires in neotropical dry forests. *Ecology* 74:140–51.

KAUPPI, P. E., K. MIELIKAINEN, and K. KUUSELA. 1992. Biomass and carbon budget of European forests, 1971 to 1990. *Science* 256:70–79.

KEELEY, J. E., and D. R. SANDQUIST. 1991. Diurnal photosynthesis cycle in CAM and non-CAM seasonal-pool aquatic macrophytes. *Ecology* 72:716–27.

KEELEY J. E., and S. C. KEELEY. 1988. Chaparral. In *North American terrestrial vegetation,* M. G. Barbour and W. D. Billings, eds., 165–208. Cambridge: Cambridge University Press.

KELLOGG, W. W., R. D. CADLE, E. R. ALLEN, A. L. LAZRUS, and E. A. MARTELL. 1972. The sulfur cycle. *Science* 75:587–96.

KENNISH, M. J. 1990. *Ecology of estuaries. Vol. II. Biological aspects.* Boca Raton, FL: CRC Press.

KENT, M. M., and A. LARSON. 1982. *Family size preferences: Evidence from world fertility surveys.* Washington, DC: Population Reference Bureau.

KERR, R. A. 1979. Global pollution: Is the Arctic haze actually industrial smog? *Science* 205:290–93.

KERR, R. A. 1981a. Mount St. Helens and a climate quandary. *Science* 211:371–74.

KERR, R. A. 1981b. Pollution of the Arctic atmosphere confirmed. *Science* 212:1013–14.

KERR, R. A. 1987a. Halocarbons linked to ozone hole. *Science* 236:1182–83.

KERR, R. A. 1987b. Winds, pollutants drive ozone hole. *Science* 238:156–58.

KERR, R. A. 1988a. The weather in the wake of El Niño. *Science* 240:883.

KERR, R. A. 1988b. Ozone hold bodes ill for the globe. *Science* 241:785–86.

KERR, R. A. 1989a. Ozone hits bottom again. *Science* 246:324.

KERR, R. A. 1989b. Arctic ozone is poised for a fall. *Science* 243:1007–1008.

KERR, R. A. 1989c. Greenhouse skeptic out in the cold. *Science* 246:1118–19.

KERR, R. A. 1990a. Ozone destruction closer to home. *Science* 247:1297.

KERR, R. A. 1990b. New Greenhouse report puts down dissenters. *Science* 249:481–82.

KERR, R. A. 1990c. Another deep Antarctic ozone hole. *Science* 250:368.

KERR, R. A. 1991. Global temperature hits record again. *Science* 251:274.

KERR, R. A. 1993. Pinatubo global cooling on target. *Science* 259:594.

KETCHUM, B. H., ed. 1983. *Estuaries and enclosed seas.* Amsterdam: Elsevier Science Publishers.

KEYFITZ, N. 1989. The growing human population. *Scientific American* 261(3):118–26.

KIEHL, J. T., and B. P. BRIEGLEB. 1993. The relative roles of sulfate aerosols and Greenhouse gases in climate forcing. *Science* 260:311–14.

KIKUCHI, W. K. 1976. Prehistoric Hawaiian fish ponds. *Science* 193:295–99.

KLEMM, E. B. et al. 1990. *The fluid Earth. Physical science and technology of the marine environment.* Curriculum Research and Development Group, College of Education, University of Hawaii, Honolulu.

KLING, G. W., M. A. CLARK, H. R. COMPTON, J. D. DEVINE, W. C. EVANS, A. M. HUMPHREY, E. D. KOENIGSBERG, J. P. LOCKWOOD, M. I. TUTTLE, and G. N. WAGNER. 1987. The 1986 Lake Nyos gas disaster in Cameroon, West Africa. *Science* 236:169–75.

KNIGHT, R. R., and L. L. EBERHARDT. 1985. Population dynamics of Yellowstone grizzly bears. *Ecology* 66:323–34.

KNIGHT, T. J., and P. J. LANGSTON-UNKEFER. 1988. Enhancement of symbiotic dinitrogen fixation by a toxin-releasing plant pathogen. *Science* 241:951–54.

KNIPLING, E. F. 1959. Sterile-male method of population control. *Science* 130:902–904.

KOEPPEN, W. 1900. Versuch einer Klassifikation der Klimate, Vorzugsweise nach ihren Beziehungen zur Pflanzenwelt. *Geographische Zeitschrift* 6:593–611.

KOEPPEN, W. 1918. Klassifikation der Klimate nach Temperature, Niederschlat, und jahres Lauf. *Petermann's Mitteilungen* 64:193–203, 243–48.

KORMONDY, E. J. 1959. The systematics of *Tetragoneuria*, based on ecological, life history, and morphological evidence (Odonata: Corduliidae). *Miscellaneous Publications of the Museum of Zoology, University of Michigan* 107:1–79.

KORMONDY, E. J. 1965. Uptake and loss of zinc-65 in the dragonfly *Plathemis lydia. Limnology and Oceanography* 10:427–33.

KORMONDY, E. J. 1969. Comparative ecology of sandspit ponds. *American Midland Naturalist* 82:28–61.

KORMONDY, E. J. 1970. Ecology and the environment of man. *BioScience* 20:751–54.

KORMONDY, E. J. 1974. Natural and human ecosystems. In *Human ecology,* F. Sargent II, ed., 27–43. Amsterdam: North Holland Publishing Company.

KORMONDY, E. J. 1981. Human intervention into natural ecosystems: The scientific background to moral choice. In *Ecological consciousness: Essays from the Earthday X Colloquium, University of Denver, April 21–24, 1980,* R. C. Schultz and J. D. Hughes, eds., 23–42. Washington, DC: University Press of America.

KORMONDY, E. J. 1982. Values and choices. In *New directions in human biology,* F. M. Hickman and J. B. Kahle, eds., 15–28. Reston, VA: National Association of Biology Teachers.

KORMONDY, E. J. 1984. Human ecology: An introduction for biology teachers. *American Biology Teacher* 46:325–29.

KORMONDY, E. J., ed. 1989. *International handbook of pollution control.* Westport, CT: Greenwood Press.

KORMONDY, E. J., and J. F. MCCORMICK. 1981. *Handbook of contemporary developments in world ecology.* Westport, CT: Greenwood Press.

KORNER, C., and J. A. ARNONE III. 1992. Response to elevated carbon dioxide in artificial tropical ecosystems. *Science* 257:1672–75.

KRATZ, T. K., and C. B. DEWITT. 1986. Internal factors controlling peatland-lake ecosystem development. *Ecology* 67:100–107.

KRAUSKOPF, K. B. 1990. Disposal of high-level nuclear waste: Is it possible? *Science* 249:1231–32.

KREBS, C. J., B. L. KELLER, and R. H. TAMARIN. 1969. *Microtus* population biology: Demographic changes in fluctuating populations of *M. ochrogaster* and *M. pennsylvanicus* in southern Indiana. *Ecology* 50:587–607.

KRUG, E. D., and C. R. FRINK. 1983. Acid rain on acid soil: A new perspective. *Science* 221:520–25.

KRUPNICK, A. J., and P. R. PORTNEY. 1991. Controlling urban air pollution: A benefit-cost assessment. *Science* 252:522–28.

KUHN, T. 1970. *The structure of scientific revolutions.* 2d ed. Chicago: University of Chicago Press.

KUNO, E. 1987. Principles of predator-prey interaction in theoretical, experimental, and natural population systems. *Advances in Ecological Research* 16:249–337.

KUSLER, J. A., W. J. MITSCH, and J. S. LARSON. 1994. Wetlands. *Scientific American* 270(1):64B–70.

LA BRECQUE, M. 1989. Detecting climate change I. *Mosaic* 20(4):2–17.

LA BRECQUE, M. 1990. Clouds and climate: A critical and unknown in the global equation. *Mosaic* 21(2):2–11.

LACHENBRUCH, A. H., and B. V. MARSHALL. 1986. Changing climate: Geothermal evidence from permafrost in the Alaskan Arctic. *Science* 234:689–96.

LAJTHA, K., and W. H. SCHLESINGER. 1988. The biogeochemistry of phosphorus cycling and phosphorus availability along a desert soil chronosequence. *Ecology* 69:24–39.

LAMARCHE, V. C., JR. 1974. Paleoclimatic inferences from long tree-ring records. *Science* 183:1043-48.

LAMARCHE, V. C., JR., D. A. GRAYBILL, H. C. FRITTS, and M. R. ROSE. 1984. Increasing atmospheric carbon dioxide: Tree ring evidence for growth enhancement in natural vegetation. *Science* 225:1019–21.

LAMB, H. F. 1985. Palynological evidence for postglacial change in the position of tree limit in Labrador. *Ecological Monographs* 55:241–258.

LANGER, W. L. 1972. Checks on population growth: 1750–1850. *Scientific American* 226(2):92–99.

LANGFORD, A. O., and F. C. FEHSENFELD. 1992. Natural vegetation as a source or sink for atmospheric ammonia: A case study. *Science* 255:581–83.

LAPORTE, T. R. 1978. Nuclear waste: Increasing scale and sociopolitical impacts. *Science* 201:22–28.

LARSON, D. 1993. The recovery of Spirit Lake. *American Scientist* 81:166–77.

LARSON, W. E., F. J. PIERCE, and R. H. DOWDY. 1983. The threat of soil erosion to long-term crop production. *Science* 219:458–65.

LAUFF, G. A., ed. 1967. *Estuaries.* Publication No. 83. Washington, DC: American Association for the Advancement of Science.

LAWSON, G. J., T. V. CALLAGHAN, and R. SCOTT. 1984. Renewable energy from plants: Bypassing fossilization. *Advances in Ecological Research* 14:57–114.

LAWTON, J. H. 1989. Food webs. In *Ecological concepts*, J. M. Cherritt, ed., 43–78. Oxford: Blackwell Scientific Publishers.

LEAN, D. R. S. 1973. Phosphorus dynamics in lake water. *Science* 179:678–79.

LEHMAN, J. T. 1986. Control of eutrophication in Lake Washington. In *Ecological knowledge and environmental problem-solving,* Committee on Applications of Ecological Theory to Environmental Problems, Commission on Life Sciences, National Research Council, ed., 301–44. Washington, DC: National Academy Press.

LEITH, H., and M. J. A. WERGER, eds. 1989. *Tropical rain forest ecosystems: Biogeographical and ecological studies. Ecosystems of the world.* vol. 14B. Amsterdam: Elsevier Science Publishers.

LEITH, H., and R. H. WHITTAKER, eds. 1975. *The primary productivity of the biosphere.* New York: Springer-Verlag.

LEMONICK, M. D. 1989. The Chernobyl cover-up. *TIME* (November 13):73.

LEMONICK, M. D. 1992. The ozone vanishes. *TIME* (February 17):60–63.

LEOPOLD, A. 1949. The land ethic. *A Sand County almanac.* New York: Oxford University Press.

LESCHINE, S. B., K. HOLWELL, and E. CANALE-PAROLE. 1988. Nitrogen fixation by anaerobic cellulolytic bacteria. *Science* 242:1157–59.

LEVERICH, W. J., and D. A. LEVIN. 1979. Age-specific survivorship and reproduction in *Phlox Drummondii. American Naturalist* 113:881–903.

LEVIN, S. A., ed. 1992. Orchestrating environmental research and assessment. *Ecological Applications* 2:103–38.

LEVIN, S. A., ed. 1993. Preserving biodiversity. *Ecological Applications* 3:201–20.

LEVITAN, D. R. 1989. Density-dependent size regulation in *Diadema antillarum*: Effects on fecundity and survivorship. *Ecology* 70:1414–24.

LEWIN, R. 1983. Finches show competition in ecology. *Science* 219:1411–12.

LEWIS, W. M., JR. 1986. Nitrogen and phosphorus runoff losses from a nutrient poor tropical moist forest. *Ecology* 67:1275–82.

LEWIS, W. M., JR. 1987. Tropical limnology. *Annual Review of Ecology and Systematics* 18:159–84.

LEWIS, W. M., JR. 1988. Primary production in the Orinoco River. *Ecology* 69:679–92.

LEWIS, W. M., JR,. and M. C. Grant. 1980. Acid precipitation in the western United States. *Science* 207:176–77.

LIGHTFOOT, D. C., and W. B. G. WHITFORD. 1987. Variation in insect densities on desert creosotebush: Is nitrogen a factor? *Ecology* 68:547–57.

LIKENS, G. E. 1981. *Some perspectives on the major biogeochemical cycles.* New York: John Wiley and Sons.

LIKENS, G. E., ed. 1985. *An ecosystem approach to aquatic ecology. Mirror Lake and its environment.* New York: Springer-Verlag.

LIKENS, G. E. 1992. *The ecosystem approach: Its use and abuse. Excellence in Ecology.* vol. 3. Oldendorf/Luhe, Germany: Ecology Institute.

LIKENS, G. E., F. H. BORMANN, R. S. PIERCE, and W. A. REINERS. 1978. Recovery of a deforested ecosystem. *Science* 199:492–96.

LIKENS, G. E., F. H. BORMANN, R. S. PIERCE, J. S. EATON, and N. M. JOHNSON. 1977. *Biogeochemistry of a forested ecosystem.* New York: Springer-Verlag.

LIKENS, G. E., R. F. WRIGHT, J. N. GALLOWAY, and T. J. BUTLER. 1979. Acid rain. *Scientific American* 241(4):43–51.

LINDBERG, S. E., R. C. HARRISS, and R. R. TURNER. 1982. Atmospheric deposition of metals to forest vegetation. *Science* 215:1609–11.

LINDEMAN, R. L. 1942. The trophic-dynamic aspect of ecology. *Ecology* 23:399–418.

LINDEN, E. 1991. Getting blacker every day. *TIME* (May 27):50–51.

LINDEN, E. 1994. Ancient creatures in a lost world. *TIME* (June 20):52–54.

LINDROTH, R. L., K. K. KINNEY, and C. L. PLATZ. 1993. Responses of deciduous trees to elevated atmospheric CO_2: Productivity, phytochemistry, and insect performance. *Ecology* 74:763–77.

LINDSTROM, E. R., H. ANDREN, P. ANGELSTAM, G. CEDERLUND, B. HORNFELDT, L. JADERBERG, P.-A. LEMNELL, B. MARTINSSON, K. SKOLD, and J. E. SWENSON. 1994. Disease reveals the predator: Sarcoptic mange, red fox predation, and prey populations. *Ecology* 75:1042–49.

LITTLE, M. A., and G. E. B. MORAN, JR. 1976. *Ecology, energetics and human variability.* Dubuque, IA: Wm. C. Brown Company, Publisher.

LIVELY, C. M., P. T. RAIMONDI, and L. F. DELPH. 1993. Intertidal community structure: Space-time interactions in the northern Gulf of California. *Ecology* 74:162–73.

LOCKE, A., and W. G. SPRULES. 1994. Effects of lake acidification and recovery on the stability of zooplankton food webs. *Ecology* 75:498–506.

LOCKERETZ, W. 1978. The lessons of the dust bowl. *American Scientist* 66:560–69.

LOGAN, W. P. D. 1953. Mortality in the London fog incident, 1952. *Lancet* 264:336–38.

LONDON, J., and J. KELLEY. 1974. Global trends in total atmospheric ozone. *Science* 184:987–89.

LOOPE, L. L. 1992. An overview of problems with introduced plant species in national parks and biosphere reserves of the United States. In *Alien plant invasions in native ecosystems of Hawaii: Management and research*, C. P. Stone, C. W. Smith and J. T. Tunison, eds., 3–28. Honolulu, HI: Cooperative National Park Resources Studies Unit.

LOPEZ, L., J. MAIER, and D. THOMPSON. 1989. Playing with fire. *TIME* (September 18):76–85.

LOTKA, A. J. 1922. The stability of normal age distribution. *Proceedings of the National Academy of Sciences* 8:339–45.

LOTKA, A. J. 1925. *Elements of physical biology.* Baltimore: Williams and Wilkins.

LOVELOCK, J. E. 1979. *Gaia.* Cambridge: Cambridge University Press.

LOVELOCK, J. E. 1988. *The ages of Gaia.* Cambridge: Cambridge University Press.

LOVELOCK, J. E., and L. MARGULIS. 1974. Atmospheric homeostasis, by and for the biosphere: The Gaia hypothesis. *Tellus* 26:1–10.

LOWRANCE, R. R., R. A. LEONARD, L. E. ASMUSSEN, and R. L. TODD. 1985. Nutrient budgets for agricultural watersheds in the Southeastern coastal plain. *Ecology* 66:287–96.

LUBCHENCO, J., A. M. OLSON, L. B. BRUBAKER, S. R. CARPENTER, M. M. HOLLAND, S. P. HUBBELL, S. A. LEVIN, J. A. MACMAHON, P. A. MATSON, J. M. MELILLO, H. A. MOONEY, C. H. PETERSON, H. R. PULLIAM, L. A. REAL, P. J. REGAL, and P. G. RISSER. 1991. The sustainable biosphere initiative: An ecological research agenda. *Ecology* 72:371–412.

LUCAS, Y., F. J. LUIZAO, A. CHAUVEL, J. ROUILLER, and D. NAHON. 1993. The relation between biological activity of the rain forest and mineral composition of soils. *Science* 260:521–23.

LUCK, R. F., and H. PODOLER. 1985. Competitive exclusion of *Aphytis lingnanensis* by *A. melinus*: Potential role of host size. *Ecology* 66:904–13.

LUCKINBILL, L. S. 1979. Regulation, stability, and diversity in a model experimental microcosm. *Ecology* 60:1098–102.

LUSSENHOP, J. 1981. Microbial and microarthropod detrital processing in a prairie soil. *Ecology* 62:964–72.

LUTHER, G. W., III, T. M. CHURCH, J. R. SCUDLARK, and M. COSMAN. 1986. Inorganic and organic sulfur cycling in salt-marsh pore waters. *Science* 232:746–49.

MACARTHUR, R. H. 1965. Patterns of species diversity. *Biological Reviews* 40:510–33.

MACARTHUR, R. H., and J. CONNELL. 1966. *The biology of populations.* New York: John Wiley and Sons.

MACDONALD, R. W., and E. C. CARMACK. 1991. Age of Canada Basin deep waters: A way to estimate primary production for the Arctic Ocean. *Science* 254:1348–50.

MACFADYEN, A. 1957. *Animal ecology: Aims and methods.* London: Sir Isaac Pitman and Sons.

MACINTYRE, F. 1974. The top millimeter of the ocean. *Scientific American* 230(5):62–77.

MACKENZIE, J. J., and M. T. EL-ASHRY. 1988. *Ill winds: Airborne pollution's toll on trees and crops.* Washington, DC: World Resources Institute.

MACKUN, I. R., D. J. LEOPOLD, and D. J. RAYNAL. 1994. Short-term responses of wetland vegetation after liming of an Adirondack watershed. *Ecological Applications* 4:535–43.

MACLULICH, D. A. 1937. Fluctuations in the numbers of the varying hare. *University of Toronto Studies, Biological Series No. 43.*

MACMAHON, J. A. 1988. Warm deserts. In *North American terrestrial vegetation,* M. G. Barbour and W. D. Billings, eds., 232–65. Cambridge: Cambridge University Press.

MADATI, P. J., and E. J. KORMONDY. 1989. Introduction. In *International handbook of pollution control,* E. J. Kormondy, ed., 1–17. Westport, CT: Greenwood Press.

MADIGAN, M. W., J. D. WALL, and H. GEST. 1979. Dark anaerobic dinitrogen fixation by a photosynthetic microorganism. *Science* 204:1428–29.

MAGNUSON, E. 1990. A cover-up on Agent Orange? *TIME* (July 23):27–28.

MAGNUSON, J. J., H. A. REGIER, W. J. CHRISTIE, and W. C. SONZONGNI. 1980. To rehabilitate and restore Great Lakes ecosystems. In *The recovery process in damaged ecosystems,* J. Cairns, Jr., ed., 95–112. Ann Arbor, MI: Ann Arbor Science Publishers.

MALTHUS, R. T. 1798. *An essay on the principle of population as it affects the future improvement of society.* London: Johnson.

MAMLOUK, M. 1982. *Knowledge and use of contraception in twenty developing countries.* Washington, DC: Population Reference Bureau.

MANN, C. 1991. Lynn Margulis: Science's unruly Earth mother. *Science* 252:378–81.

MANZER, L. E. 1990. The CFC-ozone issue: Progress on the development of alternatives to CFCs. *Science* 249:31–35.

MARES, M. A. 1992. Neotropical mammals and the myth of Amazonian biodiversity. *Science* 255:976–79.

MARSHALL, E. 1990. Radiation exposure: Hot legacy of the cold war. *Science* 249:474.

MARTIN, C. L. 1981. The American Indian as miscast ecologist. In *Ecological consciousness: Essays from the Earthday X Colloquium, University of Denver, April 21–24, 1980,* R. C. Schultz and J. D. Hughes, eds., 137–48. Washington, DC: University Press of America.

MARTIN, C. L. 1992. *In the spirit of the Earth. Rethinking history and time.* Baltimore and London: Johns Hopkins University Press.

MARTIN, L. G. 1991. Population aging policies in East Asia and the United States. *Science* 251:527–31.

MARTINEZ, L. A., M. W. SILVER, J. M. KING, and A. L. ALDREDGE. 1983. Nitrogen fixation by floating diatom mats: A source of new nitrogen to oligotrophic ocean waters. *Science* 221:152–54.

MARTINEZ, N. D. 1992. Constant connectance in community food webs. *American Naturalist* 139:1208–18.

MARX, J. L. 1977. Nitrogen fixation: Prospects for genetic manipulation. *Science* 196:638–41.

Matson, P. A., ed. 1991. The future of remote sensing in ecological studies. *Ecology* 72:1917–45.

MATSON, P. A., ed. 1992. The relative contributions of top-down and bottom-up forces in population and community ecology. *Ecology* 73:723–65.

MATSON, P. A., and A. BERRYMAN, eds. 1992. Ratio-dependent predator-prey theory. *Ecology* 73:1529–66.

MATSON, P. A., and R. D. BOONE. 1984. Natural disturbance and nitrogen mineralization: Wave-form dieback of mountain hemlock in the Oregon Cascades. *Ecology* 65:1511–16.

MATSON, P. A., and M. D. HUNTER. 1992. The relative contributions of top-down and bottom-up forces in population and community ecology. *Ecology* 73:723–65.

MATSON , P. A., P. M. VITOUSEK, J. J. EWEL, M. J. MAAZZARINO, and G. P. ROBERTSON. 1987. Nitrogen transformations following tropical forest felling and burning on a volcanic soil. *Ecology* 68:491–502.

MATTSON, W. J., Jr. 1980. Herbivory in relation to plant nitrogen content. *Annual Review of Ecology and Systematics* 11:119–61.

MAUGH, T. H., II. 1979. Restoring damaged lakes. *Science* 203:425–27.

MAUGH, T. H., II. 1980. Ozone depletion would have dire effects. *Science* 207:394–95.

MAUGH, T. H., II. 1982. New link between ozone and cancer. *Science* 216:396–97.

MAUGH, T. H., II. 1984. What is the risk from chlorofluorocarbons? *Science* 223:1051–52.

MAY, R. M. 1973. *Stability and complexity in model ecosystems.* Princeton, NJ: Princeton University Press.

MAY, R. M. 1983. Parastic infections as regulators of animal populations. *American Scientist* 71:36–45.

MAY, R. M. 1984. An overview: Real and apparent patterns in community structure. In *Ecological communities: Conceptual issues and the evidence*, D. R. Strong, Jr., D. Simberloff, L. G. Abele, and A. B. Thistle, eds., 3–16. Princeton, NJ: Princeton University Press.

MAY, R. M. 1988. How many species are there on Earth? *Science* 241:1441–49.

MAY, R. M. 1992. How many species inhabit the Earth? *Scientific American* 267(4):42–48.

MCAFEE, K. 1990. Why the Third World goes hungry. *Commonweal* (June 15):380–85.

MCCLURE, M. S. 1980. Foliar nitrogen: A basis for host suitability for elongate hemlock scale, *Fiorinia externa* (Homoptera: Diaspididae). *Ecology* 61:72–79.

MCCONNELL, E. E., G. W. LUCIER, R. C. RUMBAUGH, P. W. ALBRO, D. J. HARVIN, J. R. HASS, and M. W. HARRIS. 1984. Dioxin in soil: Bioavailability after ingestion by rats and guinea pigs. *Science* 223:1077–79.

McCORMICK, J. F., and F. B. GOLLEY. 1966. Irradiation of natural vegetation. An experimental facility, procedures and dosimetry. *Health Physics* 12:1467–74.

McCUNE, B. M., and G. COTTAM. 1985. The successional status of a southern Wisconsin oak woods. *Ecology* 66:1270–78.

McDOWELL, W. H., and G. E. LIKENS. 1988. Origin, composition, and flux of dissolved organic carbon in the Hubbard Brook Valley. *Ecological Monographs* 58:177–95.

McELROY, M. B., and R. J. SALAWITCH. 1989. Changing composition of the global stratosphere. *Science* 243:763–70.

McGOWRAN, B. 1990. Fifty million years ago. *American Scientist* 78:30–39.

McINTOSH, R. P. 1985. *The background of ecology. Concept and theory.* New York: Cambridge University Press.

McKEE, D. J., ed. 1994. *Tropospheric ozone: Human health and agricultural impacts.* Boca Raton, FL: Lewis Publishers.

McLAUGHLIN, S. B., and R. J. NORBY. 1991. Atmospheric pollution and terrestrial vegetation: Evidence of changes, linkages, and significance to selection processes. In *Ecological genetics and air pollution*, G. E. Taylor, L. F. Pitelka, and M. T. Clegg, eds., 61–101. New York: Springer-Verlag.

McLENDON, T., and E. F. REDENTE. 1991. Nitrogen and phosphorus effects on secondary succession dynamics on a semi-arid sagebrush site. *Ecology* 72:2016–24.

McLUSKY, D. 1989. *The estuarine ecosystem.* 2d ed. New York: Chapman and Hall.

McMILLAN, C. 1959. The role of ecotypic variation in the distribution of the central grassland of North America. *Ecological Monographs* 29:285–308.

McNAUGHTON, S. J. 1985. Ecology of a grazing ecosystem: The Serengeti. *Ecological Monographs* 55:259–94.

McPHERSON, J. K., and C. H. MULLER. 1969. Alleopathic effects of *Adenostoma fasciculatum*, "chamise," in the California chaparral. *Ecological Monographs* 39:177–98.

MEDVEDEV, G. 1991. *The truth about Chernobyl.* New York: Basic Books.

MEFFE, G. K. 1984. Effects of abiotic disturbance on coexistence of predator-prey fish species. *Ecology* 65:1525–34.

MERLIN, M. D., and J. O. JUVIK. 1992. Relationships among native and alien plants on Pacific islands with and without significant human disturbance and feral ungulates. In *Alien plant invasions in native ecosystems of Hawaii: Management and research*, C. P. Stone, C. W. Smith, and J. T. Tusison, eds., 597–624. Honolulu, HI: Cooperative National Park Resources Studies Unit.

MERTZ, W. 1981. The essential trace elements. *Science* 213:13.

MEYER, J. L., W. H. McDOWELL, T. L. BOTT, J. W. ELWOOD, C. ISHIZAKI, J. M. MELACK, B. L. PECKARSKY, B. J. PETERSON, and P. A. RUBLEE. 1988. Elemental dynamics in streams. *Journal of the North American Benthological Society* 7:410–32.

Meyers, J. H. 1988. Can a general hypothesis explain population cycles of forest lepidoptera? *Advances in Ecological Research* 18:179–242.

MIETTINEN, J. K. 1969. Enrichment of radioactivity by arctic ecosystems in Finnish Lapland. In *Proceedings of the 2d National Symposium on Radioecology*, D. J. Nelson and F. C. Evans, eds., 23–31. Springfield, VA: Clearinghouse of Federal Scientific and Technical Information.

MINNICH, R. A. 1983. Fire mosaics in southern California and northern Baja California. *Science* 219:1287–94.

MINNIS, P., E. F. HARRISON, L. L. STOWE, G. G. GIBSON, F. M. DENN, D. R. DOELLING, and W. L. SMITH, JR. 1993. Radiative climate forcing by the Mount Pinatubo eruption. *Science* 259:1411–15.

MITCHELL, G. J. 1991. *World on fire. Saving an endangered Earth.* New York: Charles Scribner's Sons, Macmillan Publishing Company.

MOFFAT, A. S. 1990a. Nitrogenase structure revealed. *Science* 250:1513.

MOFFAT, A. S. 1990b. Nitrogen-fixing bacteria find new partners. *Science* 250:910–12.

MOFFAT, A.S. 1993. New chemicals seek to outwit insect pests. *Science* 261:650–51.

MOHNEN, V. A. 1988. The challenge of acid rain. *Scientific American* 250(2):30–38.

MOLINA, M. J., and L. T. MOLINA. 1992. Stratospheric ozone. In *The science of global change. The impact of human activities on the environment,* D.A. Dunnette and R. J. O'Brien, eds., 24–35. Washington, DC: American Chemical Society.

MOLINA M. J., and F. S. ROWLAND. 1974. Stratospheric sink for chlorofluoromethanes: Chlorine atom-catalyzed destruction of ozone. *Nature* 249:810–12.

MONSON, R. K. 1989. On the evolutionary pathways resulting in C_4 photosynthesis and crassulacean acid metabolism (CAM). *Advances in Ecological Research* 19:58–110.

MOONEY, H. A. 1991a. Emergence of the study of global ecology: Is terrestrial ecology an impediment to progress? *Ecological Applications* 1:2–5.

MOONEY, H. A. 1991b. Biological response to climate change: An agenda for research. *Ecological Applications* 1:112–17.

MOONEY, H. A., and W. D. BILLINGS. 1961. Comparative physiological ecology of arctic and alpine populations of *Oxyria digyna. Ecological Monographs* 31:1–29.

MOONEY, H. A., T. M. BONNICKSEN, N. L. CHRISTENSEN, J. E. LOTAN, and W. A. REINERS. 1981. *Fire regimes and ecosystem properties. Proceedings of the conference.* General Technical Report WO-26. Washington, DC: U.S. Department of Agriculture, Forest Service.

MOONEY, H. A., P. M. VITOUSEK, and P. A. MATSON. 1987. Exchange of materials between terrestrial ecosystems and the atmosphere. *Science* 238:926–32.

MOORE, L. G., P. W. VAN ARSDALE, J. E. GLITTENBERG, and R. A. ALDRICH. 1980. *The biocultural basis of health.* St. Louis, MO: C. V. Mosby Company.

MOORE, T. R. 1981. Controls on the decomposition of organic matter in subarctic woodland soils. *Soil Science* 131:107–13.

MOORE, T. R. 1984. Litter decomposition in a subarctic spruce-lichen woodland, east Canada. *Ecology* 65:299–308.

MOOREHEAD, A. 1969. *Darwin and the Beagle.* New York: Harper and Row.

MORENO, J. M., and W. C. OECHEL. 1991. Fire intensity effects on germination of shrubs and herbs in southern California chaparral. *Ecology* 72:1993–2004.

MORIARTY, F. 1988. *Ecotoxicology. The study of pollutants in ecosystems.* 2d ed. New York: Academic Press.

MOUNT, G. H., S. SOLOMON, R. W. SANDERS, R. O. JAKOUBEK, and A. L. SCHMELTEKOPF. 1988. Observations of stratospheric NO_2 and O_3 at Thule, Greenland. *Science* 242:555–58.

MROZ, E. J., and W. H. ZOLLER. 1975. Composition of atmospheric particulate matter from the eruption of Heimaey, Iceland. *Science* 190:461–64.

MUELLER-DOMBOIS, D. 1988a. Perspectives for an etiology of stand-level dieback. *Annual Review of Ecology and Systematics* 17:221–44.

MUELLER-DOMBOIS, D. 1988b. Canopy dieback and ecosystem processes in the Pacific area. In *Proceedings of the XIV International Botanical Congress,* W. Greuter and B. Zimmer, eds., 445–65. Taunus: Koeltz Verlag.

MULHOLLAND, P. J. 1981. Organic carbon flow in a swamp-stream ecosystem. *Ecological Monographs* 51:307–22.

MULHOLLAND, P. J., J. D. NEWBOLD, J. W. ELWOOD, L. A. FERREN, and J. R. WEBSTER. 1985. Phosphorus spiralling in a woodland stream: Seasonal variations. *Ecology* 66:1012–23.

MULLER, C. H. 1969. The role of chemical inhibition (allelopathy) in vegetational composition. *Bulletin of the Torrey Botanical Club* 93:332–51.

MURDOCH, W. W. 1994. Population regulation in theory and practice. *Ecology* 75:271–85.

MURPHY, P. E., and A. E. LUGO. 1986. Ecology of tropical dry forest. *Annual Review of Ecology and Systematics* 17:67–88.

MURPHY, T. P., D. R. S. LEAN, and C. NALEWAJKO. 1976. Blue-green algae: Their excretion of iron-selective chelators enables them to dominate other algae. *Science* 192:900–902.

MURRAY, J. W., R. T. BARBER, M. R. ROMAN, M. P. BACON, and R. A. FREELY. 1994. Physical and biological controls on carbon cycling in the equatorial. *Pacific Science* 266:58–65.

NASH, R. 1967. *Wilderness and the American mind.* New Haven, CT: Yale University Press.

National Forum. The Summer 1988 issue (vol. 68, no. 3) is devoted entirely to the issue of renewable resources.

NATIONAL RESEARCH COUNCIL. 1981. *Atmosphere-biosphere interactions: Toward a better understanding of the ecological consequences of fossil fuel combustion.* Committee on the Atmosphere and the Biosphere, Board on Agriculture and Renewable Resources, Commission on Natural Resources. Washington, DC: National Academy Press.

NATIONAL RESEARCH COUNCIL. 1983a. *Changing climate. Report of the Carbon Dioxide Assessment Committee.* Board on Atmospheric Sciences and Climate, Commission on Physical Sciences, Mathematics, and Resources. Washington, DC: National Academy Press.

NATIONAL RESEARCH COUNCIL. 1983b. *Acid deposition. Atmospheric processes in eastern North America. A review of current scientific understanding.* Committee on Atmospheric Transport and Chemical Transformation in Acid Precipitation, Environmental Studies Board. Commission on Physical Sciences, Mathematics, and Resources. Washington, DC: National Academy Press.

NATIONAL RESEARCH COUNCIL. 1985. *The effects on the atmosphere of a major nuclear exchange.* Committee on the Atmospheric Effects of Nuclear Explosions, Commission on Physical Sciences, Mathematics, and Resources. Washington, DC: National Academy Press.

NEWBOLD, J. D., J. W. ELWOOD, R. V. O'NEILL, and A. L. SHELDON. 1983. Phosphorus dynamics in a woodland stream ecosystem: A study of nutrient spiralling. *Ecology* 64:1249–65.

NEWELL, R. E., H. G. REICHLE, JR., and W. SEILER. 1989. Carbon monoxide and the burning Earth. *Scientific American* 261(4):82–88.

NICHOLSON, A. J. 1957. The self-adjustment of populations to change. *Cold Spring Harbor Symposium of Quantitative Biology* 22:153–72.

NIXON, S. W., and M. E. Q. PILSON. 1983. Nitrogen in estuarine and coastal marine ecosystems. In *Nitrogen in the marine environment,* E. J. Carpenter and D. G. Capone, eds., 565–648. New York: Academic Press.

NORMAN, C. 1982. U.S. considers ocean dumping of radwastes. *Science* 215:1217–19.

NOVELLI, P. C., K. A. MASARIE, P. P. TANS, and P. M. LANG. 1994. Recent changes in atmospheric carbon monoxide. *Science* 263:1587–90.

NRIAGU, J. O., D. A. HOLDWAY, and R. D. COKER. 1987. Biogenic sulfur and the acidity of rainfall in remote areas of Canada. *Science* 237:1189–92.

O'BRIEN, W. J. 1979. The predator-prey interaction of planktivorous fish and zooplankton. *American Scientist* 67:572–81.

O'NEILL, R. V., D. L. DEANGELIS, J. B. WADE, and T. F. H. ALLEN. 1986. *A hierarchical concept of ecosystems.* Princeton, NJ: Princeton University Press.

ODUM, E. P. 1953. *Fundamentals in ecology.* Philadelphia: W. B. Saunders.

ODUM, E. P. 1959. *Fundamentals of ecology.* 2d ed. Philadelphia: W. B. Saunders.

ODUM, E. P. 1960. Organic production and turnover in old field succession. *Ecology* 41:34–49.

ODUM, E. P. 1962. Relationships between structure and function in ecosystems. *Japanese Journal of Ecology* 12:108–112.

ODUM, E. P. 1969. The strategy of ecosystem development. *Science* 164:262–70.

ODUM, E. P. 1984. The mesocosm. *BioScience* 34:558–62.

ODUM, E. P. 1989. *Ecology and our endangered life-support systems.* Sunderland, MA: Sinauer Associates Publishers.

ODUM, H. T. 1956. Efficiencies, size of organisms and community structure. *Ecology* 37:592–97.

ODUM, H. T., 1957. Trophic structure and productivity of Silver Springs, Florida. *Ecological Monographs* 27:55–112.

ODUM, H. T., ed. 1970. *A tropical rain forest. A study of irradiation and ecology at El Verde, Puerto Rico.* Washington, DC: U.S. Atomic Energy Commission.

ODUM, W. E., and E. J. HEALD. 1972. Trophic analysis of an estuarine mangrove community. *Bulletin of Marine Science* 22:671–738.

OLMSTED, C. E. 1944. Growth and development in range grasses, IV. Photoperiodic responses in twelve geographic strains of side-oats grama. *Botanical Gazette* 106:46–74.

OLSHANSKY, S. J., B. A. CARNES, and C. CASSEL. 1990. In search of Methuselah: Estimating the upper limits to human longevity. *Science* 250:634–40.

OOSTING, H. J. 1950. *The study of plant communities.* San Francisco: W. H. Freeman and Company.

OPPENHEIMER, M., and R. BOYLE. 1990. *Dead heat. The race against the Greenhouse Effect.* New York: Basic Books.

ORR, B. K., W. W. MURDOCH, and J. R. BENCE. 1990. Population regulation, convergence, and cannabilism in *Notonecta* (Hemiptera). *Ecology* 71:68–82.

OVINGTON, J. D. 1962. Quantitative ecology and the woodland ecosystem concept. *Advances in Ecological Research* 1:103–92.

OVINGTON, J. D. 1965. Organic production turnover and mineral cycling in woodlands. *Biological Reviews* 40:295–336.

OVINGTON, J. D. 1983. *Temperate broad-leaved evergreen forests.* Amsterdam: Elsevier Science Publishers.

PACE, M. L., and E. FUNKE. 1991. Regulation of planktonic microbial communities by nutrients and herbivores. *Ecology* 72:904–14.

PAERL, H. W., and P. E. KELLER. 1979. Nitrogen-fixing *Anabaena*: Physiological adaptations instrumental in maintaining surface blooms. *Science* 204:620–22.

PAINE, R. T. 1976. Size-limited predation: An observational and experimental approach with the *Mytilus-Pisaster* interaction. *Ecology* 57:858–73.

PAINE, R. T. 1988. Food webs: Road maps of interactions or grist for theoretical development? *Ecology* 69:1648–54.

PAINE, R. T. 1993. A salty and salutary perspective on global change. In *Biotic interactions and global change*, P. M. Kareiva, J. G. Kingsolver, and R. B. Huey, eds., 347–55. Sunderland, MA: Sinauer Associates Publishers.

PALMORE, E. 1975. *The honourable elders.* Durham, NC: Duke University Press.

PARK, T. 1933. Studies in population physiology: Factors regulating initial growth of *Tribolium confusum* populations. *Journal of Experimental Zoology* 65:17–42.

PARKER, G. G. 1983. Throughfall and stemflow in the forest nutrient cycle. *Advances in Ecological Research* 13:57–133.

PARSONS, J. W. 1973. History of salmon in the Great Lakes, 1850–1970. Technical Paper No. 68. U. S. Department of the Interior, Bureau of Sport Fisheries and Wildlife, Washington, DC.

PAStorak, R. A. 1981. The effects of predator hunger and food abundance on prey selection by *Chaoborus* larvae. *Limnology and Oceanography* 25:910–25.

PATRICK, R., V. P. BINETTI, and S. G. HALTERMAN. 1981. Acid lakes from natural and anthropogenic causes. *Science* 211:446–48.

PATTERSON, R. S., D. E. WEIDHAAS, H. R. FORD, and C. S. LOFGREN. 1970. Suppression and elimination of an island population of *Culex pipiens quinquefasciatus* with sterile males. *Science* 168:1368–70.

PAUL, E. A., F. E. CLARK, and V. O. BIEDERBECK. 1979. Natural temperate grasslands: Microorganisms. In *Grassland ecosystems of the world: Analysis of grasslands and their uses,* R. T. Coupland, ed., 87–96. London/New York: Cambridge University Press.

PAVONI, B., A. MARCOMINI, A. SFRISO, R. DONAZZOLO, and A. A. ORIO. 1992. Changes in an estuarine ecosystem: The lagoon of Venice as a case study. In *The science of global change. The impact of human activities on the environment,* D. A. Dunnette and R. J. O'Brien, eds., 287–305. Washington, DC: American Chemical Society.

PEARL, R. 1928. *The rate of living.* New York: Alfred Knopf.

PEIXOTO, J. P., and M. A. KETTANI. 1973. The control of the water cycle. *Scientific American* 228(4):46–61.

PELTIER, W. R., and A. M. TUSHINGHAM. 1989. Global sea level rise and the Greenhouse Effect: Might they be connected? *Science* 244:806–10.

PENNER, J. E., R. E. DICKINSON, and C. A. O'NEILL. 1992. Effects of aerosol from biomass burning and the global radiation budget. *Science* 256:1432–34.

PENNINGS, S. C., and R. M. CALLAWAY. 1992. Salt marsh plant zonation: The relative importance of competition and physical factors. *Ecology* 73:681–90.

PETERS, R. H. 1991. *A critique for ecology.* Cambridge: Cambridge University Press.

PETERSON, B. J. 1980. Aquatic primary productivity and the ^{14}C-CO_2 method: A history of the productivity problem. *Annual Review of Ecology and Systematics* 11:359–85.

PETERSEN R. C., JR., K. W. CUMMINS, and G. M. WARD. 1989. Microbial and animal processing of detritus in a woodland stream. *Ecological Monographs* 59:21–39.

PHILANDER, G. 1989. El Niño and La Niña. *American Scientist* 77:451–59.

PHILLIPSON, J. 1966. *Ecological energetics.* London: Edward Arnold (Publisher).

PICKETT, S. T. A., and P. S. WHITE, eds. 1985. *The ecology of natural disturbance and patch dynamics.* New York: Academic Press.

PIGOTT, C. D., and K. TAYLOR. 1964. The distribution of some woodland herbs in relation to the supply of N and P in the soil. *Journal of Ecology* 52 (Supplement):175–85.

PIMENTAL, D. 1968. Population regulation and genetic feedback. *Science* 159:1432–37.

PIMENTAL, D. 1992. Pesticides and the world food supply. In *The science of global change. The impact of human activities on the environment,* D. A. Dunnette and R. J. O'Brien, eds., 309–23. Washington, DC: American Chemical Society.

PIMENTAL, D., E. C. TERHUNE, R. DYSON-HUDSON, S. ROCHEREAU, R. SAMIS, E. A. SMITH, D. DENMAN, D. REIFSCHNIEDER, and M. SHEPARD. 1976. Land degradation: Effects on food and energy resources. *Science* 149–55.

PIMM, S. L., 1982. *Food webs.* London: Chapman and Hall.

PIMM, S. L., and J. H. LAWTON, 1978. On feeding on more than one trophic level. *Nature* 275:542–44.

PIMM, S. L., and J. L. GITTLEMAN. 1992. Biological diversity: Where is it? *Science* 255:940.

PIMM, S. L., J. H. LAWTON, and J. E. COHEN. 1991. Food web patterns and their consequences. *Nature* 350:669–74.

PIPER, D. Z., and L. A. CODISPOTI. 1975. Marine phosphorite deposits and the nitrogen cycle. *Science* 188:15–18.

PIRIE, N. W. 1967. Orthodox and unorthodox methods of meeting world food needs. *Scientific American* 216(2):27–35.

PITELKA, F. A. 1964. The nutrient recovery hypothesis for Arctic microtine cycles. I. Introduction. In *Grazing in terrestrial and marine environments,* A. J. Crisp, ed., 55–56. Oxford: Blackwell.

PITELKA, L. F., and D. J. RAYNAL. 1989. Forest decline and acidic deposition. *Ecology* 70:2–10.

PLATT, R. B. 1965. Ionizing radiation and homeostasis of ecosystems. In *Ecological effects of nuclear war,* G. M. Woodwell, ed., 39–60. Upton, NY: Biology Department, Brookhaven National Laboratory.

PLATT, T., and S. SATHYENDRANATH. 1988. Oceanic primary production: Estimation by remote sensing at local and regional scales. *Science* 241:1613–20.

PLUCKNETT, D. L., and N. J. H. SMITH. 1982. Agricultural research and Third World food production. *Science* 217:215–20.

PLUCKNETT, D. L., N. J. H. SMITH, J. T. WILLIAMS, and N. M. ANISHETTY. 1987. *Gene banks and the world's food supply.* Princeton, NJ: Princeton University Press.

POLUNIN, N. V. C. 1984. The decomposition of emergent macrophytes in fresh water. *Advances in Ecological Research* 14:115–66.

POMEROY, L. R. 1960. Residence time of dissolved phosphate in natural waters. *Science* 131:1731–32.

POMEROY, L. R., E. C. HARGROVE, and J. J. ALBERTS. 1988. The ecosystem perspective. In *Concepts of ecosystems ecology,* L. R. Pomeroy and J. J. Alberts, eds., 1–17. New York: Springer-Verlag.

PORTER, S. D., and D. A. SAVIGNANO. 1990. Invasion of polygyne fire ants decimates native ants and disrupts arthropod community. *Ecology* 71:2095–106.

POSEY, M. H. 1988. Community changes associated with the spread of an introduced seagrass, *Zostera japonica. Ecology* 69:974–83.

POST, W. M., T.-H. PENG, W. R. EMANUEL, A. W. KING, V. H. DALE, and D. L. DEANGELIS. 1990. The global carbon cycle. *American Scientist* 78:310–26.

POSTMA, H., and J. J. ZIJLISTRA, eds. 1988. *Continental shelves.* Amsterdam: Elsevier Science Publishers.

POTTER, V. 1971. *Bioethics. Bridge to the future.* Englewood Cliffs, NJ: Prentice-Hall.

POTTS, W. H., and C. H. N. JACKSON. 1952. The Shinyanga game destruction experiment. *Bulletin of Entomological Research* 43:365–74.

POWER, M. E., J. C. MARKS, and M. S. PARKER. 1992. Variation in the vulnerability of prey to different predators: Community-level consequences. *Ecology* 73:2218–23.

PRENTICE, I. C., P. J. BARTLEIN, and T. WEBB III. 1991. Vegetation and climate change in eastern North America since the last glacial maximum. *Ecology* 72:2038–56.

PRICE, P. W., M. WESTOBY, B. RICE, P. R. ATSATT, R. S. FRITZ, J. N. THOMPSON, and K. MOBLEY. 1986. Parasite mediation in ecological interactions. *Annual Review of Ecology and Systematics* 17:487–505.

PRINGLE, C., G. VELLIDIS, F. HELIOTIS, D. BANDACU, and S. CRISTOFOR. 1993. Environmental problems of the Danube delta. *American Scientist* 81:350–61.

PULLIAM, H. R., and N. M. HADDAD. 1994. Human population growth and the carrying capacity concept. *Bulletin of the Ecological Society of America* 75(3):141–57.

PULLIAM, W. M. 1993. Carbon dioxide and methane exports from a southeastern floodplain swamp. *Ecological Monographs* 3:29–53.

PUTMAN, R. J. 1994. *Community ecology.* London: Chapman and Hall.

QUENSEN, J. F., III, J. M. TIEDE, and S. A. BOYD. 1988. Reductive dechlorination of polychlorinated biphenyls by anaerobic microorganisms from sediments. *Science* 242:752–54.

RAHN, K. A., and D. H. LOWENTHAL. 1985. Pollution aerosol in the northeast: Northeastern-midwestern contributions. *Science* 228:275–84.

RAINEY, R. 1967. Natural displacement of pollution from the Great Lakes. *Science* 155:1242–43.

RAUNKIAER, C. 1934. *The life-form of plants and statistical plant geography.* Oxford: Clarendon Press.

RAVEN, P. H., and E. O. WILSON. 1992. A fifty-year plan for biodiversity studies. *Science* 258:1099–100.

RAVERA, O. 1969. Seasonal variation of the biomass and biocoenotic structure of plankton of the Bay of Ispra. *Verhandlungen der Internationalen Vereinigung für Theoretische und Angewandte Limnologie* 17:237–54.

RAYNAUD, D., J. JOUZEL, J. M. BARNOLA, J. CHAPPELLAZ, R. J. DELMAS, and C. LORIUS. 1993. The ice record of Greenhouse gases. *Science* 259:926–41.

READ, P. P. 1993. *Ablaze. The story of the heroes and victims of Chernobyl.* New York: Random House.

REDFIELD, A. C. 1958. The biological control of chemical factors in the environment. *American Scientist* 46:205–21.

REGANOLD, J. P., R. I. PAPENDICK, and J. E. PARR. 1990. Sustainable agriculture. *Scientific American* 262(6):112–20.

REGIER, H. A., and W. J. HARTMAN. 1973. Lake Erie's fish community: 150 years of cultural stress. *Science* 180:1248–55.

REICE, S. R. 1994. Nonequilibrium determinants of biological community structure. *American Scientist* 82:424–35.

REICH, J. W., E. B. RASTETTER, J. M. MELILLO, D. W. KICKLIGHTER, P. A. STEUDLER, B. J. PETERSON, A. L. GRACE, B. MOORE III, and C. J. VOROSMARTY. 1991. Potential net primary production in South America: Application of a global model. *Ecological Applications* 1:399–429.

REITER, H. 1885. *Die Consolidation der Physiognomik als Versuch einer Oekologie der Gewaechse.* Graz: Leuschner & Lubensky.

REMPEL, W. C., and E. RANDOLPH. 1983. Lethal dioxin: Monster guest in chemical lab. *Los Angeles Times* (May 9):1, 8.

REPETTO, R. 1987. Population, resources, environment: An uncertain future. *Population Bulletin* 42(2):1–44.

REPETTO, R. 1990. Deforestation in the tropics. *Scientific American* 262:36–42.

REVELLE, R. 1983. Carbon dioxide and world climate. *Scientific American* 247(2):35–43.

RICE, E. L. 1974. *Allelopathy.* New York: Academic Press.

RICHARDS, P. W. 1973. The tropical rain forest. *Scientific American* 229(6):59–67.

RICHARDSON, C. J., and P. E. MARSHALL. 1986. Processes controlling movement, storage, and export of phosphorus in a fen peatland. *Ecological Monographs* 56:279–302.

RICHARDSON, R. H., J. R. ELLISON, and W. W. AVERHOFF. 1982. Autocidal control of screwworms in North America. *Science* 215:361–70.

RICKETTS, E. F., and J. CALVIN. 1952. *Between Pacific tides.* 3d ed. Revised by J. W. Hedgpeth. Forward by John Steinbeck. Stanford, CA: Stanford University Press.

RIGGIN, P. J., S. GOODE, P. M. JACKS, and R. N. LOCKWOOD. 1998. Interaction of fire and community development in chaparral of southern California. *Ecological Monographs* 58:155–76.

RIGLER, F. H. 1964. The phosphorus fractions and turnover time of inorganic phosphorus in different types of lakes. *Limnology and Oceanography* 9:511–18.

RILEY, R. H., and P. M. VITOUSEK. 1995. Nutrient dynamics and nitrogen trace gas flux during ecosystem development in montane rain forest. *Ecology* 76:292–304.

RINGWOOD, T. 1982. Immobilization of radioactive wastes in SYNROC. *American Scientist* 70:201–207.

RISSER, R. G., and W. J. PARTON. 1982. Ecosystem analysis of the tallgrass prairie: Nitrogen cycle. *Ecology* 63:1342–51.

ROBERTS, L. 1987. Radiation accident grips Goiania. *Science* 238:1028–31.

ROBERTS, L. 1988. Is there life after climate change? *Science* 242:1010–12.

ROBERTS, L. 1989a. Does the ozone hole threaten Antarctic life? *Science* 244:288–89.

ROBERTS, L. 1989b. Global warming: Blaming the sun. *Science* 246:992–93.

ROBERTS, L. 1991a. Dioxin risks revisited. *Science* 251:624–26.

ROBERTS, L. 1991b. Report nixes "Geritol" fix for global warming. *Science* 253:1490–91.

ROBERTS, T. M., T. C. HUTCHINSON, J. PACIGA, A. CHATTOPADHYAY, R. E. JERVIS, and J. VANLOON. 1974. Lead contamination around secondary smelters: Estimation of dispersal and accumulation by humans. *Science* 186:1120–22.

ROBERTSON, D. E., E. A. CRECELIUS, J. S. FRUCHTER, and J. D. LUDWICK. 1977. Mercury emissions from geothermal power plants. *Science* 196:1094–97.

ROBERTSON, D. R. 1984. Cohabitation of competing territorial damselfishes on a Caribbean coral reef. *Ecology* 65:1121–35.

ROBERTSON, G. P., and T. ROSSWALL. 1986. Nitrogen in West Africa: The regional cycle. *Ecological Monographs* 56:43–72.

ROBERTSON, G. P., and P. M. VITOUSEK. 1981. Nitrification potentials in primary and secondary succession. *Ecology* 62:376–86.

ROBEY, B., S. O. RUTSTEIN, and L. MORRIS. 1993. The fertility decline in developing countries. *Scientific American* 269(6):60–67.

ROLSTON, H., III. 1981. What sorts of values does nature have? In *Ecological consciousness: Essays from the Earthday X Colloquium, University of Denver, April 21–24, 1980,* R. C. Schultz and J. D. Hughes, eds., 351–69. Washington, DC: University Press of America.

ROLSTON, H., III. 1988. *Environmental ethics. Duties to and values in the natural world.* Philadelphia: Temple University Press.

ROMME, W. H. 1982. Fire and landscape diversity in subalpine forests of Yellowstone National Park. *Ecological Monographs* 52:199–221.

ROMME, W. H., and D. G. DESPAIN. 1989. The Yellowstone fires. *Scientific American* 261(5):37–46.

ROOT, R. B. 1967. The niche exploitation pattern of the blue-gray gnatcatcher. *Ecological Monographs* 37:317–50.

ROOT, R. B., and A. M. OLSON. 1969. Population increase of the cabbage aphid, *Brevicoryne brassicae,* on different host plants. *Canadian Entomologist* 101:768–73.

ROSE, A. H. 1981. The microbiological production of food and drink. *Scientific American* 245(3):126–38.

ROSS, H. H. 1957. Principles of natural coexistence indicated by leafhopper populations. *Evolution* 11:113–29.

ROSS, M. S., J. J. O'BRIEN, and L. DA S. L. STERNBERG. 1994. Sea-level rise and the reduction in pine forests in the Florida Keys. *Ecological Applications* 4:144–56.

ROWLAND, F. S. 1989. Chlorofluorocarbons and the depletion of stratospheric ozone. *American Scientist* 77:36–45.

ROYAMA, T. 1984. Population dynamics of the spruce budworm *Choristoneura fumiferana. Ecological Monographs* 54:429–62.

RUBIN, E. S., R. N. COOPER, R. A. FROSCH, T. H. LEE, G. MARLAND, A. H. ROSENFELD, and D. D. STINE. 1992. Realistic mitigation options for global warming. *Science* 257:148–66.

RUDD, J. W. M., C. A. KELLY, D.W. SCHINDLER, and M. A. TURNER. 1988. Disruption of the nitrogen cycle on acidified lakes. *Science* 240:1515–17.

RUDDIMAN, W. F., and J. E. KUTZBACH. 1991. Plateau uplift and climatic change. *Scientific American* 264(3):66–75.

RUSSELL, F. S., and C. M. YONGE. 1963. *The seas.* London: Frederick Warne and Company.

RUSSELL, M. 1988. Ozone pollution: The hard choices. *Science* 241:1275–76.

RUTTNER, F. 1966. *Fundamentals of limnology.* 3d ed. Translated by D. G. Frey and F. E. J. Fry. Toronto: Toronto University Press.

RYAN, M., and C. FLAVIN. 1995. Facing China's limits. In *State of the world 1995,* L. R. Brown (project director), 113–31. New York: W. W. Norton and Company.

RYDIN, H., and S.-O. BORGEGARD. 1991. Plant characteristics over a century of primary succession on islands: Lake Hjalmaren. *Ecology* 72:1089–1101.

SAFRANY, D. R. 1974. Nitrogen fixation. *Scientific American* 231(1):64–80.

SAGAN, C., O. B. TOON, and J. B. POLLACK. 1979. Anthropogenic albedo changes and the Earth's climate. *Science* 206:1363–68.

SALA, O. E., W. J. PARTON, L. A. JOYCE, and W. K. LAUENROTH. 1988. Primary production of the central grassland region of the United States. *Ecology* 69:40–45.

SALISBURY, H. 1989. *The Great Black Dragon fire. A Chinese inferno.* Boston: Little, Brown and Company.

SALT, G. W. 1979. A comment on the use of the term *emergent properties. American Naturalist* 113:145–48.

SANDERS, H. L., and R. R. HESSLER. 1969. Ecology of the deep-sea benthos. *Science* 163:1419–24.

SANTOS, P. F., and W. G. WHITFORD. 1981. The effects of microarthropods on litter decomposition in a Chihuahuan desert ecosystem. *Ecology* 62:654–63.

SANTOS, P. F., J. PHILLIPS, and W. G. WHITFORD. 1981. The role of mites and nematodes in early stages of buried litter decomposition in a desert. *Ecology* 62:664–69.

SARNELLE, O. 1994. Inferring process from pattern: Trophic level abundances and unbedded interactions. *Ecology* 75:1835–41.

SAVAGE, M., and T. W. SWETNAM. 1990. Early 19th-century fire decline following sheep pasturing in a Navajo ponderosa pine forest. *Ecology* 71:2374–78.

SCHEINER, S. M., and J. M. REY-BENAYAS. 1994. Global patterns of plant diversity. *Evolutionary Ecology* 75:331–47.

SCHELSKE, C. L., E. F. STOERMER, D. J. CONLEY, J. A. ROBBINS, and R. M. GLOVER. 1983. Early eutrophication in the lower Great Lakes: New evidence from biogenic silica in sediments. *Science* 222:320–22.

SCHINDLER, D. W. 1977. Evolution of phosphorus limitation in lakes. *Science* 195:260–62.

SCHINDLER, D. W. 1988. Effects of acid rain on freshwater ecosystems. *Science* 239:149–57.

SCHINDLER, D. W., K. G. BEATY, E. J. FEE, D. R. CRUIKSHANK, E. R. DEBRUYN, D. L. FINDLAY, G. A. LINSEY, J. A. SHEARER, M. P. STAINTON, and M. A. TURNER. 1990. Effects of climatic warming on lakes of the central boreal forest. *Science* 250:967–70.

SCHINDLER, D. W., M. A. TURNER, M. P. STAINTON, and G. A. LINSEY. 1986. Natural sources of acid neutralizing capacity in low alkalinity lakes of the Precambrian Shield. *Science* 232:844–47.

SCHLESINGER, W. H. 1985. Decomposition of chaparral shrub foliage. *Ecology* 66:1353–59.

SCHLESINGER, W. H., and M. M. HASEY. 1981. Decomposition of chaparral shrub foliage: Losses of organic and inorganic consitituents from deciduous and evergreen leaves. *Ecology* 62:762–74.

SCHLESINGER, W. H., J. F. REYNOLDS, G. L. CUNNINGHAM, L. F. HUENNEKE, W. M. JARRELL, R. A. VIRGINIA, and W. G. WHITFORD. 1990. Biological feedbacks in global desertification. *Science* 247:1043–48.

SCHNEIDER, S. H. 1989. The Greenhouse Effect: Science and policy. *Science* 243:771–81.

SCHOEBERL, M. R., and D. L. HARTMANN. 1991. The dynamics of the stratospheric polar vortex and its relation to springtime ozone depletions. *Science* 251:46–52.

SCHOENER, W. T. 1983. The controversy over interspecific competition. *American Scientist* 70:586–95.

SCHROEDINGER, E. 1955. *What is life? The physical aspect of the living cell.* Cambridge: Cambridge University Press.

SCHULL, W. J., M. OTAKE, and J. V. NEEL. 1981. Genetic effects of the atomic bombs: A reappraisal. *Science* 213:1220–27.

SCHULZE, E. D. 1989. Air pollution and forest decline in a spruce (*Picea abies*) forest. *Science* 244:776–83.

Scientific American. 1969. The September issue (vol. 241) is devoted to the ocean ecosystem.

Scientific American. 1989. The September issue (vol. 261) is devoted to managing planet Earth.

SEARS, P. B. 1930. A record of post-glacial climate in northern Ohio. *Ohio Journal of Science* 30:205–17.

SEARS, P. B. 1932. Postglacial climate in eastern North America. *Ecology* 13:1–6.

SEARS, P. B. 1935. *Deserts on the march.* Norman, OK: University of Oklahoma Press.

SEARS, P. B. 1956. Some notes on the ecology of ecologists. *Scientific Monthly* 83:22–27.

SEASTEDT, T. R., and C. M. TATE. 1981. Decomposition rates and nutrient contents of arthropod remains in forest litter. *Ecology* 62:13–19.

SEBENS, K. P. 1985. The ecology of the rocky subtidal zone. *American Scientist* 73:548–57.

SEINFELD, J. H. 1989. Urban air pollution: State of the science. *Science* 243:745–52.

SEITZINGER, S. P., M. E. Q. PILSON, and S. W. NIXON. 1983. Nitrous oxide production in nearshore marine sediments. *Science* 222:1244–45.

SEN, A. 1993. The economics of life and death. *Scientific American* 268(5):40–47.

SHACHAK, M., C. L. JONES, and Y. GRANOT. 1987. Herbivory in rocks and the weathering of a desert. *Science* 236:1098–99.

SHANMUGAM, K. T., and R. C. VALENTINE. 1975. Molecular biology of nitrogen fixation. *Science* 187:919–24.

SHARITZ, R. R., and J. F. MCCORMICK. 1973. Population dynamics of two competing annual plant species. *Ecology* 54:723–39.

SHAW, R. W. 1987. Air pollution by particles. *Scientific American* 257(2):96–103.

SHELFORD, V. E. 1911. Physiological animal geography. *Journal of Morphology* 22:551–618.

SHELFORD, V. E. 1937. *Animal communities in temperate America.* Chicago: University of Chicago Press.

SHELFORD, V. E. 1963. *The ecology of North American.* Urbana: University of Illinois Press.

SHERMAN, K. 1991. The large marine ecosystem concept: Research and management strategy for living marine resources. *Ecological Applications* 1:349–60.

SHUKLA, J., C. NOBRE, and P. SELLERS. 1990. Amazon deforestation and climate change. *Science* 247:1322–25.

SIEGEL, B. Z., and S. M. SIEGEL. 1982. Mercury content of *Equisetum* plants around Mount St. Helens one year after the major eruption. *Science* 216:292–93.

SIGNOR, P. W. 1990. The geological history of diversity. *Annual Review of Ecology and Systematics* 21:509–39.

SIMBERLOFF, D., and T. DAYAN. 1991. The guild concept and the structure of ecological communities. *Annual Review of Ecology and Systematics* 22:115–43.

SIMMONS, L. 1970. *The role of the aged in primitive society.* Hamdon, CT: Archon Books.

SIMON, C. 1987. Hawaiian evolutionary biology: An introduction. *Trends in Ecology and Evolution* 2:175–78.

SIMON, J. L. 1990. Population growth is not bad for humanity. *National Forum* 70:12–16.

SIMS, P. L. 1988. Grasslands. In *North American terrestrial vegetation,* M. G. Barbour and W. D. Billings, eds., 265–86. Cambridge: Cambridge University Press.

SKLAR, J., and B. BERKOV. 1975. The American birth rate: Evidences of a coming rise. *Science* 189:693–700.

SKOLE, D., and C. TUCKER. 1993. Tropical deforestation and habitat fragmentation in the Amazon: Satellite data from 1978 to 1988. *Science* 260:1905–10.

SLOBODKIN, L. 1960. Ecological energy relationships at the population level. *American Naturalist* 95:213–36.

SLOVIK, P., J. H. FLYNN, and M. LAYMAN. 1991. Perceived risk, trust, and the politics of nuclear waste. *Science* 254:1603–607.

SMITH, A. P., and T. P. YOUNG. 1987. Tropical alpine plant ecology. *Annual Review of Ecology and Systematics* 18:137–58.

SMITH, F. E. 1952. Experimental methods in population dynamics: A critique. *Ecology* 33:441–50.

SMITH, J. M., J. A. BROADWAY, and A. B. STRONG. 1978. United States population dose estimates for iodine-131 in the thyroid after the Chinese atmospheric nuclear weapons tests. *Science* 200:44–46.

SMITH, R. C., B. B. PREZELIN, K. S. BAKER, R. R. BIDIGARE, N. P. BOUCHER, T. COLEY, D. KARENTZ, S. MACINTYRE, H. A. MATLICK, D. MENZIES, M. ONDRUSEK, Z. WAN, and K. J. WATERS. 1992. Ozone depletion: Ultraviolet radiation and phytoplankton biology in Antarctic waters. *Science* 255:952–59.

SMITH, R. J. 1978. Dioxins have been present since the advent of fire, says Dow. *Science* 202:1166–67.

SMITH, S. H. 1972. Factors of ecologic succession in oligotrophic fish communities of the Laurentian Great Lakes. *Journal of the Fisheries Research Board of Canada* 29:717–30.

SOLBRIG, O. 1992. The IUBS-SCOPE-UNESCO program of research in biodiversity. *Ecological Applications* 2:131–38.

SOLLINS, P., C. C. GRIER, F. M. MCCORISON, K. CROMACK, JR., R. FOGEL, and R. L. FREDRIKSEN. 1980. The internal element cycles of an old-growth Douglas-fir ecosystem in western Oregon. *Ecological Monographs* 50:261–85.

SOLOMON, S., G. H. MOUNT, R. W. SANDERS, R. O. JAKOUBEK, and A. L. SCHMELTEKOPF. 1988. Observations of the nighttime abundance of OClO in the winter stratosphere above Thule, Greenland. *Science* 242:550–55.

SOLOMON, A. M., J. R. TRABALKA, D. E. REICHLE, and L. D. VOORHEES. 1985. The global cycle of carbon. In *Atmospheric carbon dioxide and the global carbon cycle,* J. R. Trabalka, ed., 1–13. (DOE/ER-0239). Washington, DC: U. S. Department of Energy.

SOLOMON, A. M., D. C. WEST, and J. A. SOLOMON. 1981. Simulating the role of climate change and species immigration in forest succesion. In *Forest succession: Concepts and applications,* C. West, H. H. Shugart, and D. B. Botkin, eds., New York: Springer-Verlag.

SOUSA, W. P. 1984. The role of disturbance in natural communities. *Annual Review of Ecology and Systematics* 15:353–91.

SOUTHWICK, C. 1958. Population characteristics of house mice living in English corn ricks: Density relationships. *Proceedings of the Zoological Society of London* 131:163–75.

SPARROW, A. H. 1962. *The role of the cell nucleus in determining radiosensitivity.* Brookhaven Lecture Series No. 17. Upton, NY: Brookhaven National Laboratory.

SPEAR, R. W. 1989. Late-Quaternary history of high-elevation vegetation in the White Mountains of New Hampshire. *Ecological Monographs* 59:125–51.

SPENCE, D. H. N. 1982. The zonation of plants in freshwater. *Advances in Ecological Research* 12:37–125.

SPRENT, J. I. 1987. *The ecology of the nitrogen cycle.* Cambridge: Cambridge University Press.

SPRUGEL, D. G. 1984. Density, biomass, productivity, and nutrient-cycling changes during stand development in wave-regenerated balsam fir forests. *Ecological Monographs* 54:165–86.

STANTON, N. L. 1988. The underground in grasslands. *Annual Review of Ecology and Systematics* 19:573–89.

STARKE, L., ed., 1988. *State of the world 1988. Worldwatch Institute report on progress toward a sustainable society.* New York: W. W. Norton and Company.

STEINWASCHER, K. 1979. Host parasite interaction as a potential population regulating mechanism. *Ecology* 60:884–90.

STEMBERGER, R. S., and J. J. GILBERT. 1985. Body size, food concentration, and population growth in planktonic rotifers. *Ecology* 66:1151–59.

STENGEL, R. 1984. Aftermath of a nuclear spill. *TIME* (May 14):82.

STETTER, K. O., G. LAUERER, M. THOMM, and A. NEUNER. 1987. Isolation of extremely thermophilic sulfate reducers: Evidence for a novel branch of Archebacteria. *Science* 236:822–24.

STEVENS, G. C., and J. F. FOX. 1991. The causes of treeline. *Annual Review of Ecology and Systematics* 22:177–91.

STILES, F. G. 1992. Effects of a severe drought on the population biology of a tropical hummingbird. *Ecology* 73:1375–90.

STOLARSKI, R. S. 1988. The Antarctic ozone hole. *Scientific American* 258(1):30–36.

STOLARSKI, R. S., R. BOJKOV, L. BISHOP, C. ZEREFOS, J. STAEHELIN, and J. ZAWODNY. 1992. Measured trends in stratospheric ozone. *Science* 256:342–49.

STOMMEL, H., and E. STOMMEL. 1979. The year without a summer. *Scientific American* 240(6):176–84.

STONE, C. P., and J. M. SCOTT, eds. 1985. *Hawaii's terrestrial ecosystems: Preservation and management.* Honolulu, HI: University of Hawaii Press for Cooperative National Park Resources Study Unit.

STONE, R. 1994. Dioxin report faces scientific gauntlet. *Science* 265:1650.

STOPPARD, T. 1978. *Every good boy deserves favor* and *Professional foul. Two plays.* New York: Grove Press.

STRONG, D. R., JR., D. SIMBERLOFF, L. G. ABELE, and A. B. THISTLE, eds. 1984. *Ecological communities: Conceptual issues and the evidence.* Princeton, NJ: Princeton University Press.

STUIVER, M. 1978. Atmospheric carbon dioxide and carbon reservoir changes. *Science* 199:253–58.

SWETNAM, T. W., and J. L. BETANCOURT. 1990. Fire—southern oscillation relations in southwestern United States. *Science* 249:1017–20.

TABAZADEH, A., and R. P. TURCO. 1993. Stratospheric chlorine injection by volcanic eruptions: HCl scavenging and implications for ozone. *Science* 260:1082–86.

TABER, R. D., and R. F. DASMANN. 1957. The dynamics of three natural populations of the deer *Odocoileus hemionus columbianus. Ecology* 38:233–46.

TANGLEY, L. 1988. Fighting deforestation at home. *BioScience* 38:220–24.

TANSLEY, A. G. 1935. The use and abuse of vegetational concepts and terms. *Ecology* 16:284–307.

TANSLEY, A. G. 1939. *The British Isles and their vegetation.* Cambridge: Cambridge University Press.

TAUB, F. B., ed. 1984. *Lakes and reservoirs.* Amsterdam: Elsevier Science Publishing.

TEAL, J. M. 1957. Community metabolism in a temperate cold spring. *Ecological Monographs* 27:282–302.

TEGNER, M. J., and P. K. DAYTON. 1987. El Niño effects on southern California kelp forest communities. *Advances in Ecological Research* 17:243–79.

TER BRAAK, C. J. F., and I. C. PRENTICE. 1988. A theory of gradient analysis. *Advances in Ecological Research* 18:271–317.

TERRY, R.N. 1988. *The legacy of the Hawaiian cultivator in windward valleys of Hawaii.* Ph.D. dissertation, Louisiana State University and Agricultural and Mechanical College, Baton Rouge, LA.

TESTA, J. W., and D. B. SINIFF. 1987. Population dynamics of Weddell seals (*Leptonychotes weddelli*) in McMurdo Sound, Antarctica. *Ecological Monographs* 57:149–65.

THE WORLD BANK. 1986. *Poverty and hunger: Issues and options for food security in developing countries.* Washington, DC: The World Bank.

THIEMENS, M. H., and W. C. TROGLER. 1991. Nylon production: An unknown source of atmospheric nitrous oxide. *Science* 251:932–34.

THOMAS, R. B. 1973. *Human adaptation to a high Andean energy flow system.* Occasional Papers No. 7. Department of Anthropology, The Pennsylvania State University.

THOMAS, R. B., and B. P. WINTERHALDER. 1976. Physical and biotic environment of southern highland Peru. In *Man in the Andes. A multidisciplinary study of high-altitude Quechua,* P. T. Baker and M. A. Little, eds., 21–59. Stroudsburg, PA: Dowden, Hutchinson and Ross.

THOMPSON, J. N. 1978. Within-patch structure and dynamics in *Pastinaca sativa* and resource availability to a specialized herbivore. *Ecology* 59:443–48.

THOREAU, H. D. 1906. Journal. In *The Writings of Henry David Thoreau. Journals VI, X, XIII.* Ed. B. Torrey. Boston: Houghton Mifflin.

THORNTHWAITE, C. W. 1931. The climates of North America according to a new classification. *Geographical Review* 21:633–55.

THORNTHWAITE, C. W. 1933. The climates of the Earth. *Geographical Review* 23:433–40.

THURMAN, E. M. 1985. *Organic chemistry of natural waters.* Boston: Martinus Nijhoff/Dr. W. Junk.

TIEN, H. Y., Z. TIANLU, P. YU, L. JINGNENG, and L. ZHONGTANG. 1992. China's demographic dilemmas. *Population Bulletin* 47(1):1–44.

TILLY, L. J. 1968. The structure and dynamics of Cone Spring. *Ecological Monographs* 28:169–97.

TINHGEY, D. T., D. M. OSZYK, A. A. HERSTROM, and E. H. LEE. 1994. Effects of ozone on crops. In *Tropospheric ozone: Human health and agricultural impacts,* D. J. Mckee, ed., 175–206. Boca Raton, FL: Lewis Publishers.

TOLBERT, M. A., M. J. ROSSI, and D. M. GOLDEN. 1988. Antarctic ozone depletion chemistry: Reactions of N_2O_5 with H_2O and HCl on ice surfaces. *Science* 240:1018–21.

TOON, O. B., and R. P. TURCO. 1991. Polar stratospheric clouds and ozone depletion. *Scientific American* 264(6):68–74.

TOTHILL, J. C., and J. J. MOTT. 1984. *Ecology and management of the world's savannas. [International Savanna Symposium 1984.]* Canberra: Australian Academy of Sciences.

TOTSUKA, T. 1989. Japan. In *International handbook of pollution control*, E. J. Kormondy, ed., 323–35. Westport, CT: Greenwood Press.

TRABALKA, J. R., J. A. EDMONDS, J. REILLY, R. H. GARDNER, and L. D. VOORHEES. 1985. Human alterations of the global carbon cycle and the projected future. In *Atmospheric carbon dioxide and the global carbon cycle*, J. R. Trabalka, ed., 247–87. (DOE/ER-0239). Washington, DC: U.S. Department of Energy.

TRANSEAU, E. 1926. The accumulation of energy by plants. *Ohio Journal of Science* 26:1–10.

TRAVIS, J. 1994. Inside look confirms more radiation. *Science* 263:750.

TRISKA, F. J., J. R. SEDELL, K. CROMACK, JR., S. V. GREGORY, and F. M. MCCORISON. 1984. Nitrogen budget for a small coniferous forest stream. *Ecological Monographs* 54:119–40.

TURCO, R. P., O. B. TOON, T. P. ACKERMAN, J. B. POLLACK, and C. SAGAN. 1983. Nuclear winter: Global consequences of multiple nuclear explosions. *Science* 222:1283–92.

TURCO, R. P., O. B. TOON, T. P. ACKERMAN, J. B. POLLACK, and C. SAGAN. 1990. Climate and smoke: An appraisal of nuclear winter. *Science* 247:166–76.

UHL, C., and J. B. KAUFFMAN. 1990. Deforestation, fire susceptibility, and potential tree responses to fire in the Eastern Amazon. *Ecology* 71:437–49.

UNICEF. 1990. *The state of the world's children 1990.* Oxford: Oxford University Press.

UNITED NATIONS DEPARTMENT OF INTERNATIONAL ECONOMIC AND SOCIAL AFFAIRS. 1987. *Fertility behavior in the context of development: Evidence from the world fertility survey.* New York: United Nations.

UNITED STATES SOIL CONSERVATION SERVICE. 1975. *Soil taxonomy: A basic system of soil classification for making and interpreting soil surveys.* Washington, DC: U.S. Department of Agriculture, Soil Conservation Service.

UPTON, A. C. 1982. The biological effects of low-level ionizing radiation. *Scientific American* 246:41–49.

VAN SCHAIK, C. P., J. W. TERBORGH, and S. J. WRIGHT. 1993. The phenology of tropical forests: Adaptive significance and consequences for primary consumers. *Annual Review of Ecology and Systematics* 24:353–77.

VAN BUSKIRK, J., and D. C. SMITH. 1991. Density-dependent population regulation in a salamander. *Ecology* 72:1747–56.

VAN DEN ENDE, P. 1973. Predator-prey interactions in continuous culture. *Science* 181:562–64.

VAN RAALTE, C. 1982. Nitrogen fixation by non-leguminous plants. *American Biology Teacher* 44:229–32, 254.

VAN SLYKE, L. 1988. *Yangtze, nature, history and the river.* Reading, MA: Addison-Wesley Publishing Company.

VIGGIANO, A. A., R. A. MORRIS, K. GOLLINGER, and F. ARNOLD. 1995. Ozone destruction by chlorine: The impracticality of mitigation through ion chemistry. *Science* 267:82–84.

VINCENT, J. R. 1992. The tropical timber trade and sustainable development. *Science* 256:1651–55.

VINCENT, W. F. 1981. Production strategies in Antarctic inland waters: Phytoplankton ecophysiology in a permanently ice-covered lake. *Ecology* 62:1215–24.

VITOUSEK, P. M. 1992. Effects of alien plants on native ecosystems. In *Alien plant invasions in native ecosystems of Hawaii: Management and research*, C. P. Stone, C. W. Smith, and J. T. Tunison, eds., 29–41. Honolulu, HI: Cooperative National Park Resources Studies Unit.

VITOUSEK, P. M. 1994. Beyond global warming: Ecology and global change. *Ecology* 75:1861–76.

VITOUSEK, P. M., J. R. GOSZ, C. C. GRIER, J. M. MELILLO, and W. A. REINERS. 1982. A comparative analysis of potential nitrificaton and nitrate mobility in forest ecosystems. *Ecological Monographs* 52:155–77.

VITOUSEK, P. M., J. R. GOSZ, C. C. GRIER, J. M. MELILLO, W. A. REINERS, and R. L. TODD. 1979. Nitrate losses from disturbed ecosystems. *Science* 204:469–74.

VITOUSEK, P. M., L. L. LOOPE, and C. P. STONE. 1987. Introduced species in Hawaii: Biological effects and opportunities for ecological research. *Trends in Ecology and Evolution* 2:224–27.

VITOUSEK, P. M., and P. A. MATSON. 1985. Disturbance, nitrogen availability, and nitrogen losses in an intensively managed loblolloy pine plantation. *Ecology* 66:1360–76.

VITOUSEK, P. M., and W. A. REINERS. 1975. Ecosystem succession and nutrient retention: A hypothesis. *BioScience* 25:376–81.

VITOUSEK, P. M., and R. L. SANFORD, Jr. 1986. Nutrient cycling in moist tropical forest. *Annual Review of Ecology and Systematics* 17:137–67.

VOGT, K. A., C. C. GRIER, and D. J. VOGT. 1986. Production, turnover, and nutrient dynamics of above- and belowground detritus of world forests. *Advances in Ecological Research* 15:303–409.

VOLTERRA, V. 1928. Variations and fluctuations of the number of individuals in animal species living together. *Animal ecology.* Translated by R. N. Chapman. New York: Arno.

VON LIEBIG, J. 1862. *Die Naturgesetze des Feldbaues.* Braunscheig: Vieweg & Sohn.

WALKER, A. S. 1982. Deserts of China. *American Scientist* 70:366–76.

WALKER, L. R. 1987. Interactions among processes controlling successional change. *Oikos* 50:131–35.

WALKER, L. R., and F. S. CHAPIN III. 1986. Physiological controls over seedling growth in primary succession on an Alaskan floodplain. *Ecology* 67:1508–23.

WALSH, J. 1977. Seveso: The questions persist where dioxin created a watershed. *Science* 197: 1064–67.

WALSH, J. 1988. Sahel will suffer even if rains come. *Science* 224:467–71.

WALTER, H. 1973. *Vegetation of the Earth: In relation to climate and the eco-physiological conditions.* Berlin: Springer-Verlag.

WARING, R. H., and W. H. SCHLESINGER. 1985. *Forest ecosystems: Concepts and management.* New York: Academic Press.

WATSON, S. W. 1965. Characteristics of a marine nitrifying bacterium: *Nitrocystis oceanus* sp. n. *Limnology and Oceanography* 10:247–89.

WATT, K. E. F. 1965. Community stability and the strategy of biological control. *Canadian Entomologist* 97:887–95.

WATTS, W. A. 1979. Late Quaternary vegetation of Central Appalachia and the New Jersey coastal plain. *Ecological Monographs* 49:427–69.

WATTS, W. A., C. S. HANSEN, and E. C. GRIMM. 1992. Camel Lake: A 40,000-year record of vegetational and forest history from northwest Florida. *Ecology* 73:1056–66.

WATTS, W. A., and M. STUIVER. 1980. Late Wisconsin climate of northern Florida and the origin of species-rich deciduous forest. *Science* 210:325–27.

WEBB, L. J., and J. KIKKAWA, eds. 1990. *Australian tropical rainforests. Science—values—meaning.* East Melbourne, Australia: CSIRO.

WEBB, W. L., W. K. LAUENROTH, S. R. SZAREK, and R. S. KINERSON. 1983. Primary production and abiotic controls in forests, grasslands, and desert ecosystems in the United States. *Ecology* 64:134–51.

WEBSTER, J. R., and B. C. PATTEN. 1979. Effects of watershed perturbation on stream potassium and calcium dynamics. *Ecological Monographs* 19:51–72.

WEINER, J. 1985. Size hierarchies in experimental plant populations of annual plants. *Ecology* 66:743–52.

WELDEN, C. W., W. L. SLAUSON, and R. T. WARD. 1988. Competition and abiotic stress among trees and shrubs in northwest Colorado. *Ecology* 69:1566–77.

WELLS, H., and P. H. WELLS. 1992. The Monarch butterfly. *Bulletin of the Southern California Academy of Sciences* 91(1):1–25.

WELLS, P. V. 1983. Paleobiogeography of montane islands in the Great Basin since the last glaciopluvial. *Ecological Monographs* 53:341–82.

WENDORF, D. F., and J. J. HESTER. 1962. Early man's utilization of the Great Plains environment. *American Anthropologist* 18:159–71.

WEST, N. E. 1988. Intermountain deserts, shrub steppes, and woodlands. In *North American terrestrial vegetation*, M. G. Barbour and W. D. Billings, eds., 209–31. Cambridge: Cambridge University Press.

WEST, N. E. 1990. Structure and function of microphytic soil crusts in wildland ecosystems of arid to semi-arid regions. *Advances in Ecological Research* 20:179–223.

WESTLAKE, D. F. 1963. Comparisons of plant productivity. *Biological Reviews* 38:385–425.

WESTOFF, C. F. 1986. Fertility in the United States. *Science* 234:554–59.

WESTRA, L. 1994. *An environmental proposal for ethics. The principle of integrity.* Lanham, MD: Rowman and Littlefield Publishers.

WHALEN, M., C. O. KUNZ, J. M. MATUSZEK, W. E. MAHONEY, and R. C. THOMPSON. 1980. Radioactive plume from the Three Mile Island accident: Xenon-133 in air at a distance of 375 kilometers. *Science* 207:638–39.

WHITE, F. 1983. *Vegetation map of Africa (1:5,000,000).* UNESCO/AETFAT/UNSO (United Nations Educational Scientific and Cultural Organization/Association pour l'Etude Taxonomique de la Flore l'Afrique Tropicale/United Nations Sudano-Sahelian Office). Paris: UNESCO.

WHITE, L. 1967. The historical roots of our ecological crisis. *Science* 155:1203–7.

WHITEHILL, J. D. 1981. Ecological consciousness and values: Japanese perspectives. In *Ecological consciousness: Essays from the Earthday X Colloquium, University of Denver, April*

21–24, 1980, R. C. Schultz and J. D. Hughes, eds., 165–82. Washington, DC: University Press of America.

WHITFORD, W. G. 1992. Biogeochemical consequences of desertification. In *The science of global change. The impact of human activities on the environment,* D. A. Dunnette and R. J. O'Brien, eds., 352. Washington, DC: American Chemical Society.

WHITMORE, T. C. 1990. *An introduction to tropical rainforests.* Oxford: Oxford University Press.

WHITTAKER, R. H. 1953. A consideration of climax theory: The climax as a population and pattern. *Ecological Monographs* 23:41–78.

WHITTAKER, R. H. 1956. Forest dimensions and production in the Great Smoky Mountains. *Ecology* 26:1–80.

WHITTAKER, R. H., and G. E. LIKENS. 1973a. Carbon in the biota. In *Carbon and the biosphere,* G. M. Woodwell and E. V. Pecan, eds., 281–302. Washington, DC: United States Atomic Energy Commission.

WHITTAKER, R. H., and G. E. LIKENS, eds. 1973b. The primary production of the biosphere. *Human Ecology* 1:299–369.

WHITTON, B. A. 1975. *River ecology.* Berkeley, CA: University of California Press.

WILLIAMS, D. J. 1992. Great Lakes water quality: A case study. *The science of global change. The impact of human activities on the environment,* D. A. Dunnette and R. J. O'Brien, eds., 297–323. Washington, DC: American Chemical Society.

WILLIAMS, K. S., K. G. SMITH, and F. M. STEPHEN. 1993. Emergence of 13-yr periodical cicadas (Cicadidae: *Magicicada*): Phenology, mortality, and predator satiation. *Ecology* 74:1143–52.

WILSON, B. S. 1991. Latitudinal variation in activity season mortality rates of the lizard *Uta stansburiana. Ecological Monographs* 61:393–414.

WILSON, E. O. 1988. The current state of biological diversity. In *Biodiversity,* E. O. Wilson and F. M. Peter, eds., 3–18. Washington, DC: National Academy Press.

WILSON, E. O. 1992. *The diversity of life. Questions of life.* Cambridge, MA: Harvard University Press.

WILSON, E. O., and F. M. PETER, eds. 1988. *Biodiversity.* Washington, DC: National Academy Press.

WILSON, J. B., and A. D. Q. AGNEW. 1992. Positive-feedback switches in plant communities. *Advances in Ecological Research* 23:263–336.

WINOGRAD, I. J. 1981. Radioactive waste disposal in thick unsaturated zones. *Science* 212:1457–64.

WISSINGER, S. A. 1989. Seasonal variation in the intensity of competition and predation among dragonfly larvae. *Ecology* 70:1017–27.

WITHERSPOON, J. P. 1964. Cycling of cesium-134 in white oak trees. *Ecological Monographs* 34:403–20.

WITHERSPOON, P. A., N. G. W. COOK, and J. E. GALE. 1981. Geological storage of radioactive waste: Field studies in Sweden. *Science* 211:894–900.

WOLDA, H. 1988. Insect seasonality: Why? *Annual Review of Ecology and Systematics* 19:1–18.

WOLFERT, D. R. 1981. The commercial fishery for walleyes in New York waters of Lake Erie, 1959–1978. *North American Journal of Fisheries Management* 1:112–26.

WOLT, J. D., and LIETZKE, D. A. 1982. The influence of anthropogenic sulfur inputs upon soil properties in the copper region of Tennessee. *Soil Science Society of American Journal* 46:651–56.

WOOD, D. M., and R. DEL MORAL. 1987. Mechanisms of early primary succession in subalpine habitats on Mount St. Helens. *Ecology* 68:780–90.

WOOD, T., F. H. BORMANN, and G. K. VOGT. 1984. Phosphorus cycling in a northern hardwood forest: Biological and chemical control. *Science* 223:391–93.

WOODWARD, F. I. 1987. *Climate and plant distribution*. Cambridge: Cambridge University Press.

WOODWARD, F. I., ed. 1992. The ecological consequences of global climate change. *Advances in Ecological Research* 22:1–314.

WOODWELL, G. M. 1962. Effects of ionizing radiation on terrestrial ecosystems. *Science* 138:572–77.

WOODWELL, G. M. 1989. On causes of biotic impoverishment. *Ecology* 70:14–15.

WOODWELL, G. M., P. P. CRAIG, and H. A. JOHNSON. 1971. DDT in the biosphere: Where does it go? *Science* 174:1101–107.

WOODWELL, G. M., J. E. HOBBIE, R. A. HOUGHTON, J. M. MELILLO, B. MOORE, B. J. PETERSON, and G. R. SHAVER. 1983. Global deforestation: Contribution to atmospheric carbon dioxide. *Science* 222:1081–86.

WOODWELL, G. M., and A. L. REBUCK. Effects of chronic gamma radiation on the structure and diversity of an oak-pine forest. *Ecological Monographs* 37:53–69.

WOODWELL, G. M., and A. H. SPARROW. 1963. Predicted and observed effects of chronic gamma radiation on a near-climax forest ecosystem. *Radiation Botany* 3:231–37.

WOOSTER, D. 1977. *Nature's economy*. San Francisco: Sierra Club Books.

WOOTTON, J. T. 1994. The nature and consequences of indirect effects in ecological communities. *Annual Review of Ecology and Systematics* 25:443–66.

WORLD FOOD COUNCIL. 1989. T*he Cyprus initiative against hunger in the world. Introduction and part one. World hunger fifteeen years after the world food conference: The challenges ahead.* Cairo, Egypt, Fifteenth Ministerial Session. New York: United Nations.

WORLD RESOURCES INSTITUTE. 1990. *World Resources 1990–91*. Washington, DC: World Resources Institute.

WORTMAN, S. 1980. World food and nutrition: The scientific and technological base. *Science* 209:157–64.

WRIGHT, S. J. 1991. Seasonal drought and the phenology of understory shrubs in a tropical moist forest. *Ecology* 72:1643–57.

WURSTER, C. F. 1968. DDT residues and declining reproduction in the Bermuda petrel. *Science* 159:979–81.

WYMAN, R. L., ed. 1991.*Global climate change and life on earth*. Routledge, New York: Chapman and Hall.

WYNNE-EDWARDS, V. C. 1962. *Animal disperson in relation to social behavior*. Edinburgh: Oliver and Boyd.

WYNNE-EDWARDS, V. C. 1986. *Evolution through group selection*. Palo Alto, CA: Blackwell Scientific Publishers.

YANG, H. 1990. *Proceedings of the International Symposium on Grassland Vegetation, August 15–20, 1987, Hohot, The People's Republic of China*. Beijing: Science Press.

YOUNG, G. L. 1974. Human ecology as an interdisciplinary concept: A critical inquiry. *Advances in Ecological Research* 8:4–40.

YOUNG, G. L., ed. 1983. *Origins of human ecology.* Stroudsburg, PA: Hutchison Ross Publishing Company.

YUNG, Y. L., and M. B. MCELROY. 1979. Fixation of nitrogen in the prebiotic atmosphere. *Science* 203:1002–1004.

ZAMMUTO, R. M. 1987. Life histories of mammals: Analyses among and within *Spermophilus columbianus* life tables. *Ecology* 68:1351–63.

ZARET, T. M. 1982. The stability/diversity controversy: A test of hypotheses. *Ecology* 63:721–31.

ZHOU, K., and X. ZHANG. 1991. *Baiji. The Yangtze River dolphin and other endangered animals of China.* Translated by C. Luo. Washington DC: The Stone Wall Press; and Naning, China: Yilin Press.

ZILBERMAN, D., A. SCHMITZ, G. CASTERLINE, E. LICHTENBERG, and J. B. SIEBERT. 1991. The economics of pesticide use and regulation. *Science* 253:518–22.

Index